ELEMENTARY LINEAR CIRCUIT ANALYSIS

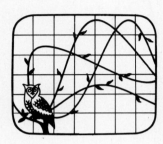

*HRW
Series in
Electrical and
Computer Engineering*

M. E. Van Valkenburg, Series Editor

Leonard S. Bobrow ELEMENTARY LINEAR CIRCUIT ANALYSIS
Benjamin C. Kuo DIGITAL CONTROL SYSTEMS
Athanasios Papoulis CIRCUITS AND SYSTEMS: A MODERN APPROACH

ELEMENTARY LINEAR CIRCUIT ANALYSIS

LEONARD S. BOBROW
University of Massachusetts, Amherst

Holt, Rinehart and Winston, Inc.

New York   Chicago   San Francisco   Atlanta
Dallas   Montreal   Toronto   London   Sydney

Editorial and production services by Cobb/Dunlop Publisher Services, Inc.

Address correspondence to:
383 Madison Avenue, New York, NY 10017

Library of Congress Cataloging in Publication Data

Bobrow, Leonard S
  Elementary linear circuit analysis.

  Includes index.
  1. Electric circuits, Linear. 2. Electric networks.
I. Title.
TK454.B58   621.319′2   80-15555
ISBN 0-03-055696-1

Printed in the United States of America

3 4 5   016   9 8 7 6 5 4

CBS COLLEGE PUBLISHING
Holt, Rinehart and Winston
The Dryden Press
Saunders College Publishing

PREFACE

This text is written for sophomore-level courses in beginning electric circuits and systems. I have found that there is sufficient material in the first seven chapters and the appendix for the first semester's course, and the next five chapters could form the basis for the second semester's course. Perhaps a school whose students allow a faster pace could include one or both of the last two chapters. As an alternative, an institution that is on the quarter system could complete the book in three quarters.

The text begins with such traditional topics as voltage, current, sources, resistors, and Ohm's law. It is assumed that the reader has had a course in high school or college freshman physics in which the concepts of the physics of electricity and magnetism (e.g., electric charge, electric potential, and magnetic fields) were introduced and discussed at an elementary level. Higher-level exposure to these topics, though illuminating, is not necessary since a more axiomatic approach to the subject matter can be taken if desired. Furthermore, it is assumed that the reader has had one or more courses in elementary calculus in which the topics of differentiation and integration were included.

Nowadays many students have already had experience with matrix notation, matrix arithmetic, and some applications of determinants (such as Cramer's rule) by the time they study this subject matter. The appendix on matrices and determinants will be a review for such people. For schools having a majority of students who are not familiar with these topics, the material in the appendix should be introduced before or during the study of Chapter 2, in which circuit analysis techniques are discussed.

Rather than devote a separate chapter to network topology, I have introduced various topological concepts throughout the text where needed as part of their application to circuit analysis.

Many circuits texts seem to neglect nonsinusoidal signals and waveforms by postponing their introduction or avoiding much of the subject. In this book, the topic is made an integral part of the discussion of inductors and capacitors. Also studied, at an even earlier stage, is the increasingly important subject of operational amplifiers. Op amp circuits appear frequently throughout the subsequent portion of the text.

The coverage of first- and second-order circuits is traditional and self-contained with regard to the solving of linear differential equations with constant coefficients. The notion of describing circuits by differential equations is expanded to matrix formulations in the chapter on state-variable analysis. There, however, it is the numerical solution via a digital computer that is stressed.

The transformation from the time domain to the frequency domain takes place in the chapter on sinusoidal analysis. The resulting concepts see application in the subsequent chapter on power, and they are later generalized with the notion of complex frequency. The subtlety of the sinusoid becomes more evident with the study of Fourier series, and its extension, the Fourier transform. It is seen that limitations on the latter can be overcome with the use of the Laplace transform.

The material in the text increases in difficulty as the subject matter progresses. A second-term student has already had the experience of one term, whereas the student in the first term is almost totally unsophisticated—and there is quite a difference between the two cases.

It is my philosophy that a text include many examples, and that these examples be worked in sufficient detail so that the reader can follow each example from beginning to end. I find phrases such as "It can be shown that…" unpleasant, for although I may be able to show it, maybe the reader cannot. Under this circumstance, if I don't show it to the reader (and the reader can't or won't show it), then the result under consideration will not be as meaningful. Of course, there are those rare occasions when, for practical reasons, I have to let a result appear without formal justification.

Rather than approach a topic initially from a general point of view, I prefer to start with an example that a student can understand, and then have it naturally evolve into something new or simply suggest something new. Generalizations are, of course, necessary, but they should appear after a student has seen some specifics (if possible). Seeing examples solved in detail gives readers a certain amount of feeling for the problem-solving process and subject material. It is then easier for them to grasp more general concepts. Inasmuch as this book has been written primarily for people to whom the material is new, and they are trying to learn it for the first time, I feel such an approach is pedagogically superior.

An extremely important part of the learning process involves solving problems. Even if a student can follow every line of every example in this book, that doesn't mean that he or she can solve problems unaided. In order to gain confidence and agility (and insight) the reader must do many exercises. In this text, numerous homework problems are given, not only at the end of chapters, but at the end of most sections as well. (In the case of fairly short sections, problems may not appear until the end of subsequent sections, so that a combination of concepts can be exercised.) Furthermore, many more problems are presented than need be given as homework assignments. Among each group of problems, certain ones are marked with an asterisk. The answers to these problems (which students can utilize for further drill or exercise) are given. The solutions for all the problems in this book are worked out in detail in the companion solutions manual. A number of problems have been chosen so that their values, as well as topology, are typical of real situations. However, since the stress of this text is on the concepts of circuit analysis and not on numerical computation, most circuits (even those with practical configurations) have numerically simple element values.

The writing style of this book may seem to be a little more casual than for a typical text. But, as the object of any book is to convey information and ideas, I feel that it is important that a text be written in a way that makes a reader (who will be spending a lot of time with it) feel comfortable. I personally think that informality tends to make students feel less intimidated.

I would appreciate hearing from any instructor who finds any typographical errors in this book or the companion solutions manual so that corrections can be made in future printings. Furthermore, I would like to receive comments and criticisms pertaining to the text.

I would like to acknowledge the assistance and support that I received while working on this project. I wish to thank Ms. Patricia Moriarty, who typed the bulk of the class notes on which this book is based, and Ms. Erica Swanson for her contribution to the typing of the class notes. I received a great deal of help from Professor Donald Scott with the FORTRAN programs given in the text, and it is very much appreciated. Thanks too, to Ms. Elizabeth M. Warriner. I am also thankful for Professor Scott's class testing of the manuscript of this text. I would also like to thank Mr. Elliott Arkin for his contribution to the cover design. I should like to acknowledge the reviewers for their comments and criticisms of the manuscript; they made many excellent suggestions. I also wish to thank Ms. Lila Gardner and my copy editor Mr. Robert Whitlock for their fine production work. Finally, I am grateful for the encouragement given to me on this project by my editor, Mr. Paul Becker. His belief in my project and my ability proves him to be not only a man of compassion and understanding, but much intelligence as well.

Leonard S. Bobrow

CONTENTS

ELEMENTARY LINEAR CIRCUIT ANALYSIS

BASIC ELEMENTS AND LAWS

Introduction

The study of electric circuits is fundamental in electrical engineering education, and can be quite valuable in other disciplines as well. The skills acquired not only are useful in such electrical engineering areas as electronics, communications, microwaves, control and power systems but also can be employed in other seemingly different fields.

By an *electric circuit* or *network* we mean a collection of electrical devices (for example, voltage and current sources, resistors, inductors, capacitors, transformers, amplifiers, and transistors) that are interconnected in some manner. The various uses of such circuits, though important, is not the major concern of this text. Instead, our prime interest will be with the process of determining the behavior of a given circuit—which is referred to as *analysis*.

We begin our study by discussing some basic electric elements and the laws that describe them. It is assumed that the reader has been introduced to the concepts of electric charge, potential, and current in various science and physics courses in high school and college.

1.1 IDEAL SOURCES

Electric charge is measured in *coulombs* (abbreviated C) in honor of the French scientist Charles de Coulomb (1736–1806), and the unit of work or energy—the *joule* (J)—is named for the British physicist James P. Joule (1818–1889). Although energy expended on electric charge has the units J/C, we give it the special name *volt* (V) in honor of the Italian physicist Alessandro Volta (1745–1827), and we say that it is a measure of *electric potential difference* or *voltage*.

An *ideal voltage source*, which is represented in Fig. 1.1, is a device that produces a voltage or potential difference of v volts across its terminals *regardless of what is connected to it.*

fig. 1.1

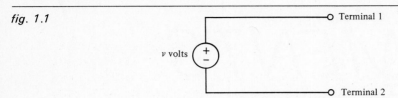

For the device shown in Fig. 1.1, terminal 1 is marked plus ($+$) and terminal 2 is marked minus ($-$). This denotes that terminal 1 is at an electric potential that is v volts higher than that of terminal 2. (Alternatively, the electric potential of terminal 2 is v volts lower than that of terminal 1.)

The quantity v can be either a positive or a negative number. For the latter case, it is possible to obtain an equivalent source with a positive value as demonstrated by the following example.

• • •

Example

Suppose that $v = -5$ volts. Then the potential at terminal 1 is -5 volts higher than that of terminal 2. However, this is equivalent to saying that terminal 1 is at a potential of $+5$ volts lower than terminal 2. Consequently, the two ideal voltage sources shown in Fig. 1.2 are equivalent.

fig. 1.2

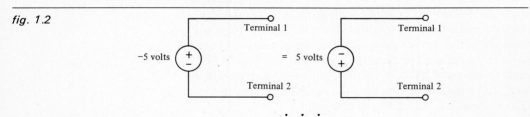

• • •

In the discussion above, we may have implied that the value of an ideal voltage source is constant, that is, it does not change with time. Such a situation is plotted in Fig. 1.3. For occasions such as this, an ideal voltage source is commonly represented by the equivalent notation shown in Fig. 1.4. We refer to such a device as an *ideal battery.*

fig. 1.3

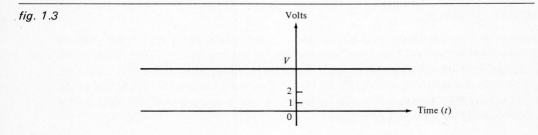

fig. 1.4

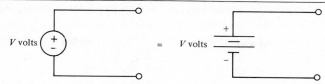

Although a "real" battery is not ideal, there are many circumstances under which an ideal battery is a very good approximation. One example is the 9-volt battery that you use for your portable transistor radio—or you may have the type that uses four or six "C" or "D" $1\frac{1}{2}$-volt batteries. More will be said about "real" or "practical" voltage sources in Chapter 3.

In general, however, the voltage (v) produced by an ideal voltage source will be a function of time (t). A few typical voltage waveforms are shown in Fig. 1.5. The waveforms shown in (a) and (b) are typical-looking AM (amplitude modulation) and

fig. 1.5

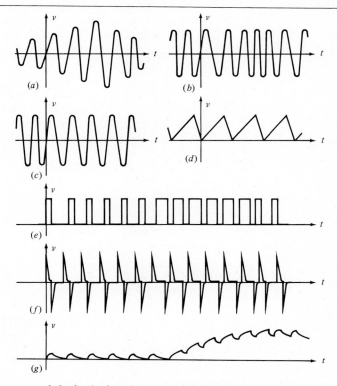

FM (frequency modulation) signals, respectively. Both types of signals are used in consumer radio communications. The sinusoid shown in (c) has a wide variety of uses; for example, this is the shape of ordinary household voltage. The "frequency" of such a waveform is often used as a reference signal, as for color detection in television. The "sawtooth" waveshape shown in (d) is used to "sweep" the electron beam in a television set (and an oscilloscope, as well). A "pulse train," such as that seen in (e), can be used to contain the horizontal and vertical synchronization information for a video signal. The waveforms shown in (f) and (g) are obtained from that given in (e). The former synchronizes the horizontal oscillator, whereas the latter synchronizes the vertical oscillator.

Since the voltage produced by a source is in general a function of time, say $v(t)$, then the most general representation of an ideal voltage source is as shown in Fig. 1.6.

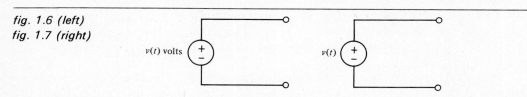

fig. 1.6 (left)
fig. 1.7 (right)

There should be no confusion if the units "volts" are not included in the representation of the source. Thus, the ideal voltage source shown in Fig. 1.7 is identical to the previous one—the units "volts" being understood. For the sake of illumination, consider an ideal voltage source (Fig. 1.7) as described by the function $v(t)$, shown in Fig. 1.8. The

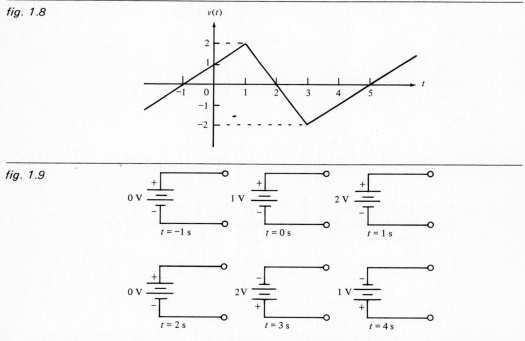

fig. 1.8

fig. 1.9

behavior of the source at six specific instances of time is shown in Fig. 1.9. Here, for example, at time $t = 1$ second (s) the source behaves as an ideal 2-volt battery.

Placing an electric potential difference across some material generally results in a flow of electric charge. Negative charge (in the form of electrons) flows from a given electric potential to a higher potential. Conversely, positive charge tends to flow from a given potential to a lower potential. Charge is usually denoted by q, and since this quantity is generally time-dependent, the total amount of charge that flows is designated by $q(t)$.

We define *current*, denoted $i(t)$, to be the flow rate of the charge; that is,

$$i(t) = \frac{dq(t)}{dt}$$

and the units of current (coulombs per second or C/s) are referred to as *amperes* (abbreviated A) in honor of the French physicist André Ampère (1775–1836). Due to Benjamin Franklin (a positive thinker), the direction of current has been chosen to be that direction in which positive charge would flow.

An *ideal current source* (Fig. 1.10) is a device that, *when connected to anything*, will always pull I amperes into terminal 2 and push I amperes out of terminal 1 (vice versa if the arrow is pointed in the opposite direction). As a consequence of the definition, it should be quite clear that the ideal current sources in Fig. 1.11 are equivalent.

fig. 1.10

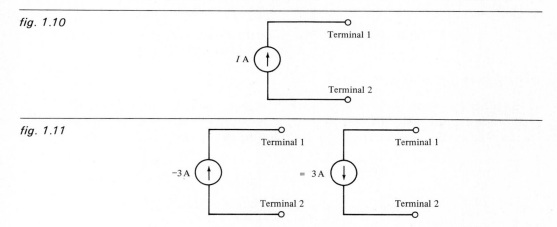

Again, in general, the amount of current produced by an ideal source will be a function of time. Thus, the general representation of an ideal current source is given by Fig. 1.12, where the units "amperes" are understood.

fig. 1.12

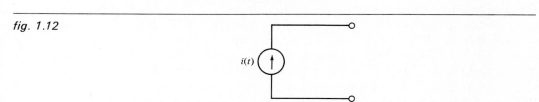

PROBLEMS   *1.1.   An ideal voltage source has a value of $v(t) = 10e^{-t}$ V. What is the value of this voltage source when (a) $t = 0$ s, (b) $t = 1$ s, (c) $t = 2$ s, (d) $t = 3$ s, (e) $t = 4$ s?
Ans.   (a) 10 V, (c) 1.35 V, (e) 0.183 V

1.2.   Repeat Problem 1.1 for $v(t) = 5 \sin(\pi/2)t$ V.

1.3.   Repeat Problem 1.1 for $v(t) = 3 \cos(\pi/2)t$ V.

*1.4.   The total charge in some material is described by the function $q(t) = 4e^{-2t}$ C. Find the current through this material.

     Ans. $-8e^{-2t}$ A

1.5.   Repeat Problem 1.4 for the case that $q(t) = 3 \sin \pi t$ C.

1.6.   Repeat Problem 1.4 for the case that $q(t) = 6 \cos 2\pi t$ C.

---

fig. P1.7

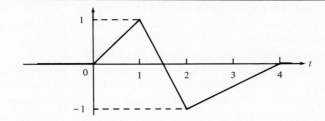

1.7.   Repeat Problem 1.1 for the case that $v(t)$ is described by the function shown in Fig. P1.7.

1.8.   The total charge $q(t)$ in some material is described by the function given in Problem 1.7. Sketch the current $i(t)$ through the material.

1.2   RESISTORS AND OHM'S LAW

Suppose that some material is connected to the terminals of an ideal voltage source $v(t)$ as shown in Fig. 1.13. Suppose also that $v(t)$ is described by the waveform given in Fig. 1.8. Then at time $t = 0$ s, $v(t) = 1$ V, and thus the electric potential at the top of

---

fig. 1.13

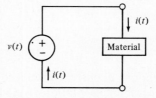

the material is 1 V above the potential at the bottom. Electrons in the material, therefore, will tend to flow from bottom to top, so we say that the current tends to go from top to bottom. Hence, for the given polarity, when $v(t)$ is a positive number, then $i(t)$ will be a positive number with the direction indicated. At time $t = 1$ s, $v(t) = 2$ V. Again, the potential at the top is greater than at the bottom, so $i(t)$ will again be positive. However, because the potential at time $t = 1$ s is twice as much than at time $t = 0$ s, the current at $t = 1$ s will be greater than at time $t = 0$ s. (If the material is a "linear" element, the current will be twice as great.) At time $t = 2$ s, $v(t) = 0$ V. Thus the potential at the top and the bottom of the material is the same. The result is no flow of electrons, and hence, no current. In this case $i(t) = 0$. At time $t = 3$ s, $v(t) = -2$ V. Consequently, the top of the material will be at a potential lower than that at the bottom of the material. Hence a current from bottom to top will result, and $i(t)$ will be a negative number. Note that the current $i(t)$ through the material must also go through the voltage source, since there is nowhere else for it to go.

If in Fig. 1.13, the resulting current $i(t)$ is always directly proportional to the voltage, for any function $v(t)$, then the material is called a *linear resistor*—or *resistor*, for short.

Since voltage and current are directly proportional for a resistor, there exists a proportionality constant R, called *resistance*, such that

$$v(t) = Ri(t)$$

Dividing both sides of this equation by $i(t)$, we obtain

$$R = \frac{v(t)}{i(t)}$$

The unit of resistance (volts per ampere) is referred to as the *ohm** and is denoted by the capital Greek letter omega, Ω. The accepted schematic representation of a resistor whose resistance is R ohms is shown in Fig. 1.14. A plot of voltage versus current for a (linear) resistor is shown in Fig. 1.15.

fig. 1.14

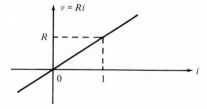

fig. 1.15

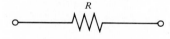

An example of a nonlinear resistor is an incandescent lamp. A typical volt-ampere characteristic curve is illustrated in Fig. 1.16. Even though such an element has this nonlinear characteristic, for small variations in current (or voltage) such an element

fig. 1.16

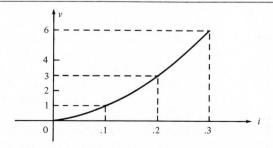

_____
* Named for the German physicist Georg Ohm (1787–1854).

approximately behaves as a (linear) resistor. For example, if the current only varies between 0 and 0.1 A, then the lamp is approximately a resistor whose value is

$$R = \frac{1.0}{0.1} = 10 \ \Omega$$

whereas if the current only varies between 0.2 and 0.3 A, then the resistance of the lamp is approximately

$$R = \frac{6.0 - 3.0}{0.3 - 0.2} = \frac{3.0}{0.1} = 30 \ \Omega$$

It was Ohm who discovered that if a resistor R has a voltage $v(t)$ *across* it and a current $i(t)$ going *through* it, then if one is the cause, the other is the effect. Furthermore, if the polarity of the voltage and the direction of the current are as shown in Fig. 1.17,

fig. 1.17

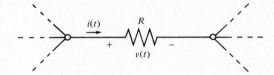

then it is true that

$$v(t) = Ri(t)$$

This equation is often called *Ohm's law*. From it, we may immediately deduce that

$$R = \frac{v(t)}{i(t)}$$

and

$$i(t) = \frac{v(t)}{R}$$

It should be pointed out here that directions of currents and polarities of voltages are crucial when writing Ohm's law; the accepted convention is given in Fig. 1.17. For example, in the situation shown in Fig. 1.18, before writing Ohm's law note that the current's direction is opposite to that dictated by convention. However, this difficulty is easily remedied by redrawing Fig. 1.18 in the equivalent form shown in Fig. 1.19. (A current of 5 A going through the resistor to the left is the same as a current of -5 A

fig. 1.18

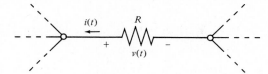

fig. 1.19

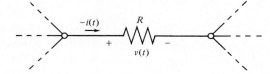

going to the right.) Since the direction of the current and the polarity of the voltage now conform to convention, we use Ohm's law to write

$$v(t) = R[-i(t)] \qquad \text{or} \qquad v(t) = -Ri(t)$$

Alternatively, we could have redrawn the given figure in another equivalent form (Fig. 1.20) and again used Ohm's law to write

$$-v(t) = Ri(t)$$

from which we get

$$v(t) = -Ri(t)$$

fig. 1.20

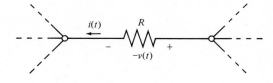

. . .

Examples
Assume the circuit shown in Fig. 1.21. We use the letter k to represent the prefix "kilo," which indicates a value of 10^3. Some of the more common symbols are as

fig. 1.21

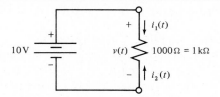

shown in the table:

Value	Prefix	Symbol
10^{-12}	pico	p
10^{-9}	nano	n
10^{-6}	micro	μ
10^{-3}	milli	m
10^{3}	kilo	k
10^{6}	mega	M
10^{9}	giga	G

In Fig. 1.21 the voltage across the 1k-Ω resistor is, by the definition of an ideal voltage source, $v(t) = 10$ V. Thus, by Ohm's law, we get

$$i_1(t) = \frac{v(t)}{R} = \frac{10}{1000} = \frac{1}{100} = 0.01 \text{ A} = 10 \text{ mA}$$

and

$$i_2(t) = \frac{-v(t)}{R} = \frac{-10}{1000} = \frac{-1}{100} = -0.01 \text{ A} = -10 \text{ mA}$$

Note that $i_2(t) = -i_1(t)$, as expected.

Let us now consider the circuit shown in Fig. 1.22. Since $v(t) = 163$ cos $120\pi t$ V, then

$$i(t) = \frac{v(t)}{R} = 1.63 \cos 120\pi t \text{ A}$$

fig. 1.22

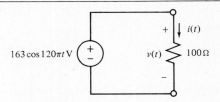

To find the value of the current at any time t_0, we merely substitute t_0 into the expression for $i(t)$. For example, at time $t = 0$, the value of the current is

$$i(0) = 1.63 \cos[120\pi(0)]$$

$$= 1.63 \cos 0$$

$$= 1.63 \text{ A}$$

At time $t = 1/240$ s, the current is

$$i(1/240) = 1.63 \cos[120\pi(1/240)]$$

$$= 1.63 \cos \pi/2$$

$$= 0 \text{ A}$$

At time $t = 1/120$ s, the current is

$$i(1/120) = 1.63 \cos[120\pi(1/120)]$$

$$= 1.63 \cos \pi$$

$$= -1.63 \text{ A}$$

The above sinusoidal voltage source actually represents the ordinary household electrical supply. Why such a source is usually designated "115 volts ac/60 hertz" will be discussed in Chapter 9.

Assume we wish to find the voltage in the circuit of Fig. 1.23. By the definition

fig. 1.23

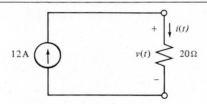

of an ideal current source, the current flowing through the 20-Ω resistor is $i(t) = 12$ A. Thus, by Ohm's law, we get

$$v(t) = Ri(t)$$

$$= (20)(12)$$

$$= 240 \text{ V}$$

For the circuit in Fig. 1.24, Ohm's law yields

$$v(t) = -Ri(t)$$

$$= -(50)(3 \sin \omega t)$$

$$= -150 \sin \omega t \text{ V}$$

fig. 1.24

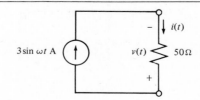

Suppose we wish to find the resistance in the circuit in Fig. 1.25. By Ohm's law we find that

$$R = \frac{v(t)}{i(t)}$$

$$= \frac{50}{5}$$

$$= 10 \; \Omega$$

fig. 1.25

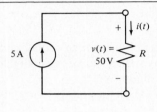

· · ·

If a resistor R is connected to an ideal voltage source $v(t)$ as shown in Fig. 1.26, we conclude that since

$$i(t) = \frac{v(t)}{R}$$

fig. 1.26

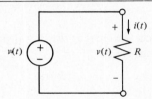

for any particular ideal source $v(t)$, the amount of current $i(t)$ that results can be made to be any finite value by choosing the appropriate value for R (for example, to make $i(t)$ large, make R small). Thus, we see that an ideal voltage source is capable of supplying *any* amount of current, and that the amount depends on what is connected to the source: Only the voltage is constrained to be $v(t)$ volts at the terminals.

Connecting a resistor R to an ideal current source (see Fig. 1.27), we know that

$$v(t) = Ri(t)$$

fig. 1.27

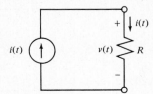

Therefore, for a given current source $i(t)$, the voltage $v(t)$ that results can be made to be any finite value by appropriately choosing R. (For example, to make $v(t)$ large, make R large.) Hence, we conclude that an ideal current source is capable of producing *any* amount of voltage across its terminals, and that amount depends on what is connected to the source: Only the current is constrained to be $i(t)$ amperes through the source.

Physical (nonideal) sources do not have the ability to produce unlimited currents and voltages. As a matter of fact, an actual source may approximate an ideal source only for a limited range of values.

Now consider the two ideal voltage sources whose terminals are connected as shown in Fig. 1.28. By definition of an ideal 3-V source, $v(t)$ must be 3 V. However, by the definition of a 5-V ideal voltage source, $v(t) = 5$ V. Clearly, both conditions cannot

fig. 1.28

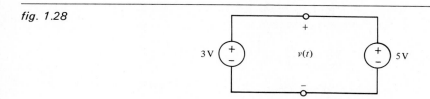

be satisfied simultaneously. (Note that connecting a resistor—or anything else, for that matter—to the terminals will not in any way alleviate this matter.) Therefore, to avoid this paradoxical situation, we will insist that two ideal voltage sources never have their terminals connected together as above. (The only exception is the connection of two sources with the same value and the same polarity.)

Now consider a resistor whose value is zero ohms. An equivalent representation (called a *short circuit*) of such a resistance is shown in Fig. 1.29. By Ohm's law,

$$v(t) = Ri(t)$$

$$= 0i(t)$$

$$= 0 \text{ V}$$

fig. 1.29

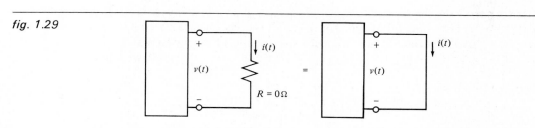

Thus, no matter what finite value $i(t)$ has, $v(t)$ will be zero. Hence we see that *a zero-ohm resistor is equivalent to an ideal voltage source whose value is zero volts*, provided that the current through it is finite. Therefore, in order for a zero resistance to be synonymous with a constraint of zero volts (and to avoid the unpleasantness of infinite currents), we shall insist that we are never allowed to place a short circuit directly across a voltage source as shown in Fig. 1.30. In actuality, the reader will be spared a lot of grief if he or she never attempts this in a laboratory or field situation either.

fig. 1.30

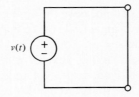

Next consider a resistor having infinite resistance. An equivalent representation (called an *open circuit*) of such a situation is shown in Fig. 1.31. By Ohm's law,

$$i(t) = \frac{v(t)}{R}$$

$$i(t) = 0 \text{ A}$$

fig. 1.31

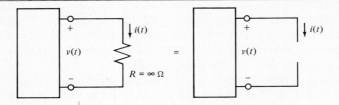

as long as $v(t)$ has a finite value. Thus, we may conclude that *an infinite resistance is equivalent to an ideal current source whose value is zero amperes.* Furthermore, we shall always assume that an ideal current source has something connected to its terminals.

PROBLEMS   *1.9. Consider the circuit shown in Fig. P1.9.
(a) If $i_1 = 4$ A, find v_1.   (c) If $i_3 = 2$ A, find v_3.
(b) If $i_2 = -2$ A, find v_2.   (d) If $i_4 = -2$ A, find v_4.
Ans.   (a) 20 V

fig. P1.9

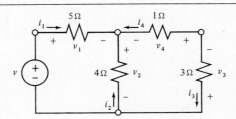

fig. P1.10

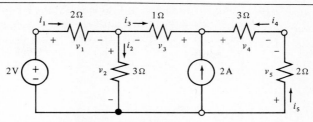

*1.10. Consider the circuit shown in Fig. P1.10.
(a) If $v_1 = -\frac{2}{3}$ V, find i_1.   (d) If $v_4 = \frac{4}{3}$ V, find i_4.
(b) If $v_2 = \frac{8}{3}$ V, find i_2.   (e) If $v_5 = -\frac{14}{9}$ V, find i_5.
(c) If $v_3 = \frac{11}{9}$ V, find i_3.
Ans.   (a) $-\frac{1}{3}$ A

fig. P1.11

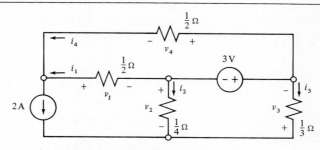

1.11. Consider the circuit shown in Fig. P1.11.
(a) If $i_1 = -2$ A, find v_1.   (c) If $i_3 = \frac{30}{7}$ A, find v_3.
(b) If $v_2 = -\frac{11}{7}$ V, find i_2.   (d) If $v_4 = 2$ V, find i_4.

1.12. Consider the circuit in Problem 1.9.
(a) Find $i_1(t)$ if $v_1(t) = 20$ V.
(b) Find $i_2(t)$ if $v_2(t) = 3e^{-2t}$ V.
(c) Find $v_3(t)$ if $i_3(t) = 6 \sin 2t$ A.
(d) Find $v_4(t)$ if $i_4(t) = -e^{-t} \cos 5t$ A.

fig. P1.13

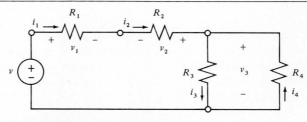

*1.13. Consider the circuit shown in Fig. P1.13.
(a) If $i_1 = 2$ A and $v_1 = 4$ V, find R_1.
(b) If $i_2 = 2$ A and $v_2 = -12$ V, find R_2.
(c) If $i_3 = \frac{4}{3}$ A and $v_3 = 4$ V, find R_3.
(d) If $i_4 = -\frac{2}{3}$ A and $v_3 = 4$ V, find R_4.
Ans. (a) 2 Ω

*1.14. Consider the circuit in Problem 1.13.
(a) If $i_1(t) = -e^{-2t}$ and $v_1(t) = -2e^{-2t}$, find R_1.
(b) If $i_2(t) = -e^{-2t}$ and $v_2(t) = 6e^{-2t}$, find R_2.
(c) If $i_3(t) = -\frac{2}{3}e^{-2t}$ and $v_3(t) = -2e^{-2t}$, find R_3.
(d) If $i_4(t) = \frac{1}{3}e^{-2t}$ and $v_3(t) = -2e^{-2t}$, find R_4.
Ans. (a) $2\,\Omega$

1.3   KIRCHHOFF'S LAWS

As a consequence of the work of the German physicist Gustav Kirchhoff (1824–1887), we are able to analyze an interconnection of any number of elements (voltage sources, current sources, resistors, and elements not even discussed yet). We shall refer to any such system as a *circuit* or a *network*.

For a given circuit, a connection of two or more elements is called a *node* (or *vertex*). An example of a node is depicted in the partial circuit shown in Fig. 1.32. Even if this figure is redrawn to make it appear that there may be more than one node as in Fig. 1.33, the connection of the six elements actually constitutes only one

fig. 1.32

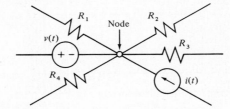

fig. 1.33

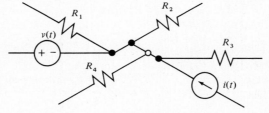

node. As depicted here, we use both a solid dot and a hollow dot to indicate a connection. The hollow dot explicitly represents a node. In cases in which we want to specifically consider a short circuit we will use nodes at both ends of the short circuit.

We now present the first of Kirchhoff's two laws, which is essentially the law of conservation of electric charge.

Kirchhoff's Current Law (KCL)

Kirchhoff's current law may be stated as follows:

KCL: At any node n of a network, at every instant of time, the sum of the currents into n is equal to the sum of the currents out of n.

As an example of KCL, consider a portion of some network, as shown in Fig. 1.34. Applying KCL at the node labeled n, we obtain the equation

$$i_1(t) + i_4(t) + i_5(t) = i_2(t) + i_3(t) + i(t)$$

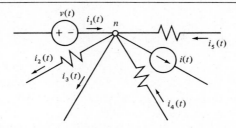

Note that even if one of the elements—the one in which $i_3(t)$ flows—is a short circuit, KCL holds. In other words, KCL applies regardless of the nature of the elements in the circuit.

An alternative, but equivalent, form of KCL can be obtained by considering currents directed into a node to be positive in sense, and currents directed out of a node to be negative in sense (or vice versa). Under this circumstance, the alternative form of KCL can be stated as follows:

KCL: At any node of a circuit, the currents sum to zero.

Applying this form of KCL to node n in Fig. 1.34, selecting currents directed in to be positive in sense, we get

$$i_1(t) - i_2(t) - i_3(t) + i_4(t) - i(t) + i_5(t) = 0$$

A close inspection of the last two equations, however, reveals that they are the same!

From this point on, we shall simplify our notation somewhat by often abbreviating functions of time t such as $v(t)$ and $i(t)$ by v and i, respectively. For instance, we may rewrite the last two equations, respectively, as

$$i_1 + i_4 + i_5 = i_2 + i_3 + i$$

and

$$i_1 - i_2 - i_3 + i_4 - i + i_5 = 0$$

It should always be understood, however, that lowercase letters such as v and i, in general, represent time-varying quantities.*

· · ·

Example

Let us find the voltage v in the two-node circuit shown in Fig. 1.35, in which the directions of i_1, i_2, and i_3 and the polarity of v were chosen arbitrarily. (The directions of the 2-A and 13-A sources are given.)

---

* A constant is a special case of a function of time.

fig. 1.35

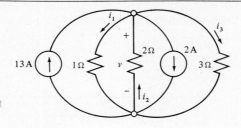

By KCL (at either of the two nodes), we have

$$13 - i_1 + i_2 - 2 - i_3 = 0$$

From this we can write

$$i_1 - i_2 + i_3 = 11$$

By Ohm's law,

$$i_1 = \frac{v}{1} \qquad i_2 = \frac{-v}{2} \qquad i_3 = \frac{v}{3}$$

Substituting these into the previous equation yields

$$\left(\frac{v}{1}\right) - \left(-\frac{v}{2}\right) + \left(\frac{v}{3}\right) = 11$$

$$v + \frac{v}{2} + \frac{v}{3} = 11$$

$$\frac{6v + 3v + 2v}{6} = 11$$

$$\frac{11v}{6} = 11$$

$$v = 6 \text{ V}$$

Having solved for v, we now find that

$$i_1 = \frac{v}{1} = \frac{6}{1} = 6 \text{ A}$$

$$i_2 = -\frac{v}{2} = -\frac{6}{2} = -3 \text{ A}$$

$$i_3 = \frac{v}{3} = \frac{6}{3} = 2 \text{ A}$$

Note that a reordering of the above circuit elements, as shown in Fig. 1.36, will result in the same equation when KCL is applied. Since Ohm's law remains unchanged, the same answers are obtained.

fig. 1.36

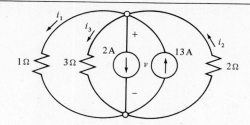

Although the given circuit was drawn in a manner that displayed the nodes explicitly, it could have equally as well been drawn as shown in Fig. 1.37. Do not be confused because the curved lines have been made straight. This is exactly the

fig. 1.37

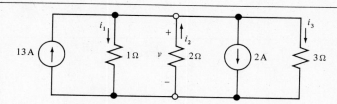

same circuit as the one shown in Fig. 1.35. (It may be helpful to think of the two nodes as being stretched out.)

. . .

The Parallel Connection

Let us now consider the circuit shown in Fig. 1.38. (*Note:* Unless otherwise specified, the value of any resistor will always be assumed to be given in ohms.) For this circuit, let

fig. 1.38

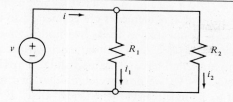

us determine the value of the current i. This circuit clearly has two nodes, and by the definition of an ideal voltage source, the voltage across each resistor is v. By Ohm's law we can write

$$i_1 = \frac{v}{R_1} \quad \text{and} \quad i_2 = \frac{v}{R_2}$$

Applying KCL (at either node) we get

$$i = i_1 + i_2$$

Substituting the above values for i_1 and i_2 in the previous equation, we get

$$i = \frac{v}{R_1} + \frac{v}{R_2}$$

or

$$i = v\left(\frac{1}{R_1} + \frac{1}{R_2}\right)$$

Now consider the simple circuit shown in Fig. 1.39. Clearly,

$$i = \frac{v}{R} = v\left(\frac{1}{R}\right)$$

fig. 1.39

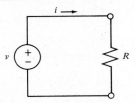

For the same source v in each of the previous two circuits (Figs. 1.38 and 1.39) we now see that we can make the resulting current i the same by choosing R such that

$$\frac{1}{R} = \frac{1}{R_1} + \frac{1}{R_2}$$

Note that the resistors R_1 and R_2 are connected to the same pair of nodes. We call this a *parallel connection*. In general, if m elements are connected to the same pair of nodes (regardless of what else is connected to these two nodes), then we say that these m elements are *connected in parallel*. As a consequence of this definition, we see that elements connected in parallel all have the same voltage across them.

From the discussion above, we now conclude that two resistors, R_1 and R_2, in parallel are equivalent to a single resistor R provided that

$$\boxed{\frac{1}{R} = \frac{1}{R_1} + \frac{1}{R_2}}$$

Therefore,

$$\frac{1}{R} = \frac{R_2 + R_1}{R_1 R_2}$$

or

$$R = \frac{R_1 R_2}{R_1 + R_2}$$

This is a very useful and important formula; so you might as well take 10 seconds now and memorize it!

In concluding that the parallel connection of R_1 and R_2 is equivalent to a single resistance of value $R_1 R_2 / (R_1 + R_2)$, we employed an ideal source to produce the voltage v. However, it really does not matter how the voltage v is produced. For example, even if the voltage results from connecting R_1 in parallel with R_2 to an arbitrary circuit (Fig. 1.40), we can connect the equivalent resistance to the same

fig. 1.40

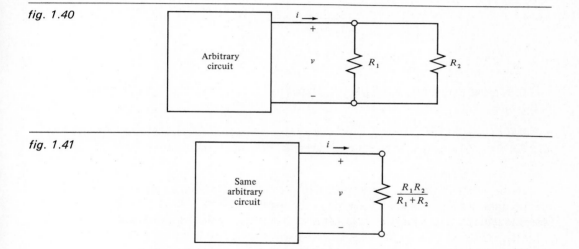

fig. 1.41

arbitrary circuit (Fig. 1.41) and the relationship between v and i will remain the same. In both cases it is

$$i = v\left(\frac{1}{R_1} + \frac{1}{R_2}\right)$$

or

$$i = v\left(\frac{R_1 + R_2}{R_1 R_2}\right)$$

and the effect on the arbitrary circuit is identical.

In summary, in a given circuit, two resistors in parallel can be replaced by an equivalent resistor without affecting the remainder of the circuit. This equivalence is depicted in Fig. 1.42.

fig. 1.42

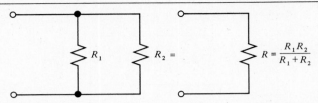

Completely analogous to the discussion above, three resistors R_1, R_2, and R_3 in parallel are equivalent to a single resistance R, provided that

$$\frac{1}{R} = \frac{1}{R_1} + \frac{1}{R_2} + \frac{1}{R_3}$$

This situation is shown in Fig. 1.43.

fig. 1.43

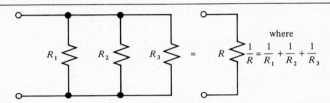

In general, m resistors $R_1, R_2, \ldots, R_m$ in parallel are equivalent to a single resistor R, provided that

$$\frac{1}{R} = \frac{1}{R_1} + \frac{1}{R_2} + \cdots + \frac{1}{R_m}$$

In the discussion above, the reciprocal of a resistance appeared a number of times. It is because of the frequent appearance of the quantity $1/R$ that we denote this by a separate symbol. Given an R-ohm resistor, we define its *conductance* G to be

$$G = \frac{1}{R}$$

Furthermore, since the units of R are ohms, denoted by Ω, we define the units of G to be *mhos* and denote this by $\mho$. Hence, we have two equivalent ways of representing the same element (see Fig. 1.44). The equivalent forms of Ohm's law are as follows:

$$v = Ri = \frac{i}{G}$$

$$i = \frac{v}{R} = Gv$$

$$R = \frac{v}{i} = \frac{1}{G}$$

$$G = \frac{i}{v} = \frac{1}{R}$$

fig. 1.44

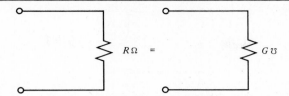

Now note that when combining resistors in parallel, in order to obtain an equivalent resistor, we add conductances, as in Fig. 1.45. If G_1 and G_2 are positive numbers, then from

$$G = G_1 + G_2$$

fig. 1.45

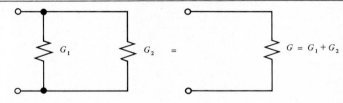

we may deduce that

$$G > G_1 \quad \text{and} \quad G > G_2$$

Hence

$$\frac{1}{R} > \frac{1}{R_1} \quad \text{and} \quad \frac{1}{R} > \frac{1}{R_2}$$

or

$$R < R_1 \quad \text{and} \quad R < R_2$$

Thus the equivalent resistance of two resistors in parallel is less than the value of either of the two resistors.

Now consider the circuit shown in Fig. 1.46. We now ask the question that has been plaguing students of circuit theory for generations: "If I were an ampere, where

fig. 1.46

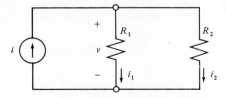

would I go?" To answer this question, we simply replace the parallel connection of R_1 and R_2 by its equivalent resistance as shown in Fig. 1.47. Thus

$$v = Ri = \frac{R_1 R_2 i}{R_1 + R_2}$$

fig. 1.47

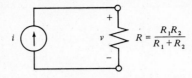

Since this is the same v that appears in the original circuit, we conclude that

$$i_1 = \frac{v}{R_1}$$

and

$$i_1 = \frac{R_2 i}{R_1 + R_2}$$

as well as

$$i_2 = \frac{v}{R_2}$$

and

$$i_2 = \frac{R_1 i}{R_1 + R_2}$$

Since, by KCL, we know that $i_1 + i_2 = i$, the above two boxed formulas describe how the current i is divided by the resistors. For this reason, a pair of resistors in parallel is often referred to as a *current divider*. Note that if R_1 and R_2 are both positive, and if R_1 is greater than R_2, then i_2 is greater than i_1. In other words, a larger amount of current will go through the smaller resistor; thus amperes tend to take the path of least resistance!

In the discussion above, the current i that was divided was produced by an ideal current source. However, it makes no difference where the current comes from, a given current divider will always divide it the same way. For example, the same two boxed current divider formulas given above hold for the circuit of Fig. 1.48, in which a current i is produced.

fig. 1.48

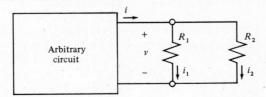

• • •

Example

For the circuit of Fig. 1.49, let us determine the currents indicated by using current division.

fig. 1.49

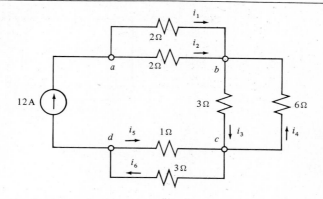

The current into node *a* that is to be divided by the two 2-Ω resistors is 12 A. By the current division formulas,

$$i_1 = i_2 = \frac{2(12)}{2 + 2} = 6 \text{ A}$$

The current into node *b* that is to be divided by the 3-Ω and 6-Ω resistors is $i_1 + i_2$ = 12 A. By the current division formulas, we have

$$i_3 = \frac{6(12)}{3 + 6} = 8 \text{ A} \qquad \text{and} \qquad -i_4 = \frac{3(12)}{3 + 6} = 4 \text{ A}$$

and therefore

$$i_4 = -4 \text{ A}$$

Finally, the current into node *d* that is to be divided by the 1-Ω and 3-Ω resistors is − 12 A. Thus, by current division,

$$i_5 = \frac{3(-12)}{1 + 3} = -9 \text{ A} \qquad \text{and} \qquad -i_6 = \frac{1(-12)}{1 + 3} = -3 \text{ A}$$

or

$$i_6 = 3 \text{ A}$$

• • •

Just as KCL applies to any node of a circuit (i.e., to satisfy the physical law of conservation of charge, the current in must equal the current out) so must KCL hold for any closed region.

In Fig. 1.50 three regions have been identified. Applying KCL to region I, we get

$$i_2 + i_5 = i_4$$

fig. 1.50

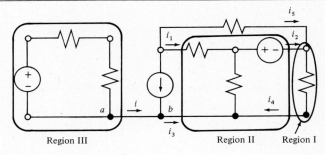

Region III   Region II   Region I

For region II,

$$i_1 + i_3 + i_4 = i_2$$

For region III,

$$i = 0$$

You now may ask, "Since no current flows from point *a* to point *b* (or vice versa) why is the connection (a short circuit) between the points there?" If the connection between the two points is removed, two separate circuits result. The voltages and currents within each individual circuit remain the same as before. Having the connection present constrains points *a* and *b* to be the same node, and hence be at the same potential. It also indicates that the two separate portions are physically connected (even though no current flows between them).

PROBLEMS *1.15.   For the circuit shown in Fig. P1.15, find *v* if (a) $i = 1$ A, (b) $i = 2$ A, (c) $i = 3$ A.
Ans.   (a) 5 V, (b) 0 V, (c) −5 V

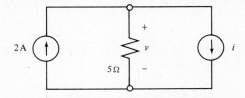

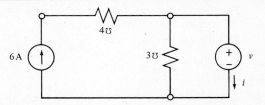

1.16.   For the circuit shown in Fig. P1.16, find *i* if (a) $v = 1$ V, (b) $v = 2$ V, (c) $v = 3$ V.

fig. P1.17

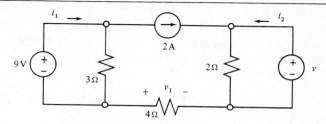

1.17. For the circuit shown in Fig. P1.17, find v_1 if (a) $v = 2$ V, (b) $v = 4$ V, (c) $v = 6$ V.

fig. P1.18

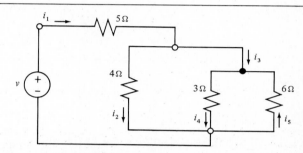

*1.18. For the circuit shown in Fig. P1.18, if $i_1 = 5$ A, determine i_2, i_3, i_4, and i_5.
Ans. $\frac{5}{3}$ A, $\frac{10}{3}$ A, $\frac{20}{9}$ A, $-\frac{10}{9}$ A

1.19. For the circuit of Problem 1.18 assume that $i_4 = 4$ A. Find i_1, i_2, i_3, and i_5.

1.20. For the same circuit, if $i_5 = 4$ A, find i_1, i_2, i_3, and i_4.

1.21. For the same circuit, if $i_2(t) = 6 \cos 5t$, find $i_1(t)$, $i_3(t)$, $i_4(t)$, and $i_5(t)$.

fig. P1.22

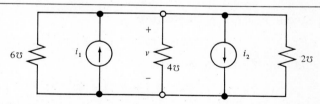

*1.22. Suppose that $i_1 = 2$ A in the circuit shown in Fig. P1.22. Find v if (a) i_2 = 1 A, (b) $i_2 = 2$ A, (c) $i_2 = 3$ A.
Ans.   (a) $\frac{1}{12}$ V, (b) 0 V, (c) $-\frac{1}{12}$ V

1.23. For the circuit given in Problem 1.22, assume that $i_1(t) = 3 \cos 2t$. Find $v(t)$ if $i_2(t) = 5 \cos 2t$.

1.24. Repeat Problem 1.23 for $i_2(t) = 5 \sin 2t$.

Kirchhoff's Voltage Law (KVL)

We shall now present the second of Kirchhoff's laws—the voltage law. To do this, we must introduce the concept of a "loop." Starting at any node n in a circuit, we form a *loop* by traversing through elements (open circuits included!) and returning to the starting node *n*, never encountering any other node more than once. As an example, consider the partial circuit shown in Fig. 1.51. In this circuit the 1-Ω, 2-Ω, 3-Ω, and 6-Ω resistors, along with the 2-A current source constitute the loop *a, b, c, e, f, a*. A few (but not all) of the other loops are: (1) *a, b, c, d, e, f, a*; (2) *c, d, e, c*; (3) *d, e, f, d*; (4) *a, b, c, e, d, f, a*; (5) *a, b, e, f, a*; (6) *b, c, e, b*.

fig. 1.51

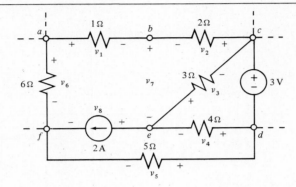

We now can state Kirchhoff's voltage law, as follows:

KVL: If any loop in any circuit is traversed, at every instant of time, the sum of the voltages of one polarity equals the sum of the voltages of the other polarity.

In Fig. 1.51, suppose that the voltages across the elements are as shown. Then by KVL around loop *a, b, c, e, f, a* we have

$$v_1 + v_8 = v_2 + v_3 + v_6$$

and around loop *b, c, d, e, b*, we have

$$v_2 + v_7 = 3 + v_4$$

In this last loop, one of the elements traversed (the element between nodes *b* and *e*) is an open circuit; however, KVL holds regardless of the nature of the elements in the circuit.

Since voltage is energy (or work) per unit charge, then KVL is another way of stating the physical law of the conservation of energy.

An alternative statement of KVL can be obtained by considering voltages across elements that are traversed from + to − to be positive in sense and voltages across elements that are traversed from − to + to be negative in sense (or vice versa). Under this circumstance, KVL has the following alternative form.

KVL: Around any loop in a circuit, the voltages sum to zero.

Applying this form of KVL to the partial circuit shown in Fig. 1.51, selecting a traversal from + to − to be positive in sense, around loop a, b, c, e, f, a we get

$$v_1 - v_2 - v_3 + v_8 - v_6 = 0$$

and around loop b, c, d, e, b we get

$$-v_2 + 3 + v_4 - v_7 = 0$$

These two equations are, respectively, the same as the previous two equations.

. . .

Example
Let us find the current i in Fig. 1.52, where the polarities of v_1, v_2, v_3 and the direction of i were chosen arbitrarily.

fig. 1.52

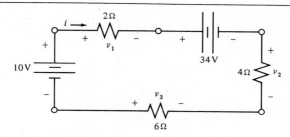

Applying KVL, we get

$$v_1 + 34 + v_2 - v_3 - 10 = 0$$

Thus

$$v_1 + v_2 - v_3 = -24$$

From Ohm's law,

$$v_1 = 2i \qquad v_2 = 4i \qquad v_3 = -6i$$

Substituting these into the previous equation yields

$$(2i) + (4i) - (-6i) = -24$$
$$2i + 4i + 6i = -24$$
$$12i = -24$$
$$i = -2 \text{ A}$$

Having solved for i, we now find that

$$v_1 = 2i = 2(-2) = -4 \text{ V}$$
$$v_2 = 4i = 4(-2) = -8 \text{ V}$$
$$v_3 = -6i = (-6)(-2) = 12 \text{ V}$$

Note that a reordering of the Fig. 1.52 circuit elements, as shown in Fig. 1.53, will result in the same equation when KVL is applied. Since Ohm's law remains unchanged, the same answers are obtained.

fig. 1.53

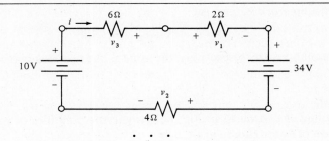

· · ·

The Series Connection

Let us consider the circuit shown in Fig. 1.54 and determine the value of the voltage v.

fig. 1.54

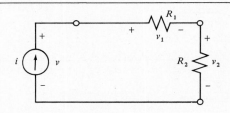

By Ohm's law, we can write

$$v_1 = R_1 i \quad \text{and} \quad v_2 = R_2 i$$

Applying KVL we get

$$v = v_1 + v_2$$

Thus

$$v = R_1 i + R_2 i$$

or

$$v = (R_1 + R_2)i$$

Next consider the simple circuit shown in Fig. 1.55. Clearly,

$$v = Ri$$

fig. 1.55

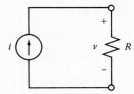

For the same source i in each of the previous two circuits (Figs. 1.54 and 1.55), we now see that we can make the resulting voltage v the same by choosing R such that

$$R = R_1 + R_2$$

Note that the resistors R_1 and R_2 have a node in common, and no other element is connected to this node. We call this a *series connection*. In general, if m elements are connected together such that each resulting node joins no more than two of the elements, then we say that these m elements are *connected in series*. As a consequence of this definition, we see that elements connected in series all have the same current through them.

From the above discussion, we now conclude that two resistors R_1 and R_2 in series are equivalent to a single resistor R provided that

$$R = R_1 + R_2$$

Needless to say, this simple formula should also be committed to memory.

In concluding that the series connection of R_1 and R_2 is equivalent to a single resistance of value $R_1 + R_2$ we employed an ideal source to produce the current i. However, it makes no difference how the current i is produced. For instance, even if the current is caused by connecting the series connection of R_1 and R_2 to an arbitrary circuit as shown in Fig. 1.56, then by connecting the equivalent resistance to the same

fig. 1.56

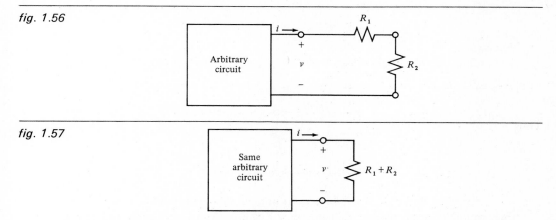

fig. 1.57

arbitrary circuit (Fig. 1.57) the relationship between v and i remains the same. In both cases it is

$$v = i(R_1 + R_2)$$

and the effect on the arbitrary circuit is identical.

In summary, two resistors in series in a given circuit can be replaced by an equivalent resistor without affecting the remainder of the circuit. This equivalence is depicted in Fig. 1.58.

fig. 1.58

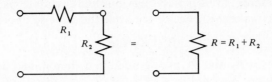

In a manner similar to the above discussion, we can show that three resistors R_1, R_2, and R_3 in series are equivalent to a single resistance R provided that

$$R = R_1 + R_2 + R_3$$

This situation is shown in Fig. 1.59.

fig. 1.59

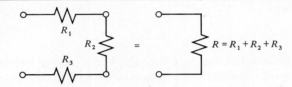

In general, m resistors $R_1, R_2, \ldots, R_m$ in series are equivalent to a single resistor R provided that

$$R = R_1 + R_2 + \cdots + R_m$$

$\cdot \quad \cdot \quad \cdot$

Example
Let us find i in the circuit of Fig. 1.60.

fig. 1.60

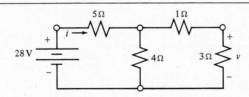

In order to find i, we can replace series and parallel connections of resistors by their equivalent resistances. We begin by noting that the 1-Ω and 3-Ω resistors are in series. Combining them we obtain Fig. 1.61. Note that it is not possible to

fig. 1.61

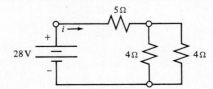

display the original voltage v in this figure. Since the two 4-Ω resistors are connected in parallel, we can further simplify the circuit as shown in Fig. 1.62. Here, the 5-Ω and 2-Ω resistors are in series, so we may combine them and obtain the circuit of Fig. 1.63. Then, from Ohm's law, we have

$$i = \frac{28}{7} = 4 \text{ A}$$

fig. 1.62 (left)
fig. 1.63 (right)

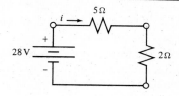 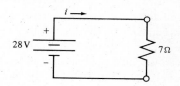

· · ·

Now consider the circuit shown in Fig. 1.64. In order to determine how the voltage v is divided between R_1 and R_2, we can calculate i by replacing the series connection of the resistors by its equivalent resistance as shown in Fig. 1.65. By Ohm's law,

$$i = \frac{v}{R}$$

$$i = \frac{v}{R_1 + R_2}$$

fig. 1.64 (left)
fig. 1.65 (right)

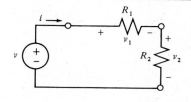 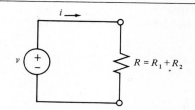

Thus, for the original circuit, we find that

$$v_1 = R_1 i$$

and therefore

$$\boxed{v_1 = \frac{R_1 v}{R_1 + R_2}}$$

Moreover,

$$v_2 = R_2 i$$

and therefore

$$v_2 = \frac{R_2 v}{R_1 + R_2}$$

The above two boxed equations describe how the voltage v is divided by the resistors. Because of this, a pair of resistors in series is often called a *voltage divider*. Note that for positive valued resistors, if R_1 is greater than R_2, then v_1 is greater than v_2. In other words, a larger voltage will be maintained across the larger resistor.

In the foregoing discussion the divided voltage v was produced by an ideal voltage source. However, how the voltage v is produced is irrelevant as far as the division of the voltage is concerned. For instance, the same two boxed voltage divider equations above hold for the circuit of Fig. 1.66, in which a voltage v is produced.

fig. 1.66

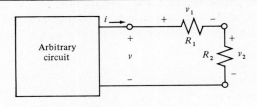

· · ·

Example
Let us find v in the circuit shown in Fig. 1.67.

fig. 1.67

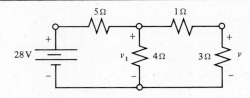

Combining the series connection of the 1-Ω and 3-Ω resistors, we obtain the circuit of Fig. 1.68. Now the pair of 4-Ω resistors in parallel can be combined as shown in Fig. 1.69. By voltage division,

$$v_1 = \frac{2(28)}{2 + 5}$$

$$v_1 = \frac{56}{7} = 8 \text{ V}$$

fig. 1.68

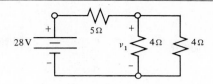

fig. 1.69

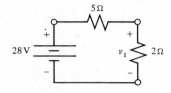

Returning to the original circuit and applying voltage division again yields

$$v = \frac{3v_1}{3 + 1}$$

$$v = \frac{3(8)}{4} = 6 \text{ V}$$

. . .

There is generally more than one way to analyze a given circuit and as long as the different techniques are correctly employed, the same results will be obtained. To demonstrate this, let us again consider the circuit given in the preceding example and take a slightly different approach.

. . .

Example
Consider the circuit shown in Fig. 1.70.

fig. 1.70

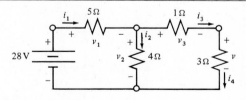

By using Ohm's law we obtain

$$i_4 = \frac{v}{3}$$

By KCL,

$$i_3 = i_4 = \frac{v}{3}$$

By Ohm's law,

$$v_3 = 1i_3 = \frac{v}{3}$$

By KVL,

$$v_2 = v_3 + v = \frac{v}{3} + v = \frac{4v}{3}$$

By Ohm's law,

$$i_2 = \frac{v_2}{4} = \frac{v}{3}$$

By KCL,

$$i_1 = i_2 + i_3$$

$$i_1 = \frac{v}{3} + \frac{v}{3} = \frac{2v}{3}$$

By Ohm's law,

$$v_1 = 5i_1 = \frac{10v}{3}$$

By KVL,

$$v_1 + v_2 = 28$$

$$\frac{10v}{3} + \frac{4v}{3} = 28$$

$$\frac{14v}{3} = 28$$

Hence

$$v = 28\left(\frac{3}{14}\right) = 6 \text{ V}$$

· · ·

Combining Sources

We mentioned before that it is not permissible to connect ideal voltage sources in parallel. However, consider the series connection of two ideal voltage sources (Fig. 1.71). From KVL we know that $v = v_1 + v_2$, and by the definition of an ideal voltage source, this must always be the voltage between nodes a and b, regardless of what is

fig. 1.71

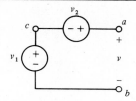

connected to them. Thus, a series connection of two voltage sources as shown in Fig. 1.71 (remember, no other element can be connected to node *c*!) is equivalent to the ideal voltage source shown in Fig. 1.72.

fig. 1.72

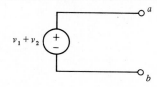

Clearly, the obvious generalization to *m* voltage sources in series holds.

· · ·

Example

In a previous example we determined *i* in the circuit shown in Fig. 1.73. By rearranging the order in this one loop circuit (of course this does not affect *i*), we obtain the circuit shown in Fig. 1.74, from which we obtain Fig. 1.75. By Ohm's law,

$$i = \frac{-24}{2 + 4 + 6}$$

$$i = \frac{-24}{12} = -2 \text{ A}$$

fig. 1.73

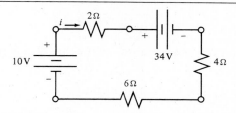

fig. 1.74

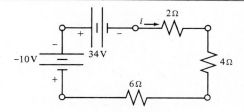

fig. 1.75

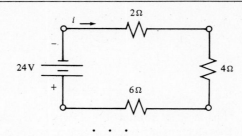

. . .

Let us now consider a parallel connection of ideal current sources as shown in Fig. 1.76. From KCL we find that $i = i_1 + i_2$, and by the definition of an ideal current source, this must always be the current into the arbitrary circuit. Therefore, a parallel

fig. 1.76

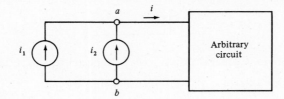

connection of two current sources as shown in Fig. 1.76 (remember, other elements *can* be connected to nodes *a* and *b*!) is equivalent to the ideal current source shown in Fig. 1.77.

fig. 1.77

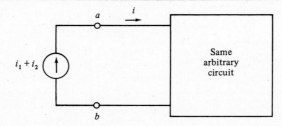

Again, this result can be generalized to the case of *m* current sources in parallel.

. . .

Example
In a previous example, we determined v in the circuit shown in Fig. 1.78.

fig. 1.78

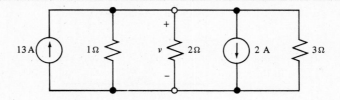

Combining the current sources in parallel we obtain the circuit in Fig. 1.79.

fig. 1.79

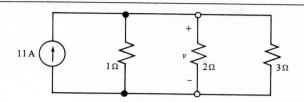

Since the equivalent resistance of the three resistors in parallel is R, where

$$\frac{1}{R} = \frac{1}{1} + \frac{1}{2} + \frac{1}{3} = \frac{6 + 3 + 2}{6} = \frac{11}{6}$$

we obtain

$$R = \frac{6}{11} \, \Omega$$

Then, from Ohm's law,

$$v = \frac{6}{11} \, (11) = 6 \text{ V}$$

. . .

PROBLEMS   1.25.   In the circuits shown in Fig. P1.25, what values of v are permissible?

fig. P1.25

(a) 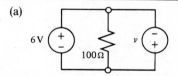   (b)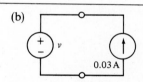

(c)   Is the circuit of part (b) a series or parallel connection?

fig. P1.26

(a) 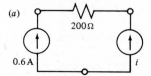   (b)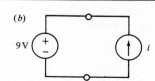

1.26.   In the circuits shown in Fig. P1.26, what values of i are permissible?
*1.27.   In the circuits shown in Fig. P1.27, find the variables indicated.
   Ans.   (a) 2 A, (b) 8 V, (c) 10 V, 19 V, (d) -5 A, 2 A

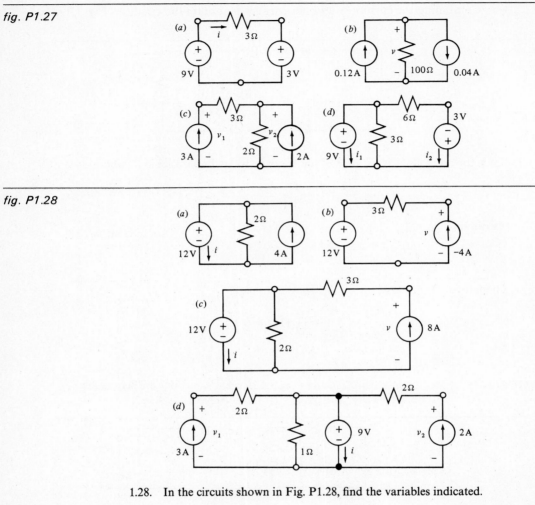

fig. P1.27

fig. P1.28

1.28.   In the circuits shown in Fig. P1.28, find the variables indicated.

fig. P1.29

1.29.   In the circuits shown in Fig. P1.29(a)–(c), find the variables indicated.

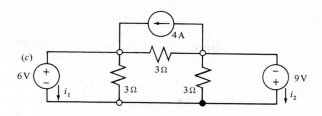

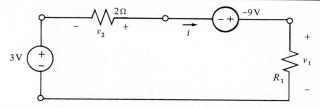

*1.30. In the circuit shown in Fig. P1.30, find the variables indicated for
(a) $R_1 = 2\ \Omega$ and (b) $R_1 = 4\ \Omega$.
Ans. (a) 9 V, $\frac{9}{2}$ A, $-\frac{9}{2}$ A

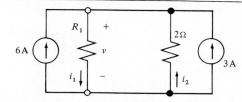

1.31. Repeat Problem 1.30 for the circuit shown in Fig. P1.31.

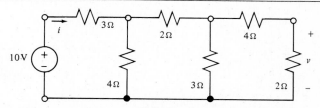

*1.32. In the series-parallel circuit shown in Fig. P1.32, find v and i.
Ans. $\frac{2}{3}$ V, 2 A

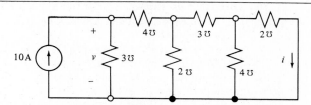

1.33. Repeat Problem 1.32 for the circuit in Fig. P1.33.

fig. P1.34

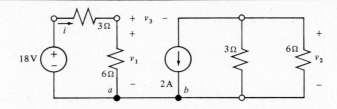

1.34.   Given the circuit in Fig. P1.34,
   (a)   Find i, v_1, v_2, and v_3.
   (b)   Remove the short circuit between a and b. Find i, v_1, and v_2. (Don't try to find v_3—it can't be done!)

fig. P1.35

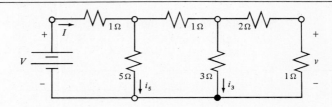

*1.35.   Given the series-parallel circuit in Fig. P1.35.
   (a)   If $v = 2$ volts, what is V?
   (b)   If $i_3 = 3$ amperes, what is V?
   (c)   If $i_5 = 4$ amperes, what is V?
   (d)   What is the resistance $R_{eq} = V/I$ "seen" by the battery for part (a)? part (b)? part (c)?
Ans.   (a) 16 V, (b) 24 V, (c) 32 V, (d) $\frac{8}{3}$ Ω

fig. P1.36

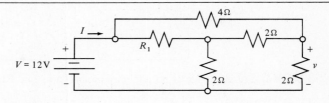

1.36.   Given the non-series-parallel circuit in Fig. P1.36.
   (a)   When $R_1 = \frac{1}{2}$ Ω, then $v = 6$ V. What is the resistance $R_{eq} = V/I$ seen by the battery?
   (b)   When $R_1 = 4$ Ω, then $v = 4$ V. What is R_{eq}?
   (c)   When $v = 3$ V, what are the values of R_1 and R_{eq}?
1.37.   The non-series-parallel circuit in Fig. P1.37 is known as a *twin-T network*.
   (a)   Suppose that $R_1 = 1$ Ω, $R_2 = 3$ Ω, and $v = 3$ V. Find the resistance $R_{eq} = V/I$ seen by the battery.
   (b)   Suppose that $R_2 = \frac{2}{7}$ Ω and $v = 1$ V. Find R_1 and R_{eq}.

fig. P1.37

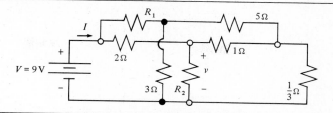

fig. P1.38

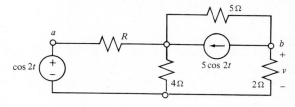

1.38. In the circuit in Fig. P1.38, $v(t) = -6 \cos 2t$. Find R.
1.39. For the circuit given in Problem 1.38, place a 7-Ω resistor between nodes a and b. What value of R results in the same value for $v(t)$?

fig. P1.40

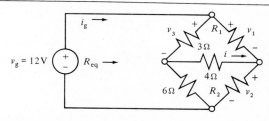

1.40. The circuit in Fig. P1.40 is a non-series-parallel connection known as a *bridge circuit*.
  (a) Given that $R_1 = 6 \, \Omega$, $R_2 = 3 \, \Omega$, and $v_1 = 7$ V, find v_2, i, v_3, and the resistance $R_{eq} = v_g/i_g$ seen by the voltage source.
  (b) Repeat part (a) for $R_1 = 3 \, \Omega$, $R_2 = 6 \, \Omega$, and $v_1 = 4$ V.
  (c) When the current $i = 0$, we say that the bridge is *balanced*. Under what condition (find an expression relating R_1 and R_2) will this bridge be balanced?

1.4 DEPENDENT SOURCES

Up to this point we have been considering voltage and current sources whose values were, in general, time-dependent. However, these values were given to be independent of the behavior of the circuits to which the sources belonged. For this reason, we say that such sources are *independent*.

  We now wish to consider an ideal source, either voltage or current, whose value depends upon some parameter (usually a voltage or current) in the circuit to which the source belongs. We call such an ideal source a *dependent* or *controlled* source, and represent it as shown in Fig. 1.80. Note that the dependent source is represented by a diamond-shaped symbol so as not to confuse it with an independent source.

fig. 1.80

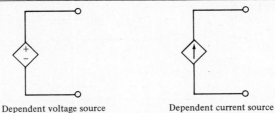

Dependent voltage source   Dependent current source

· · ·

Example

Let us begin with the circuit shown in Fig. 1.81, in which the value of the dependent current source depends on the current i_1 in the 3-Ω resistor. Thus, we say that it is a *current-dependent current source*. Specifically, the value of the dependent

fig. 1.81

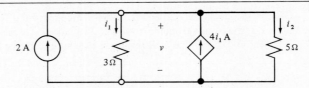

source is $4i_1$, and the units are amperes, of course. Hence, if the current through the 3-Ω resistor is $i_1 = 2$ A, then the dependent current source has a value of $4i_1 = 8$ A; if the current is $i_1 = -3$ A, then the dependent current source has a value of $4i_1 = -12$ A; and so on. To find the actual quantity i_1, we proceed as follows:

Applying KCL (at either node) we can write

$$i_1 + i_2 = 4i_1 + 2$$

Simplifying this equation, we have

$$-3i_1 + i_2 = 2$$

By Ohm's law,

$$i_1 = \frac{v}{3} \quad \text{and} \quad i_2 = \frac{v}{5}$$

Therefore,

$$-3\left(\frac{v}{3}\right) + \frac{v}{5} = 2$$

$$-v + \frac{v}{5} = 2$$

$$\frac{-4v}{5} = 2$$

$$v = \frac{-10}{4} = -\tfrac{5}{2} \text{ V}$$

Hence, we see that in actuality

$$i_1 = \frac{v}{3} = -\tfrac{5}{6} \text{ A}$$

and thus the value of the dependent current source is

$$4i_1 = \frac{-20}{6} = -\tfrac{10}{3} \text{ A}$$

Remember, if the independent source has a different value, then the quantity $4i_1$ will in general be different from $-\tfrac{10}{3}$ A.

. . .

Example

Now let us consider the circuit shown in Fig. 1.82. In this circuit the value of the dependent current source is specified by a voltage! In other words, the value of the

fig. 1.82

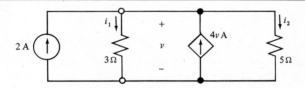

source is $4v$ *amperes*, where v is the *amount* of voltage across the 3-Ω resistor (also the 5-Ω resistor in this case). Such a device is a *voltage-dependent current source*.

To solve for v, we apply KCL and obtain

$$i_1 + i_2 = 4v + 2$$

Thus,

$$\frac{v}{3} + \frac{v}{5} = 4v + 2$$

$$-4v + \frac{8v}{15} = 2$$

$$\frac{-52v}{15} = 2$$

$$v = \frac{-30}{52} = -\tfrac{15}{26} \text{ V}$$

Consequently,

$$4v = -\tfrac{30}{13}$$

and this is the value of the dependent current source, in amperes.

Again, we point out that if the independent source has a value other than 2 amperes, then the dependent source's value $4v$ amperes will be something other than $-\frac{30}{13}$ amperes. The other variables in this circuit are

$$i_1 = \frac{v}{3} = -\tfrac{5}{26}\ \text{A} \qquad \text{and} \qquad i_2 = \frac{v}{5} = -\tfrac{3}{26}\ \text{A}$$

$$\bullet \quad \bullet \quad \bullet$$

Example

The circuit shown in Fig. 1.83 contains a *current-dependent voltage source*. The value of this dependent source is determined by the loop current i.

fig. 1.83

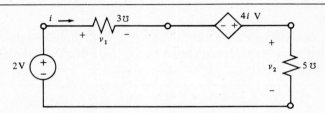

Before we analyze this circuit, let us compare it with the circuit given in the preceding example. What is a parallel connection there is a series connection here, what is a voltage (v) there is a current (i) here, what is a current there (2 A, i_1, i_2, $4v$ A) is a voltage here (2 V, v_1, v_2, $4i$ V, respectively), and what is a resistance there (3 Ω, 5 Ω) is a conductance here (3 ℧, 5 ℧, respectively). For this reason, we say that this circuit is the *dual* of the previous circuit (and vice versa). Once a circuit has been analyzed, its dual is automatically analyzed. A more formal discussion of the concept of duality is given in Chapter 4. However, for the time being, let us verify it for this particular case.

By KVL,

$$v_1 - 4i + v_2 = 2$$

By Ohm's law,

$$v_1 = \frac{i}{3} \qquad \text{and} \qquad v_2 = \frac{i}{5}$$

Thus

$$\frac{i}{3} - 4i + \frac{i}{5} = 2$$

from which

$$\left(\frac{5 - 60 + 3}{15}\right)i = 2$$

or

$$i = \frac{-30}{52} = -\tfrac{15}{26}\ \text{A}$$

which is the value for v in the previous (dual) circuit. The value of the dependent source is

$$4i = 4\left(-\frac{15}{26}\right) = -\frac{30}{13}$$

and the units of this value are volts. Furthermore,

$$v_1 = \frac{i}{3} = -\tfrac{5}{26} \text{ V} \qquad \text{and} \qquad v_2 = \frac{i}{5} = -\tfrac{3}{26} \text{ V}$$

which are the respective values for i_1 and i_2 (in amperes) in the dual circuit.

· · ·

Example

Finally, the circuit shown in Fig. 1.84 contains a *voltage-dependent voltage source*, whose value in this case depends on the voltage across the 4-Ω resistor. To analyze

fig. 1.84

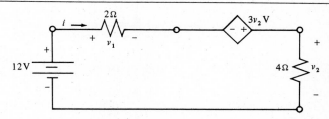

this circuit we proceed as follows:
    By KVL,

$$v_1 - 3v_2 + v_2 = 12$$

or

$$v_1 - 2v_2 = 12$$

By Ohm's law,

$$v_1 = 2i \qquad \text{and} \qquad v_2 = 4i$$

Therefore

$$2i - 2(4i) = 12$$
$$2i - 8i = 12$$
$$-6i = 12$$
$$i = -2 \text{ A}$$

Hence

$$v_2 = 4i = -8 \text{ V}$$

and the value of the dependent voltage source is

$$3v_2 = -24 \text{ V}$$

· · ·

The concept of dependent sources is not introduced simply for abstraction; rather, it is precisely this type of source that models the behavior of such important electronic devices as transistors, amplifiers, and vacuum tubes.

. . .

Example
The circuit of Fig. 1.85 is a simple field-effect transistor (FET) amplifier in which the input is v_1 and the output is v_2. The portion of the circuit in the box formed by the dashed lines is an approximate model of an FET.

fig. 1.85

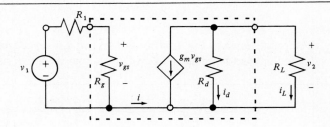

To analyze this circuit, by voltage division,

$$v_{gs} = \frac{R_g v_1}{R_g + R_1}$$

Applying KCL to the right-hand portion of the circuit, we get

$$g_m v_{gs} + i_d + i_L = 0$$

or, substituting,

$$g_m \left(\frac{R_g v_1}{R_g + R_1} \right) + \frac{v_2}{R_d} + \frac{v_2}{R_L} = 0$$

Putting v_1 and v_2 on different sides of the equal sign, we obtain

$$v_2 \left(\frac{1}{R_d} + \frac{1}{R_L} \right) = \frac{-g_m R_g v_1}{R_g + R_1}$$

or

$$v_2 \left(\frac{R_L + R_d}{R_d R_L} \right) = \frac{-g_m R_g v_1}{R_g + R_1}$$

from which

$$v_2 = \frac{-g_m R_d R_g R_L v_1}{(R_g + R_1)(R_L + R_d)}$$

Typical values for an actual amplifier are $g_m = 5 \times 10^{-3}$ ℧, $R_1 = 600$ Ω, $R_g = 1$ MΩ, $R_d = 10$ kΩ, and $R_L = 10$ kΩ. Substituting these into the previous expression, we obtain*

$$v_2 \approx -25v_1$$

so, for example, if the input voltage is $v_1(t) = 0.1 \cos 120\pi t$, then the output voltage is $v_2(t) \approx -2.5 \cos 120\pi t$.

. . .

From this example, you may think that circuit analysis is simple. It is—especially when you analyze a simple circuit.

. . .

Example
The circuit shown in Fig. 1.86 is a simple bipolar transistor amplifier. The portion in the dashed box is an approximate T-model of a transistor in the common-emitter configuration.

fig. 1.86

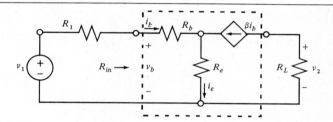

By KVL,

$$v_1 = R_1 i_b + R_b i_b + R_e i_e$$

By KCL,

$$i_e = i_b + \beta i_b = (1 + \beta)i_b$$

Thus,

$$v_1 = R_1 i_b + R_b i_b + R_e(1 + \beta)i_b$$
$$= [R_1 + R_b + (1 + \beta)R_e]i_b$$

so

$$i_b = \frac{v_1}{R_1 + R_b + (1 + \beta)R_e}$$

But, by Ohm's law,

$$v_2 = -R_L \beta i_b$$

---

*The symbol $\approx$ means "approximately equal to."

Thus

$$v_2 = \frac{-R_L \beta v_1}{R_1 + R_b + (1 + \beta)R_e}$$

Typical values for such a circuit are $\beta = 50$, $R_L = 1.5\,\text{k}\Omega$, $R_e = 25\,\Omega$, $R_b = 150\,\Omega$, and $R_1 = 75\,\Omega$. Using these values, we get

$$v_2 = \frac{-(1500)(50)v_1}{75 + 150 + (51)25}$$

$$= -50v_1$$

The voltage across the input of the transistor is

$$v_b = R_b i_b + R_e i_e$$

$$= R_b i_b + R_e(1 + \beta)i_b$$

$$= [R_b + (1 + \beta)R_e]i_b$$

Thus, the resistance R_{in} looking into the input of the transistor is

$$R_{\text{in}} = \frac{v_b}{i_b} = R_b + (1 + \beta)R_e$$

$$= 150 + (51)25 = 1425\,\Omega$$

The resistance R_{eq} seen by the voltage source is

$$R_{\text{eq}} = R_1 + R_{\text{in}}$$

$$= 75 + 1425 = 1500\,\Omega$$

The voltage v_b across the input of the transistor is, by voltage division,

$$v_b = \frac{R_{\text{in}}v_1}{R_{\text{in}} + R_1}$$

$$= \frac{1425v_1}{1425 + 75} = 0.95v_1$$

From the fact that $v_2 = -50v_1$, then $v_1 = -v_2/50$. Thus

$$v_b = 0.95\left(\frac{-v_2}{50}\right)$$

or

$$v_2 = -\frac{50}{0.95}\, v_b = -52.63v_b$$

This result can also be obtained from the fact that

$$i_b = \frac{v_b}{R_b + (1 + \beta)R_e}$$

and

$$v_2 = -R_L\beta i_b = \frac{-R_L\beta v_b}{R_b + (1 + \beta)R_e}$$

$$= \frac{-(1500)(50)v_b}{150 + (51)25} = -52.63v_b$$

. . .

Comments about Circuit Elements

Certain types of nonideal circuit elements, such as positive-valued resistors, come in a wide variety of values and are readily available in the form of discrete components. For many situations it is reasonable to assume that an actual resistor behaves as an ideal resistor.

With few exceptions, however, this is not the case with independent voltage and current sources. Although an actual battery can often be thought of as an ideal voltage source, other nonideal independent sources are approximated by a combination of circuit elements (some of which will be discussed later). Among these elements are the dependent sources, which are not discrete components as are many resistors and batteries but are in a sense part of electronic devices like transistors and operational amplifiers (which consist of numerous transistors and resistors). But don't try to peel open a transistor's metal can (for those that are constructed that way) so that you can see a little diamond-shaped object. The dependent source is a theoretical element that is used to help describe or model the behavior of various electrical devices.

In summary, although ideal circuit elements are not "off-the-shelf" circuit components, their importance lies in the fact that they can be interconnected (on paper or on a computer) to approximate actual circuits that are comprised of nonideal elements and assorted electrical components—thus allowing for the analysis of such circuits.

PROBLEMS *1.41. Given the circuit shown in Fig. P1.41, find i_1 for (a) $K = 2$, (b) $K = 3$, (c) $K = 4$.
Ans.   (a) 2 A, (b) 0 A, (c) −2 A

fig. P1.41

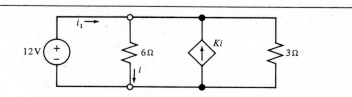

fig. P1.42

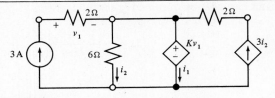

1.42.   Repeat Problem 1.41 for the circuit in Fig. P1.42.

fig. P1.43

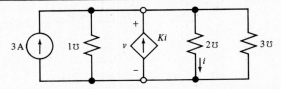

1.43.   Consider the circuit shown in Fig. P1.43. Find v for (a) $K = 2$, (b) $K = 4$.

fig. P1.44

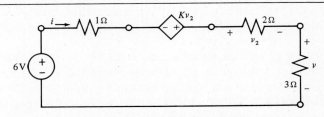

*1.44.   Repeat Problem 1.43 for the circuit in Fig. P1.44.
       Ans.   (a) 9 V, (b) -9 V

fig. P1.45

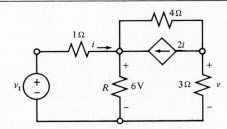

1.45.   For the circuit in Fig. P1.45, find v when (a) $R = 12\ \Omega$, (b) $R = 2\ \Omega$, (c) $R = 8\ \Omega$.

fig. P1.46

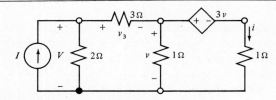

1.46.   Consider the circuit shown in Fig. P1.46.

(a) If $i = 4$ A, what is I?
(b) If $v_3 = 9$ V, what is I?
(c) Find $R_{eq} = V/I$ for parts (a) and (b).

fig. P1.47

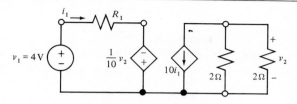

1.47. Consider the circuit in Fig. P1.47.
(a) If $v_2 = -10$ V, what is R_1?
(b) If $v_2 = -20$ V, what is R_1?
(c) Find $R_{eq} = v_1/i_1$ for parts (a) and (b).

fig. P1.48

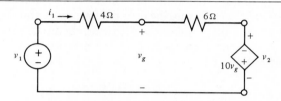

1.48. In the circuit shown in Fig. P1.48, find the output voltage v_2 in terms of the input voltage v_1. Also find $R_{eq} = v_1/i_1$.

fig. P1.49

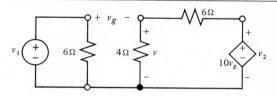

*1.49. Consider the circuit in Fig. P1.49.
(a) Use voltage division to find v in terms of v_g.
(b) Find the output voltage v_2 in terms of the input voltage v_1.
Ans. (a) $4v_g$, (b) $2v_1$,

1.5 POWER

Recall that electrons flow through a resistor from a given potential to a higher potential, and hence current flows from a given potential to a lower one. Potential difference or voltage is a measure of work per unit charge (that is, joules per coulomb). To obtain a current through an element as shown in Fig. 1.87 it takes a certain amount of work or energy—and we say that this energy is absorbed by the element. Taking the product of voltage (energy per unit charge) and current (charge per unit

fig. 1.87

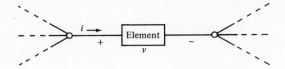

time) we get a quantity that measures energy per unit time. Such a term is known as *power*. It's for this reason that we define $p(t)$ the *instantaneous power absorbed* by an element as in Fig. 1.87 to be the product of voltage and current. That is,

$$p(t) = v(t)i(t)$$

The unit of power (joules per second) is called the *watt* (abbreviated W).* Given an expression for instantaneous power absorbed $p(t)$, to determine the power absorbed at time t_0, simply substitute the number t_0 into the expression.

As with Ohm's law, when using the formula for power absorbed, we must always be conscious of both the voltage polarity and the current direction. For example, for the situation shown in Fig. 1.88, going back to the definition of instantaneous power absorbed, we find in this case that

$$p(t) = -v(t)i(t)$$

fig. 1.88

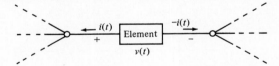

Since power absorbed in a given element can be either a positive or negative quantity (depending on the relationship between voltage and current for the given element), we can say that the element absorbs x watts or, equivalently, that it *supplies* or *delivers* $-x$ watts.

· · ·

Example
Consider the circuit shown in Fig. 1.89.

fig. 1.89

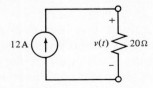

* Named for the Scottish inventor James Watt (1736–1819).

By Ohm's law,

$$v(t) = 20(12) = 240 \text{ V}$$

By definition, the instantaneous power absorbed by the 20-Ω resistor is

$$p_R(t) = v(t)i(t)$$
$$= (240)(12) = 2880 \text{ W} = 2.88 \text{ kW}$$

The instantaneous power absorbed by the current source (observing voltage polarity and current direction) is

$$p_s(t) = -v(t)i(t)$$
$$= -(240)(12) = -2880 \text{ W} = -2.88 \text{ kW}$$

or we may say that the current source delivers $-p_s(t) = 2.88$ kW of power.

· · ·

In the preceding example the expressions for (instantaneous) power absorbed and delivered were not time-dependent; that is, they were constant with respect to time. This is a consequence of the fact that the value of the current source did not vary with time.

· · ·

Example
As a second example, consider the circuit shown in Fig. 1.90.

fig. 1.90

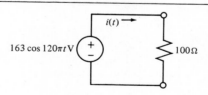

By Ohm's law,

$$i(t) = \frac{v(t)}{R} = \frac{163}{100} \cos 120\pi t \text{ A}$$

By definition, the power absorbed by the resistor is

$$p_R(t) = v(t)i(t)$$
$$p_R(t) = \frac{(163)^2}{100} \cos^2 120\pi t$$
$$p_R(t) = 265.69 \cos^2 120\pi t \text{ W}$$

The power absorbed by the voltage source is

$$p_s(t) = -v(t)i(t)$$

$$p_s(t) = -\frac{(163)^2}{100} \cos^2 120\pi t$$

$$p_s(t) = -265.69 \cos^2 120\pi t \text{ W}$$

In particular, at time $t = 0$ the power absorbed by the resistor is

$$p_R(0) = 265.69 \cos[120\pi(0)]$$

$$= 265.69 \text{ W}$$

whereas the source delivers

$$-p_s(0) = 265.69 \cos[120\pi(0)]$$

$$= 265.69 \text{ W}$$

At time $t = \frac{1}{240}$ second, however, since

$$\cos 120\pi\left(\frac{1}{240}\right) = \cos \frac{\pi}{2} = 0$$

then both the resistor and the source absorb 0 watts.

·   ·   ·

Example
Consider the circuit shown in Fig. 1.91. Let us determine the power absorbed in each of the elements.

fig. 1.91

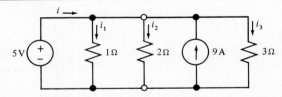

Note that the voltage across each of the elements is 5 V since all the elements are in parallel. Therefore, by Ohm's law,

$$i_1 = \frac{5}{1} = 5 \text{ A} \qquad i_2 = \tfrac{5}{2} \text{ A} \qquad i_3 = \tfrac{5}{3} \text{ A}$$

and the power absorbed in the 1-Ω, 2-Ω, and 3-Ω resistors are

$$p_1 = 5i_1 = 5(5) = 25 \text{ W}$$

$$p_2 = 5i_2 = 5\left(\frac{5}{2}\right) = \tfrac{25}{2} \text{ W}$$

$$p_3 = 5i_3 = 5\left(\frac{5}{3}\right) = \tfrac{25}{3} \text{ W}$$

respectively, for a total of

$$25 + \frac{25}{2} + \frac{25}{3} = \frac{150 + 75 + 50}{6} = \frac{275}{6} \text{ W}$$

absorbed by the resistors.

By KCL,

$$i + 9 = i_1 + i_2 + i_3 = \frac{5}{1} + \frac{5}{2} + \frac{5}{3}$$

or

$$i = \frac{30 + 15 + 10}{6} - 9 = \tfrac{1}{6} \text{ A}$$

Thus the power absorbed by the voltage source is

$$p_s = -5i = -\tfrac{5}{6} \text{ W}$$

Also note that the power absorbed by the current source is

$$p_t = -5(9) = -45 \text{ W}$$

Hence, the total power absorbed is

$$p_1 + p_2 + p_3 + p_s + p_t = 5 + \frac{5}{2} + \frac{5}{3} - \frac{5}{6} - 45$$

$$= \frac{275}{6} - \frac{5}{6} - \frac{270}{6} = 0 \text{ W}$$

Recalling from freshman physics that power is work (energy) per unit time, we see that the fact that the total power absorbed is zero is equivalent to saying that the principle of the conservation of energy is satisfied in this circuit (as it is in any circuit). In this case, the sources supply all the power and the resistors absorb it all.

. . .

Example

Now, however, consider the circuit shown in Fig. 1.92 which is nearly identical to Fig. 1.91.

fig. 1.92

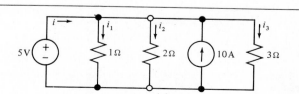

In this case,

$$i_1 = 5 \text{ A} \qquad i_2 = \tfrac{5}{2} \text{ A} \qquad i_3 = \tfrac{5}{3} \text{ A}$$

and

$$p_1 = 25 \text{ W} \qquad p_2 = \tfrac{25}{2} \text{ W} \qquad p_3 = \tfrac{25}{3} \text{ W}$$

as before. By KCL, however,

$$i + 10 = i_1 + i_2 + i_3$$

and thus

$$i = \frac{55}{6} - 10 = -\tfrac{5}{6} \text{ A}$$

Therefore

$$p_s = 5\left(\frac{5}{6}\right) = \tfrac{25}{6} \text{ W}$$

and

$$p_t = -5(10) = -50 \text{ W}$$

Hence the total power absorbed is

$$p_1 + p_2 + p_3 + p_s + p_t = \frac{275}{6} + \frac{25}{6} - 50$$

$$= \frac{300}{6} - 50 = 0 \text{ W}$$

and, again energy (power) is conserved. However, in this case not only do the resistors absorb power, but so does the voltage source. It is the current source that supplies all the power absorbed in the circuit.

· · ·

In all the examples worked so far, the reader may have noted that the power absorbed in every resistor was a nonnegative* number. As we now shall see, this is always the case, for consider Fig. 1.93. By definition, the power absorbed in the resistor is

$$p = vi$$

---

fig. 1.93

---

*A *positive number* is a number that is greater than zero, whereas a *nonnegative number* is a number that is greater than or equal to zero.

But by Ohm's law

$$v = Ri$$

Thus

$$p = (Ri)i$$

or

$$p = Ri^2$$

Also,

$$i = \frac{v}{R}$$

so that

$$p = v\left(\frac{v}{R}\right)$$

or

$$p = \frac{v^2}{R}$$

and both formulas for calculating power absorbed in a resistor R demonstrate that p is always a nonnegative number when R is positive.

· · ·

Example
In the previously given simple transistor amplifier circuit (see Fig. 1.86), the power absorbed by the load resistor R_L is

$$p_2 = \frac{v_2^2}{R_L} = \frac{v_2^2}{1500}$$

and the power supplied by the independent voltage source is

$$p_1 = v_1 i_b$$

$$= v_1\left(\frac{v_1}{R_1 + R_b + (1 + \beta)R_e}\right)$$

or

$$p_1 = \frac{v_1^2}{R_1 + R_b + (1 + \beta)R_e} = \frac{v_1^2}{1500}$$

(Also note that p_1 is the power absorbed by R_{eq}.)

However, since

$$v_2 = -50v_1$$

then

$$p_2 = \frac{(-50v_1)^2}{1500} = \frac{2500v_1^2}{1500}$$

We call the ratio p_2/p_1 the *power gain* from source to load. For this example it is

$$\frac{p_2}{p_1} = \frac{2500v_1^2}{1500}\left(\frac{1500}{v_1^2}\right) = 2500$$

$$\cdot \quad \cdot \quad \cdot$$

A resistor always absorbs power. In a physical resistor, this power is dissipated as heat. In some types of resistors (such as an incandescent lamp or bulb, a toaster, or an electric space heater), this property is desirable in that the net result may be light or warmth. In other types of resistors, as those in electronic circuits, the heat dissipated in a resistor may be a problem that cannot be ignored.

The common carbon resistor comes in values that range from less than 10 Ω to more than 10 MΩ. The physical size of such resistors will determine the amount of power they can safely dissipate. Typical power ratings of carbon resistors are $\frac{1}{8}, \frac{1}{4}, \frac{1}{2}$, 1, and 2 watts. A power dissipation that exceeds the rating of a resistor can physically damage the resistor. When an application requires the use of a resistor that must dissipate more than 2 W of heat, another type—the wire-wound resistor—is often used. Metal film resistors, though more expensive to construct, can have higher power ratings than carbon resistors and are more reliable and stable. In many electronic applications, resistors need dissipate only small amounts of power. This fact, together with present-day technology, allows for the use of resistors that are part of integrated circuits, in addition to the "lumped" circuit elements like the carbon, wire-wound, and metal film resistors mentioned above.

PROBLEMS   *1.50.   A flashlight incandescent lamp is rated at $\frac{1}{2}$ W when used with a (series) connection of two $1\frac{1}{2}$-V batteries. Assuming ideal batteries, what current is drawn by the lamp?
Ans. $\frac{1}{6}$ A

1.51.   For the circuit given in Problem 1.45, find the power absorbed by each element if $R = 1$ Ω.

1.52.   For each circuit given in the first four examples in the section on dependent sources (Figs. 1.81–1.84), find the power absorbed by each element.

fig. P1.53

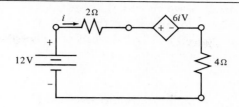

*1.53.   Find the power absorbed by each element in the circuit shown in Fig.
P1.53.

Ans. -12 W, 2 W, 4 W, 6 W

fig. P1.54

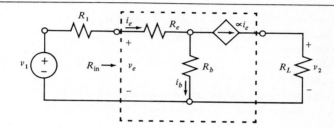

1.54.   The circuit in Fig. P1.54 is a simple bipolar transistor amplifier. The
portion of the circuit in the dashed box is an approximate T-model of a
transistor in the common-base connection.
   (a)   Find i_e in terms of the various resistors, α, and the input voltage v_1.
   (b)   Find v_2 in terms of the resistors, α, and v_1.
   (c)   Given that $R_1 = 40\ \Omega$, $R_e = 27\ \Omega$, $R_b = 150\ \Omega$, $R_L = 1500\ \Omega$, and
$\alpha = 0.98$, find the voltage gain v_2/v_1.
   (d)   Find the numerical power gain p_2/p_1, where p_1 is the power supplied
by v_1 and p_2 is the power absorbed by R_L.
1.55.   Consider the circuit given in Problem 1.54.
   (a)   Find the resistance $R_{eq} = v_1/i_e$ seen by the voltage source.
   (b)   Find the resistance $R_{in} = v_e/i_e$ at the input of the amplifier.
   (c)   Find the voltage gain v_2/v_e.

Summary

1.   An ideal voltage source is a device that produces a
specific (not necessarily constant with time) electric
potential difference across its terminals regardless of
what is connected to it.
2.   An ideal current source is a device that produces a
specific (not necessarily constant with time) current
through it regardless of what is connected to it.

3. A resistor is a device in which the voltage across it is directly proportional to the current through it. If we write the equation (Ohm's law) describing this as $v = Ri$, the constant of proportionality R is called resistance. If we write $i = Gv$, then G is called conductance. Thus, $R = 1/G$.

4. An open circuit is an infinite resistance, and a short circuit is a zero resistance. The former is equivalent to an ideal current source whose value is zero, and the latter is equivalent to an ideal voltage source whose value is zero.

5. At any node of a circuit, the currents sum to zero; this is known as Kirchhoff's current law (KCL).

6. Around any loop in a circuit, the voltages sum to zero; this is known as Kirchhoff's voltage law (KVL).

7. Current going into a parallel connection of resistors divides among them; the smallest resistor has the most current through it.

8. Voltage across a series connection of resistors divides among them; the largest resistor has the most voltage across it.

9. Resistances in series behave as a single resistance whose value equals the sum of the component values.

10. Conductances in parallel behave as a single conductance whose value equals the sum of the component values.

11. Voltage sources in series can be combined into a single voltage source, and current sources in parallel can be combined into a single current source.

12. Voltage and current sources can be dependent as well as independent.

13. A dependent source is a voltage or current source whose value depends on some other circuit parameter.

14. The power absorbed by an element is the product of the current through it and the voltage across it.

CIRCUIT ANALYSIS TECHNIQUES   2

Introduction

The process by which we determine a parameter (either voltage or current) of a circuit is called *analysis*. If all the element voltages and currents are found, we say that the analysis is *complete*. In analyzing circuits throughout this book, the analysis may or may not be complete. Although in practice complete analysis is not usually required, the analysis techniques we shall study will enable us to determine all the parameters of a circuit if we so desire.

Up to this point we have been dealing with circuits that were not very complicated. Don't be fooled, though; some very simple circuits can be quite useful. We have already come across some simple single-stage amplifier circuits whose analysis was accomplished by applying the basic principles covered to date—Ohm's law, KCL, and KVL. Although the fortuitous use of approximate transistor circuit models made this possible, as we shall verify in the future, the results obtained are surprisingly accurate when compared to those obtained by using more sophisticated models. Nonetheless, we must also be able to analyze more complicated circuits, those in which it is simply not possible to conveniently write and solve a single equation having only one unknown. Instead, we shall have to write and solve a set of linear algebraic equations. To obtain such equations, we again utilize Ohm's law, KCL, and KVL.

There are several distinct approaches that we can take. In one we write a set of simultaneous equations in which the variables are voltages; this is known as *nodal analysis*. In another we write a set of simultaneous equations in which the variables are currents; this is known as *mesh analysis*. Although nodal analysis can be used for all circuits, this is not the case for mesh analysis. A class of circuits known as "nonplanar networks" cannot be handled with mesh analysis. However, a similar approach—*loop analysis*—can be used. Here the variables are also currents.

Although in this chapter we shall consider *resistive circuits*, those containing only resistors and sources, both independent and dependent, we shall see later that the same techniques are applicable to networks that contain other types of elements as well.

2.1   NODAL ANALYSIS

Suppose that we wish to analyze the circuit of Fig. 2.1. If we can determine v_1 and v_2, then any other parameter (that is, voltage or current) can be determined as follows. By KVL,

$$v_{12} = v_1 - v_2$$

fig. 2.1

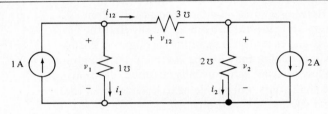

By Ohm's law,

$$i_1 = 1v_1 = v_1$$

$$i_2 = 2v_2$$

$$i_{12} = 3v_{12}$$

One way to determine v_1 and v_2 in this circuit is by a technique known as *nodal analysis*, wherein we write a set of simultaneous equations in which the unknowns are voltages.

The first step in the procedure is the arbitrary choice of a node in the circuit known as the *reference* (or *datum*) *node*. This node is frequently indicated by any of the symbols shown in Fig. 2.2. In this text we shall use the symbol on the left to signify the reference node.

fig. 2.2

A node in a circuit that is labeled v denotes that the potential at that node is v volts above the potential at the reference node. In other words, the voltage between the node labeled v and the reference node is v volts, where the plus is at the former and the minus is at the latter. (Of course, the reference node is at a potential of zero volts with respect to itself.)

· · ·

Example

In the circuit shown in Fig. 2.3 all of the element voltages, as well as the current flowing through each, are indicated.

fig. 2.3

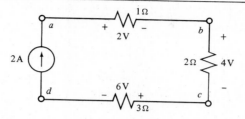

If we select node *d* as the reference, then node *c* is at a potential of 6 V with respect to the reference. Node *b* is at a potential of 10 V, and node *a* is at 12 V with respect to the reference. These facts are summarized in Fig. 2.4. Note that if all the

fig. 2.4

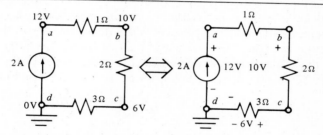

node voltages (with respect to the reference) are known, then the voltage across any element or between any pair of nodes is easily obtained by KVL. For suppose that v_{xy} is the voltage between nodes x and y, with the plus at node x and the minus at node y. Then

$$v_{ac} = v_{ad} - v_{cd} = 12 - 6 = 6 \text{ V}$$

Also

$$v_{ba} = v_{bd} - v_{ad} = 10 - 12 = -2 \text{ V}$$

Suppose now that node *c* is chosen as the reference. Then we have the equivalence shown in Fig. 2.5. Therefore, we have

$$v_{bd} = v_{bc} - v_{dc} = 4 - (-6) = 10 \text{ V}$$

fig. 2.5

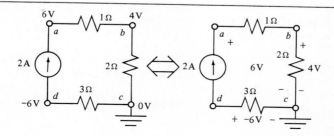

and

$$v_{ba} = v_{bc} - v_{ac} = 4 - 6 = -2 \text{ V}$$

Choosing node b or a as the reference, we have the respective situations shown in Fig. 2.6.

fig. 2.6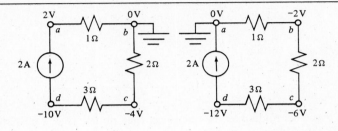

The important thing to remember is that the choice of a reference node does not in any way affect the operation of a circuit. It merely indicates to which node the various node voltages are referenced. In particular, the voltage across any element or between any pair of nodes is the same regardless of which node is chosen as the reference.

Returning now to the circuit given in Fig. 2.1, let us select as the reference the node common to the 1-℧ and 2-℧ conductances. The other two nodes are v_1 and v_2, as shown in Fig. 2.7. For this circuit let us apply KCL at the node labeled v_1. We obtain

$$i_1 + i_{12} = 1$$

fig. 2.7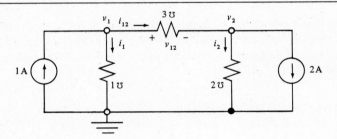

At the node labeled v_2, applying KCL yields

$$i_{12} = i_2 + 2$$

Thus we have two equations (one for each nonreference node) in the three variables i_1, i_2, and i_{12}. However, by Ohm's law, we can express these currents in terms of the voltages v_1, v_2, and v_{12} as follows:

$$i_1 = 1v_1 = v_1$$

$$i_2 = 2v_2$$

$$i_{12} = 3v_{12}$$

Substituting these into the above two equations we get

$$v_1 + 3v_{12} = 1 \quad \text{and} \quad 3v_{12} = 2v_2 + 2$$

But note that by KVL we have

$$v_{12} = v_1 - v_2$$

Hence, substituting this into the previous two equations yields

$$v_1 + 3(v_1 - v_2) = 1 \quad \text{and} \quad 3(v_1 - v_2) = 2v_2 + 2$$

Combining terms, we get the pair of simultaneous equations whose variables are the nonreference node voltages, as follows:

$$4v_1 - 3v_2 = 1$$

and

$$3v_1 - 5v_2 = 2$$

From the first equation, we can write

$$v_1 = \frac{1 + 3v_2}{4}$$

and substituting this into the second equation we get

$$3\left(\frac{1 + 3v_2}{4}\right) - 5v_2 = 2$$

$$\frac{3 + 9v_2}{4} - 5v_2 = 2$$

$$3 + 9v_2 - 20v_2 = 8$$

$$-11v_2 = 5$$

$$v_2 = -\tfrac{5}{11} \text{ V}$$

from which we obtain

$$v_1 = \frac{1 + 3(-5/11)}{4} = \frac{1 - (15/11)}{4} = -\tfrac{1}{11} \text{ V}$$

Alternatively, expressing the two equations as a single matrix equation, we have

$$\begin{bmatrix} 4 & -3 \\ 3 & -5 \end{bmatrix} \begin{bmatrix} v_1 \\ v_2 \end{bmatrix} = \begin{bmatrix} 1 \\ 2 \end{bmatrix}$$

Calculating the following determinants,

$$\Delta = \begin{vmatrix} 4 & -3 \\ 3 & -5 \end{vmatrix} = (-20) - (-9) = -11$$

$$\Delta_1 = \begin{vmatrix} 1 & -3 \\ 2 & -5 \end{vmatrix} = (-5) - (-6) = 1$$

$$\Delta_2 = \begin{vmatrix} 4 & 1 \\ 3 & 2 \end{vmatrix} = 8 - 3 = 5$$

and applying Cramer's rule, we obtain

$$v_1 = \frac{\Delta_1}{\Delta} = -\tfrac{1}{11} \text{ V}$$

and

$$v_2 = \frac{\Delta_2}{\Delta} = -\tfrac{5}{11} \text{ V}$$

Note that the above matrix equation can be written as

$$\mathbf{Gv} = \mathbf{i}$$

where

$$\mathbf{G} = \begin{bmatrix} 4 & -3 \\ 3 & -5 \end{bmatrix} \qquad \mathbf{v} = \begin{bmatrix} v_1 \\ v_2 \end{bmatrix} \qquad \mathbf{i} = \begin{bmatrix} 1 \\ 2 \end{bmatrix}$$

The 2×2 matrix $\mathbf{G}$ has entries that were derived from the conductances of the circuit. (Just trace over the steps in the derivation of the voltage equations to see this fact.) Hence $\mathbf{G}$ is called a *conductance matrix*. Obviously, $\mathbf{v}$ is a voltage matrix. A short search indicates that the entries of $\mathbf{i}$ come from the current sources in the circuit. Thus $\mathbf{i}$ is a current matrix. Therefore we see that the matrix equation

$$\mathbf{Gv} = \mathbf{i}$$

is just Ohm's law in matrix form.

In the foregoing paragraphs we analyzed a circuit by the technique of *nodal analysis*, a method that is described by the more general rules given in the accompanying box.

Nodal Analysis with No Voltage Sources

Given a circuit with n nodes and no voltage sources (we shall discuss the case of voltage sources very shortly), proceed as follows:

1. Select any node as the reference node.
2. Label the remaining $n - 1$ nodes (e.g., $v_1, v_2, \ldots, v_{n-1}$).
3. Arbitrarily assign currents $i_1, i_2, i_3, \ldots$ to the elements in which no current is designated.
4. Apply KCL at each nonreference node.
5. Employ Ohm's law to express the currents $i_1, i_2, i_3, \ldots$ in terms of the node voltages $v_1, v_2, \ldots, v_{n-1}$, and substitute these values into the current equations obtained in step 4.
6. Solve the resulting set of $n - 1$ simultaneous voltage equations to determine the node voltages.

· · ·

Example

In the circuit of Fig. 2.8 we have already selected the reference node (as indicated), labeled the nonreference nodes, and assigned arbitrary currents through the conductances.

fig. 2.8

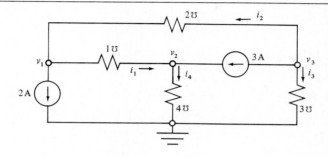

Now let us apply KCL at each nonreference node. We get, at node v_1,

$$i_1 + 2 = i_2$$

at node v_2,

$$i_1 + 3 = i_4$$

and, at node v_3,

$$i_2 + i_3 + 3 = 0$$

Next, by Ohm's law, we find that

$$i_1 = 1(v_1 - v_2)$$
$$i_2 = 2(v_3 - v_1)$$
$$i_3 = 3v_3$$
$$i_4 = 4v_2$$

where we have used KVL in writing the equations for i_1 and i_2. In other words, the voltage across the 1-℧ conductance is $v_1 - v_2$ (the plus on the left) and the voltage across the 2-℧ conductance is $v_3 - v_1$ (the plus on the right). Substituting these expressions for i_1, i_2, i_3, i_4 into the above node equations, we get, for the first equation,

$$1(v_1 - v_2) + 2 = 2(v_3 - v_1)$$
$$v_1 - v_2 + 2 = 2v_3 - 2v_1$$
$$3v_1 - v_2 - 2v_3 = -2$$

for the second equation,

$$1(v_1 - v_2) + 3 = 4v_2$$

$$v_1 - v_2 + 3 = 4v_2$$

$$v_1 - 5v_2 = -3$$

and, for the third equation,

$$2(v_3 - v_1) + 3v_3 + 3 = 0$$

$$-2v_1 + 5v_3 = -3$$

$$2v_1 - 5v_3 = 3$$

Expressing these equations in matrix form, we have

$$\begin{bmatrix} 3 & -1 & -2 \\ 1 & -5 & 0 \\ 2 & 0 & -5 \end{bmatrix} \begin{bmatrix} v_1 \\ v_2 \\ v_3 \end{bmatrix} = \begin{bmatrix} -2 \\ -3 \\ 3 \end{bmatrix}$$

which has the form

$$\mathbf{Gv = i}$$

and hence is the matrix form of Ohm's law.

Thus

$$\Delta = \begin{vmatrix} 3 & -1 & -2 \\ 1 & -5 & 0 \\ 2 & 0 & -5 \end{vmatrix} = 75 - 20 - 5 = 50$$

$$\Delta_1 = \begin{vmatrix} -2 & -1 & -2 \\ -3 & -5 & 0 \\ 3 & 0 & -5 \end{vmatrix} = -50 - 30 + 15 = -65$$

$$\Delta_2 = \begin{vmatrix} 3 & -2 & -2 \\ 1 & -3 & 0 \\ 2 & 3 & -5 \end{vmatrix} = 45 - 6 - 12 - 10 = 17$$

$$\Delta_3 = \begin{vmatrix} 3 & -1 & -2 \\ 1 & -5 & -3 \\ 2 & 0 & 3 \end{vmatrix} = -45 + 6 - 20 + 3 = -56$$

and from Cramer's rule,

$$v_1 = \frac{\Delta_1}{\Delta} = -\frac{65}{50} = -1.30 \text{ V}$$

$$v_2 = \frac{\Delta_2}{\Delta} = \frac{17}{50} = 0.34 \text{ V}$$

$$v_3 = \frac{\Delta_3}{\Delta} = -\frac{56}{50} = -1.12 \text{ V}$$

. . .

Example

Now let us consider a circuit that contains a dependent current source. The configuration in Fig. 2.9 is a single-stage transistor amplifier that, in essence, we previously analyzed without the 15k-Ω resistor in parallel with the dependent current source. To see how this additional resistance affects the previous results, we proceed as follows

fig. 2.9

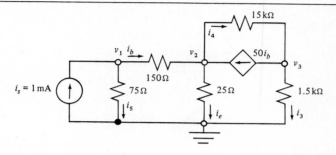

Applying KCL at the nonreference nodes, we get, at node v_1,

$$i_b + i_5 = i_s$$

at node v_2,

$$i_b + 50i_b = i_e + i_4$$

or

$$51i_b = i_e + i_4$$

at node v_3,

$$50i_b + i_3 = i_4$$

and, from Ohm's law,

$$i_b = \frac{v_1 - v_2}{150}$$

$$i_e = \frac{v_2}{25}$$

$$i_3 = \frac{v_3}{1.5\ k}$$

$$i_4 = \frac{v_2 - v_3}{15\ k}$$

$$i_5 = \frac{v_1}{75}$$

Substituting these into the node equations yields, for the first equation,

$$\frac{v_1 - v_2}{150} + \frac{v_1}{75} = 10^{-3}$$

from which

$$3v_1 - v_2 = 0.15$$

For the second equation,

$$51\left(\frac{v_1 - v_2}{150}\right) = \frac{v_2}{25} + \frac{v_2 - v_3}{15 \text{ k}}$$

Simplifying yields

$$5100v_1 - 5701v_2 + v_3 = 0$$

For the third equation,

$$50\left(\frac{v_1 - v_2}{150}\right) + \frac{v_3}{1.5 \text{ k}} = \frac{v_2 - v_3}{15 \text{ k}}$$

from which we obtain

$$5000v_1 - 5001v_2 + 11v_3 = 0$$

Writing these equations in matrix form, we have

$$\begin{bmatrix} 3 & -1 & 0 \\ 5100 & -5701 & 1 \\ 5000 & -5001 & 11 \end{bmatrix} \begin{bmatrix} v_1 \\ v_2 \\ v_3 \end{bmatrix} = \begin{bmatrix} 0.15 \\ 0 \\ 0 \end{bmatrix}$$

Calculating determinants, we obtain

$$\Delta = \begin{vmatrix} 3 & -1 & 0 \\ 5100 & -5701 & 1 \\ 5000 & -5001 & 11 \end{vmatrix} = -188,133 - 5000 + 15,003 + 56,100 = -122,030$$

$$\Delta_1 = \begin{vmatrix} 0.15 & -1 & 0 \\ 0 & -5701 & 1 \\ 0 & -5001 & 11 \end{vmatrix} = -9406.65 + 750.15 = -8656.5$$

$$\Delta_2 = \begin{vmatrix} 3 & 0.15 & 0 \\ 5100 & 0 & 1 \\ 5000 & 0 & 11 \end{vmatrix} = 750 - 8415 = -7665$$

$$\Delta_3 = \begin{vmatrix} 3 & -1 & 0.15 \\ 5100 & -5701 & 0 \\ 5000 & -5001 & 0 \end{vmatrix} = -3,825,765 + 4,275,750 = 449,985$$

and by Cramer's rule,

$$v_1 = \frac{\Delta_1}{\Delta} = \frac{-8656.5}{-122{,}030} = 0.071 \text{ V}$$

$$v_2 = \frac{\Delta_2}{\Delta} = \frac{-7665}{-122{,}030} = 0.063 \text{ V}$$

$$v_3 = \frac{\Delta_3}{\Delta} = \frac{449{,}985}{-122{,}030} = -3.69 \text{ V}$$

The voltage gain for this amplifier is

$$\frac{v_3}{v_1} = -51.98$$

The gain we calculated previously by ignoring the 15k-Ω resistor (removing it) was -52.63, which is roughly a 1 percent difference from this result. Thus the inclusion of the 15k-Ω resistor (i.e., using a more accurate model of the transistor) does not significantly change the result obtained.

The resistance R_{eq} seen by the independent source is

$$R_{eq} = \frac{v_1}{i_s} = \frac{0.071}{0.001} = 71 \ \Omega$$

In the above analysis, exact numerical values were used. In practice, however, approximations are the rule rather than the exception. For this example, 5701 and 5001 would be replaced by 5700 and 5000, respectively. Typically, 5100 may be approximated by 5000 as well. Using the first two approximations we get

$$\Delta = -122{,}000$$

$$\Delta_1 = -8655$$

$$\Delta_2 = -7665$$

$$\Delta_3 = 450{,}000$$

from which

$$\frac{v_3}{v_1} \approx -52$$

The additional use of the last approximation results in

$$\Delta = -123{,}100$$

$$\Delta_1 = -8655$$

$$\Delta_2 = -7500$$

$$\Delta_3 = -525{,}000$$

from which

$$\frac{v_3}{v_1} \approx -60.7$$

Depending on the degree of accuracy desired, therefore, the last approximation may not be an acceptable one.

. . .

We shall now investigate nodal analysis for the case when a circuit contains one or more voltage sources. We begin with an example.

. . .

Example
Consider the circuit shown in Fig. 2.10.

fig. 2.10

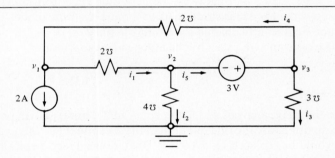

Applying KCL at node v_1, we get

$$i_1 + 2 = i_4$$

and, by Ohm's law,

$$i_1 = 2(v_1 - v_2)$$
$$i_4 = 2(v_3 - v_1)$$

Substituting these expressions for i_1 and i_4 into the previous equation we get

$$2(v_1 - v_2) + 2 = 2(v_3 - v_1)$$
$$2v_1 - 2v_2 + 2 = 2v_3 - 2v_1$$
$$4v_1 - 2v_2 - 2v_3 = -2$$
$$2v_1 - v_2 - v_3 = -1$$

Next applying KCL at node v_2, we get

$$i_1 = i_2 + i_5 \tag{2.1}$$

Since

$$i_1 = 2(v_1 - v_2) \quad \text{and} \quad i_2 = 4v_2$$

we can substitute these expressions into Equation (2.1). However, can we express i_5 in terms of the node voltages? The answer is NO, since Ohm's law does not apply to voltage sources. (Remember that an ideal voltage source can pass any amount of current—it depends on what is connected to the source.) But note that a similar situation will occur when we apply KCL at node v_3. For there

$$i_5 = i_3 + i_4 \tag{2.2}$$

and we can write

$$i_3 = 3v_3 \quad \text{and} \quad i_4 = 2(v_3 - v_1)$$

and we can substitute these values into Equation (2.2), but not an expression in terms of the node voltages for i_5. However, if we combine Equations (2.1) and (2.2) as follows,

$$i_1 = i_2 + i_5$$
$$i_1 = i_2 + (i_3 + i_4)$$
$$i_1 = i_2 + i_3 + i_4 \tag{2.3}$$

we can eliminate i_5 and then substitute the expressions obtained from Ohm's law into Equation (2.3) to obtain

$$2(v_1 - v_2) = 4v_2 + 3v_3 + 2(v_3 - v_1)$$
$$2v_1 - 2v_2 = 4v_2 + 3v_3 + 2v_3 - 2v_1$$
$$4v_1 - 6v_2 - 5v_3 = 0$$

which is only the second equation of node voltages that we have obtained. Thus, so far we have applied KCL at all three of the nonreference nodes and only have two node-voltage equations to show for it. But, the cause of this—the voltage source—actually makes amends, since by KVL, we have that

$$v_3 - v_2 = 3$$

or

$$-v_2 + v_3 = 3$$

which is the missing third equation.

What we effectively did, by combining the two current equations containing i_5, was to think of the voltage source (and its associated nodes) as being one big node, called a *supernode*, as shown in Fig. 2.11.

fig. 2.11

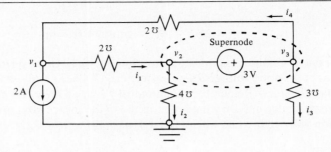

Applying KCL to this supernode, we get

$$i_1 = i_2 + i_3 + i_4$$

from which

$$4v_1 - 6v_2 - 5v_3 = 0$$

as above.

In summary, the three simultaneous node equations that we obtained for the circuit above are

$$2v_1 - v_2 - v_3 = -1$$

$$4v_1 - 6v_2 - 5v_3 = 0$$

$$-v_2 + v_3 = 3$$

which, when written in matrix form, is

$$\begin{bmatrix} 2 & -1 & -1 \\ 4 & -6 & -5 \\ 0 & -1 & 1 \end{bmatrix} \begin{bmatrix} v_1 \\ v_2 \\ v_3 \end{bmatrix} = \begin{bmatrix} -1 \\ 0 \\ 3 \end{bmatrix}$$

Again, note that this equation has the form

$$\mathbf{Gv} = \mathbf{i}$$

From this equation, we obtain

$$\Delta = \begin{vmatrix} 2 & -1 & -1 \\ 4 & -6 & -5 \\ 0 & -1 & 1 \end{vmatrix} = -12 + 4 - 10 + 4 = -14$$

$$\Delta_1 = \begin{vmatrix} -1 & -1 & -1 \\ 0 & -6 & -5 \\ 3 & -1 & 1 \end{vmatrix} = 6 + 15 - 18 + 5 = 8$$

$$\Delta_2 = \begin{vmatrix} 2 & -1 & -1 \\ 4 & 0 & -5 \\ 0 & 3 & 1 \end{vmatrix} = -12 + 30 + 4 = 22$$

$$\Delta_3 = \begin{vmatrix} 2 & -1 & -1 \\ 4 & -6 & 0 \\ 0 & -1 & 3 \end{vmatrix} = -36 + 4 + 12 = -20$$

so that

$$v_1 = \frac{\Delta_1}{\Delta} = \frac{8}{-14} = -\tfrac{4}{7} \text{ V}$$

$$v_2 = \frac{\Delta_2}{\Delta} = \frac{22}{-14} = -\tfrac{11}{7} \text{ V}$$

$$v_3 = \frac{\Delta_3}{\Delta} = \frac{-20}{-14} = \tfrac{10}{7} \text{ V}$$

· · ·

Example
Consider the circuit shown in Fig. 2.12.

fig. 2.12

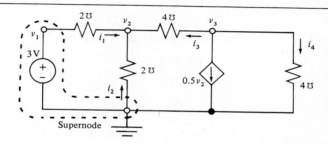

In this case the supernode determined by the voltage source contains the reference node. Therefore, let us not apply KCL at node v_1, but instead we see that, by KVL,

$$v_1 = 3 \text{ V}$$

This is our first node equation! Applying KCL at node v_2, we get

$$i_1 + i_2 + i_3 = 0$$

and since

$$i_1 = 2(v_1 - v_2)$$
$$i_2 = -2v_2$$
$$i_3 = 4(v_3 - v_2)$$

by substitution we find that

$$2(v_1 - v_2) - 2v_2 + 4(v_3 - v_2) = 0$$

or

$$2v_1 - 8v_2 + 4v_3 = 0 \qquad (2.4)$$

Applying KCL at node v_3, we get

$$i_3 + 0.5v_2 + i_4 = 0$$

and hence

$$4(v_3 - v_2) + 0.5v_2 + 4v_3 = 0$$

or

$$-\tfrac{7}{2}v_2 + 8v_3 = 0 \tag{2.5}$$

Of course, we could write these three equations in matrix form as follows:

$$\begin{bmatrix} 1 & 0 & 0 \\ 2 & -8 & 4 \\ 0 & -\tfrac{7}{2} & 8 \end{bmatrix} \begin{bmatrix} v_1 \\ v_2 \\ v_3 \end{bmatrix} = \begin{bmatrix} 3 \\ 0 \\ 0 \end{bmatrix}$$

However, since one of the unknowns is actually known $(v_1 = 3 \text{ V})$, we can substitute this fact into Equation (2.4) to obtain

$$-8v_2 + 4v_3 = -6$$

which, together with Equation (2.5), in matrix form is

$$\begin{bmatrix} -8 & 4 \\ -\tfrac{7}{2} & 8 \end{bmatrix} \begin{bmatrix} v_2 \\ v_3 \end{bmatrix} = \begin{bmatrix} -6 \\ 0 \end{bmatrix}$$

In order to solve this equation, we form

$$\Delta = \begin{vmatrix} -8 & 4 \\ -\tfrac{7}{2} & 8 \end{vmatrix} = -64 + 14 = -50$$

$$\Delta_2 = \begin{vmatrix} -6 & 4 \\ 0 & 8 \end{vmatrix} = -48$$

$$\Delta_3 = \begin{vmatrix} -8 & -6 \\ -\tfrac{7}{2} & 0 \end{vmatrix} = -21$$

from which we obtain

$$v_2 = \frac{\Delta_2}{\Delta} = \frac{-48}{-50} = 0.96 \text{ V}$$

$$v_3 = \frac{\Delta_3}{\Delta} = \frac{-21}{-50} = 0.42 \text{ V}$$

and, since $v_1 = 3$ V, the conductance G_{eq} seen by the independent voltage source is

$$G_{eq} = \frac{i_1}{v_1} = \frac{2(v_1 - v_2)}{v_1}$$

$$= \frac{2(3 - 0.96)}{3}$$

$$= 1.36 \text{ } \mho$$

· · ·

Example

Now let us consider the circuit of Fig. 2.13, in which a dependent voltage source is connected to the reference node. For this circuit, the supernode contains the reference node and the two voltage sources. Because the reference node is included,

fig. 2.13

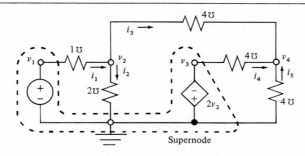

Supernode

we do not apply KCL at the supernode. However, the independent voltage source constrains node voltage v_1. Let's not even be specific as to the function v_1 and, instead, express our answers in terms of v_1.

At node v_2,

$$i_1 = i_2 + i_3$$

and, by Ohm's law,

$$1(v_1 - v_2) = 2v_2 + 4(v_2 - v_4)$$

or

$$7v_2 - 4v_4 = v_1$$

The dependent voltage source gives us the equation

$$v_3 = -2v_2$$

or

$$2v_2 + v_3 = 0$$

Finally, summing currents at node v_4, we get

$$i_3 + i_4 + i_5 = 0$$

so that

$$4(v_2 - v_4) + 4(v_3 - v_4) - 4v_4 = 0$$

and, dividing by 4, we obtain

$$v_2 + v_3 - 3v_4 = 0$$

We now write the node-voltage equations in matrix form

$$\begin{bmatrix} 7 & 0 & -4 \\ 2 & 1 & 0 \\ 1 & 1 & -3 \end{bmatrix} \begin{bmatrix} v_2 \\ v_3 \\ v_4 \end{bmatrix} = \begin{bmatrix} v_1 \\ 0 \\ 0 \end{bmatrix}$$

from which we obtain

$$\Delta = \begin{vmatrix} 7 & 0 & -4 \\ 2 & 1 & 0 \\ 1 & 1 & -3 \end{vmatrix} = -21 - 8 + 4 = -25$$

$$\Delta_2 = \begin{vmatrix} v_1 & 0 & -4 \\ 0 & 1 & 0 \\ 0 & 1 & -3 \end{vmatrix} = -3v_1$$

$$\Delta_3 = \begin{vmatrix} 7 & v_1 & -4 \\ 2 & 0 & 0 \\ 1 & 0 & -3 \end{vmatrix} = 6v_1$$

$$\Delta_4 = \begin{vmatrix} 7 & 0 & v_1 \\ 2 & 1 & 0 \\ 1 & 1 & 0 \end{vmatrix} = 2v_1 - v_1 = v_1$$

and thus

$$v_2 = \frac{\Delta_1}{\Delta} = \frac{-3v_1}{-25} = 0.12v_1$$

$$v_3 = \frac{\Delta_2}{\Delta} = \frac{6v_1}{-25} = -0.24v_1$$

$$v_4 = \frac{\Delta_3}{\Delta} = \frac{v_1}{-25} = -0.04v_1$$

The conductance G_{eq} seen by the independent voltage source is

$$G_{eq} = \frac{i_1}{v_1} = \frac{1(v_1 - v_2)}{v_1}$$

$$= \frac{v_1 - 0.12v_1}{v_1} = 0.88 \; \mho$$

$$\bullet \quad \bullet \quad \bullet$$

In the preceding examples we saw how to deal with voltage sources in a circuit when utilizing nodal analysis. The technique is summarized in the accompanying box.

Nodal Analysis with Voltage Sources

1. Form supernodes corresponding to the voltage sources and apply KCL to each nonreference node that is not part of a supernode.
2. Apply KCL to each supernode not containing the reference node.
3. For each voltage source apply KVL to obtain an equation in which the nonreference node voltages are the variables.

Note that in applying nodal analysis to a circuit, we get the matrix equation form of Ohm's law

$$Gv = i$$

. . .

Example

For demonstration purposes, we finally consider the circuit shown in Fig. 2.14. In this case, since the two voltage sources have a node in common, the supernode

fig. 2.14

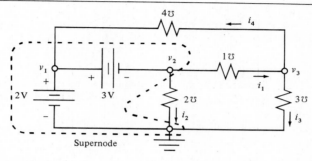

formed is actually composed of the apparent two supernodes determined by each of the voltage sources.

Again, since the supernode contains the reference node, we do not apply KCL, but instead we know that

$$v_1 = 2 \text{ V}$$

and

$$v_2 = -3 + v_1$$
$$v_2 = -3 + 2$$

and hence

$$v_2 = -1 \text{ V}$$

By KCL, at node v_3,

$$i_1 = i_3 + i_4$$

and thus

$$1(v_2 - v_3) = 3v_3 + 4(v_3 - v_1)$$
$$4v_1 + v_2 - 8v_3 = 0$$

By substituting the values for v_1 and v_2 into this equation, we get

$$4(2) + (-1) - 8v_3 = 0$$
$$-8v_3 = -7$$
$$v_3 = \tfrac{7}{8} \text{ V}$$

. . .

PROBLEMS *2.1.   Using nodal analysis on the circuit in Fig. P2.1, with *a* as the reference node, find i_1, i_2, i_3, i_4. (Note that this circuit is identical to Fig. 2.8.)
Ans. $-\frac{41}{25}$ A, $\frac{9}{25}$ A, $-\frac{84}{25}$ A, $\frac{34}{25}$ A

fig. P2.1

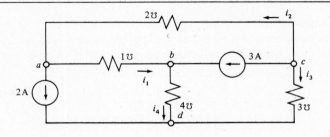

2.2.   Repeat Problem 2.1 using node *c* as the reference node.
2.3.   Repeat Problem 2.1 using node *b* as the reference node.

fig. P2.4

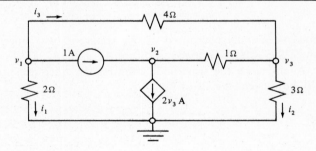

2.4.   Use nodal analysis to find i_1, i_2, i_3 in the circuit shown in Fig. P2.4.

fig. P2.5

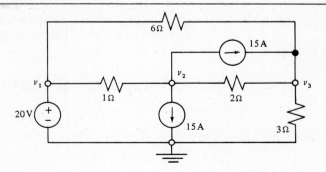

*2.5.   Use nodal analysis to find the node voltages for the circuit in Fig. P2.5.
Ans.   20 V, $-\frac{2}{3}$ V, 18 V
2.6.   Use nodal analysis to find i_1, i_2, i_3 in the circuit shown in Fig. P2.6.

fig. P2.6

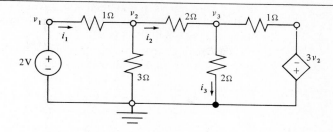

fig. P2.7

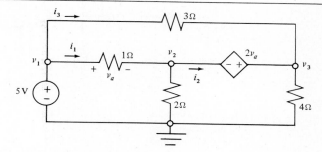

2.7. For the circuit in Fig. P2.7, use nodal analysis to find i_1, i_2, i_3.

fig. P2.8

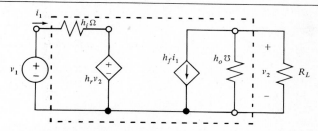

2.8. In the simple transistor amplifier circuit shown in Fig. P2.8, the portion in the dashed box is the *hybrid* or *h-parameter model* of a bipolar transistor. Note that h_i is a resistance and h_o is a conductance. Use nodal analysis to find (a) the voltage gain v_2/v_1 and (b) the resistance $R_{eq} = v_1/i_1$.

fig. P2.9

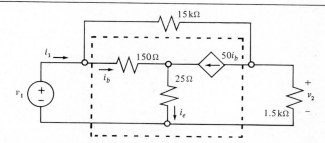

*2.9. The circuit in Fig. P2.9 is a simple bipolar transistor amplifier with "feedback." The portion of the circuit in the dashed box is an approximate

T-model of a transistor in the common-emitter configuration. Use nodal analysis to find (a) the voltage gain v_2/v_1 and (b) the resistance $R_{eq} = v_1/i_1$.
Ans.   (a) -47.76, (b) 253 Ω

2.10.   For the circuit given in Problem 2.9, place a 75-Ω resistor in series with voltage source v_1 and repeat parts (a) and (b).

fig. P2.11

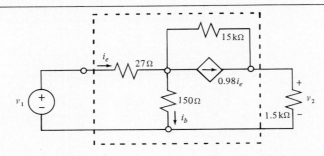

2.11.   The circuit shown in Fig. P2.11 is a simple bipolar transistor amplifier. The portion of the circuit in the dashed box is a T-model of a transistor in the common-base configuration. Use nodal analysis to find (a) the voltage gain v_2/v_1 and (b) the resistance $R_{eq} = v_1/i_e$.

2.12.   For the circuit given in Problem 2.11, place a 40-Ω resistor in series with voltage source v_1 and repeat parts (a) and (b).

fig. P2.13

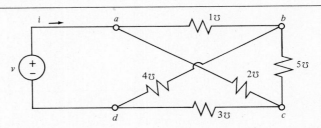

*2.13.   For the circuit shown in Fig. P2.13, use nodal analysis to determine $G = i/v$.
Ans. $\frac{31}{15}$ $\mho$

fig. P2.14a

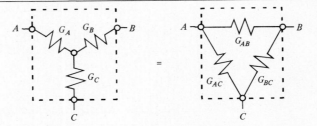

2.14.   Suppose that a circuit contains the dashed box shown on the left in Fig. P2.14a. Without affecting the remainder of the circuit, the box on the left can be replaced by the box on the right provided that

$$G_{AB} = \frac{G_A G_B}{G_A + G_B + G_C} \qquad G_{AC} = \frac{G_A G_C}{G_A + G_B + G_C}$$

$$G_{BC} = \frac{G_B G_C}{G_A + G_B + G_C}$$

Such a process is called a *Y-Δ (wye-delta) transformation*.

fig. P2.14b

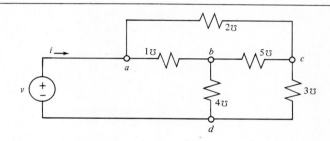

The circuit in Fig. P2.14b is identical to the circuit given in Problem 2.13. Use a Y-Δ transformation on the 1-℧, 4-℧, and 5-℧ conductances, and then combine elements in series and parallel to determine $G = i/v$.

fig. P2.15

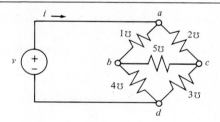

2.15.   With reference to Problem 2.14, the formulas for a Δ-Y (*delta-wye*) *transformation* are

$$R_A = \frac{R_{AB} R_{AC}}{R_{AB} + R_{AC} + R_{BC}} \qquad R_B = \frac{R_{AB} R_{BC}}{R_{AB} + R_{AC} + R_{BC}}$$

$$R_C = \frac{R_{AC} R_{BC}}{R_{AB} + R_{AC} + R_{BC}}$$

where $R = 1/G$.

The circuit shown in Fig. P2.15 is identical to the circuit given in Problem 2.13. Use a Δ-Y transformation on the 1-℧, 2-℧, and 5-℧ conductances, and then combine elements in series and parallel to determine $G = i/v$.

2.16.   The circuit shown in Fig. P2.16 is a connection of 1-Ω resistors. Find the resistance R.

fig. P2.16

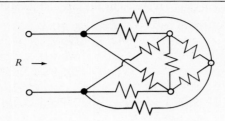

fig. P2.17

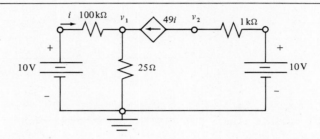

*2.17.   Use nodal analysis to find v_1 and v_2 in the circuit shown in Fig. P2.17.
Ans.   0.123 V, 5.16 V

2.2   *MESH ANALYSIS*

Having discussed one technique for analyzing a circuit, that being nodal analysis, we are now in a position to introduce another. Suppose that we wish to analyze the circuit of Fig. 2.15. Clearly, it would be a simple matter to proceed by nodal analysis. Instead,

fig. 2.15

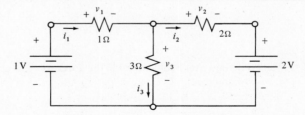

however, note that if we can determine i_1 and i_2, then any other parameter can be determined. By KCL,

$$i_3 = i_1 - i_2$$

By Ohm's law,

$$v_1 = 1i_1 = i_1$$

$$v_2 = 2i_2$$

$$v_3 = 3i_3$$

Since the current flowing "down" through the 3-Ω resistor is i_3 amperes, and since $i_3 = i_1 - i_2$, this current can be broken up into its component parts: i_1 amperes flowing down and $-i_2$ amperes flowing down. However, since $-i_2$ amperes flowing down is identical to the condition of i_2 amperes flowing up, we can represent i_3 equivalently by the two currents in the circuit as shown in Fig. 2.16. Note that the actual current flowing

fig. 2.16

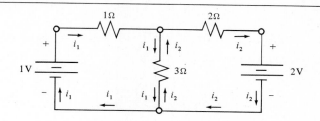

down through the 3-Ω resistor is still $i_1 - i_2 = i_3$. By applying KCL at the other nodes in the circuit, we see that the current flowing up through the 1-V battery is i_1, whereas the current flowing down through the 2-V source is i_2. These situations are also indicated in the circuit. It is because of this that in the loop on the left we can imagine that there is a current i_1 circulating clockwise, and in the loop on the right there is a current i_2 also circulating clockwise. We represent this pictorially in Fig. 2.17. We refer

fig. 2.17

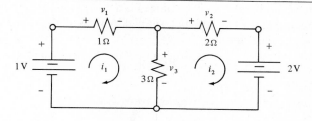

to i_1 and i_2 as *mesh currents*. For this circuit, let us apply KVL to loop i_1—the loop labeled with the mesh current i_1. We have

$$v_1 + v_3 = 1$$

For loop i_2, applying KVL yields

$$-v_2 + v_3 = 2$$

Note that it does not matter in what direction you traverse a loop when applying KVL; the result is always the same.

By Ohm's law,

$$v_1 = 1i_1 = i_1$$
$$v_2 = 2i_2$$
$$v_3 = 3(i_1 - i_2)$$

Substituting these quantities into the two loop equations obtained above, we get

$$i_1 + 3(i_1 - i_2) = 1$$
$$4i_1 - 3i_2 = 1$$

and

$$-2i_2 + 3(i_1 - i_2) = 2$$
$$3i_1 - 5i_2 = 2$$

In matrix form, these equations can be written as

$$\begin{bmatrix} 4 & -3 \\ 3 & -5 \end{bmatrix} \begin{bmatrix} i_1 \\ i_2 \end{bmatrix} = \begin{bmatrix} 1 \\ 2 \end{bmatrix}$$

This matrix equation has the form

$$\mathbf{Ri} = \mathbf{v}$$

where

$$\mathbf{R} = \begin{bmatrix} 4 & -3 \\ 3 & -5 \end{bmatrix} \qquad \mathbf{i} = \begin{bmatrix} i_1 \\ i_2 \end{bmatrix} \qquad \mathbf{v} = \begin{bmatrix} 1 \\ 2 \end{bmatrix}$$

The 2 × 2 matrix **R** has entries that were derived from the resistances of circuit, and is consequently called a *resistance matrix*. Since **i** is a current matrix, and since it can be easily verified that the entries of **v** are determined by the voltage sources, **v** is a voltage matrix. Thus, the matrix equation

$$\mathbf{Ri} = \mathbf{v}$$

is also Ohm's law in matrix form.

For the specific matrix equation given above, we obtain

$$\Delta = \begin{vmatrix} 4 & -3 \\ 3 & -5 \end{vmatrix} = -20 + 9 = -11$$

$$\Delta_1 = \begin{vmatrix} 1 & -3 \\ 2 & -5 \end{vmatrix} = -5 + 6 = 1$$

$$\Delta_2 = \begin{vmatrix} 4 & 1 \\ 3 & 2 \end{vmatrix} = 8 - 3 = 5$$

so by Cramer's rule,

$$i_1 = \frac{\Delta_1}{\Delta} = \frac{1}{-11} = -\tfrac{1}{11} \text{ A}$$

$$i_2 = \frac{\Delta_2}{\Delta} = \frac{5}{-11} = -\tfrac{5}{11} \text{ A}$$

What we have just seen is an example of the technique known as *mesh analysis*. Before listing the rules for applying this method of analysis, we must mention that the set of rules that will follow shortly, unlike that for nodal analysis, will not be valid for all types of circuits. A general procedure, which can be applied to any circuit, will be discussed in the section on loop analysis. We shall now specify those types of circuits to be considered at this point.

In the circuit shown in Fig. 2.18, note that the 4-Ω and 5-Ω resistors are not connected, but that one "crosses over" the other. This situation, however, can be avoided simply by redrawing the circuit as in Fig. 2.19. This circuit, which is identical to the previous one, is displayed in a manner in which no element crosses over another.

fig. 2.18

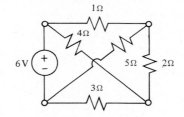

fig. 2.19

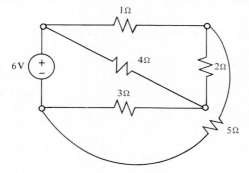

Such a circuit is said to be *planar*; that is, a planar circuit is one which may be drawn in the plane such that there is no element that crosses over another.

For the ten-element circuit (one voltage source and nine resistors) shown in Fig. 2.20, no matter how you try, there is no way of redrawing this circuit so that none of the elements crosses. This, therefore, is an example of a *nonplanar* circuit.

A second example of a nonplanar circuit is shown in Fig. 2.21. This circuit has nine elements: one voltage source and eight resistors. There is no nonplanar circuit having less than nine elements; that is, any circuit with eight or fewer elements must be planar.

fig. 2.20

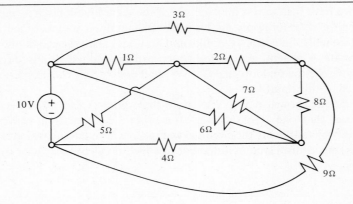

fig. 2.21

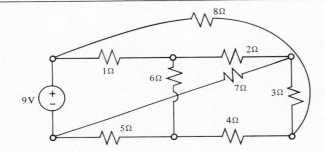

In the discussion of mesh analysis that follows immediately, we shall deal only with planar circuits. Note that in a planar circuit, the elements partition the plane into regions called *meshes*. For example, in Fig. 2.22 the three finite meshes are explicitly displayed. The infinite region that surrounds the circuit will be ignored. With this in mind, we present in the accompanying box the rules for analyzing a planar circuit having *m* meshes and no current sources via mesh analysis. The situation for circuits containing current sources will be discussed immediately afterward.

fig. 2.22

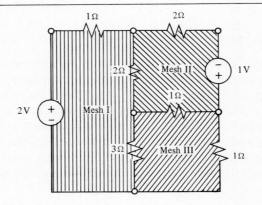

Mesh Analysis for Planar Circuits (No Current Sources)

1. Arbitrarily assign mesh currents (e.g., $i_1, i_2, \ldots, i_m$) to the finite meshes of the circuit.
2. For the elements not assigned a voltage, arbitrarily assign the voltages $v_1, v_2, v_3, \ldots$.
3. Apply KVL to each of the m meshes.
4. For each equation obtained in step 3, express the voltages in terms of the mesh currents $i_1, i_2, \ldots, i_m$, using Ohm's law.
5. Solve the resulting set of m simultaneous mesh equations for the mesh currents $i_1, i_2, \ldots, i_m$.

・ ・ ・

Example

Let us analyze the previous circuit (Fig. 2.22) via mesh analysis. In Fig. 2.23 we have assigned the mesh currents i_1, i_2, i_3 to the three finite meshes in an arbitrary manner. Furthermore, the resistances have been arbitrarily labeled with the voltages $v_1, v_2, v_3, v_4, v_5, v_6$.

fig. 2.23

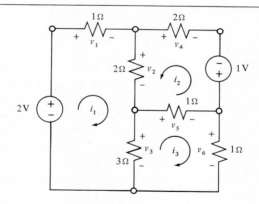

Let us now apply KVL to the three meshes (note that the direction in which a mesh is traversed is irrelevant). For the mesh containing the current i_1 (called mesh i_1 for short),

$$v_1 + v_2 + v_3 - 2 = 0$$

for mesh i_2,

$$v_4 - 1 - v_5 - v_2 = 0$$

and, for mesh i_3,

$$v_3 - v_6 - v_5 = 0$$

By Ohm's law,

$$v_1 = 1i_1 = i_1$$

$$v_2 = 2(i_1 + i_2) = 2i_1 + 2i_2$$

$$v_3 = 3(i_1 - i_3) = 3i_1 - 3i_3$$

$$v_4 = -2i_2$$

$$v_5 = 1(i_2 + i_3) = i_2 + i_3$$

$$v_6 = 1i_3 = i_3$$

Substituting these expressions into the three mesh equations given above yields, for mesh i_1,

$$i_1 + (2i_1 + 2i_2) + (3i_1 - 3i_3) - 2 = 0$$

or

$$6i_1 + 2i_2 - 3i_3 = 2$$

for mesh i_2,

$$-2i_2 - 1 - (i_2 + i_3) - (2i_1 + 2i_2) = 0$$

or

$$-2i_1 - 5i_2 - i_3 = 1$$

and, for mesh i_3,

$$(3i_1 - 3i_3) - i_3 - (i_2 + i_3) = 0$$

or

$$3i_1 - i_2 - 5i_3 = 0$$

Expressing these equations in matrix form, the result is

$$\begin{bmatrix} 6 & 2 & -3 \\ -2 & -5 & -1 \\ 3 & -1 & -5 \end{bmatrix} \begin{bmatrix} i_1 \\ i_2 \\ i_3 \end{bmatrix} = \begin{bmatrix} 2 \\ 1 \\ 0 \end{bmatrix}$$

which is Ohm's law:

$$\mathbf{Ri = v}$$

Calculating determinants,

$$\Delta = \begin{vmatrix} 6 & 2 & -3 \\ -2 & -5 & -1 \\ 3 & -1 & -5 \end{vmatrix} = 150 - 6 - 6 - 45 - 6 - 20 = 150 - 83 = 67$$

$$\Delta_1 = \begin{vmatrix} 2 & 2 & -3 \\ 1 & -5 & -1 \\ 0 & -1 & -5 \end{vmatrix} = 50 + 3 - 2 + 10 = 61$$

$$\Delta_2 = \begin{vmatrix} 6 & 2 & -3 \\ -2 & 1 & -1 \\ 3 & 0 & -5 \end{vmatrix} = -30 - 6 + 9 - 20 = -47$$

$$\Delta_3 = \begin{vmatrix} 6 & 2 & 2 \\ -2 & -5 & 1 \\ 3 & -1 & 0 \end{vmatrix} = 6 + 4 + 30 + 6 = 46$$

Hence, by Cramer's rule,

$$i_1 = \frac{\Delta_1}{\Delta} = \tfrac{61}{67} \text{ A}$$

$$i_2 = \frac{\Delta_2}{\Delta} = \tfrac{-47}{67} \text{ A}$$

$$i_3 = \frac{\Delta_3}{\Delta} = \tfrac{46}{67} \text{ A}$$

. . .

Example
As a second example of mesh analysis, let us consider the circuit shown in Fig. 2.24.

fig. 2.24

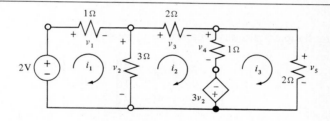

Applying KVL to each of the three meshes we get, for mesh i_1,

$$v_1 + v_2 = 2$$

for mesh i_2,

$$-v_2 + v_3 + v_4 - 3v_2 = 0$$

and, for mesh i_3,

$$3v_2 - v_4 + v_5 = 0$$

and, by Ohm's law,

$$v_1 = 1i_1 = i_1$$
$$v_2 = 3(i_1 - i_2)$$
$$v_3 = 2i_2$$
$$v_4 = 1(i_2 - i_3) = i_2 - i_3$$
$$v_5 = 2i_3$$

Substituting these values into the above mesh equations yields, for the first equation,

$$i_1 + 3(i_1 - i_2) = 2$$
$$i_1 + 3i_1 - 3i_2 = 2$$
$$4i_1 - 3i_2 = 2$$

for the second equation,

$$-4[3(i_1 - i_2)] + 2i_2 + (i_2 - i_3) = 0$$
$$-12i_1 + 12i_2 + 2i_2 + i_2 - i_3 = 0$$
$$-12i_1 + 15i_2 - i_3 = 0$$

and, for the third equation,

$$3[3(i_1 - i_2)] - (i_2 - i_3) + 2i_3 = 0$$
$$9i_1 - 9i_2 - i_2 + i_3 + 2i_3 = 0$$
$$9i_1 - 10i_2 + 3i_3 = 0$$

Writing these equations in matrix form, we obtain

$$\begin{bmatrix} 4 & -3 & 0 \\ -12 & 15 & -1 \\ 9 & -10 & 3 \end{bmatrix} \begin{bmatrix} i_1 \\ i_2 \\ i_3 \end{bmatrix} = \begin{bmatrix} 2 \\ 0 \\ 0 \end{bmatrix}$$

Calculating determinants,

$$\Delta = \begin{vmatrix} 4 & -3 & 0 \\ -12 & 15 & -1 \\ 9 & -10 & 3 \end{vmatrix} = 180 + 27 - 40 - 108 = 59$$

$$\Delta_1 = \begin{vmatrix} 2 & -3 & 0 \\ 0 & 15 & -1 \\ 0 & -10 & 3 \end{vmatrix} = 90 - 20 = 70$$

$$\Delta_2 = \begin{vmatrix} 4 & 2 & 0 \\ -12 & 0 & -1 \\ 9 & 0 & 3 \end{vmatrix} = -18 + 72 = 54$$

$$\Delta_3 = \begin{vmatrix} 4 & -3 & 2 \\ -12 & 15 & 0 \\ 9 & -10 & 0 \end{vmatrix} = 240 - 270 = -30$$

and, by Cramer's rule,

$$i_1 = \frac{\Delta_1}{\Delta} = \frac{70}{59} = 1.19 \text{ A}$$

$$i_2 = \frac{\Delta_2}{\Delta} = \frac{54}{59} = 0.92 \text{ A}$$

$$i_3 = \frac{\Delta_3}{\Delta} = -\frac{30}{59} = -0.51 \text{ A}$$

$\cdot$ $\cdot$ $\cdot$

We shall now consider mesh analysis for the case when a circuit contains one or more current sources, either independent or dependent. We begin with an example.

$\cdot$ $\cdot$ $\cdot$

Example
Consider the circuit shown in Fig. 2.25

fig. 2.25

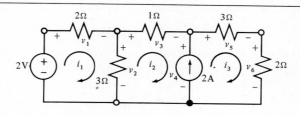

Applying KVL to mesh i_1, we get

$$v_1 + v_2 = 2$$

and, by Ohm's law,

$$v_1 = 2i_1$$

$$v_2 = 3(i_1 - i_2)$$

Substituting these expressions for v_1 and v_2 into the previous equation we get

$$2i_1 + 3(i_1 - i_2) = 2$$

$$5i_1 - 3i_2 = 2$$

Next, applying KVL to mesh i_2, we get

$$-v_2 + v_3 + v_4 = 0$$

Since

$$v_2 = 3(i_1 - i_2)$$

$$v_3 = 1i_2 = i_2$$

we can substitute these expressions into the previous equation. However, we cannot express v_4 in terms of the mesh currents, since Ohm's law does not apply to a current source. (Recall that an ideal current source is able to have any voltage across it; this depends on what is connected to the source.) Note too that the same apparent problem will appear when KVL is applied to mesh i_3. For there we find that

$$-v_4 + v_5 + v_6 = 0$$

and, by Ohm's law,

$$v_5 = 3i_3$$

$$v_6 = 2i_3$$

We now rewrite the voltage equation from mesh i_2 as

$$v_4 = v_2 - v_3$$

and the voltage equation from mesh i_3 as

$$v_4 = v_5 + v_6$$

We can combine these two equations to eliminate v_4. The result is

$$v_2 - v_3 = v_5 + v_6$$

This equation is the equation that results from applying KVL to the *supermesh* which is formed from combining the two meshes having the current source as one of its sides—that is, mesh i_2 and mesh i_3. Traversing this supermesh, we have

$$-v_2 + v_3 + v_5 + v_6 = 0$$

which is the same as the previous equation. Substituting the expressions we obtained from Ohm's law into this equation we get

$$-3(i_1 - i_2) + i_2 + 3i_3 + 2i_3 = 0$$

$$-3i_1 + 4i_2 + 5i_3 = 0$$

Having eliminated the apparent difficulty with v_4, by applying KVL to all three meshes we have obtained only the two equations

$$5i_1 - 3i_2 = 2$$

$$-3i_1 + 4i_2 + 5i_3 = 0$$

in the three variables i_1, i_2, and i_3. The third equation we need to uniquely determine the unknowns i_1, i_2, and i_3 is obtained when we realize that the current flowing up through the current source, which by definition is 2 A, is expressible in terms of the mesh currents as $i_3 - i_2$. We finally have the third equation:

$$i_3 - i_2 = 2$$

or

$$-i_2 + i_3 = 2$$

Writing the three mesh current equations in matrix form, we have

$$\begin{bmatrix} 5 & -3 & 0 \\ -3 & 4 & 5 \\ 0 & -1 & 1 \end{bmatrix} \begin{bmatrix} i_1 \\ i_2 \\ i_3 \end{bmatrix} = \begin{bmatrix} 2 \\ 0 \\ 2 \end{bmatrix}$$

Again, note that this equation has the form

$$\mathbf{Ri} = \mathbf{v}$$

From this equation we obtain

$$\Delta = \begin{vmatrix} 5 & -3 & 0 \\ -3 & 4 & 5 \\ 0 & -1 & 1 \end{vmatrix} = 20 + 25 - 9 = 36$$

$$\Delta_1 = \begin{vmatrix} 2 & -3 & 0 \\ 0 & 4 & 5 \\ 2 & -1 & 1 \end{vmatrix} = 8 - 30 + 10 = -12$$

$$\Delta_2 = \begin{vmatrix} 5 & 2 & 0 \\ -3 & 0 & 5 \\ 0 & 2 & 1 \end{vmatrix} = -50 + 6 = -44$$

$$\Delta_3 = \begin{vmatrix} 5 & -3 & 2 \\ -3 & 4 & 0 \\ 0 & -1 & 2 \end{vmatrix} = 40 + 6 - 18 = 28$$

and thus

$$i_1 = \frac{\Delta_1}{\Delta} = \frac{-12}{36} = -\tfrac{1}{3} \text{ A}$$

$$i_2 = \frac{\Delta_2}{\Delta} = \frac{-44}{36} = -\tfrac{11}{9} \text{ A}$$

$$i_3 = \frac{\Delta_3}{\Delta} = \frac{28}{36} = \tfrac{7}{9} \text{ A}$$

$$\cdot \quad \cdot \quad \cdot$$

Example

Now let us consider the circuit shown in Fig. 2.26. We have already analyzed this by nodal analysis (see Fig. 2.10).

fig. 2.26

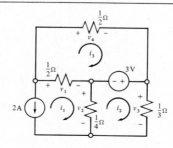

In this case the supermesh resulting from the 2-A current source does not consist of two *finite* meshes. However, the source current is directly expressible in terms of a mesh current; that is,

$$i_1 = -2 \text{ A}$$

This, then is our first mesh equation. Applying KVL to mesh i_2, we have

$$-v_2 - 3 + v_3 = 0$$
$$-v_2 + v_3 = 3$$

and since

$$v_2 = \frac{1}{4}(i_1 - i_2)$$

$$v_3 = \frac{1}{3}i_2$$

by substitution we obtain

$$-\frac{1}{4}(i_1 - i_2) + \frac{1}{3}i_2 = 3$$

$$-\frac{1}{4}i_1 + \frac{7}{12}i_2 = 3$$

Applying KVL to mesh i_3 yields

$$-v_1 + v_4 + 3 = 0$$
$$-v_1 + v_4 = -3$$

and since

$$v_1 = \frac{1}{2}(i_1 - i_3)$$

$$v_4 = \frac{1}{2}i_3$$

we have

$$-\frac{1}{2}(i_1 - i_3) + \frac{1}{2}i_3 = 3$$

$$-\frac{1}{2}i_1 + i_3 = -3$$

In matrix form, the above three mesh-current equations can be written as

$$\begin{bmatrix} 1 & 0 & 0 \\ -\frac{1}{4} & \frac{7}{12} & 0 \\ -\frac{1}{2} & 0 & 1 \end{bmatrix} \begin{bmatrix} i_1 \\ i_2 \\ i_3 \end{bmatrix} = \begin{bmatrix} -2 \\ 3 \\ -3 \end{bmatrix}$$

Instead of solving this matrix equation, however, let us substitute the fact that $i_1 = -2$ A into the second and third equations. The result is:

$$-\frac{1}{4}(-2) + \frac{7}{12}i_2 = 3$$

$$\frac{1}{2} + \frac{7}{12}i_2 = 3$$

$$\frac{7}{12}i_2 = \frac{5}{2}$$

$$i_2 = \frac{30}{7} \text{ A}$$

and

$$-\frac{1}{2}(-2) + i_3 = -3$$

$$1 + i_3 = -3$$

$$i_3 = -4 \text{ A}$$

With these values of i_1, i_2, and i_3, we can now determine the various voltages in the circuit, as follows:

$$v_1 = \frac{1}{2}(i_1 - i_3) = \frac{1}{2}(-2 + 4) = 1 \text{ V}$$

$$v_2 = \frac{1}{4}(i_1 - i_2) = \frac{1}{4}\left(-2 - \frac{30}{7}\right) = -\frac{11}{7} \text{ V}$$

$$v_3 = \frac{1}{3}i_2 = \frac{1}{3}\left(\frac{30}{7}\right) = \frac{10}{7} \text{ V}$$

$$v_4 = \frac{1}{2}i_3 = \frac{1}{2}(-4) = -2 \text{ V}$$

$$\cdot \quad \cdot \quad \cdot$$

Naturally, these values are identical to the results we obtained when we employed nodal analysis on this circuit.

. . .

Example
Let us now consider a circuit that has more than one current source (Fig. 2.27).

fig. 2.27

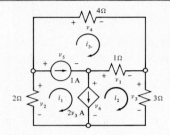

We can of course apply KVL to each mesh and obtain, for mesh i_1,

$$-v_2 + v_5 + v_6 = 0 \tag{2.6}$$

for mesh i_2,

$$-v_6 + v_1 + v_3 = 0 \tag{2.7}$$

and, for mesh i_3,

$$-v_5 + v_4 - v_1 = 0 \tag{2.8}$$

Since v_5 and v_6 cannot be expressed in terms of the mesh currents i_1, i_2, and i_3, we can combine the above equations to obtain a single equation not containing either v_5 or v_6 as follows: From Equation (2.7),

$$v_6 = v_1 + v_3$$

and, from Equation (2.8),

$$v_5 = v_4 - v_1$$

Substituting these into Equation (2.6), we obtain

$$-v_2 + (v_4 - v_1) + (v_1 + v_3) = 0$$

or

$$-v_2 + v_3 + v_4 = 0 \tag{2.9}$$

which corresponds to the loop consisting of the 2-Ω, 3-Ω, and 4-Ω resistors. Note, however, that this loop is the supermesh obtained by combining the meshes containing the current sources.

Since

$$v_2 = -2i_1$$

$$v_3 = 3i_2$$

$$v_4 = 4i_3$$

and substituting these values into Equation (2.9), we get

$$-(-2i_1) + 3i_2 + 4i_3 = 0$$

or

$$2i_1 + 3i_2 + 4i_3 = 0$$

To get two more equations in the variables i_1, i_2, and i_3, we just express the source currents in terms of the mesh currents as follows:

$$i_1 - i_3 = 1$$

and

$$i_1 - i_2 = 2v_3 = 2(3i_2)$$

or

$$i_1 - 7i_2 = 0$$

Expressing the three mesh equations in matrix form, we have

$$\begin{bmatrix} 2 & 3 & 4 \\ 1 & 0 & -1 \\ 1 & -7 & 0 \end{bmatrix} \begin{bmatrix} i_1 \\ i_2 \\ i_3 \end{bmatrix} = \begin{bmatrix} 0 \\ 1 \\ 0 \end{bmatrix}$$

Thus

$$\Delta = \begin{vmatrix} 2 & 3 & 4 \\ 1 & 0 & -1 \\ 1 & -7 & 0 \end{vmatrix} = -3 - 28 - 14 = -45$$

$$\Delta_1 = \begin{vmatrix} 0 & 3 & 4 \\ 1 & 0 & -1 \\ 0 & -7 & 0 \end{vmatrix} = -28$$

$$\Delta_2 = \begin{vmatrix} 2 & 0 & 4 \\ 1 & 1 & -1 \\ 1 & 0 & 0 \end{vmatrix} = -4$$

$$\Delta_3 = \begin{vmatrix} 2 & 3 & 0 \\ 1 & 0 & 1 \\ 1 & -7 & 0 \end{vmatrix} = 3 + 14 = 17$$

and hence

$$i_1 = \frac{\Delta_1}{\Delta} = \frac{-28}{-45} = 0.62 \text{ A}$$

$$i_2 = \frac{\Delta_2}{\Delta} = \frac{-4}{-45} = 0.089 \text{ A}$$

$$i_3 = \frac{\Delta_3}{\Delta} = \frac{17}{-45} = -0.38 \text{ A}$$

$$\cdot \quad \cdot \quad \cdot$$

This last example has demonstrated that in analyzing (planar) circuits having current sources by mesh analysis we can proceed in two ways. First, we can apply KVL to each mesh and then manipulate the resulting equations in order to eliminate the appearance of voltages across current sources. Second, we can instead apply KVL to the meshes (supermeshes) obtained by considering the current sources as open circuits; thinking of a current source as an open circuit is just a temporary convenience for applying KVL.

In both methods, expressing the source currents in terms of the mesh currents results in the remaining necessary equations.

PROBLEMS *2.18. Use mesh analysis to find v in the circuit in Fig. P2.18.
          Ans. $\frac{7}{8}$ V

fig. P2.18

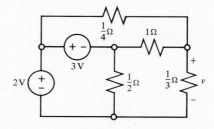

fig. P2.19

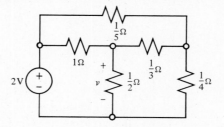

2.19. Use mesh analysis to find v in the circuit shown in Fig. P2.19.

fig. P2.20

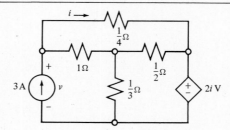

2.20. Use mesh analysis to find v in the circuit shown in Fig. P2.20.
*2.21. Use mesh analysis to find v in the circuit shown in Fig. P2.21.
        Ans. -0.18 V

fig. P2.21

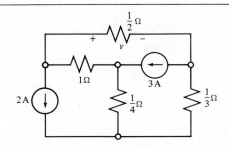

fig. P2.22

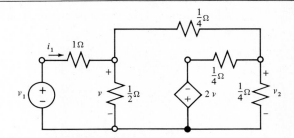

2.22.   For the circuit shown in Fig. P2.22, use mesh analysis to find (a) the voltage gain v_2/v_1 and (b) the resistance $R_{eq} = v_1/i_1$.

fig. P2.23

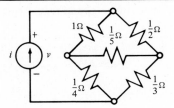

*2.23.   For the circuit shown in Fig. P2.23, use mesh analysis to determine $R = v/i$.
    Ans. $\frac{15}{31}\,\Omega$

fig. P2.24

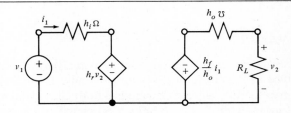

2.24.   The simple transistor amplifier in Fig. P2.24 incorporates an alternative h-parameter equivalent circuit of the transistor. Use mesh analysis to find (a) the voltage gain v_2/v_1 and (b) the resistance $R_{eq} = v_1/i_1$.

fig. P2.25

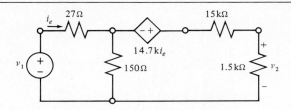

2.25. For the transistor amplifier circuit in Fig. P2.25, use mesh analysis to find (a) the voltage gain v_2/v_1 and (b) the resistance $R_{eq} = v_1/i_e$.

2.26. For the circuit given in Problem 2.25, place a 40-Ω resistor in series with voltage source v_1 and repeat parts (a) and (b).

fig. P2.27

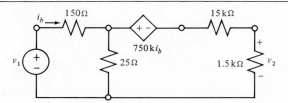

2.27. For the transistor amplifier circuit shown in Fig. P2.27, use mesh analysis to find (a) the voltage gain v_2/v_1 and (b) the resistance $R_{eq} = v_1/i_b$.

2.28. For the circuit given in Problem 2.27, place a 75-Ω resistor in series with voltage source v_1 and repeat parts (a) and (b).

fig. P2.29

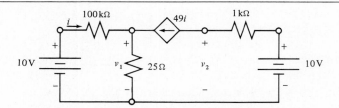

*2.29. Use mesh analysis to find v_1 and v_2 in the circuit in Fig. P2.29.
   Ans.   0.123 V, 5.16 V

2.3   LOOP ANALYSIS

In our discussions of nodal analysis and mesh analysis, we established rules to follow which enabled us to systematically analyze circuits. In the case of nodal analysis, the technique presented is valid for all circuits—both planar and nonplanar. However, for the case of mesh analysis, the technique presented is valid only for planar circuits. Nonplanar circuits can be analyzed via a technique known as loop analysis, which is quite similar to mesh analysis. In order to study this technique, we must first introduce some concepts from the mathematical discipline known as *graph theory*.

Perhaps you have referred to a plot of some function $f(x)$ versus x as a "graph." This is precisely the notion that we are *not* referring to when we speak about a graph. A *graph* is merely a collection of points and lines. The points are called *nodes*, and the lines are called *edges*. Each end of an edge is connected to a node, and both ends of an edge may be connected to the same node. As a matter of convention, the edges are drawn by smooth—but not necessarily straight—lines. A few examples of graphs are shown in Fig. 2.28. In a graph, two or more edges can meet only at a node. Edges can be drawn "crossing over" each other, but this does not mean that they intersect.

fig. 2.28

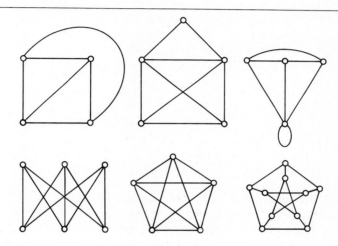

Having introduced the concept of a graph, we shall now proceed to represent circuits as graphs so that we may discuss loop analysis.

As we mentioned earlier, a node in a circuit is determined by the connection of two or more circuit elements—excluding short circuits and open circuits. Given a circuit, we can form a graph by replacing each circuit element (of course, excluding short circuits and open circuits) by an edge.

As an example, for the circuit shown in Fig. 2.29 the nodes are labeled $a, b, c,$ and d. Replacing each element by a single edge, we get the corresponding graph shown in Fig. 2.30. In this graph, edges e_3 and e_4 correspond to the voltage and current sources,

fig. 2.29

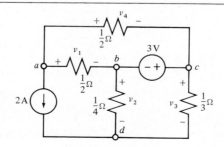

fig. 2.30

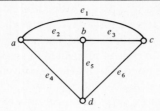

respectively. The remaining edges correspond to the resistors. Figure 2.30 is an example of a *connected graph*, that is, a graph in which any node can be reached from any other node by traversing (walking along) the edges. An example of an *unconnected* or *disconnected graph* is shown in Fig. 2.31. In this graph, for example, there is no way of getting from node b to node g by traversing the edges.

fig. 2.31

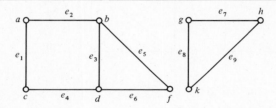

Given a set of n nodes, what is the fewest number of edges required to connect all the nodes? Clearly, one edge placed between any pair of nodes, say n_1 and n_2, will connect them. A second edge placed between n_3 and either n_1 or n_2 will connect the three nodes: n_1, n_2, n_3. A third edge placed between n_4 and either n_1 or n_2 or n_3 will connect the four nodes: n_1, n_2, n_3, n_4. Continuing in this manner, we see that $n - 1$ edges (and no fewer) can be used to connect n nodes. We call a connected graph with n nodes and $n - 1$ edges a *tree*. Examples of trees with four, five, and six nodes are shown in Fig. 2.32.

fig. 2.32

Since $n - 1$ is the fewest number of edges required to connect n nodes, a connected graph having n nodes contains a minimum of $n - 1$ edges. In other words, a connected n-node graph that is not a tree will contain a number of n-node trees called *spanning trees*. Each edge of a spanning tree is called a *branch*.

· · ·

Example
Consider the four-node connected graph in Fig. 2.33, which was discussed earlier. This graph contains 16 spanning trees. They are shown in Fig. 2.34.

fig. 2.33

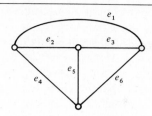

fig. 2.34

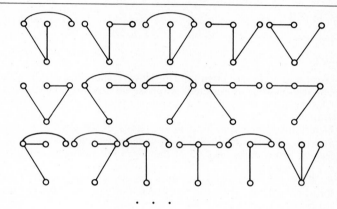

· · ·

Given an *n*-node connected graph, unfortunately there is no simple way of determining the number of spanning trees contained in the graph. Given a graph, let us select some node n_1. Suppose that we start at n_1 and traverse some of the edges of the graph such that we finish at n_1. If we do not traverse any edge more than once and if we do not encounter any node except n_1 more than once and n_1 exactly twice, then the resulting set of edges traversed is called a *loop*.

· · ·

Example

For the graph given in Fig. 2.33, the set of edges e_1, e_2, e_3 forms a loop, as does the set e_3, e_5, e_6. Actually, all the loops of this graph are depicted in Fig. 2.35.

fig. 2.35

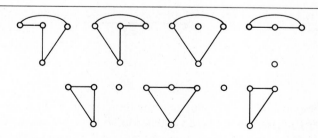

· · ·

Given a graph, unfortunately there is no simple way of determining the number of loops contained in the graph. Note that a tree cannot contain any loops, since any edge in a loop can be removed without disconnecting the nodes connected by that loop.

In our previous discussion of mesh analysis, we placed a mesh current in each window or mesh of a planar circuit. However, we could not extend this idea to nonplanar circuits for it is not apparent where to place the appropriate mesh currents in such cases. Loop analysis, though, tells us how to select appropriate loop currents for nonplanar, as well as planar, circuits. The technique is outlined in the accompanying box.

Loop Analysis

1. Given a (connected) circuit, form the corresponding graph of the circuit.
2. For the resulting (connected) graph, select any spanning tree.
3. Choose any edge not in the spanning tree—that is, any edge that is not a branch.
4. Determine the unique loop that consists of the edge chosen in step 3 and appropriate tree branches. This is called a *fundamental loop.*
5. Repeat this process until all of the edges not in the spanning tree have been chosen in step 3.
6. Establish loop currents in the given circuit that correspond to the fundamental loops determined above. (The direction in which a loop is traversed is again arbitrary.) Then proceed as for mesh analysis.

Following these steps, for each edge not in the selected spanning tree, we get a unique loop. If the graph has n nodes and e edges, since there are $n - 1$ branches in a spanning tree, there are $e - (n - 1) = e - n + 1$ edges not in the tree. Hence, for loop analysis, we write $e - n + 1$ loop equations.

· · ·

Example
Suppose that the corresponding graph of a circuit is the graph considered in Fig. 2.33. Furthermore, suppose that the tree selected is the one indicated by the bold edges (e_2, e_3, e_5) in Fig. 2.36.

---

fig. 2.36

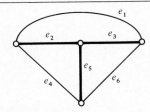

Choosing edge e_1 results in the fundamental loop e_1, e_2, e_3, whereas choosing e_4 results in the fundamental loop e_4, e_2, e_5. Finally, choosing e_6 yields e_6, e_3, e_5. Thus, for the spanning tree selected, we obtain the loops that would result by using mesh analysis. Since we have already analyzed this circuit with mesh analysis in Section 2.2, let us select the spanning tree depicted in Fig. 2.37. Choosing e_1, we get (fundamental) loop e_1, e_2, e_5, e_6; choosing e_3 yields loop e_3, e_5, e_6; and choosing e_4 results in loop e_2, e_4, e_5. Referring to the original circuit (Fig. 2.29) that corresponds to Fig. 2.37, the loop currents resulting from the given spanning tree are shown in Fig. 2.38.

fig. 2.37

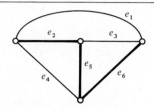

fig. 2.38

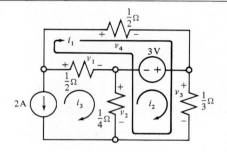

For loop i_1,

$$-v_1 - v_2 + v_3 + v_4 = 0$$

and, for loop i_2,

$$-v_2 + v_3 - 3 = 0$$

Since

$$v_1 = \frac{1}{2}(-i_1 + i_3)$$

$$v_2 = \frac{1}{4}(-i_1 - i_2 + i_3)$$

$$v_3 = \frac{1}{3}(i_1 + i_2)$$

$$v_4 = \frac{1}{2}i_1$$

substituting these voltage expressions into the previous two loop equations, we get

$$-\frac{1}{2}(-i_1 + i_3) - \frac{1}{4}(-i_1 - i_2 + i_3) + \frac{1}{3}(i_1 + i_2) + \frac{1}{2}i_1 = 0$$

and

$$-\frac{1}{4}(-i_1 - i_2 + i_3) + \frac{1}{3}(i_1 + i_2) - 3 = 0$$

Multiplying both of these equations by 12 and simplifying, we obtain

$$19i_1 + 7i_2 - 9i_3 = 0$$

and

$$7i_1 + 7i_2 - 3i_3 = 36$$

However, from loop i_3, we deduce that $i_3 = -2$ A. Thus, the previous two equations become

$$19i_1 + 7i_2 = -18$$

and

$$7i_1 + 7i_2 = 30$$

Subtracting the second equation from the first, we get

$$12i_1 = -48 \Rightarrow i_1 = -4 \text{ A}$$

and hence

$$7i_2 = 30 - 7i_1 = 58 \Rightarrow i_2 = \tfrac{58}{7} \text{ A}$$

Having determined the loop currents, we can now find

$$v_1 = \frac{1}{2}(-i_1 + i_3) = \frac{1}{2}(4 - 2) = 1 \text{ V}$$

$$v_2 = \frac{1}{4}(-i_1 - i_2 + i_3) = \frac{1}{4}\left(4 - \frac{58}{7} - 2\right) = -\tfrac{11}{7} \text{ V}$$

$$v_3 = \frac{1}{3}(i_1 + i_2) = \frac{1}{3}\left(-4 + \frac{58}{7}\right) = \tfrac{10}{7} \text{ V}$$

$$v_4 = \frac{1}{2}i_1 = -2 \text{ V}$$

Of course, the values agree with those obtained when we used mesh analysis to analyze the circuit. The different values for i_1, i_2, i_3 is due to the fact that these variables represent different currents in the two analyses.

$$\cdot \quad \cdot \quad \cdot$$

Example
Instead of the previous spanning tree, let us now select the one indicated in Fig. 2.39.

---

fig. 2.39

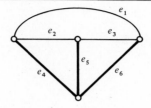

Choosing e_1 yields the loop e_1, e_4, e_6. Choosing e_2 results in the loop e_2, e_4, e_5, and choosing e_3 gives the loop e_3, e_5, e_6. With respect to the original circuit, the loop currents, therefore, are as shown in Fig. 2.40.

---

fig. 2.40

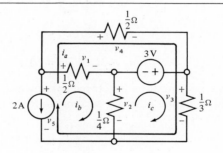

For loop i_a,

$$v_3 + v_4 - v_5 = 0$$

for loop i_b,

$$v_1 + v_2 - v_5 = 0$$

and, for loop i_c,

$$-v_2 + v_3 - 3 = 0$$

Also,

$$v_1 = \frac{1}{2} i_b$$

$$v_2 = \frac{1}{4} (i_b - i_c)$$

$$v_3 = \frac{1}{3} (i_a + i_c)$$

$$v_4 = \frac{1}{2} i_a$$

However, because of the current source, no similar simple expression exists for v_5. Thus, combining the equations for loops i_a and i_b, we have

$$v_1 + v_2 - v_3 - v_4 = 0$$

so that

$$\frac{1}{2}i_b + \frac{1}{4}(i_b - i_c) - \frac{1}{3}(i_a + i_c) - \frac{1}{2}i_a = 0$$

or

$$-10i_a + 9i_b - 7i_c = 0$$

From the equation for loop i_c,

$$-\frac{1}{4}(i_b - i_c) + \frac{1}{3}(i_a + i_c) - 3 = 0$$

or

$$4i_a - 3i_b + 7i_c = 36$$

Thus far, we have two equations and three unknowns: i_a, i_b, i_c. The third equation is obtained when we express the 2-A current source in terms of the loop currents. We have

$$i_a + i_b = -2$$

Writing the three loop equations in matrix form

$$\begin{bmatrix} -10 & 9 & -7 \\ 4 & -3 & 7 \\ 1 & 1 & 0 \end{bmatrix} \begin{bmatrix} i_a \\ i_b \\ i_c \end{bmatrix} = \begin{bmatrix} 0 \\ 36 \\ -2 \end{bmatrix}$$

and

$$\Delta = \begin{vmatrix} -10 & 9 & -7 \\ 4 & -3 & 7 \\ 1 & 1 & 0 \end{vmatrix} = 63 - 28 - 21 + 70 = 84$$

$$\Delta_a = \begin{vmatrix} 0 & 9 & -7 \\ 36 & -3 & 7 \\ -2 & 1 & 0 \end{vmatrix} = -126 - 252 + 42 = -336$$

$$\Delta_b = \begin{vmatrix} -10 & 0 & -7 \\ 4 & 36 & 7 \\ 1 & -2 & 0 \end{vmatrix} = 56 + 252 - 140 = 168$$

$$\Delta_c = \begin{vmatrix} -10 & 9 & 0 \\ 4 & -3 & 36 \\ 1 & 1 & -2 \end{vmatrix} = -60 + 324 + 360 + 72 = 696$$

Thus,

$$i_a = \frac{\Delta_a}{\Delta} = \frac{-336}{84} = -4 \text{ A}$$

$$i_b = \frac{\Delta_b}{\Delta} = \frac{168}{84} = 2 \text{ A}$$

$$i_c = \frac{\Delta_c}{\Delta} = \frac{696}{84} = \tfrac{58}{7} \text{ A}$$

from which

$$v_1 = \frac{1}{2}i_b = \frac{1}{2}(2) = 1 \text{ V}$$

$$v_2 = \frac{1}{4}(i_b - i_c) = \frac{1}{4}\left(2 - \frac{58}{7}\right) = -\tfrac{11}{7} \text{ V}$$

$$v_3 = \frac{1}{3}(i_a + i_c) = \frac{1}{3}\left(-4 + \frac{58}{7}\right) = \tfrac{10}{7} \text{ V}$$

$$v_4 = \frac{1}{2}i_a = \frac{1}{2}(-4) = -2 \text{ V}$$

as was obtained previously.

$\bullet \quad \bullet \quad \bullet$

PROBLEMS   2.30.   In the circuit shown in Fig. P2.30, the $\frac{1}{2}$-Ω, $\frac{1}{3}$-Ω, and $\frac{1}{4}$-Ω resistors form a spanning tree. Use loop analysis with respect to this tree in order to determine v.

fig. P2.30

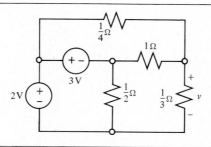

fig. P2.31

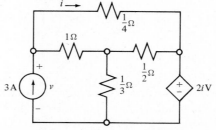

2.31.   Repeat Problem 2.30 for the circuit shown in Fig. P2.31.

fig. P2.32

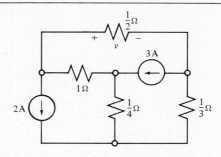

*2.32.   Repeat Problem 2.30 for the circuit in Fig. P2.32.
        Ans.   −0.18 V

fig. P2.33

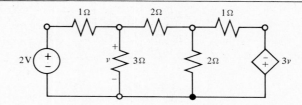

2.33.   In the circuit shown in Fig. P2.33, the two 1-Ω resistors and the two 2-Ω resistors form a spanning tree. Use loop analysis with respect to this tree in order to determine v.

*2.34.   Repeat Problem 2.33 using the spanning tree consisting of the two 1-Ω resistors and the two voltage sources.
        Ans. $\frac{48}{59}$ V

fig. P2.35

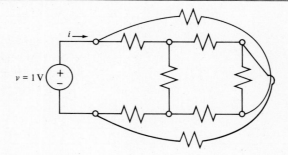

2.35.   Use loop analysis to find i and hence $R = v/i = 1/i$ for the nonplanar circuit in Fig. P2.35, which consists of 1-Ω resistors.

1.   To use nodal analysis, apply KCL at each nonreference node, express the resulting equations in terms of the node voltages, and solve.

Summary

2. A set of node equations can be expressed as a single matrix equation of the form $\mathbf{i} = \mathbf{Gv}$.

3. Using nodal analysis for circuits with voltage sources necessitates the concept of a supernode.

4. To use mesh analysis (on planar circuits), apply KVL around each finite mesh, express the resulting equations in terms of the mesh currents, and solve.

5. A set of mesh equations can be expressed as a single matrix equation of the form $\mathbf{v} = \mathbf{Ri}$.

6. Using mesh analysis for circuits with current sources necessitates the concept of a supermesh.

7. To use loop analysis (on planar or nonplanar circuits) apply KVL around each loop formed from one nonbranch edge and branches, express the resulting equations in terms of the loop currents, and solve.

8. A set of loop equations can be expressed as a single matrix equation of the form $\mathbf{v} = \mathbf{Ri}$.

IMPORTANT 3
CIRCUIT
CONCEPTS

Up to this point we have assumed that all the elements under consideration were ideal. Although nothing is ideal, including voltage and current sources, a combination of ideal elements may accurately describe the behavior of an actual source. The subject of nonideal sources will be discussed in this chapter. Included is the condition for which a practical voltage source and a practical current source are equivalent.

If an ideal voltage source is connected to some resistive load, all of the power supplied by the source is absorbed by the load. However, when a load is connected to a nonideal voltage source, not all of the power produced will be delivered to the load; the proportion depends on the value (resistance) of the load. In this chapter we shall determine which load receives the maximum power for a given practical voltage source.

In order to determine the maximum power transfer to a load when it is connected to an arbitrary circuit, we shall study Thévenin's theorem—an extremely important concept.

We shall see that a circuit containing more than one independent source can be analyzed by considering each independent source separately, and then adding the results. This is known as the principle of superposition.

The chapter concludes with procedures for analyzing circuits that contain the important and useful circuit element known as an operational amplifier.

3.1 NONIDEAL SOURCES

We know that an ideal voltage source is capable of supplying any amount of current. The amount supplied depends on what is connected to the voltage source. However, actual physical voltage sources (practical voltage sources) do not have this capability in

116

that there is a limit as to what amount of current they can supply. A more practical representation of a real voltage source is shown in Fig. 3.1. That is, an actual voltage source *behaves* more as an ideal voltage source v_s in series with a resistance R_s than as an ideal source. Do not in any way infer that a practical voltage source is constructed from an ideal source and a resistor. (Practical sources may be constructed from transistors, resistors, capacitors, diodes, and so on.) It is simply that the behavior of certain practical sources may be better described or modeled by an ideal voltage source in series with a resistor.

fig. 3.1

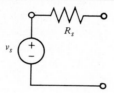

Suppose that we connect some load resistance R_L to the practical source shown in Fig. 3.1. The result is shown in Fig. 3.2. We see that, by Ohm's law,

$$i_L = \frac{v_s}{R_s + R_L}$$

and, by voltage division,

$$v_L = \frac{R_L v_s}{R_s + R_L}$$

fig. 3.2

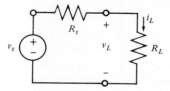

Thus, we find that the current and the voltage reaching the load is affected by the resistance R_s—called the *internal resistance* or the *source resistance*. Furthermore, the amount of load current and load voltage is less than would be the case if R_s were equal to zero, assuming of course that R_s is nonnegative.

Question: Why is Fig. 3.3 not a model of a practical source? The answer to the question is as follows: Placing a resistance R_s in parallel with an ideal voltage source simply establishes a current of v_s/R_s in R_s. The voltage across the terminals *a* and *b* will not be affected—that is, $v = v_s$—and, regardless of what is connected to the terminals (with the exception of another ideal voltage source, of course), the voltage will remain v_s. Thus, the model given below describes an ideal voltage source, not a practical voltage source.

fig. 3.3

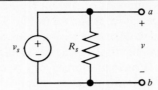

If we revise the previous discussion of voltage sources to the case of current sources, we would conclude that a more realistic representation of a practical current source is as shown in Fig. 3.4. In this case the internal resistance of the source is labeled R_g. For a practical voltage source, the closer the internal resistance is to zero, the closer the source is to being ideal. However, for a practical current source, the larger R_g is, the closer the source is to being ideal. Furthermore, Fig. 3.5 is not a model of a practical current source but actually describes an ideal current source.

fig. 3.4 (left)
fig. 3.5 (right)

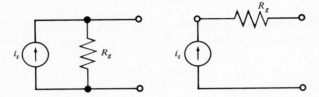

We may now ask whether or not a voltage source and a current source can exhibit the same behavior. For the case of an ideal voltage source and an ideal current source, due to their definitions, the answer is no. However, this is not the case for practical voltage and current sources. To see this, let us indicate precisely what we mean by the phrase "exhibit the same behavior."

Suppose that two sources have the same load. If the effect on that load is the same, then the two sources are said to be equivalent with respect to that given load. If two sources are equivalent with respect to all possible loads, then they are said to be *equivalent*.

With this definition of equivalence, let us consider the two circuits shown in Fig. 3.6. For the circuit with the practical voltage source,

$$i_1 = \frac{v}{R_s} = \frac{v_s - v_1}{R_s}$$

whereas for the circuit with the practical current source,

$$i_2 = i_s - i = i_s - \frac{v_2}{R_g} = \frac{i_s R_g - v_2}{R_g}$$

The effect on the load is the same when

$$i_1 = i_2$$

fig. 3.6

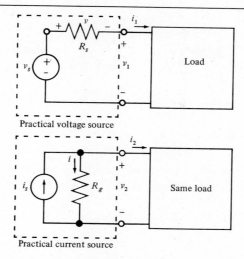

Practical voltage source

Practical current source

which means that the resulting voltages are equal (i.e., $v_1 = v_2$), or vice versa. In other words, the effect is the same when

$$\frac{v_s - v_1}{R_s} = \frac{i_s R_g - v_2}{R_g}$$

If $v_1 = v_2$, we can obtain this equality by letting

$$v_s = i_s R_g \qquad \text{or} \qquad i_s = \frac{v_s}{R_g}$$

and

$$R_s = R_g$$

Thus, these two conditions will result in the practical voltage source and the practical current source having the same effect on any arbitrary load. (If you're not convinced of this, assume that the load is the resistor R_L, and calculate the resulting load current given that $R_s = R_g$ and $v_s = i_s R_g$.) To summarize, the sources shown in Fig. 3.7 are equivalent. Alternatively, we have the equivalence shown in Fig. 3.8. Replacing one source by an equivalent source is often called a *source transformation*.

fig. 3.7

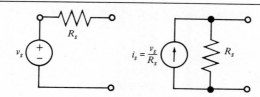

fig. 3.8

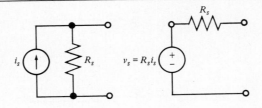

. . .

Example

By performing a source transformation on the practical voltage source given in Fig. 3.9 we obtain the equivalent source shown in Fig. 3.10.

fig. 3.9 (left)
fig. 3.10 (right)

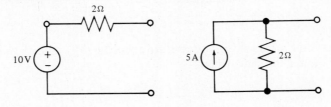

. . .

Example

The circuit shown in Fig. 3.11 has been analyzed previously with mesh analysis (see Fig. 2.25). Performing a source transformation on the 2-V source in series with the 2-Ω resistor, we obtain the circuit shown in Fig. 3.12. The 2-Ω and 3-Ω resistors in parallel can be combined into a $\frac{6}{5}$-Ω resistor. However, a source transformation can then be performed on the parallel combination of the 1-A source and $\frac{6}{5}$-Ω resistor. The result is shown in Fig. 3.13.

fig. 3.11

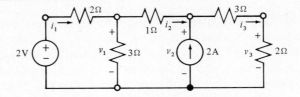

fig. 3.12

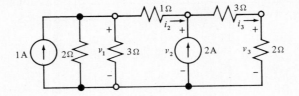

fig. 3.13

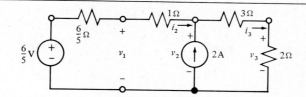

The $\frac{6}{5}$-Ω and 1-Ω resistors in series can be combined into a $\frac{11}{5}$-Ω resistor. Performing a source transformation on the $\frac{11}{5}$-Ω resistor in series with the $\frac{6}{5}$-V source yields the circuit of Fig. 3.14. The $\frac{6}{11}$-A and 2-A current sources are connected in parallel so they may be combined into a single $\frac{28}{11}$-A current source, which will be in parallel with the $\frac{11}{5}$-Ω resistor. The resulting circuit is shown in Fig. 3.15. By current division, we have that

$$i_3 = \frac{\dfrac{11}{5}\left(\dfrac{28}{11}\right)}{\dfrac{11}{5} + 3 + 2} = \frac{\dfrac{28}{5}}{\dfrac{36}{5}}$$

$$= \tfrac{7}{9} \text{ A}$$

and

$$v_3 = 2i_3 = \tfrac{14}{9} \text{ V}$$

fig. 3.14

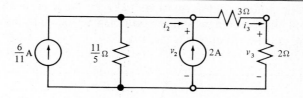

fig. 3.15

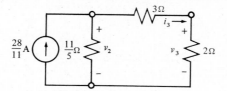

Returning to previous circuits, we have that

$$v_2 = 3i_3 + 2i_3 = 5i_3 = \tfrac{35}{9} \text{ V}$$

and

$$i_2 = i_3 - 2 = -\tfrac{11}{9} \text{ A}$$

Also,

$$v_1 = 1i_2 + v_2 = \tfrac{24}{9} \text{ V}$$

so

$$i_1 = \frac{2 - v_1}{2} = -\tfrac{1}{3} \text{ A}$$

Note that when a source transformation is performed, a variable (or variables) may disappear. For instance, in the original circuit i_1 is the current through the series 2-Ω resistor; note, however, that after the first transformation, this is not the same as the current through the parallel 2-Ω resistor. Although a source transformation does not change the behavior of the remainder of the circuit, whatever was transformed is, needless to say, no longer the same.

· · ·

Not only can source transformations be used for independent sources, they can be used for dependent sources as well. Care must be taken, however, to keep the dependent variable intact.

· · ·

Example
The circuit of Fig. 3.16 was analyzed previously via nodal analysis (see Fig. 2.13). Performing source transformations, we get the circuit shown in Fig. 3.17. Combining the parallel resistances, and then performing source transformations again, we get the circuit shown in Fig. 3.18.

fig. 3.16

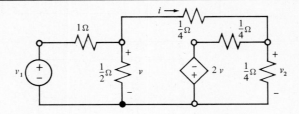

fig. 3.17

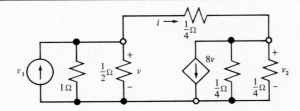

fig. 3.18

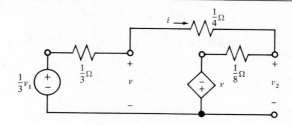

Writing a mesh equation we get

$$\frac{1}{3}v_1 = \frac{1}{3}i + \frac{1}{4}i + \frac{1}{8}i - v$$

$$= \frac{17}{24}i - v$$

Also,

$$\frac{1}{3}v_1 = \frac{1}{3}i + v$$

Adding these two equations, we get

$$\frac{2}{3}v_1 = \frac{25}{24}i$$

from which

$$i = \frac{16}{25}v_1$$

Thus,

$$v_2 = -\frac{1}{4}i - \frac{1}{3}i + \frac{1}{3}v_1$$

$$= -\frac{7}{12}\left(\frac{16}{25}v_1\right) + \frac{1}{3}v_1$$

$$= -\frac{1}{25}v_1$$

$\bullet$ $\bullet$ $\bullet$

PROBLEMS   *3.1.   In the circuit shown in Fig. P3.1, use source transformations to determine the current i.

Ans.   2 A

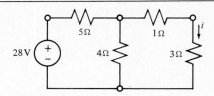

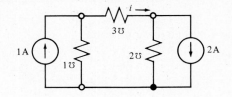

3.2.   Repeat Problem 3.1 for the circuit in Fig. P3.2.

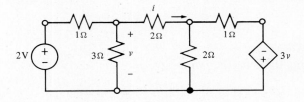

3.3.   Repeat Problem 3.1 for the circuit in Fig. P3.3.

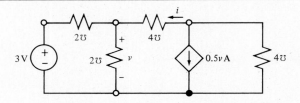

3.4.   Repeat Problem 3.1 for the circuit in Fig. P3.4.

3.2   MAXIMUM POWER TRANSFER

Consider the practical voltage source shown in Fig. 3.19 with a resistive load R_L. For a practical voltage source having the parameters v_s and R_s, what value of R_L will result in the maximum amount of power absorbed by the load resistance?

Since the power absorbed by the load is $p = i_L^2 R_L$, it may be tempting to believe that to increase the power absorbed, we simply increase the load resistance R_L. This reasoning, however, is invalid since an increase in R_L will result in a decrease in i_L. To

fig. 3.19

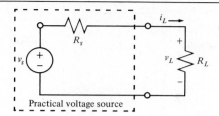

Practical voltage source

find the value for which maximum power is delivered to the load, we proceed as follows:

By Ohm's law,

$$i_L = \frac{v_s}{R_s + R_L}$$

The (instantaneous) power absorbed by the load is

$$p = i_L^2 R_L$$

$$= \left(\frac{v_s}{R_s + R_L}\right)^2 R_L$$

$$= \frac{v_s^2 R_L}{(R_s + R_L)^2}$$

Thus we have an expression for the power in terms of the variable R_L. For what value of R_L is the power a maximum? From freshman calculus we know that if we have a function $f(x)$ of the real variable x, then maxima occur for the values of x where the derivative of $f(x)$ is equal to zero. Thus let us take the derivative of p with respect to the variable R_L. Using the fact that

$$\frac{d}{dx}\left[\frac{f(x)}{g(x)}\right] = \frac{g(x)\dfrac{d}{dx}f(x) - f(x)\dfrac{d}{dx}g(x)}{[g(x)]^2}$$

we obtain

$$\frac{dp}{dR_L} = \frac{(R_s + R_L)^2 \dfrac{d}{dR_L}(v_s^2 R_L) - v_s^2 R_L \dfrac{d}{dR_L}(R_s + R_L)^2}{[(R_s + R_L)^2]^2}$$

$$= \frac{(R_s + R_L)^2 v_s^2 - v_s^2 R_L[2(R_s + R_L)]}{(R_s + R_L)^4}$$

$$= \frac{(R_s + R_L)[(R_s + R_L)v_s^2 - 2R_L v_s^2]}{(R_s + R_L)^4}$$

$$= \frac{R_s v_s^2 - R_L v_s^2}{(R_s + R_L)^3} = \frac{(R_s - R_L)v_s^2}{(R_s + R_L)^3}$$

Clearly, $dp/dR_L = 0$ when the numerator $(R_s - R_L)v_s^2 = 0$. Furthermore, given that v_s is a nonzero source, then $dp/dR_L = 0$ when

$$R_s - R_L = 0$$

or

$$R_s = R_L$$

Therefore, when $R_s = R_L$, the power

$$p = \frac{v_s^2 R_L}{(R_s + R_L)^2}$$

becomes

$$p_m = \frac{v_s^2 R_s}{(2R_s)^2} = \frac{v_s^2}{4R_s} = \frac{v_s^2}{4R_L}$$

We may verify that $p_m = v_s^2/4R_s$ is maximum by taking the second derivative of p, setting $R_s = R_L$, and obtaining a negative quantity.

In summary, given a practical voltage source with an internal resistance R_s, the maximum power that can be delivered to a resistive load R_L is obtained when $R_L = R_s$, and that power is

$$\boxed{p_m = \frac{v_s^2}{4R_L}}$$

3.3   THÉVENIN'S THEOREM

We have just seen what value load resistance absorbs the maximum amount of power when it is connected to a simple series connection of a voltage source and a resistance. However, suppose that a load R_L is connected to an arbitrary circuit (arbitrary in the sense that it contains only elements discussed previously), as shown in Fig. 3.20.

fig. 3.20

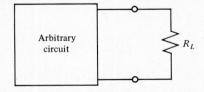

What value of R_L will absorb the maximum amount of power? Knowing the particular circuit, we can use one of the analysis techniques studied previously to obtain an expression for the power absorbed by R_L, then take the derivative of this expression in order to determine what value of R_L results in maximum power. This was the approach taken when the arbitrary circuit consisted of a single independent voltage source in series with a single resistance—and the amount of effort required was not inconsequential. Can you imagine how distasteful such an approach will be in general? Fortunately, though, a remarkable and important circuit theory concept states that as far as R_L is concerned, the arbitrary circuit shown above behaves as though it is a single independent voltage source in series with a single resistance. Once we determine the values of the source and the resistance, we simply apply the previously obtained results on maximum power transfer in order to find the appropriate value of R_L.

Given an arbitrary circuit containing any or all of the following elements: resistors, voltage sources, current sources. (The sources can be dependent as well as independent.) Let us identify a pair of nodes, say node a and node b, such that the circuit can be partitioned into two parts shown in Fig. 3.21.

fig. 3.21

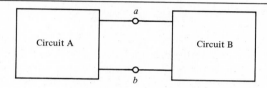

Suppose that circuit A contains no dependent source that is dependent on a parameter in circuit B, and vice versa. Then we can replace circuit A by an appropriate independent voltage source, call it v_{oc}, in series with an appropriate resistance, call it R_0, and the effect on circuit B is the same as that produced by circuit A. This series voltage source and resistance is called the *Thévenin equivalent* of circuit A. In other words, circuit A and the contents of the dashed box shown in Fig. 3.22 have the same effect on circuit B. This result, known as *Thévenin's theorem*,* is one of the more useful and significant concepts in circuit theory. A proof of this theorem will be given later when it is again discussed in its general form.

fig. 3.22

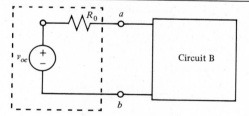

To obtain the voltage v_{oc}, called the *open-circuit voltage*, remove circuit B from circuit A and determine the voltage between nodes a and b. This voltage, as shown in Fig. 3.23, is v_{oc}.

To obtain the resistance R_0, called the *Thévenin equivalent resistance*, again remove circuit B from circuit A. Next, set all independent sources in circuit A to zero.

---

* Named for the French engineer M. L. Thévenin (1857–1926).

fig. 3.23

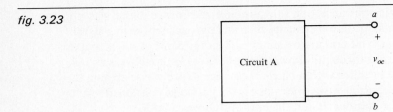

Leave the dependent sources as is! A zero-voltage source is equivalent to a short circuit and a zero-current source is equivalent to an open circuit. Now determine the resistance between nodes a and b (this is R_0) as shown in Fig. 3.24. If a circuit A contains

fig. 3.24

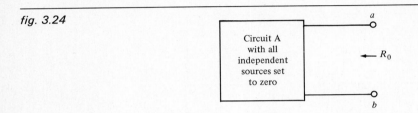

no dependent sources, when all independent sources are set to zero, the result may be simply a series-parallel resistive network. In this case, R_0 can be found by appropriately combining resistors in series and parallel. In general, however, R_0 can be found by applying an independent source between nodes a and b and then by taking the ratio of voltage to current. This procedure is depicted in Fig. 3.25. For the most part, it doesn't matter whether v_0 is the source and i_0 is the response or vice versa. (Exceptions to this rule are special cases which will be discussed later.)

fig. 3.25

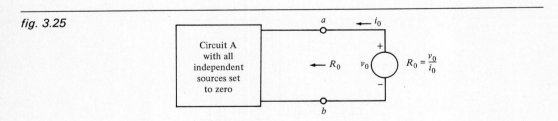

In applying Thévenin's theorem, circuit B (which is often called the *load*) may consist of many circuit elements, a single element, or no elements (that is, circuit B already may be an open circuit).

· · ·

Example
Consider the circuit shown in Fig. 3.26.

To find the Thévenin equivalent of circuit A, first remove the 3-Ω load resistor

fig. 4.16

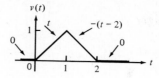

Analytically,

$$v(t) = \begin{cases} 0 & \text{for } -\infty < t < 0 \\ t & \text{for } 0 \le t < 1 \\ -(t-2) & \text{for } 1 \le t < 2 \\ 0 & \text{for } 2 \le t < \infty \end{cases}$$

We then have that

$$i(t) = C\frac{dv(t)}{dt} = 2\frac{dv(t)}{dt} = \begin{cases} 0 & \text{for } -\infty < t < 0 \\ 2 & \text{for } 0 \le t < 1 \\ -2 & \text{for } 1 \le t < 2 \\ 0 & \text{for } 2 \le t < \infty \end{cases}$$

A sketch of the current is shown in Fig. 4.17.

fig. 4.17

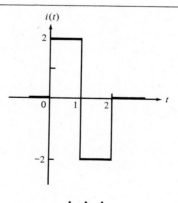

$\cdot \quad \cdot \quad \cdot$

At this point, compare these results with the example on page 174 in the preceding section, in which the current and voltage functions are described by $v(t)$ and $i(t)$, respectively. Calculating the instantaneous power absorbed by the capacitor in this example results in the same function $p(t)$ that was obtained in the inductor example. As a consequence, a similar subsequent discussion leads us to the conclusion that a capacitor, like an inductor, is an energy storage element.

. . .

Example

In the op amp circuit shown in Fig. 4.18, because the inverting input is at a potential of zero volts, then

$$v_C = v$$

fig. 4.18

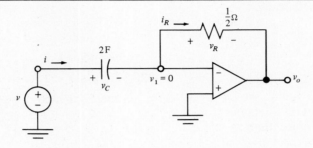

Thus

$$i = C\frac{dv_C}{dt} = 2\frac{dv}{dt}$$

Since the op amp draws no current,

$$i_R = i$$

so

$$v_R = \frac{1}{2}i_R = \frac{1}{2}i = \frac{1}{2}\left(2\frac{dv}{dt}\right) = \frac{dv}{dt}$$

Finally, by KVL,

$$v_o = -v_R$$

Hence,

$$v_o = -\frac{dv}{dt}$$

In other words, the output voltage $v_o(t)$ is the derivative of the input voltage $v(t)$ (multiplied by the constant -1). Because of this, we call such a circuit a *differentiator*.

For the case that $v(t)$ is the function given in the last example, the output voltage $v_o(t)$ is given by

$$v_o(t) = -\frac{dv(t)}{dt} = \begin{cases} 0 & \text{for } -\infty < t < 0 \\ -1 & \text{for } 0 \le t < 1 \\ 1 & \text{for } 1 \le t < 2 \\ 0 & \text{for } 2 \le t < \infty \end{cases}$$

a sketch of which is shown in Fig. 4.19.

fig. 4.19

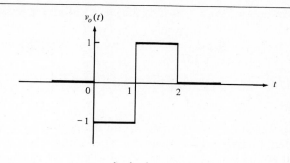

For a capacitor, the energy is stored in the electric field that exists between its plates. To obtain an expression for the energy stored in a capacitor, we proceed as follows.

For the capacitor shown in Fig. 4.20,

$$p(t) = v(t)i(t)$$

$$= v(t)\left[C\,\frac{dv(t)}{dt} \right]$$

$$= Cv(t)\,\frac{dv(t)}{dt}$$

$$w_c(t) = \int_{-\infty}^{t} p(t)\,dt$$

$$= \int_{-\infty}^{t} Cv(t)\,\frac{dv(t)}{dt}\,dt$$

$$= \int_{v(-\infty)}^{v(t)} Cv(t)\,dv(t)$$

$$= C\,\frac{v^2(t)}{2}\bigg|_{v(-\infty)}^{v(t)}$$

$$= \frac{1}{2}\,C[v^2(t) - v^2(-\infty)]$$

fig. 4.20

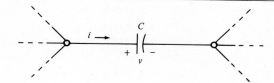

Again, by convention, $v(-\infty) = 0$; thus the energy stored in a capacitor of C farads at time t is

$$w_C(t) = \frac{1}{2} Cv^2(t)$$

Since no physical dielectric is perfect, an actual capacitor will allow a certain amount of current (maybe extremely small), called *leakage current*, to flow through it. For this reason we model a physical capacitor as an ideal capacitor in parallel with a resistance. This "leakage resistance" is inversely proportional to the capacitance. Depending on the construction of capacitors, values can range from a few picofarads (pF, or 10^{-12} F) to 10,000 μF and more. The product of leakage resistance and capacitance typically lie between 10 and 10^6 $\Omega \cdot$ F. The *working voltage* of a capacitor—that is, the maximum voltage that can be applied to the capacitor without damaging it or breaking down the dielectric—generally varies from a few volts to hundreds or even thousands of volts.

The capacitor, like the operational amplifier and the resistor, has the valuable ability for miniaturization, so that it can be made part of integrated circuits. On the contrary, because semiconductors do not possess the appropriate magnetic properties and because of size limitations, inductors are not in general readily adaptable to integrated-circuit form. This, however, does not diminish the importance of the inductor—your radio, television set, and various kinds of telecommunications equipment employ many of them.

Since the expressions for the energy stored in an inductor and a capacitor are, respectively,

$$w_L(t) = \frac{1}{2} Li^2(t) \quad \text{and} \quad w_C(t) = \frac{1}{2} Cv^2(t)$$

we see that regardless of the current through the inductor or the voltage across the capacitor, the energy stored is never negative. That is, these elements can supply energy, but never more than has been previously delivered to them. In other words, they can store and return energy, but not produce it. We refer to such elements as being *passive*. Since the power absorbed by a resistor is $i^2R = v^2/R$, the energy absorbed is always positive, and consequently a resistor is also a passive element.

An element that is capable of producing energy (e.g., an independent or dependent source) is called an *active* element. A circuit that consists only of passive elements is called a *passive circuit* or *network*, whereas one that contains one or more active elements is known as an *active circuit* or *network*.

In the foregoing discussion about resistors, inductors, and capacitors being passive, it was assumed that these elements were positive-valued. For the case that such elements have negative values, we see that they must necessarily be active elements. The remainder of this book will deal almost exclusively with positive-valued inductors, capacitors, and resistors.

PROBLEMS 4.1. Consider the circuit shown in Fig. P4.1a. Suppose that the current $i(t)$, shown in Fig. P4.1b, is given by

$$i(t) = \begin{cases} 0 & \text{for } -\infty < t < 0 \\ t & \text{for } 0 \le t < 1 \\ 1 & \text{for } 1 \le t < \infty \end{cases}$$

Find $v(t)$, $w_L(t)$, $p_R(t)$, $v_R(t)$, and $v_s(t)$, and sketch these functions.

fig. P4.1a

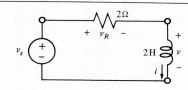

fig. P4.1b

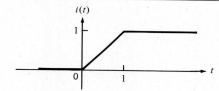

fig. P4.2

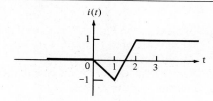

4.2. Repeat Problem 4.1 for the current in Fig. P4.2.

$$i(t) = \begin{cases} 0 & \text{for } -\infty < t < 0 \\ -t & \text{for } 0 \le t < 1 \\ 2t - 3 & \text{for } 1 \le t < 2 \\ 1 & \text{for } 2 \le t < \infty \end{cases}$$

fig. P4.3

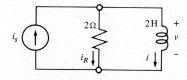

4.3. In the circuit shown in Fig. P4.3, suppose that the current $i(t)$ is given by the function in Problem 4.1. Find $v(t)$, $w_L(t)$, $p_R(t)$, $i_R(t)$, and $i_s(t)$, and sketch these functions.

4.4. Repeat Problem 4.3 for the current given in Problem 4.2.

fig. P4.5

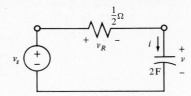

4.5.   Consider the circuit shown in Fig. P4.5 and suppose that the voltage $v(t)$ is described by the function given in Problem 4.1. Find $i(t)$, $w_C(t)$, $p_R(t)$, $v_R(t)$, and $v_s(t)$, and sketch these functions.

4.6.   Repeat Problem 4.5 for the case that the voltage $v(t)$ is described by the function given in Problem 4.2.

fig. P4.7

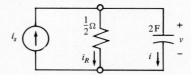

4.7.   In the circuit in Fig. P4.7, suppose that the voltage $v(t)$ is described by the function given in Problem 4.1. Find $i(t)$, $w_C(t)$, $p_R(t)$, $i_R(t)$, and $i_s(t)$, and sketch these functions.

4.8.   Repeat Problem 4.7 for the case that the voltage $v(t)$ is described by the function given in Problem 4.2.

fig. P4.9

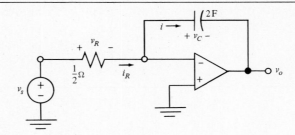

4.9.   Consider the op amp circuit in Fig. P4.9. Suppose that $v_C(t)$ is described by the function given in Problem 4.1. Find $i(t)$, $i_R(t)$, $v_R(t)$, $v_s(t)$, and $v_o(t)$.

4.10.   Repeat Problem 4.9 for the case that there is an additional $\frac{1}{2}$-Ω resistor in parallel with the capacitor.

4.11.   Repeat Problem 4.9 for the op amp circuit in Fig. P4.11.

fig. P4.11

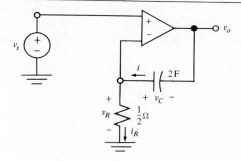

4.12.   For the op amp circuit given in Problem 4.11, place an additional $\frac{1}{2}$-Ω resistor in parallel with the capacitor and repeat Problem 4.9.

fig. P4.13

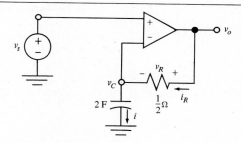

4.13.   Repeat Problem 4.9 for the op amp circuit shown in Fig. P4.13.
4.14.   For the op amp circuit given in Problem 4.13, place an additional $\frac{1}{2}$-Ω resistor in parallel with the capacitor and repeat Problem 4.9.

4.3   RAMPS, STEPS, AND IMPULSES

In previous examples we encountered the function diagrammed in Fig. 4.21, which is described as follows:

$$f(t) = \begin{cases} 0 & \text{for } -\infty < t < 0 \\ t & \text{for } 0 \le t < 1 \\ -(t - 2) & \text{for } 1 \le t < 2 \\ 0 & \text{for } 2 \le t < \infty \end{cases}$$

fig. 4.21

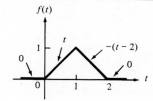

What we are doing here is specifying the function $f(t)$ by partitioning time into a number of intervals, and then describing $f(t)$ for each interval. However, this is not the only way in which a function can be specified. It is possible to express the above and many other functions as a sum of elementary functions which will be defined shortly. There are several reasons for doing this : (1) We get a one-line expression for $f(t)$ that is often more convenient to handle ; (2) the mathematical juggling can be less involved ; (3) we frequently may take analytical advantage of the linearity properties of circuits ; and (4) we can easily decompose the response of a circuit, thereby getting a better understanding of its behavior.

We begin by considering the simple function $f(t) = t$. Clearly, $df(t)/dt = 1$. Plots of this function and its derivative are shown in Fig. 4.22.

fig. 4.22

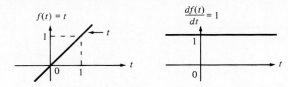

We now define a similar function, the difference being that we want to constrain this new function to be zero for all $t < 0$. The *unit ramp function* $r(t)$ is defined as follows :

$$r(t) = \begin{cases} 0 & \text{for } -\infty < t < 0 \\ t & \text{for } 0 \le t < \infty \end{cases}$$

Pictorially, the unit ramp function $r(t)$ is shown in Fig. 4.23. To take the derivative of this function, we proceed in two steps as follows :

$$\text{For } t < 0, \quad \frac{d[r(t)]}{dt} = \frac{d[0]}{dt} = 0$$

$$\text{For } t \ge 0, \quad \frac{d[r(t)]}{dt} = \frac{d(t)}{dt} = 1$$

fig. 4.23

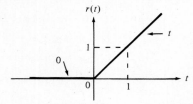

The result is a function, called the *unit step function*, that is denoted by $u(t)$. In other words,

$$\frac{d}{dt}[r(t)] = u(t) = \begin{cases} 0 & \text{for } -\infty < t < 0 \\ 1 & \text{for } 0 \le t < \infty \end{cases}$$

A sketch of the unit step function is shown in Fig. 4.24.

fig. 4.24

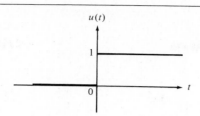

Now let us go one step further and differentiate the unit step function. In this case,

$$\frac{d[u(t)]}{dt} = \frac{d[0]}{dt} = 0 \qquad \text{for } -\infty < t < 0$$

$$\frac{d[u(t)]}{dt} = \frac{d[1]}{dt} = 0 \qquad \text{for } 0 < t < \infty$$

However, at $t = 0$ the step function is not continuous. In rigorous mathematical terms, therefore, the derivative of $u(t)$ does not exist at $t = 0$. Yet, let us say, instead, that the derivative of a step function at $t = 0$ is infinite; although it is undefined, it corresponds to the intuitive concept of slope.

We pictorially describe the derivative of the unit step function in Fig. 4.25 and refer to this "function," denoted $\delta(t)$, as the *unit impulse function* or the *unit delta function.* Note that the unit delta function $\delta(t)$ is equal to zero for all values of t, except for $t = 0$, where it is infinite!

fig. 4.25

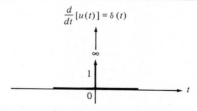

Needless to say, because of its zero width and infinite height, it is not possible physically to obtain a unit impulse function as either a voltage or a current. But don't get upset—it's not possible to construct ideal sources, ideal resistors, ideal inductors, and ideal capacitors either. Nor is it possible to get a unit step function—it is not physically possible to have a function (either voltage or current) change a specified amount in zero time.* However, we shall find the unit impulse function to be a convenient and very useful mathematical tool.

Suppose that K is a constant. If we multiply the unit ramp, step, and impulse

*Physically, a unit step function can be approximated more accurately than can a unit impulse function.

functions by K, we get the following more general forms of the ramp, step, and impulse, respectively.

$$Kr(t) = \begin{cases} 0 & \text{for } -\infty < t < 0 \\ Kt & \text{for } 0 \le t < \infty \end{cases}$$

$$\frac{d}{dt}[Kr(t)] = Ku(t) = \begin{cases} 0 & \text{for } -\infty < t < 0 \\ K & \text{for } 0 \le t < \infty \end{cases}$$

$$\frac{d}{dt}[Ku(t)] = K\delta(t)$$

Sketches of these functions are shown in Fig. 4.26.

fig. 4.26

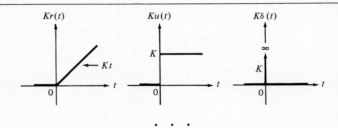

$\cdot$ $\cdot$ $\cdot$

Example
For the case that $K = -5$, the corresponding ramp, step, and impulse functions are shown in Fig. 4.27.

fig. 4.27

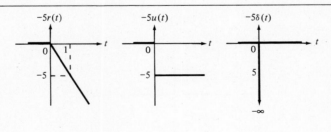

$\cdot$ $\cdot$ $\cdot$

Now let us further generalize the ramp, step, and impulse functions by choosing some real number a. Then we have, from the definition of a ramp function, that

$$Kr(t-a) = \begin{cases} 0 & \text{for } -\infty < t - a < 0 \\ K(t-a) & \text{for } 0 \le t - a < \infty \end{cases}$$

Since $t - a < 0$ if and only if $t < a$; and $t - a \ge 0$ if and only if $t \ge a$, then

$$Kr(t-a) = \begin{cases} 0 & \text{for } -\infty < t < a \\ K(t-a) & \text{for } a \le t < \infty \end{cases}$$

Plotting the ramp function $Kr(t - a)$ versus t, we therefore obtain the plot in Fig. 4.28 when $a > 0$. Note that the plot of this ramp function can be obtained from the plot of the ramp function $Kr(t)$ simply by "shifting" the latter to the right by the amount a.

fig. 4.28

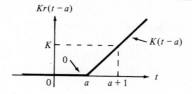

Similarly, if $a < 0$, then the shift would be to the left by the amount a. In other words, if $a = -b$, where $b > 0$, then a sketch of $Kr(t - a) = Kr(t + b)$ would have the form shown in Fig. 4.29.

fig. 4.29

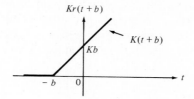

The discussion above is based on the fact that a plot of $f(t)$ versus t is identical to a plot of $f(t - a)$ versus $t - a$; and to obtain a plot of $f(t - a)$ versus t we merely add the amount a to every point on the horizontal axis in the plot of $f(t - a)$ versus $t - a$.

We may now differentiate the ramp function in the general form $Kr(t - a)$ to obtain $Ku(t - a)$, the general form of the step function. In other words,

$$\frac{d}{dt}[Kr(t - a)] = Ku(t - a) = \begin{cases} 0 & \text{for } -\infty < t < a \\ K & \text{for } a \le t < \infty \end{cases}$$

a sketch of which is shown in Fig. 4.30. As before, the derivative of the step function $Ku(t - a)$ is the delta function $K\delta(t - a)$, which is depicted in Fig. 4.31.

fig. 4.30 (left)
fig. 4.31 (right)

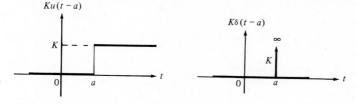

One way we can construct an independent source that produces a step function is with the use of a switch. A *normally open ideal switch* can be represented as in Fig. 4.32.

fig. 4.32

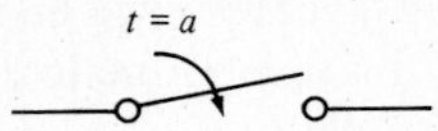

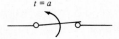

The resistance between the two terminals of the switch is infinite for time $t < a$ seconds; during this time interval we say that the switch is *open*. For time $t \geq a$ seconds, the resistance between the two terminals is zero ohms, and we say that the switch is *closed* for this time interval.

A *normally closed ideal switch*, represented by Fig. 4.33, is an ideal switch which is closed for $t < a$ seconds and open for $t \geq a$ seconds.

fig. 4.33

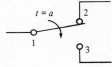

Both switches described above are examples of *single-pole, single-throw switches*. The "single-pole" refers to the fact that only one pair of terminals is connected. The "single-throw" means that there is only one pair of terminals. An example of a *single-pole, double-throw* switch is shown in Fig. 4.34. In this case the switch connects terminals 1 and 2 for time $t < a$ seconds, and connects terminals 1 and 3 for time $t \geq a$ seconds.

fig. 4.34

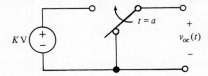

fig. 4.35

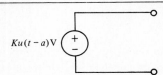

Now consider the connection of an ideal constant voltage source and a switch, as shown in Fig. 4.35. Let us find the Thévenin equivalent for this simple circuit. For $t < a$ then $v_{oc}(t) = 0$; while for $t \geq a$ we have that $v_{oc}(t) = K$ volts. The result is that $v_{oc}(t) = Ku(t - a)$. Clearly, $R_0 = 0$ for both $t < a$ and $t \geq a$. The Thévenin equivalent of this simple circuit is shown in Fig. 4.36.

fig. 4.36

$Ku(t-a)$V

fig. 4.37

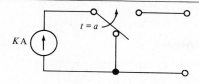

fig. 4.38

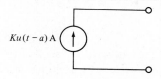

For the constant current source and switch shown in Fig. 4.37, finding the Norton equivalent reveals that this circuit is equivalent to the ideal current source in Fig. 4.38.

Let us now employ ramps, steps, and impulses in the analysis of some simple circuits.

· · ·

Example

Reconsider the circuit and current shown in Fig. 4.39. Let us analyze this circuit again by first expressing $i(t)$ as a sum of the elemental functions just discussed.

fig. 4.39

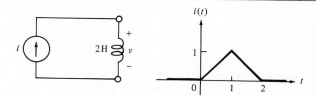

To write an analytical expression for $i(t)$, note that for $t \leq 1$, $i(t)$ is described by the unit ramp function $r(t)$. However, for $t > 1$ this is no longer the case. Thus, to negate the effect of $r(t)$ for $t > 1$, we subtract a ramp of value $K = 1$ that is zero for $t < 1$; that is, we subtract $r(t-1)$ from [add $-r(t-1)$ to] $r(t)$. The result is shown in Fig. 4.40. Now, however, we would like to add a function to $r(t) - r(t-1)$ such that the result is unchanged for $t \leq 1$ and has a slope of -1 when $t > 1$. This is accomplished simply by adding $-r(t-1)$. The result, $r(t) - r(t-1) - r(t-1) = r(t) - 2r(t-1)$, is shown in Fig. 4.41.

fig. 4.40

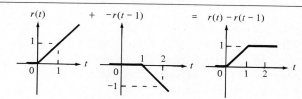

fig. 4.41

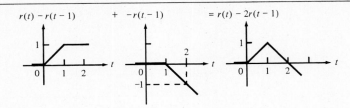

We now see that $r(t) - 2r(t - 1)$ describes $i(t)$ so long as $t \leq 2$. To obtain the function that describes $i(t)$ for all t, we simply add $r(t - 2)$ to $r(t) - 2r(t - 1)$. The result, $i(t) = r(t) - 2r(t - 1) + r(t - 2)$, is shown in Fig. 4.42.

fig. 4.42

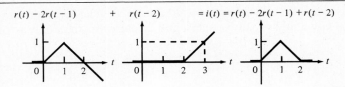

Since

$$v = L \frac{di}{dt}$$

we find that

$$v(t) = 2 \frac{d}{dt} [r(t) - 2r(t - 1) + r(t - 2)]$$

However, since the derivative of a sum of functions is equal to the sum of the derivatives of the functions, we have

$$v(t) = 2 \left(\frac{d}{dt} [r(t)] + \frac{d}{dt} [-2r(t - 1)] + \frac{d}{dt} [r(t - 2)] \right)$$

and since the derivative of a ramp is a step,

$$v(t) = 2u(t) - 4u(t - 1) + 2u(t - 2)$$

A sketch of $v(t)$ is obtained by adding the component step functions as shown in Fig. 4.43. Thus, we get the same sketch (naturally) of $v(t)$ as was obtained previously (see Fig. 4.6) by expressing $i(t)$ in terms of ramp functions and working directly with this analytical expression. In general, the approach that is simpler to employ will depend on the nature of the problem.

fig. 4.43

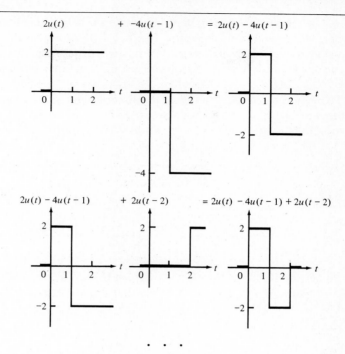

2u(t)   +   −4u(t − 1)   = 2u(t) − 4u(t − 1)

2u(t) − 4u(t − 1)   + 2u(t − 2)   = 2u(t) − 4u(t − 1) + 2u(t − 2)

· · ·

Example

For the circuit in Fig. 4.44 let us find the current $i(t)$, where $v(t)$ is described by the plot in Fig. 4.45.

fig. 4.44 (left)
fig. 4.45 (right)

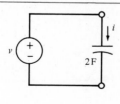 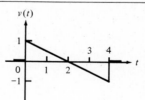

Since

$$i = C \frac{dv}{dt} = 2 \frac{dv}{dt}$$

then

$$i = 2 \frac{d}{dt} [0] = 0 \quad \text{for } -\infty < t < 0$$

$$i = 2 \frac{d}{dt} \left[-\frac{1}{2} t + 1 \right] = -1 \quad \text{for } 0 < t < 4$$

$$i = 2 \frac{d}{dt} [0] = 0 \quad \text{for } 4 < t < \infty$$

We may be tempted to believe that the solution is complete. However, note that there are two discontinuities in $v(t)$—one at $t = 0$ and one at $t = 4$. Thus, in the expression (or plot) for $i(t)$ there will be two delta functions. To verify this analytically, let us first write an expression for $v(t)$ in terms of ramps and steps as follows.

To begin with, in order to obtain the discontinuity at $t = 0$, we start off with a step function $u(t)$. Since we do not want $v(t)$ to be equal to one for $t > 0$, but rather be a straight line with a slope of $-\frac{1}{2}$, we add to $u(t)$ a ramp function with coefficient $K = -\frac{1}{2}$. The result is shown in Fig. 4.46.

fig. 4.46

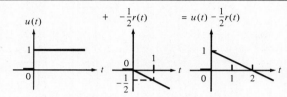

To get the resulting waveform to return to zero at $t = 4$ and stay there for $t > 4$, we first add a ramp having $K = \frac{1}{2}$ that is zero for $t < 4$. This is shown in Fig. 4.47.

fig. 4.47

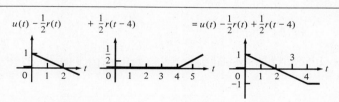

fig. 4.48

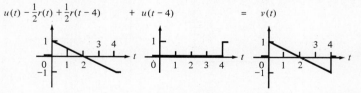

Finally, we need only add a unit step function that is zero for $t < 4$. The result, $v(t) = u(t) - \frac{1}{2}r(t) + \frac{1}{2}r(t - 4) + u(t - 4)$, is shown in Fig. 4.48. Since

$$i = C \frac{dv}{dt}$$

then

$$i(t) = 2 \frac{d}{dt}\left[u(t) - \frac{1}{2} r(t) + \frac{1}{2} r(t - 4) + u(t - 4) \right]$$

$$i(t) = 2 \left(\frac{d}{dt}\left[u(t) \right] + \frac{d}{dt}\left[-\frac{1}{2} r(t) \right] + \frac{d}{dt}\left[\frac{1}{2} r(t - 4) \right] + \frac{d}{dt}\left[u(t - 4) \right] \right)$$

But the derivative of a ramp is a step, and the derivative of a step is an impulse. Thus

$$i(t) = 2\left[\delta(t) - \frac{1}{2}u(t) + \frac{1}{2}u(t-4) + \delta(t-4)\right]$$

$$i(t) = 2\delta(t) - u(t) + u(t-4) + 2\delta(t-4)$$

To obtain a sketch of $i(t)$, we simply add the component waveforms, as shown in Fig. 4.49.

fig. 4.49

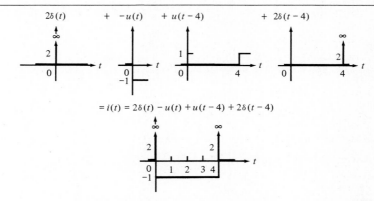

$$= i(t) = 2\delta(t) - u(t) + u(t-4) + 2\delta(t-4)$$

An alternative, and often simpler, approach to expressing a function in terms of steps (and ramps) uses the "sampling pulse" shown in Fig. 4.50. If $f(t)$ is an arbitrary function, then the product of $f(t)$ and the sampling pulse is

$$f(t)[u(t-a) - u(t-b)] = \begin{cases} 0 & \text{for } -\infty < t < a \\ f(t) & \text{for } a \le t < b \\ 0 & \text{for } b \le t < \infty \end{cases}$$

fig. 4.50

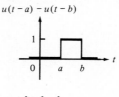

$u(t-a) - u(t-b)$

• • •

Let us now demonstrate the use of the sampling pulse.

• • •

Example
Reconsider the current $i(t)$ shown in Fig. 4.51, which we encountered previously.
Since we can express $i(t)$ as the sum

$$i(t) = i_1(t) + i_2(t)$$

fig. 4.51

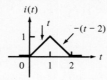

where

$$i_1(t) = \begin{cases} 0 & \text{for } -\infty < t < 0 \\ t & \text{for } 0 \leq t < 1 \\ 0 & \text{for } 1 \leq t < \infty \end{cases} \qquad i_2(t) = \begin{cases} 0 & \text{for } -\infty < t < 1 \\ -(t-2) & \text{for } 1 \leq t < 2 \\ 0 & \text{for } 2 \leq t < \infty \end{cases}$$

then

$$i(t) = t[u(t) - u(t-1)] - (t-2)[u(t-1) - u(t-2)]$$

$$= tu(t) - tu(t-1) - (t-2)u(t-1) + (t-2)u(t-2)$$

$$= tu(t) - (2t-2)u(t-1) + (t-2)u(t-2)$$

$$= tu(t) - 2(t-1)u(t-1) + (t-2)u(t-2)$$

However, from the definition of a unit ramp function,

$$r(t-a) = \begin{cases} 0 & \text{for } -\infty < t < a \\ t-a & \text{for } a \leq t < \infty \end{cases}$$

since

$$(t-a)u(t-a) = \begin{cases} 0 & \text{for } -\infty < t < a \\ t-a & \text{for } a \leq t < \infty \end{cases}$$

we see that

$$r(t-a) = (t-a)u(t-a)$$

Thus, we can also express $i(t)$ as

$$i(t) = r(t) - 2r(t-1) + r(t-2)$$

as was obtained previously.

. . .

Example

Again consider the function $v(t)$ shown in Fig. 4.52.

fig. 4.52

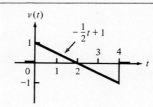

Clearly,

$$v(t) = \left(-\frac{1}{2}t + 1\right)[u(t) - u(t-4)]$$

$$= \left(-\frac{1}{2}t + 1\right)u(t) - \left(-\frac{1}{2}t + 1\right)u(t-4)$$

$$= -\frac{1}{2}tu(t) + u(t) + \frac{1}{2}tu(t-4) - u(t-4)$$

But since

$$\frac{1}{2}tu(t-4) = \frac{1}{2}(t-4)u(t-4) + 2u(t-4)$$

then

$$v(t) = -\frac{1}{2}tu(t) + u(t) + \frac{1}{2}(t-4)u(t-4) + u(t-4)$$

$$= -\frac{1}{2}r(t) + u(t) + \frac{1}{2}r(t-4) + u(t-4)$$

as was obtained before.

• • •

PROBLEMS *4.15.   Express the functions shown in Fig. P4.15 in terms of ramps and step functions.
   Ans.   (a) $r(t) - r(t-1)$; (b) $-r(t) + 3r(t-1) - 2r(t-2)$

fig. P4.15

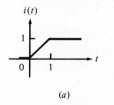

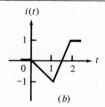

(a)

(b)

fig. P4.16

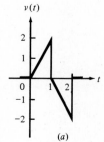

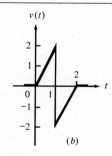

(a)

(b)

4.16.   Repeat Problem 4.15 for the functions shown in Fig. P4.16.

4.17.   The current through a 2-H inductor is described by function (a) in Problem 4.15. Express the resulting voltage in terms of step and impulse functions. Sketch this voltage.

4.18.   Repeat Problem 4.17 for the current given in part (b) of Problem 4.15.

4.19.   The voltage across a 2-F capacitor is described by function (a) in Problem 4.16. Express the resulting current in terms of step and impulse functions. Sketch this current.

*4.20.   Repeat Problem 4.19 for the voltage given in part (b) of Problem 4.16.
   Ans. $4u(t) - 8\delta(t - 1) - 4u(t - 2)$

4.21.   Sketch the function $f(t) = r(t) - r(t - 1) + u(t - 1) - r(t - 2) + r(t - 4)$. Determine $df(t)/dt$ and sketch this function.

fig. P4.22

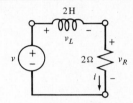

4.22.   For the circuit in Fig. P4.22, suppose that $i(t) = r(t) - r(t - 1)$. Find $v_R(t)$, $v_L(t)$, and $v(t)$. Sketch these functions.

fig. P4.23

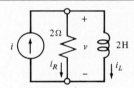

4.23.   In the circuit in Fig. P4.23, the current through the inductor is $i_L(t) = r(t) - r(t - 1)$. Find $v(t)$, $i_R(t)$, and $i(t)$. Sketch these functions.

fig. P4.24

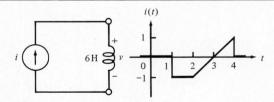

*4.24.   Assume the circuit and current shown in Fig. P4.24.
   (a)   Find an expression for $v(t)$ in terms of ramps and steps.
   (b)   Sketch $v(t)$.
   Ans. $-6\delta(t - 1) + 6u(t - 2) - 6u(t - 4) - 6\delta(t - 4)$

4.25.   Consider the circuit and input voltage $v(t)$ shown in Fig. P4.25.
   (a)   Write an expression for $v(t)$.
   (b)   Find $i_R(t)$, $i_C(t)$, and $i(t)$. Sketch these functions.

fig. P4.25

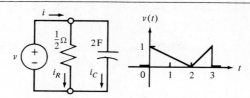

fig. P4.26

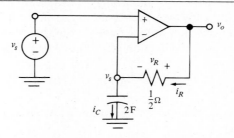

4.26. Consider the circuit and input current $i(t)$ shown in Fig. P4.26.
(a) Write an expression for $i(t)$.
(b) Find $v_R(t)$, $v_L(t)$, and $v(t)$. Sketch these functions.

fig. P4.27

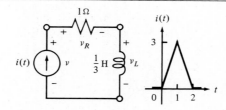

4.27. For the op amp circuit in Fig. P4.27, find $i_C(t)$, $i_R(t)$, $v_R(t)$, and $v_o(t)$, and sketch these functions, when
(a) $v_s(t) = u(t) - u(t - 1)$
(b) $v_s(t) = r(t) - 2r(t - 1) + r(t - 2)$

fig. P4.28

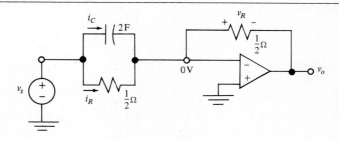

4.28. Repeat Problem 4.27 for the op amp circuit in Fig. P4.28.

*4.29. For the circuit in Fig. P4.29, suppose that $v_s(t) = -r(t) + 3r(t - 1)$
$- 2r(t - 2)$.
(a)   Sketch $v_s(t)$.
(b)   Find $i(t)$ and $v(t)$. Sketch these functions.
Ans. $-u(t) + 3u(t - 1) - 2u(t - 2)$, $6u(t) - 18u(t - 1) + 12u(t - 2)$

fig. P4.29

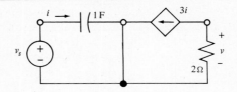

4.4   INTEGRAL RELATIONSHIPS

Given any fixed value t_0, the integral

$$\int_{-\infty}^{t_0} f(t)\, dt$$

is a real number that represents the net area under the curve $f(t)$ for the segment of the t-axis which extends from $-\infty$ to t_0. Since t_0 can be any real number between $-\infty$ and ∞, let us instead use for the upper limit on the integral the variable t. In this way, the integral will not result in a specific number, but rather it will result in a function of the variable t. Let us call this function $g(t)$; that is, we define

$$g(t) = \int_{-\infty}^{t} f(t)\, dt$$

Specifically then, the area under the curve $f(t)$ between $-\infty$ and t_0 is

$$g(t_0) = \int_{-\infty}^{t_0} f(t)\, dt$$

Therefore, for all $t > t_0$, we have that

$$g(t) = \int_{-\infty}^{t} f(t)\, dt$$

$$g(t) = \int_{-\infty}^{t_0} f(t)\, dt + \int_{t_0}^{t} f(t)\, dt$$

or

$$g(t) = g(t_0) + \int_{t_0}^{t} f(t)\, dt$$

Now let us return to the inductor. With reference to Fig. 4.53, we know that

$$v = L\frac{di}{dt}$$

fig. 4.53

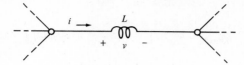

This is the differential relationship between voltage and current for an inductor. From this we may now obtain an integral relationship between voltage and current for an inductor. To do this, first integrate both sides of this equation with respect to time, choosing t_0 and t as the lower and upper limits, respectively. Doing this we get

$$\int_{t_0}^{t} v(t)\, dt = \int_{t_0}^{t} L\, \frac{di(t)}{dt}\, dt$$

By the chain rule of calculus, we can change the variable of integration for the term on the right. Since $t = t_0$ implies that $i(t) = i(t_0)$, we have

$$\int_{t_0}^{t} v(t)\, dt = \int_{i(t_0)}^{i(t)} L\, di(t) = Li(t)\Big|_{i(t_0)}^{i(t)}$$
$$= L[i(t) - i(t_0)]$$

Hence,

$$i(t) - i(t_0) = \frac{1}{L} \int_{t_0}^{t} v(t)\, dt$$

or

$$\boxed{\; i(t) = i(t_0) + \frac{1}{L} \int_{t_0}^{t} v(t)\, dt \;}$$

But as we have seen above, for $t > t_0$ we can write this last equation as

$$\boxed{\; i(t) = \frac{1}{L} \int_{-\infty}^{t} v(t)\, dt \;}$$

Since these last two equations are equivalent, either one is referred to as the integral relationship between voltage and current for an inductor.

· · ·

Example
Consider the step function $Ku(t - a)$. For $t < a$, we have that

$$\int_{-\infty}^{t} Ku(t - a)\, dt = \int_{-\infty}^{t} 0\, dt = 0$$

For $t \geq a$, however, we have that

$$\int_{-\infty}^{t} Ku(t-a) \, dt = \int_{-\infty}^{a} Ku(t-a) \, dt + \int_{a}^{t} Ku(t-a) \, dt$$

$$= \int_{-\infty}^{a} 0 \, dt + \int_{a}^{t} K \, dt$$

$$= 0 + Kt \Big|_{a}^{t}$$

$$= K(t-a)$$

Thus

$$\int_{-\infty}^{t} Ku(t-a) \, dt = K(t-a)u(t-a)$$

$$= Kr(t-a)$$

That is, the integral of a step is a ramp.

This result should be no surprise since the derivative of a ramp is a step; that is,

$$\frac{d}{dt}\left[Kr(t-a)\right] = Ku(t-a)$$

Hence, since the derivative of a step function is a delta function,

$$\frac{d}{dt}\left[Ku(t-a)\right] = K\delta(t-a)$$

then it follows that

$$\int_{-\infty}^{t} K\delta(t-a) \, dt = Ku(t-a)$$

These results are sketched in Fig. 4.54.

 fig. 4.54

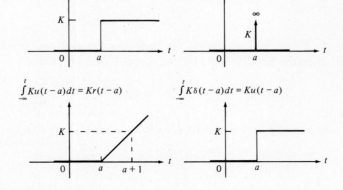

We see from the sketch of the integral of the impulse function $K\delta(t - a)$ that although at the point $t = a$ the function is infinite, the area under it is finite and is equal to K.

$$\cdot \quad \cdot \quad \cdot$$

We now derive an important and useful property that we shall make use of many times in the subsequent material. The property involves the product of a unit impulse function $\delta(t - a)$ and an arbitrary function $f(t)$.

Since $\delta(t - a) = 0$ for $t \neq a$, then

$$f(t)\delta(t - a) = 0 \qquad \text{for } t \neq a$$

as is

$$f(a)\delta(t - a) = 0 \qquad \text{for } t \neq a$$

However, when $t = a$, we have

$$f(t)\delta(t - a) = f(a)\delta(t - a) \qquad \text{for } t = a$$

provided that $f(a)$ exists. Combining these facts, we find that

$$\boxed{f(t)\delta(t - a) = f(a)\delta(t - a) \qquad \text{for all } t}$$

This result is known as the *sampling property* of the impulse function.

Examples of the sampling property are as follows:

$$(\cos 3t)\delta(t - \pi) = (\cos 3\pi)\delta(t - \pi) = -\delta(t - \pi)$$

$$e^{-2t}\delta(t) = e^{-2(0)}\delta(t) = \delta(t)$$

$$(1 - e^{-4t})\delta(t) = (1 - e^{0})\delta(t) = 0\delta(t)$$

In this last example, we obtained an impulse function whose area is zero! Taking the integral of the zero impulse function $0\delta(t - a)$ we get the zero step function $0u(t - a)$. The derivative of this step function, which again is $0\delta(t - a)$, is clearly zero. Our conclusion is that an impulse function whose area is zero is equivalent to zero; that is,

$$0\delta(t - a) = 0$$

$$\cdot \quad \cdot \quad \cdot$$

Example

In the circuit shown in Fig. 4.55 let us find the inductor current i given that the voltage v is described in Fig. 4.56.

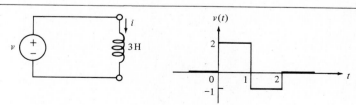

fig. 4.55 (left)
fig. 4.56 (right)

One way to determine i is to first express v in terms of step functions. To do this, we note that $v(t) = 2u(t) - 3u(t - 1) + u(t - 2)$. The justification of this expression for v is shown in Fig. 4.57. Since

$$i(t) = \frac{1}{L} \int_{-\infty}^{t} v(t) \, dt$$

 fig. 4.57

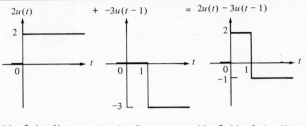

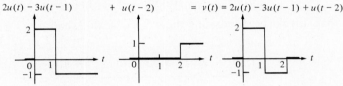

then

$$i(t) = \frac{1}{3} \int_{-\infty}^{t} [2u(t) - 3u(t - 1) + u(t - 2)] \, dt$$

$$= \frac{1}{3} \left[\int_{-\infty}^{t} 2u(t) \, dt + \int_{-\infty}^{t} -3u(t - 1) \, dt + \int_{-\infty}^{t} u(t - 2) \, dt \right]$$

$$= \frac{1}{3} [2r(t) - 3r(t - 1) + r(t - 2)]$$

$$= \frac{2}{3} r(t) - r(t - 1) + \frac{1}{3} r(t - 2)$$

A sketch of $i(t)$ is developed as shown in Fig. 4.58.

An alternative scheme to determine i is to partition the t-axis into segments and then use the integral formula for $i(t)$. This is done as follows:

For $t \leq 0$,

$$i(t) = \frac{1}{L} \int_{-\infty}^{t} v(t) \, dt$$

$$= \frac{1}{3} \int_{-\infty}^{t} 0 \, dt$$

$$= 0 \text{ A}$$

fig. 4.58

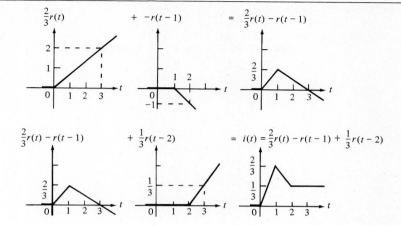

For $0 < t \le 1$,

$$i(t) = \frac{1}{L} \int_{-\infty}^{t} v(t)\, dt = i(t_0) + \frac{1}{L} \int_{t_0}^{t} v(t)\, dt$$

$$= i(0) + \frac{1}{L} \int_{0}^{t} v(t)\, dt$$

and, since $i(t) = 0$ for $t \le 0$,

$$i(t) = 0 + \frac{1}{3} \int_{0}^{t} 2\, dt$$

$$= \frac{1}{3} (2t) \Big|_{0}^{t}$$

$$= \frac{2}{3} t$$

For $1 < t \le 2$,

$$i(t) = \frac{1}{L} \int_{-\infty}^{t} v(t)\, dt$$

$$= i(t_0) + \frac{1}{L} \int_{t_0}^{t} v(t)\, dt$$

$$= i(1) + \frac{1}{L} \int_{1}^{t} v(t)\, dt$$

and since $i(t) = \frac{2}{3}t$ for $0 < t \leq 1$,

$$i(t) = \frac{2}{3}(1) + \frac{1}{3}\int_1^t (-1)\, dt$$

$$= \frac{2}{3} - \frac{1}{3}t \Big|_1^t$$

$$= \frac{2}{3} - \frac{1}{3}(t - 1)$$

$$= -\frac{1}{3}t + 1$$

For $t > 2$,

$$i(t) = i(t_0) + \frac{1}{L}\int_{t_0}^t v(t)\, dt$$

$$= i(2) + \frac{1}{L}\int_2^t v(t)\, dt$$

and, since $i(t) = -\frac{1}{3}t + 1$ for $1 < t \leq 2$,

$$i(t) = \left[-\frac{1}{3}(2) + 1 \right] + \frac{1}{3}\int_2^t 0\, dt = \frac{1}{3}$$

Thus, in summary, we have that

$$i(t) = \begin{cases} 0 & \text{for } t \leq 0 \\ \frac{2}{3}t & \text{for } 0 < t \leq 1 \\ -\frac{1}{3}t + 1 & \text{for } 1 < t \leq 2 \\ \frac{1}{3} & \text{for } t > 2 \end{cases}$$

A sketch of $i(t)$ is shown in Fig. 4.59.

fig. 4.59

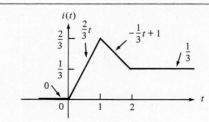

We can also express $i(t)$ in terms of ramps by using sampling pulses as follows:

$$i(t) = \frac{2}{3}t[u(t) - u(t-1)] + \left(-\frac{1}{3}t + 1\right)[u(t-1) - u(t-2)] + \frac{1}{3}u(t-2)$$

$$= \frac{2}{3}tu(t) - \frac{2}{3}tu(t-1) - \frac{1}{3}tu(t-1) + u(t-1)$$

$$+ \frac{1}{3}tu(t-2) - u(t-2) + \frac{1}{3}u(t-2)$$

$$= \frac{2}{3}tu(t) - (t-1)u(t-1) + \frac{1}{3}(t-2)u(t-2)$$

$$= \frac{2}{3}r(t) - r(t-1) + \frac{1}{3}r(t-2)$$

which confirms the previous calculations.

In this example, we see that for time $t < 0$ the voltage applied to the inductor is zero, and as we would expect, the resulting current through the inductor is zero. For time $0 \leq t < 2$ the voltage is nonzero, and as our analysis tells us, so is the current. But look—for time $t > 2$, the voltage applied is again zero, but the current is not. It remains constant forever! Why?

For time $t > 2$, the voltage source has a value of zero volts, and thus behaves as a short circuit. Unlike a resistor, an inductor can have zero volts across it and at the same time have a nonzero current through it (remember that an inductor acts as a short circuit to dc). So there is no paradox.

The energy stored in the inductor is

$$w_L(t) = \frac{1}{2} Li^2(t) = \frac{3}{2} i^2(t) = \begin{cases} 0 & \text{for } -\infty < t < 0 \\ \frac{2}{3}t^2 & \text{for } 0 \leq t < 1 \\ \frac{3}{2}(-\frac{1}{3}t + 1)^2 & \text{for } 1 \leq t < 2 \\ \frac{1}{6} & \text{for } 2 \leq t < \infty \end{cases}$$

A sketch of this function is shown in Fig. 4.60. We see that for $t < 0$, since the current is zero, the energy stored is zero. When the input voltage is positive, for $0 \leq t < 1$, the energy stored in the inductor increases with time. When the input

fig. 4.60

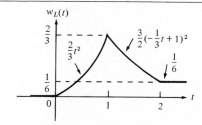

voltage goes negative, for $1 \leq t < 2$, the energy stored decreases until time $t = 2$ where it is $\frac{1}{6}$ J. For $t \geq 2$, however, when the voltage applied is zero, we have the equivalent situation shown in Fig. 4.61. Since the inductor and the voltage source are ideal, no energy is dissipated, and the energy stays constant for $t \geq 2$. This causes the current to stay constant for $t \geq 2$.

fig. 4.61

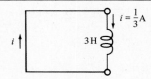

Since all physical inductors and voltage sources have some associated resistance, in a corresponding actual circuit the current $i(t)$ and the energy stored $w_L(t)$ would eventually become zero. The more resistance, the sooner these zero values would be reached.

. . .

Returning to the capacitor with the voltage and current indicated as in Fig. 4.62, we know that

$$i = C \frac{dv}{dt}$$

fig. 4.62

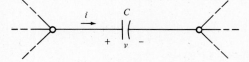

and this is the differential relationship between current and voltage. In a manner identical to that discussed for the inductor, we can obtain the equivalent integral relationships between voltage and current for a capacitor, which are as follows:

$$v(t) = v(t_0) + \frac{1}{C} \int_{t_0}^{t} i(t) \, dt$$

and

$$v(t) = \frac{1}{C} \int_{-\infty}^{t} i(t) \, dt$$

. . .

Example

For the circuit of Fig. 4.63 let us determine v given that i is described by the waveform in Fig. 4.64.

fig. 4.63 (left)
fig. 4.64 (right)

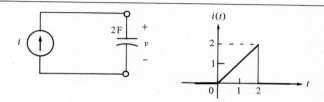

In this case let us consider three intervals of time. For $t \leq 0$,

$$v(t) = \frac{1}{C} \int_{-\infty}^{t} i(t)\, dt$$

$$= \frac{1}{2} \int_{-\infty}^{t} 0\, dt$$

$$= 0$$

For $0 < t \leq 2$,

$$v(t) = \frac{1}{C} \int_{-\infty}^{t} i(t)\, dt$$

$$= v(0) + \frac{1}{C} \int_{0}^{t} i(t)\, dt$$

$$= 0 + \frac{1}{2} \int_{0}^{t} t\, dt$$

$$= \frac{1}{2}\left(\frac{t^2}{2}\right)\Big|_{0}^{t}$$

$$= \frac{t^2}{4}$$

For $t > 2$,

$$v(t) = \frac{1}{C} \int_{-\infty}^{t} i(t)\, dt$$

$$= v(2) + \frac{1}{2} \int_{2}^{t} i(t)\, dt$$

Since $v(t) = t^2/4$ for $0 < t \leq 2$,

$$v(t) = \frac{2^2}{4} + \frac{1}{2} \int_2^t 0 \, dt$$

$$= 1$$

In summary, then,

$$v(t) = \begin{cases} 0 & \text{for } -\infty < t < 0 \\ t^2/4 & \text{for } 0 \leq t < 2 \\ 1 & \text{for } 2 \leq t < \infty \end{cases}$$

A sketch of v is shown in Fig. 4.65.

---

fig. 4.65

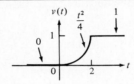

In this example, for time $t \leq 0$ the current applied to the capacitor is zero, and the resuiting voltage across it is therefore zero. For $0 < t < 2$ when the current is nonzero, so is the voltage. For $t \geq 2$, the current is again zero, but the voltage remains nonzero; in particular, it stays a constant 1 V for all time.

The energy stored in the capacitor is

$$w_C(t) = \frac{1}{2}Cv^2(t) = v^2(t) = \begin{cases} 0 & \text{for } -\infty < t < 0 \\ t^4/16 & \text{for } 0 \leq t < 2 \\ 1 & \text{for } 2 \leq t < \infty \end{cases}$$

Thus this energy increases from 0 to 1 J in the time interval $0 \leq t < 2$. But for $t \geq 2$, since the current source has a value of zero, it behaves as an open circuit. Hence there is no way for the capacitor to discharge; that is, there is no element that can absorb the energy stored in the capacitor. Consequently, the energy stored and the voltage across the capacitor remain constant for $t \geq 2$.

· · ·

Since a physical capacitor has some associated leakage resistance (it may be very large) and a practical current source has an internal resistance, in a corresponding actual circuit the energy stored and the voltage across the capacitor would eventually become zero. The smaller the effective resistance, the sooner zero is reached.

Previously we discussed two approaches for describing functions: One required the partitioning of time into intervals and in the other we expressed a function as a sum of elementary functions (e.g., ramps, steps, and impulses). In the example just presented the former approach was taken because for this particular example it seems to be more convenient. In taking the alternative approach, we first obtain an expression for $i(t)$. From the plot of $i(t)$ given in Fig. 4.64,

$$i(t) = t[u(t) - u(t - 2)]$$
$$= tu(t) - tu(t - 2)$$
$$= tu(t) - (t - 2)u(t - 2) - 2u(t - 2)$$

Thus we have

$$i(t) = r(t) - r(t - 2) - 2u(t - 2)$$

Since

$$v(t) = \frac{1}{C} \int_{-\infty}^{t} i(t)\, dt$$

in order to obtain $v(t)$, we must know what is the integral of a ramp. Once this is known, an analytical expression for $v(t)$ is readily obtained. (Getting the plot of $v(t)$ versus t with this approach can be a little tricky for this example — try it if you like.) To take this approach requires the facts that

$$r(t - a) = (t - a)u(t - a)$$

and

$$\int_{-\infty}^{t} r(t - a)\, dt = \frac{1}{2}(t - a)^2 u(t - a) = \frac{1}{2}[r(t - a)]^2$$

We shall leave the details of the last result as an exercise for the reader (see Problem 4.31).

The following example demonstrates the convenience of expressing a function as a sum of elementary functions.

· · ·

Example

Consider the op amp circuit shown in Fig. 4.66.

fig. 4.66

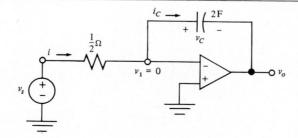

Since $v_1 = 0$, then

$$i = 2v_s$$

But, the op amp draws no current, so

$$i_C = i = 2v_s$$

Thus

$$v_C = \frac{1}{C} \int_{-\infty}^{t} i_C(t) \, dt = \frac{1}{2} \int_{-\infty}^{t} 2v_s(t) \, dt = \int_{-\infty}^{t} v_s(t) \, dt$$

By KVL,

$$v_o = -v_C$$

Hence

$$v_o(t) = -\int_{-\infty}^{t} v_s(t) \, dt$$

and the output voltage is the integral of the input voltage (multiplied by the constant -1). For this reason, such a circuit is known as an *integrator*. This type of circuit is extremely useful; it is the backbone of the analog computer.

Suppose that the input voltage is $v_s(t) = 2u(t) - 3u(t - 1) + u(t - 2)$; for a sketch see Fig. 4.56. Then the output voltage is

$$v_o(t) = -\int_{-\infty}^{t} [2u(t) - 3u(t - 1) + u(t - 2)] \, dt$$

$$= -2r(t) + 3r(t - 1) - r(t - 2)$$

a sketch of which is shown in Fig. 4.67.

---

fig. 4.67

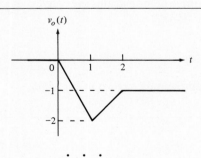

$$\cdot \ \cdot \ \cdot$$

PROBLEMS   4.30.   Show that (a) $[u(t)]^2 = u(t)$ and (b) $[u(t - a)]^2 = u(t - a)$.

4.31.   The integral of a ramp function is called a *parabola*. Show that

$$\int_{-\infty}^{t} r(t - a) \, dt = \frac{1}{2}(t - a)^2 u(t - a) = \frac{1}{2}[r(t - a)]^2$$

4.32. (a) The voltage $v(t)$ across a 2-F capacitor at time $t = 1$ s is $\frac{1}{4}$ V. If the current through the capacitor is

$$i(t) = \begin{cases} t & \text{for } 1 \le t < 2 \\ 0 & \text{for } 2 \le t < \infty \end{cases}$$

find $v(t)$ for $t \ge 1$.

(b) The current $i(t)$ through a 3-H inductor at time $t = 1$ s is $\frac{2}{3}$ A. If the voltage across the inductor is

$$v(t) = \begin{cases} -1 & \text{for } 1 \le t < 2 \\ 0 & \text{for } 2 \le t < \infty \end{cases}$$

find $i(t)$ for $t \ge 1$.

fig. P4.33

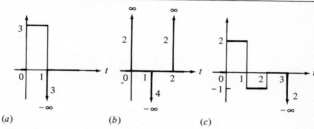

(a)   (b)   (c)

*4.33. Express the three functions shown in Fig. P4.33 in terms of step and impulse functions.

Ans. (a) $3u(t) - 3u(t-1) - 3\delta(t-1)$; (b) $2\delta(t) - 4\delta(t-1) + 2\delta(t-2)$; (c) $2u(t) - 3u(t-1) + u(t-2) - 2\delta(t-3)$

fig. P4.34

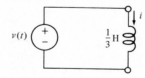

4.34. For the circuit shown in Fig. P4.34, let $v(t)$ be described by the functions given in Problem 4.33. Find $i(t)$ for parts (a), (b), and (c) and sketch these functions.

fig. P4.35

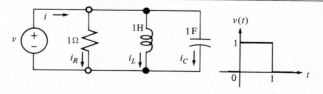

4.35. For the circuit and input voltage $v(t)$ shown in Fig. P4.35, find $i_R(t)$, $i_L(t)$, $i_C(t)$, and $i(t)$. Sketch these functions.

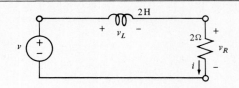

*4.36.   For the circuit in Fig. P4.36, suppose that $v_L(t) = u(t) - u(t - 1)$. Find
$i(t)$, $v_R(t)$, and $v(t)$. Sketch these functions.
Ans. $\frac{1}{2}r(t) - \frac{1}{2}r(t - 1)$; $r(t) - r(t - 1)$; $u(t) - u(t - 1) + r(t) - r(t - 1)$

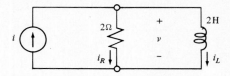

4.37.   For the circuit in Fig. P4.37, suppose that $v(t) = u(t) - u(t - 1)$. Find $i_R(t)$,
$i_L(t)$, and $i(t)$. Sketch these functions.

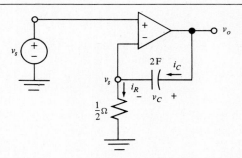

4.38.   For the op amp circuit in Fig. P4.38, suppose that $v_s(t) = u(t) - u(t - 1)$.
Find $i_R(t)$, $i_C(t)$, $v_C(t)$, and $v_o(t)$. Sketch these functions.

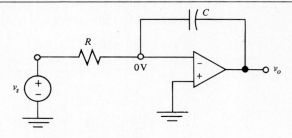

4.39.   Show that for the integrator circuit in Fig. P4.39,

$$v_o(t) = -\frac{1}{RC} \int_{-\infty}^{t} v_s(t) \, dt$$

4.5   *SERIES AND PARALLEL CONNECTIONS*

Just as resistors connected in series or parallel can be combined into an equivalent resistor, so can inductors and capacitors be combined. To see specifically how, we begin with a series connection of inductors as shown in Fig. 4.68.

fig. 4.68

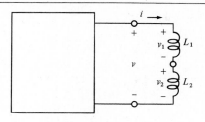

We find that

$$v = v_1 + v_2$$

$$= L_1 \frac{di}{dt} + L_2 \frac{di}{dt}$$

$$= (L_1 + L_2) \frac{di}{dt}$$

$$= L \frac{di}{dt}$$

where $L = L_1 + L_2$. Since the relationship between v and i is not changed, the series connection of L_1 and L_2 can be replaced by a single inductor of value $L = L_1 + L_2$ and the effect on the remainder of the network is the same. We depict this condition in Fig. 4.69. Similarly, the situation for three inductors in series is shown in Fig. 4.70.

fig. 4.69

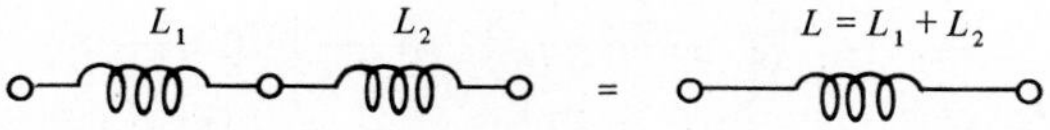

$$\underset{L_1}{\text{—}000\text{—}} \quad \underset{L_2}{\text{—}000\text{—}} \quad = \quad \underset{L = L_1 + L_2}{\text{—}000\text{—}}$$

fig. 4.70

$$\underset{L_1}{\text{—}000\text{—}} \quad \underset{L_2}{\text{—}000\text{—}} \quad \underset{L_3}{\text{—}000\text{—}} \quad = \quad \underset{L = L_1 + L_2 + L_3}{\text{—}000\text{—}}$$

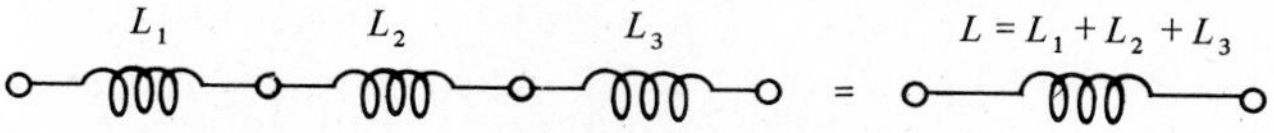

For the case of two inductors in parallel, as shown in Fig. 4.71, by KCL,

$$i = i_1 + i_2$$

$$= \frac{1}{L_1} \int_{t_0}^{t} v \, dt + i_1(t_0) + \frac{1}{L_2} \int_{t_0}^{t} v \, dt + i_2(t_0)$$

$$= \left(\frac{1}{L_1} + \frac{1}{L_2} \right) \int_{t_0}^{t} v \, dt + i_1(t_0) + i_2(t_0)$$

$$= \frac{1}{L} \int_{t_0}^{t} v \, dt + i(t_0)$$

fig. 4.71

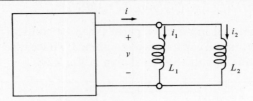

where

$$\frac{1}{L} = \frac{1}{L_1} + \frac{1}{L_2} \quad \text{or} \quad L = \frac{L_1 L_2}{L_1 + L_2}$$

and

$$i(t_0) = i_1(t_0) + i_2(t_0)$$

Thus, the parallel combination of inductors L_1 and L_2 can be replaced by a single inductor L provided that $1/L = (1/L_1) + (1/L_2)$ and the initial current in L is equal to the sum of the initial currents in L_1 and L_2. This result is shown in Fig. 4.72. Similarly, the situation for three inductors in parallel is shown in Fig. 4.73.

fig. 4.72

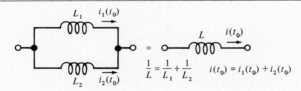

fig. 4.73

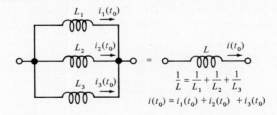

Let us now consider capacitors in parallel. Specifically, for two capacitors in parallel (Fig. 4.74), by KCL,

$$i = i_1 + i_2$$

$$= C_1 \frac{dv}{dt} + C_2 \frac{dv}{dt}$$

$$= (C_1 + C_2) \frac{dv}{dt}$$

$$= C \frac{dv}{dt}$$

fig. 4.74

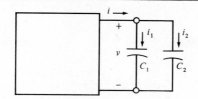

so that we obtain the result shown in Fig. 4.75. Similarly, for three capacitors in parallel, we obtain the result in Fig. 4.76.

For capacitors in series, analogously to inductors in parallel, we get the result shown in Fig. 4.77.

fig. 4.75

fig. 4.76

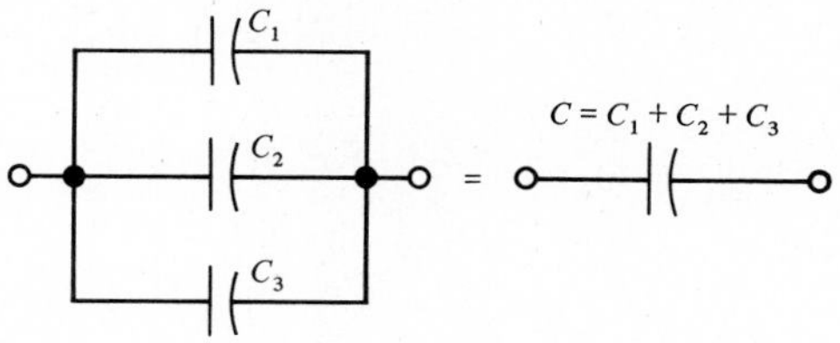

fig. 4.77

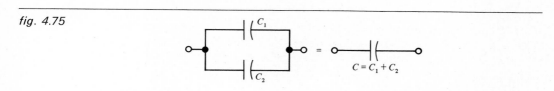

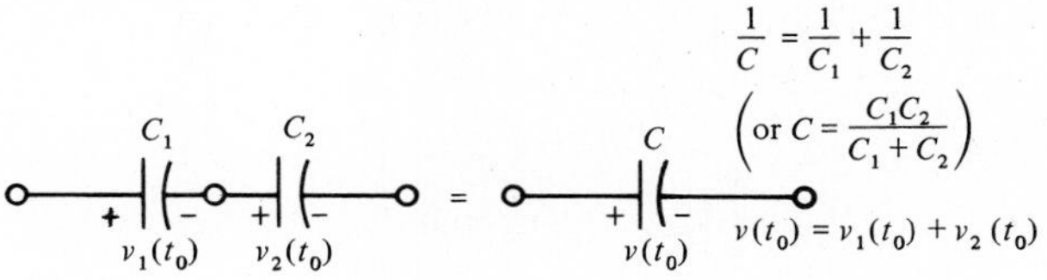

In summary, inductors in series and parallel are treated like resistors, whereas capacitors are treated like conductances.

4.6   MESH AND NODAL ANALYSIS

Just as we analyzed resistive circuits with the use of node, mesh, or loop equations, we can write a set of equations for circuits that contain capacitors and inductors in addition to resistors and sources. The procedure is similar to that described for the resistive case—the difference being that for inductors and capacitors the appropriate relationship between voltage and current is used in place of Ohm's law.

· · ·

Example
Consider the circuit shown in Fig. 4.78.

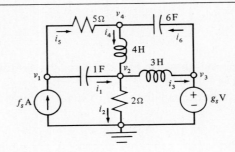

fig. 4.78

By nodal analysis, at node v_1,

$$f_s = i_1 + i_5$$

from which

$$f_s = 1\frac{d}{dt}(v_1 - v_2) + \frac{1}{5}(v_1 - v_4)$$

At node v_2,

$$i_1 + i_4 = i_2 + i_3$$

or

$$1\frac{d}{dt}(v_1 - v_2) + \frac{1}{4}\int_{-\infty}^{t}(v_4 - v_2)\,dt = \frac{1}{2}v_2 + \frac{1}{3}\int_{-\infty}^{t}(v_2 - v_3)\,dt$$

At node v_3,

$$v_3 = g_s$$

At node v_4,

$$i_4 = i_5 + i_6$$

so

$$\frac{1}{4}\int_{-\infty}^{t}(v_4 - v_2)\,dt = \frac{1}{5}(v_1 - v_4) + 6\frac{d}{dt}(v_3 - v_4)$$

The equation, whose variables are voltages, obtained at node v_1 is called a *differential equation* since it contains variables and their derivatives. The equations obtained at nodes v_2 and v_4 are called *integrodifferential equations* since they contain integrals as well as derivatives.

Referring to Fig. 4.79 we can write a set of mesh equations for the circuit as follows:

fig. 4.79

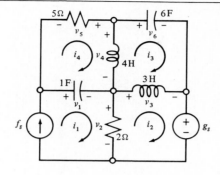

For mesh i_1,

$$f_s = i_1$$

For mesh i_2,

$$g_s = v_2 - v_3$$

$$g_s = 2(i_1 - i_2) - 3\frac{d}{dt}(i_2 - i_3)$$

For mesh i_3,

$$-v_3 - v_4 + v_6 = 0$$

$$-3\frac{d}{dt}(i_2 - i_3) - 4\frac{d}{dt}(i_4 - i_3) + \frac{1}{6}\int_{-\infty}^{t} i_3 \, dt = 0$$

For mesh i_4,

$$-v_1 + v_4 - v_5 = 0$$

$$-\frac{1}{1}\int_{-\infty}^{t}(i_1 - i_4)\,dt + 4\frac{d}{dt}(i_4 - i_3) - (-5i_4) = 0$$

$$\cdot \quad \cdot \quad \cdot$$

Writing the equations for a circuit, as in the preceding example, is not difficult. Finding the solution of equations like these, however, is another matter—it is no simple task. Thus, with the exception of some very simple circuits, we shall have to resort to additional concepts and techniques to be discussed later.

4.7   DUALITY

Let us formalize our discussion of "duality," which was mentioned briefly earlier. Consider the two circuits shown in Fig. 4.80.

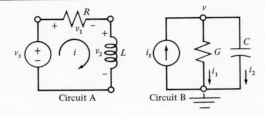

fig. 4.80

Circuit A   Circuit B

Using mesh analysis for circuit A and nodal analysis for circuit B, we get the following results:

Circuit A	Circuit B
$v_s = v_1 + v_2$	$i_s = i_1 + i_2$
$v_s = Ri + L\dfrac{di}{dt}$	$i_s = Gv + C\dfrac{dv}{dt}$

In other words, both circuits are described by the same equations

$$x_s = x_1 + x_2$$

$$x_s = a_1 y + a_2\,\frac{dy}{dt}$$

except that a variable that is a current in one circuit is a voltage in the other and vice versa. For these two circuits this result is not a coincidence, but rather is due to a concept known as *duality*, which has its roots in the subject of graph theory—a topic we discussed briefly previously.

In actuality, circuits A and B are *dual circuits* when the values $R = G$ and $L = C$ and the functions $v_s = i_s$ since the following roles are interchanged:

Series ↔ Parallel
Voltage ↔ Current
Resistance ↔ Conductance
Capacitance ↔ Inductance

In summary, the relationships between a circuit and its dual are as follows:

Circuit	Dual Circuit
Series connection	Parallel connection
Parallel connection	Series connection
Voltage of x volts	Current of x amperes
Current of x amperes	Voltage of x volts
Resistance of x ohms	Conductance of x mhos
Conductance of x mhos	Resistance of x ohms
Capacitance of x farads	Inductance of x henries
Inductance of x henries	Capacitance of x farads

The usefulness of duality lies in the fact that once a circuit is analyzed, its dual is in essence analyzed also—two for the price of one! Note that if circuit B is the dual of circuit A, then taking the dual of circuit B in essence results in circuit A. Not every circuit, however, has a dual. With the aid of graph theory it can be shown that a circuit has a dual if and only if it is a planar network.

Given a series-parallel network, its dual can be found by inspection using the list above. However, a circuit may not be series-parallel. We now present a technique for obtaining the dual of a planar circuit—regardless of whether or not it is a series-parallel network.

1. Inside of each mesh, including the infinite region surrounding the circuit, place a node.
2. Suppose two of these nodes, say nodes a and b, are in adjacent meshes. Then there is at least one element in the boundary common to these two meshes. Place the dual of each common element between nodes a and b.

When using this procedure to obtain the dual circuit, in order to get the mesh equations of the original circuit to correspond to the node equations of the dual circuit, place clockwise mesh currents $i_1, i_2, \ldots, i_n$ in the finite regions. The corresponding nodes in the dual are labeled with the voltages $v_1, v_2, \ldots, v_n$, respectively. The reference node of the dual circuit corresponds to the infinite region of the original circuit. As far as the polarities of the voltages and the directions of the currents are concerned, first consider a current of x amperes flowing through some element. This element is common to one or two finite meshes, call one of them i_a. If the x amperes and the clockwise mesh current i_a have the same direction through the element, then in the dual circuit the polarity at node a is positive for the corresponding voltage; otherwise it is negative. For the case of y volts across the element, if mesh current i_a traverses the voltage from $+$ to $-$, then in the dual the corresponding current has direction out of node a; otherwise it is directed in. These situations are depicted in Fig. 4.81.

fig. 4.81

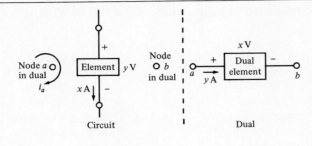

Circuit   Dual

. . .

Example

For the circuit given in the previous example (Fig. 4.79), the dual is shown in heavy lines in Fig. 4.82. The dual network is redrawn in Fig. 4.83 for simplicity.

fig. 4.82

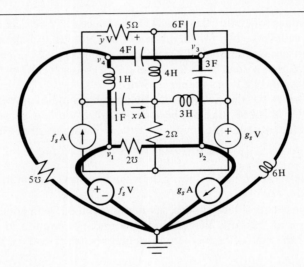

fig. 4.83

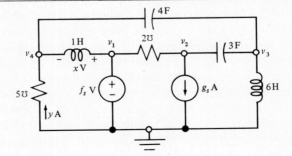

By nodal analysis, at node v_1,

$$v_1 = f_s$$

at node v_2,

$$g_s = 2(v_1 - v_2) + 3 \frac{d}{dt}(v_3 - v_2)$$

at node v_3,

$$3 \frac{d}{dt}(v_3 - v_2) + 4 \frac{d}{dt}(v_3 - v_4) + \frac{1}{6} \int_{-\infty}^{t} v_3 \, dt = 0$$

and, at node v_4,

$$\frac{1}{1} \int_{-\infty}^{t} (v_4 - v_1) \, dt + 4 \frac{d}{dt}(v_4 - v_3) + 5v_4 = 0$$

Note that these are the duals of the mesh equations that we obtained earlier for the circuit given in Fig. 4.79.

· · ·

4.8 LINEARITY AND SUPERPOSITION

For the capacitor and the inductor and their associated descriptions (see Fig. 4.84), all of the relationships between voltage and current are linear relationships. For instance, suppose that for either element, the voltage v is determined from the current i. Then when $i = i_1 + i_2$, the resulting voltage $v = v_1 + v_2$, where v_1 is the voltage due solely to

fig. 4.84

$$i_C = C \frac{dv_C}{dt}$$

$$v_C = \frac{1}{C} \int_{-\infty}^{t} i_C \, dt$$

$$v_L = L \frac{di_L}{dt}$$

$$i_L = \frac{1}{L} \int_{-\infty}^{t} v_L \, dt$$

current i_1 and v_2 is the voltage due solely to current i_2. Also, the voltage due to the current Ki (that is the current i scaled by the constant factor K) is Kv (the voltage due to current i scaled by the same constant K). Consequently, the principle of superposition can also be used for analyzing circuits with capacitors and inductors if the above current-voltage relationships are employed.

· · ·

Example

The circuit in Fig. 4.85 contains two independent sources. We shall use the principle of superposition to determine $v(t)$ given that $v_s(t) = -r(t) + 3r(t-1) - 2r(t-2)$.

First, let us determine $v_1(t)$, the portion of $v(t)$ that is due to the 6-V battery. Setting $v_s(t) = 0$ we have the circuit of Fig. 4.86. The reason why $i_1(t) = 0$ is twofold: The voltage across the capacitor is zero and a capacitor acts as an open

fig. 4.85

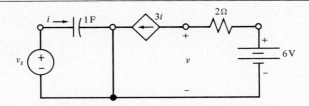

fig. 4.86

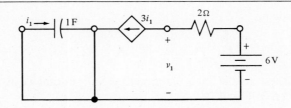

circuit to dc. In either case, we find that $3i_1(t) = 0$, so the voltage across the 2-Ω resistor is zero since the current through it is zero. By KVL,

$$v_1(t) = 6 \text{ V}$$

To find $v_2(t)$, the portion of $v(t)$ due to $v_s(t)$, set the 6-V battery to zero. The resulting circuit is shown in Fig. 4.87. By KVL, the voltage across the capacitor is $v_C(t) = v_s(t)$. Thus,

$$i_2(t) = C \frac{dv_s(t)}{dt} = 1 \frac{d}{dt} [-r(t) + 3r(t-1) - 2r(t-2)]$$

$$= -u(t) + 3u(t-1) - 2u(t-2)$$

fig. 4.87

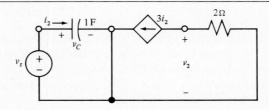

By Ohm's law,

$$v_2(t) = 2(-3i_2) = 6u(t) - 18u(t-1) + 12u(t-2)$$

By the principle of superposition, therefore,

$$v(t) = v_1(t) + v_2(t) = 6 + 6u(t) - 18u(t-1) + 12u(t-2)$$

Sketches of $v_s(t)$ and $v(t)$ are shown in Fig. 4.88.

fig. 4.88

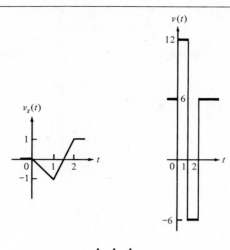

$\cdots$

Although the previously mentioned linear relationships between voltage and current for inductors and capacitors allow us to use the principle of superposition, we must be very careful when employing the integral relationships that incorporate initial conditions at time t_0.

For the case that a capacitor has an initial voltage $v_C(t_0)$ or an inductor has an initial current $i_L(t_0)$, the current-voltage relationships that can be used are

$$v_C = \frac{1}{C} \int_{t_0}^{t} i_C \, dt + v_C(t_0)$$

and

$$i_L = \frac{1}{L} \int_{t_0}^{t} v_L \, dt + i_L(t_0)$$

respectively. In both cases, however, the relationships are nonlinear. If an initial condition is zero, the result is a linear relationship.

Since $v_C(t_0)$ is a number whose units are volts, and $i_L(t_0)$ is a number whose units are amperes, for $t \geq t_0$ we can represent a capacitor with the initial condition $v_C(t_0)$ as the series connection of a capacitor with zero initial voltage and a voltage source whose value (for $t \geq t_0$) equals the initial condition as shown in Fig. 4.89. For $t \geq t_0$ we can represent an inductor with the initial condition $i_L(t_0)$ as the parallel connection of an inductor with zero initial current and a current source whose value (for $t \geq t_0$) equals the initial condition as shown in Fig. 4.90.

fig. 4.89 (left)
fig. 4.90 (right)

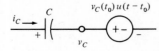
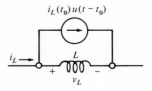

With these representations of capacitors and inductors having initial conditions, we can then utilize the principle of superposition.

$\cdot$ $\cdot$ $\cdot$

Example

Again consider the simple inductor circuit shown in Fig. 4.91, which was analyzed previously (see Figs. 4.55 and 4.56).

fig. 4.91

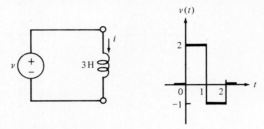

We have already determined that

$$i(t) = \begin{cases} 0 & \text{for } -\infty < t \le 0 \\ \frac{2}{3}t & \text{for } 0 < t \le 1 \end{cases}$$

Of course we can determine $i(t)$ for $t > 1$ as was done before; however, let us now take a different approach. Although this alternative scheme is more involved in this particular case, we shall demonstrate it here inasmuch as there are situations in which it can be employed advantageously.

It is that part of the applied voltage $v(t)$ from $t = -\infty$ to $t = 1$ that gives us the initial condition ($t_0 = 1$) that

$$i(1) = \tfrac{2}{3}(1) = \tfrac{2}{3} \text{ A}$$

For $t \ge 1$, the applied voltage behaves as $-u(t - 1) + u(t - 2)$. This, together with the fact that $i(1) = \frac{2}{3}$ A, allows us to determine $i(t)$ for $t \ge 1$ from the circuit shown in Fig. 4.92. In this circuit $i_L(1) = 0$. To use the principle of superposition,

fig. 4.92

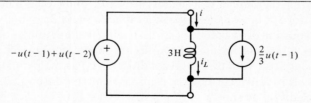

set the current source to zero. We then have the circuit shown in Fig. 4.93. Thus the current due to the voltage source is

$$i_1(t) = \frac{1}{3} \int_{-\infty}^{t} [-u(t - 1) + u(t - 2)] \, dt$$

$$= -\frac{1}{3}r(t - 1) + \frac{1}{3}r(t - 2)$$

fig. 4.93

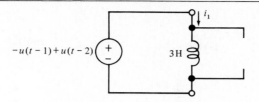

fig. 4.94

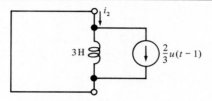

When the voltage source is set to zero, we have the circuit shown in Fig. 4.94. Since the voltage across the inductor is zero, the current through it is zero, and thus the current due to the current source is

$$i_2(t) = \frac{2}{3}u(t-1)$$

By the principle of superposition,

$$i(t) = i_1(t) + i_2(t) = -\frac{1}{3}r(t-1) + \frac{1}{3}r(t-2) + \frac{2}{3}u(t-1)$$

and for $1 \leq t < 2$ this expression becomes

$$i(t) = -\frac{1}{3}(t-1) + \frac{2}{3} = -\frac{1}{3}t + 1$$

while for $t \geq 2$ it becomes

$$i(t) = -\frac{1}{3}(t-1) + \frac{1}{3}(t-2) + \frac{2}{3}$$

$$= -\frac{1}{3}t + \frac{1}{3} + \frac{1}{3}t - \frac{2}{3} + \frac{2}{3} = \frac{1}{3}$$

and these results agree with those obtained previously.

• • •

The circuits containing inductors and capacitors encountered so far have been relatively simple. In the next chapter we shall begin to deal with the general case. The first approach we come across will utilize the solving of more complicated equations known as differential equations. In the analysis of circuits in the future, we shall repeatedly make use of two very important properties—one for inductors and one for capacitors. Let us now derive these properties.

fig. 4.95

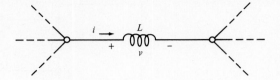

Again, consider the simple representation of an inductor, as shown in Fig. 4.95. Since

$$i(t) = \frac{1}{L} \int_{-\infty}^{t} v(t)\, dt$$

then the current through the inductor at time $t = a$ is

$$i(a) = \frac{1}{L} \int_{-\infty}^{a} v(t)\, dt$$

For any positive real number ε, the current through the inductor at time $t = a + \varepsilon$ is

$$i(a + \varepsilon) = \frac{1}{L} \int_{-\infty}^{a+\varepsilon} v(t)\, dt$$

$$= \frac{1}{L} \int_{-\infty}^{a} v(t)\, dt + \frac{1}{L} \int_{a}^{a+\varepsilon} v(t)\, dt$$

$$= i(a) + \frac{1}{L} \int_{a}^{a+\varepsilon} v(t)\, dt$$

If ε gets arbitrarily small, then

$$\frac{1}{L} \int_{a}^{a+\varepsilon} v(t)\, dt$$

gets arbitrarily small and $i(a + \varepsilon)$ gets arbitrarily close to $i(a)$—provided that $v(t)$ does not contain an impulse at time $t = a$. We therefore conclude that *the current through an inductor cannot change instantaneously* unless there is an impulse of voltage present.

Since for a capacitor (Fig. 4.96) it is true that

$$v(t) = \frac{1}{C} \int_{-\infty}^{t} i(t)\, dt$$

fig. 4.96

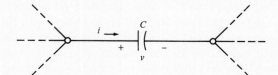

proceeding in a manner as above we can deduce that *the voltage across a capacitor cannot change instantaneously* except for an impulse of current.

In practical terms, there is no such thing as an impulse function. So, practically speaking, we can say that the current through an inductor cannot change instantaneously nor can the voltage across a capacitor. However, the voltage across an inductor and the current through a capacitor *can* change instantaneously (theoretically) or in a very short period of time (practically) without the presence of an impulse function.

PROBLEMS   4.40.   For the circuit shown in Fig. P4.40, derive the voltage divider formula

$$v_2(t) = \frac{L_2}{L_1 + L_2} \, v_s(t)$$

fig. P4.40

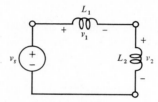

fig. P4.41

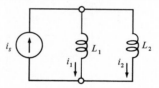

4.41.   For the circuit in Fig. P4.41, assume that $i_1(t_0) = i_2(t_0) = 0$. Derive the current divider formula

$$i_2(t) = \frac{L_1}{L_1 + L_2} \, i_s(t)$$

fig. P4.42

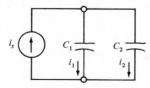

4.42.   With reference to Problem 4.40, use duality to obtain a current division formula for the circuit in Fig. P4.42.

4.43.   With reference to Problem 4.41, use duality to obtain a voltage division formula for the circuit in Fig. P4.43.

fig. P4.43

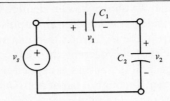

fig. P4.44

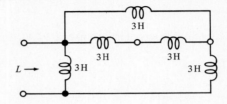

*4.44. For the connection of inductors in Fig. P4.44, find the equivalent inductance L.

Ans. $\frac{15}{8}$ H

fig. P4.45

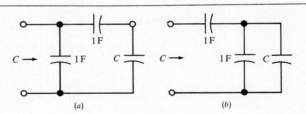

(a)   (b)

4.45. For each of the circuits shown in Fig. P4.45, what value of C results in an equivalent capacitance of C farads?

fig. P4.46

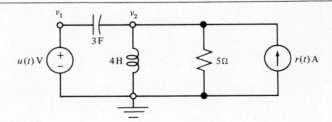

*4.46. Write the node equations for the circuit shown in Fig. P4.46.

Ans. $v_1 = u(t); 3\dfrac{d}{dt}(v_2 - v_1) + \dfrac{1}{4}\displaystyle\int_{-\infty}^{t} v_2\, dt + \dfrac{1}{5}v_2 = r(t)$

4.47. Write the node equations for the circuit shown in Fig. P4.47.

fig. P4.47

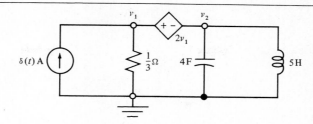

fig. P4.48

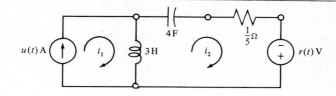

*4.48.   For the circuit in Fig. P4.48, write the mesh equations.

Ans. $i_1 = u(t); 3\dfrac{d}{dt}(i_2 - i_1) + \dfrac{1}{4}\displaystyle\int_{-\infty}^{t} i_2\, dt + \tfrac{1}{5}i_2 = r(t)$

fig. P4.49

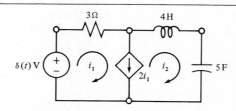

4.49.   For the circuit shown in Fig. P4.49, write the mesh equations.
*4.50.   Find the dual of the circuit given in Problem 4.46.
       Ans.   Circuit in Problem 4.48
4.51.   Find the dual of the circuit given in Problem 4.47.
4.52.   Find the dual of the circuit given in Problem 4.48.
4.53.   Find the dual of the circuit given in Problem 4.49.

fig. P4.54

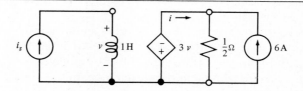

4.54.   For the circuit in Fig. P4.54, suppose that $i_s(t) = r(t) - 3r(t - 1)$
       $+ 2r(t - 2)$. Use the principle of superposition to find $i(t)$. Sketch this
       function.

4.55. For the circuit in Fig. P4.55, suppose that $i_s(t) = 2u(t) - 3u(t - 1)$ $+ u(t - 2)$. Use the principle of superposition to find $i(t)$, and sketch this function, given that (a) $v_s(t) = 6$ V and (b) $v_s(t) = -3u(t - 2)$ V.

fig. P4.55

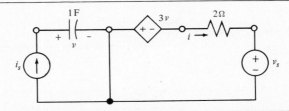

Summary

1. The voltage across an inductor is directly proportional to the derivative of the current through it.
2. An inductor behaves as a short circuit to direct current.
3. The current through a capacitor is directly proportional to the derivative of the voltage across it.
4. A capacitor behaves as an open circuit to direct current.
5. The derivative of a ramp function is a step function. The derivative of a step function is an impulse function.
6. The integral of an impulse function is a step function. The integral of a step function is a ramp function.
7. The current through an inductor is directly proportional to the integral of the voltage across it.
8. The voltage across a capacitor is directly proportional to the integral of the current through it.
9. The energy stored in an inductor is $Li^2/2$ joules; the energy stored in a capacitor is $Cv^2/2$ joules.
10. Inductors in series and parallel are combined in the same way as are resistances. Capacitors in series and parallel are combined in the same way as are conductances.
11. Writing node and mesh (or loop) equations for circuits containing inductors and capacitors are done as were for resistive circuits. Except for simple circuits, the solutions of equations in this form will be avoided.
12. A planar circuit and its dual are in essence described by the same equations.
13. A charged capacitor can be modeled as an uncharged capacitor in series with a voltage source. An inductor with an initial current through it can be modeled as an inductor with no initial current in parallel with a current source.
14. The current through an inductor cannot change instantaneously except for an impulse of voltage. The voltage across a capacitor cannot change instantaneously except for an impulse of current.

FIRST-ORDER CIRCUITS

5

Introduction

To analyze a resistive circuit, we can write a set of node or mesh or loop equations—which are algebraic equations—and solve them. However, for circuits with inductors and/or capacitors, not all the equations will be algebraic. To see how to analyze such circuits, we begin by considering simple networks that in addition to resistors contain a single inductor or capacitor.

We shall first consider the situation where an initial condition is present but an independent source is not (the zero-input case), and then the situation where an independent source is present but the initial condition is zero (the zero-state case). We shall also consider circuits that have both nonzero initial conditions and nonzero inputs. The analysis of such circuits requires the solution of simple "differential equations"—this topic being part of our study.

Once a few simple circuits have been analyzed we shall see that the use of various principles such as duality, linearity, and "time invariance" will enable us to handle certain more complicated situations without having to start from scratch.

5.1 THE ZERO-INPUT RESPONSE

We begin this section by considering the simple RC circuit shown in Fig. 5.1.

Suppose that at time $t = 0$, the voltage across the capacitor is known to be $v_C(0)$.

fig. 5.1

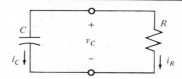

235

(We shall see how this can occur shortly.) We can then determine $v_C(t)$ for all $t \geq 0$ as follows: By KCL,

$$i_C + i_R = 0$$

from which

$$C \frac{dv_C}{dt} + \frac{v_C}{R} = 0$$

or, dividing both sides of the equation by C, we get

$$\frac{dv_C}{dt} + \frac{1}{RC} v_C = 0$$

In this equation there appears the function $v_C(t)$ we wish to determine, as well as its first derivative $dv_C(t)/dt$. For this reason we say that this equation is a *first-order differential equation*. Since the circuit shown in Fig. 5.1 is described by a first-order differential equation, we call it a *first-order circuit*. Since the variable v_C and its derivative are of the first power, we call the differential equation *linear*. Furthermore, since the right side is zero, we say that the equation is *homogeneous*. Note that the coefficients of v_C and its derivative are constants. In order to solve this equation—that is, find the function $v_C(t)$ that satisfies the equation and the initial (or boundary) condition $v_C(0)$, we proceed as follows:

From the given differential equation we have

$$\frac{dv_C}{dt} = -\frac{1}{RC} v_C$$

and dividing both sides by v_C (called *separating variables*), we get

$$\frac{1}{v_C} \frac{dv_C}{dt} = -\frac{1}{RC}$$

Integrating both sides of this equation with respect to time, we have

$$\int \frac{1}{v_C} \frac{dv_C}{dt} \, dt = \int -\frac{1}{RC} \, dt$$

By the chain rule, we can write

$$\int \frac{1}{v_C} \frac{dv_C}{dt} \, dt = \int \frac{dv_C}{v_C} = \int -\frac{1}{RC} \, dt$$

Integrating we get

$$\ln v_C(t) = -\frac{t}{RC} + K$$

where K is a constant of integration. From this equation, taking powers of e, the result is

$$v_C(t) = e^{-t/RC + K} = e^{-t/RC} e^K \tag{5.1}$$

Setting $t = 0$, we obtain

$$v_C(0) = e^0 e^K = e^K$$

Substituting this value of e^K into Equation (5.1) yields

$$v_C(t) = v_C(0)e^{-t/RC} \qquad \text{for } t \geq 0$$

Thus, this is the solution to the differential equation

$$\frac{dv_C(t)}{dt} + \frac{1}{RC} v_C(t) = 0$$

subject to the constraint of the initial condition that the voltage across the capacitor at $t = 0$ is $v_C(0)$.

Of course, a variable is a variable is a variable—so even if we change the variable's name from $v_C(t)$ to $x(t)$, by going through the same routine we get the same results. In other words, the solution to the homogeneous, first-order, linear differential equation (with constant coefficients)

$$\frac{dx(t)}{dt} + ax(t) = 0$$

subject to the initial (or boundary) condition $x(0)$ is

$$\boxed{x(t) = x(0)e^{-at}}$$

and don't you forget it!

A sketch of the function $v_C(t)$ versus t for $t \geq 0$ is shown in Fig. 5.2. Note that the resistor current can now be determined for $t \geq 0$ since

$$i_R(t) = \frac{v_C(t)}{R} = \frac{v_C(0)}{R} e^{-t/RC}$$

fig. 5.2

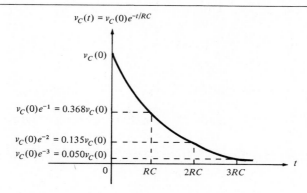

$$v_C(t) = v_C(0)e^{-t/RC}$$

Furthermore, the capacitor current for $t \geq 0$ is

$$i_C(t) = -i_R(t) = -\frac{v_C(0)}{R} e^{-t/RC}$$

or, alternatively,

$$i_C(t) = C \frac{dv_C(t)}{dt} = C \frac{d}{dt} [v_C(0)e^{-t/RC}]$$

$$= C\left(\frac{-1}{RC}\right)v_C(0)e^{-t/RC}$$

$$= -\frac{v_C(0)}{R} e^{-t/RC}$$

From these results we see that initially (at $t = 0$) the capacitor is charged to $v_C(0)$ volts, and for $t > 0$, the capacitor discharges through the resistor exponentially. From the formula

$$w_C(t) = \frac{1}{2} Cv_C^2(t)$$

we see that the energy stored in the capacitor at $t = 0$ is

$$w_C(0) = \frac{1}{2} Cv_C^2(0)$$

Since the voltage goes to zero as time goes to infinity, the energy stored in the capacitor also goes to zero as time goes to infinity. Where does this energy go?

The power absorbed by the resistor is

$$p_R(t) = Ri_R^2(t)$$

$$= R\left(\frac{v_C(0)}{R} e^{-t/RC}\right)^2 = \frac{v_C^2(0)}{R} e^{-2t/RC}$$

and therefore the total energy absorbed by the resistor is

$$w_R = \int_0^\infty p_R(t) \, dt$$

$$= \int_0^\infty \frac{v_C^2(0)}{R} e^{-2t/RC} \, dt = -\frac{RC}{2} \frac{v_C^2(0)}{R} e^{-2t/RC} \Big|_0^\infty$$

$$= -\frac{C}{2} v_C^2(0) [0 - 1] = \frac{1}{2} Cv_C^2(0) = w_C(0)$$

Thus, we see that the energy initially stored in the capacitor is eventually dissipated as heat by the resistor.

We see that the voltage and current in the above circuit have the form

$$f(t) = Ke^{-t/\tau}$$

where K and τ are constants. The constant τ is called the *time constant* and, since the power of e should be a dimensionless number, the units of τ therefore are seconds. For the voltage and current expressions above, we have that the time constant is

$$\tau = RC$$

Thus we see that the product of resistance and capacitance has as units seconds; that is, ohms × farads = seconds. This fact can be verified from $R = v/i$ and $C = q/v$, which have the respective units volts/ampere = volts/coulombs/second and coulombs/volt. Hence, the product RC has as its unit the second. For the example above, making R or C or both larger, increases the time constant. Conversely, making R or C or both smaller decreases the time constant.

A sketch of $f(t)$ versus t is shown in Fig. 5.3, in which we can see that as t goes to infinity, $f(t)$ goes to zero. Although $f(t)$ never identically equals zero for any finite value of t, when $t = 3\tau$ the value of $f(t)$ is down to only about 5 percent of the value of $f(t)$ at $t = 0$. Hence, after just a few time constants, the value of the function $f(t)$ is practically zero.

fig. 5.3

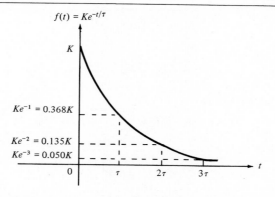

$$f(t) = Ke^{-t/\tau}$$

K

$Ke^{-1} = 0.368K$

$Ke^{-2} = 0.135K$
$Ke^{-3} = 0.050K$

0 τ 2τ 3τ t

In the circuit shown in Fig. 5.1 there was no independent source (that is, no input), only an initial condition (or initial state) of the circuit. For this reason, we say that the voltage and current that we determined are *zero-input* (or *natural*) *responses*.

Now let us look at a situation in which a zero-input response occurs.

· · ·

Example

Let us begin this example by considering the circuit in Fig. 5.4, in which there is a switch that is closed for time $t < 0$, is opened at time $t = 0$, and stays open for all time $t > 0$. For this circuit, let us determine $v_C(t)$ for all values of t.

For $t < 0$, since the capacitor behaves as an open circuit for dc, we can determine $v_C(t)$ by simply using the voltage division formula

$$v_C(t) = \frac{RV}{R_1 + R} \qquad \text{for } t < 0$$

fig. 5.4

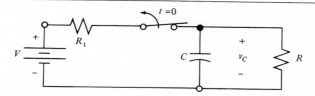

At time $t = 0$, the switch is opened. Since the voltage across a capacitor cannot change instantaneously (except for an impulse of current), we must have that

$$v_C(0) = \frac{RV}{R_1 + R}$$

Next, for time $t \geq 0$, since the switch is then open, to determine $v_C(t)$ we need only consider the simple RC circuit shown in Fig. 5.5, where

$$v_C(0) = \frac{RV}{R_1 + R}$$

fig. 5.5

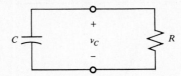

But we have just seen that

$$v_C(t) = v_C(0)e^{-t/RC} \qquad \text{for } t \geq 0$$

Thus, we have

$$v_C(t) = \frac{RV}{R_1 + R} e^{-t/RC} \qquad \text{for } t \geq 0$$

A sketch of $v_C(t)$ versus t, for all values of t, is given in Fig. 5.6. Note that instead of writing the two-part expression

$$v_C(t) = \begin{cases} \dfrac{RV}{R_1 + R} & \text{for } t < 0 \\[2ex] \dfrac{RV}{R_1 + R} e^{-t/RC} & \text{for } t \geq 0 \end{cases}$$

fig. 5.6

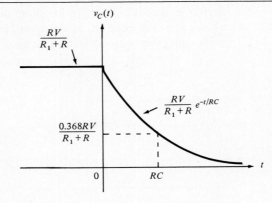

we can equivalently write the single expression

$$v_C(t) = \frac{RV}{R_1 + R} - \frac{RV}{R_1 + R} \, u(t) + \frac{RV}{R_1 + R} \, e^{-t/RC} u(t)$$

$$= \frac{RV}{R_1 + R} - \frac{RV}{R_1 + R} \, (1 - e^{-t/RC}) u(t)$$

$$= \frac{RV}{R_1 + R} \, [1 - (1 - e^{-t/RC}) u(t)]$$

$$\cdot \quad \cdot \quad \cdot$$

Now suppose that in the preceding example the switch was opened at some arbitrary time $t = t_0$. To find $v_C(t)$ we would proceed in a similar manner as before. In the process, we again would have the simple RC circuit shown in Fig. 5.7, which is described by the differential equation

$$\frac{dv_C}{dt} + \frac{1}{RC} \, v_C = 0$$

fig. 5.7

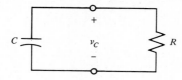

This time, however, the solution must satisfy the boundary condition $v_C(t_0)$. To obtain $v_C(t)$ for $t \geq t_0$, we could proceed as before. However, let us take a slightly different approach this time. We do this as follows:

The given differential equation can be written as

$$\frac{dv_C}{dt} = -\frac{1}{RC} \, v_C$$

Separating variables, we get

$$\frac{1}{v_C} \frac{dv_C}{dt} = -\frac{1}{RC}$$

Integrating both sides with respect to time and choosing t_0 and t as the lower and upper limits, respectively, we have

$$\int_{t_0}^{t} \frac{1}{v_C} \frac{dv_C}{dt} \, dt = \int_{t_0}^{t} -\frac{1}{RC} \, dt$$

Using the chain rule to change the variable of integration, we can write

$$\int_{v_C(t_0)}^{v_C(t)} \frac{dv_C}{v_C} = -\frac{1}{RC} \int_{t_0}^{t} dt$$

Integrating, we get

$$\ln v_C \Big|_{v_C(t_0)}^{v_C(t)} = -\frac{1}{RC} t \Big|_{t_0}^{t}$$

from which we obtain

$$\ln v_C(t) - \ln v_C(t_0) = -\frac{1}{RC}(t - t_0)$$

and since $\ln a - \ln b = \ln(a/b)$, then

$$\ln \frac{v_C(t)}{v_C(t_0)} = \frac{-(t - t_0)}{RC}$$

Taking powers of e, we get

$$\frac{v_C(t)}{v_C(t_0)} = e^{-(t - t_0)/RC}$$

from which

$$v_C(t) = v_C(t_0)e^{-(t - t_0)/RC} \qquad \text{for } t \geq t_0$$

If $t_0 = 0$, then we obtain the previous expression for $v_C(t)$. A plot of $v_C(t)$ versus t is shown in Fig. 5.8.

---

fig. 5.8

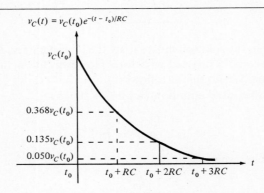

$$v_C(t) = v_C(t_0)e^{-(t - t_0)/RC}$$

We now conclude that the solution to the general linear, homogeneous, first-order differential equation

$$\frac{dx(t)}{dt} + ax(t) = 0$$

subject to the boundary condition $x(t_0)$ is

$$x(t) = x(t_0)e^{-a(t - t_0)} \qquad \text{for } t \geq t_0$$

Now let us look at the simple RL circuit shown in Fig. 5.9, where at time $t = 0$ the inductor current is $i_L(0)$.

fig. 5.9

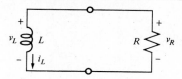

We can determine $i_L(t)$ for $t \geq 0$ as follows: By KVL,

$$v_L - v_R = 0$$

from which

$$L \frac{di_L}{dt} + Ri_L = 0$$

so that

$$\frac{di_L}{dt} + \frac{R}{L} i_L = 0$$

From our previous discussion we know the solution to this differential equation. Hence we can write

$$i_L(t) = i_L(0)e^{-Rt/L} \qquad \text{for } t \geq 0$$

A sketch of $i_L(t)$ versus t for $t \geq 0$ is shown in Fig. 5.10. Note that in this case the time constant is $\tau = L/R$. Thus, dividing inductance by resistance yields seconds.

fig. 5.10

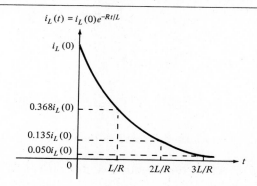

In the simple RL circuit in Fig. 5.9 the inductor voltage is

$$v_L(t) = L \frac{di_L(t)}{dt}$$

$$= L \frac{d}{dt} [i_L(0)e^{-Rt/L}]$$

$$= L \left(\frac{-R}{L} \right) i_L(0)e^{-Rt/L}$$

$$= -Ri_L(0)e^{-Rt/L} \qquad \text{for } t \geq 0$$

and the resistor voltage is

$$v_R(t) = -Ri_L(t)$$
$$= -Ri_L(0)e^{-Rt/L} \qquad \text{for } t \geq 0$$

This, of course, verifies the fact that $v_L(t) = v_R(t)$, which we know by KVL.

The energy stored in the inductor initially (at $t = 0$) is determined from

$$w_L(t) = \frac{1}{2} Li_L^2(t)$$

to be

$$w_L(0) = \frac{1}{2} Li_L^2(0)$$

and as time goes to infinity, since the current in the inductor goes to zero exponentially, the energy stored in the inductor also goes to zero.

The power absorbed by the resistor is

$$p_R(t) = Ri_L^2(t) = R[i_L(0)e^{-Rt/L}]^2 = Ri_L^2(0)e^{-2Rt/L}$$

and therefore the total energy absorbed by the resistor is

$$w_R = \int_0^\infty p_R(t) \, dt = \int_0^\infty Ri_L^2(0)e^{-2Rt/L} \, dt$$
$$= -\frac{L}{2R} Ri_L^2(0)e^{-2Rt/L} \Big|_0^\infty$$
$$= -\frac{L}{2}i_L^2(0)[0 - 1] = \frac{1}{2}Li_L^2(0) = w_L(0)$$

Thus, we see that the energy initially stored in the inductor is eventually dissipated as heat by the resistor.

For the case we are given the boundary condition $i_L(t_0)$ instead of the initial condition $i_L(0)$ for the simple RL circuit in Fig. 5.9, as was done for the simple RC circuit given previously, we can show that

$$i_L(t) = i_L(t_0)e^{-R(t - t_0)/L} \qquad \text{for } t \geq t_0$$

We now present a situation in which a zero-input response like the one just discussed occurs.

$$\cdot \quad \cdot \quad \cdot$$

Example

In the circuit shown in Fig. 5.11 the switch is closed for time $t < 0$ and is open for time $t \geq 0$. Let us determine $i_L(t)$ for all t.

fig. 5.11

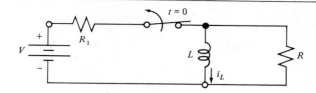

For $t < 0$, since the inductor acts as a short circuit to dc, the inductor current is simply

$$i_L(t) = \frac{V}{R_1} \qquad \text{for } t < 0$$

At time $t = 0$, the switch is opened. Since the current through an inductor cannot change instantaneously (except for an impulse of voltage), we must have that

$$i_L(0) = \frac{V}{R_1}$$

Next for time $t \geq 0$, since the switch is then open, to determine $i_L(t)$ we need only consider the simple RL circuit, shown in Fig. 5.12, where $i_L(0) = V/R_1$. But we have just seen that

$$i_L(t) = i_L(0)e^{-Rt/L} \qquad \text{for } t \geq 0$$

fig. 5.12

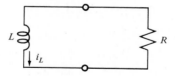

Thus, we have that

$$i_L(t) = \frac{V}{R_1}e^{-Rt/L} \qquad \text{for } t \geq 0$$

A sketch of $i_L(t)$ versus t for all t is identical to the sketch of $v_C(t)$ versus t in Fig. 5.6; only the constants have been changed.

Again, we may write either

$$i_L(t) = \begin{cases} \dfrac{V}{R_1} & \text{for } t < 0 \\[2ex] \dfrac{V}{R_1}e^{-Rt/L} & \text{for } t \geq 0 \end{cases}$$

or

$$i_L(t) = \frac{V}{R_1} - \frac{V}{R_1} u(t) + \frac{V}{R_1} e^{-Rt/L} u(t)$$

$$= \frac{V}{R_1} - \frac{V}{R_1} (1 - e^{-Rt/L}) u(t)$$

$$= \frac{V}{R_1} [1 - (1 - e^{-Rt/L}) u(t)]$$

$$\bullet \quad \bullet \quad \bullet$$

Now let us consider an example of a zero-input response for a first-order circuit that contains a dependent source.

$$\bullet \quad \bullet \quad \bullet$$

Example
The circuit shown in Fig. 5.13 has no independent source, but it does have a dependent current source.

fig. 5.13

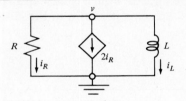

Suppose that $i_L(0) = 5$ A. By KCL,

$$i_R + 2i_R + i_L = 0$$

or

$$3i_R + i_L = 0$$

Since $i_R = v/R$, then

$$\frac{3v}{R} + i_L = 0$$

However, $v = L \, di_L/dt$. Thus

$$\frac{3L}{R} \frac{di_L}{dt} + i_L = 0$$

or

$$\frac{di_L}{dt} + \frac{R}{3L} i_L = 0$$

The solution of this differential equation is

$$i_L(t) = i_L(0)e^{-Rt/3L}$$

$$= 5e^{-Rt/3L} \qquad \text{for } t \geq 0$$

Note that because of the presence of the dependent source, the time constant is not L/R but rather $3L/R$.

An alternative for analyzing this circuit is to utilize Thévenin's theorem. We do this by removing the inductor (this can be done since the dependent source does not depend on the inductor's voltage or current) and replacing the resistor and dependent source with its Thévenin equivalent circuit.

First, we calculate v_{oc} from the circuit of Fig. 5.14. By KCL,

$$i_R + 2i_R = 0$$

$$3i_R = 0$$

$$i_R = 0$$

fig. 5.14

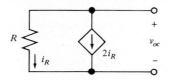

Thus,

$$v_{oc} = Ri_R$$

$$= 0 \text{ V}$$

Next, we calculate R_0 from Fig. 5.15. By KCL,

$$i_R + 2i_R = i_0$$

$$3i_R = i_0$$

fig. 5.15

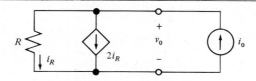

But

$$i_R = \frac{v_0}{R}$$

so

$$\frac{3v_0}{R} = i_0$$

and

$$\frac{v_0}{i_0} = \frac{R}{3} = R_0$$

Hence, the Thévenin equivalent circuit is a resistor whose value is $R/3$ ohms. Connecting this equivalent circuit to the inductor we get the circuit of Fig. 5.16, and we have already seen that for such a circuit

$$i_L(t) = i_L(0)e^{-Rt/3L}$$

$$= 5e^{-Rt/3L} \qquad \text{for } t \geq 0$$

fig. 5.16

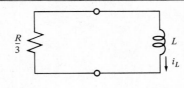

. . .

Next, let us consider a zero-input response for a more complicated first-order circuit.

. . .

Example

In the circuit shown in Fig. 5.17 there are two switches—one that opens at time $t = 0$ and one that closes at time $t = 1$ s. For this circuit, let us determine $v_C(t)$ for all t.

fig. 5.17

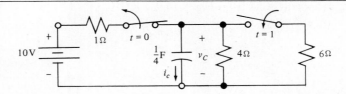

Since a capacitor acts as an open circuit to dc, for $t < 0$, by voltage division we have that

$$v_C(t) = \frac{4}{4+1}(10) = 8 \text{ V}$$

Since the voltage across a capacitor cannot change instantaneously, when the first switch is opened at time $t = 0$, we have

$$v_C(0) = 8 \text{ V}$$

For $0 \leq t < 1$, the voltage across the capacitor is determined by the simple RC circuit in Fig. 5.18, in which $v_C(0) = 8$ V.

fig. 5.18

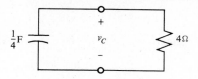

Thus we have that

$$v_C(t) = v_C(0)e^{-t/RC}$$

$$= 8e^{-t} \quad \text{for } 0 \le t < 1$$

However, at time $t = 1$ s the other switch is closed, and since the voltage across a capacitor cannot change instantaneously, we have

$$v_C(1) = 8e^{-1}$$

$$= 8(0.368) = 2.943 \text{ V}$$

For $t \ge 1$, the 4-Ω and 6-Ω resistors are connected in parallel, and the effective resistance is

$$\frac{(4)(6)}{4 + 6} = 2.4 \; \Omega$$

We now therefore have the simple RC circuit shown in Fig. 5.19, in which $v_C(1) = 2.943$ V.

fig. 5.19

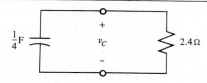

From our previous discussion, we know that

$$v_C(t) = v_C(1)e^{-(t-1)/RC}$$

$$= 2.943e^{-(t-1)/0.6}$$

$$= 2.943e^{-5(t-1)/3} \quad \text{for } t \ge 1$$

In summary,

$$v_C(t) = \begin{cases} 8 & \text{for } t < 0 \\ 8e^{-t} & \text{for } 0 \le t < 1 \\ 2.943e^{-5(t-1)/3} & \text{for } t \ge 1 \end{cases}$$

Furthermore, since $i_C = C \, dv_C/dt$, we also have that for $t < 0$,

$$i_C(t) = \frac{1}{4} \frac{d}{dt} (8) = 0$$

for $0 \leq t < 1$,

$$i_C(t) = \frac{1}{4}\frac{d}{dt}(8e^{-t}) = -2e^{-t}$$

and, for $t \geq 1$,

$$i_C(t) = \frac{1}{4}\frac{d}{dt}[2.94e^{-5(t-1)/3}]$$

$$= \frac{1}{4}(2.94)\left(-\frac{5}{3}\right)e^{-5(t-1)/3}$$

$$= -1.226e^{-5(t-1)/3}$$

In other words,

$$i_C(t) = \begin{cases} 0 & \text{for } t < 0 \\ -2e^{-t} & \text{for } 0 \leq t < 1 \\ -1.226e^{-5(t-1)/3} & \text{for } t \geq 1 \end{cases}$$

Sketches of $v_C(t)$ and $i_C(t)$ are shown in Fig. 5.20. Note that although the voltage across the capacitor does not change instantaneously, the current through it does at time $t = 0$ and $t = 1$ s. In this example, for $t < 0$ the capacitor is charged to 8 V.

fig. 5.20

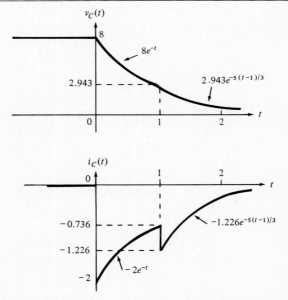

After the first switch is opened at $t = 0$, the capacitor begins to discharge through the 4-Ω resistor with a time constant of $\tau = RC = 4(1/4) = 1$ s. At $t = 1$ s, the capacitor has not completely discharged; there is still 2.943 V across it. At this time the second switch is closed, and the capacitor begins to discharge through the parallel combination of the 4-Ω and 6-Ω resistors (effectively 2.4 Ω) with a time constant of $\tau = (2.4)(1/4) = \frac{3}{5}$ s.

. . .

PROBLEMS *5.1.   In the circuit shown in Fig. P5.1, the switch is opened when $t = 0$. Find $v(t)$ and $i(t)$ for all time. Sketch these functions.
    Ans. $-6e^{-2t}u(t),\ -2e^{-2t}u(t)$

fig. P5.1

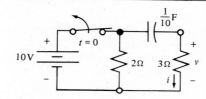

fig. P5.2

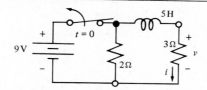

5.2.   Repeat Problem 5.1 for the circuit in Fig. P5.2.

fig. P5.3

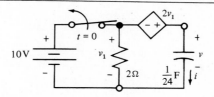

5.3.   Repeat Problem 5.1 for the circuit in Fig. P5.3.

fig. P5.4

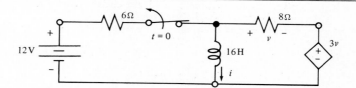

*5.4.   Repeat Problem 5.1 for the circuit in Fig. P5.4.
    Ans. $-16e^{-2t}u(t),\ 2 + 2(e^{-2t} - 1)u(t)$
5.5.   Repeat Problem 5.1 for the circuit in Fig. P5.5.

fig. P5.5

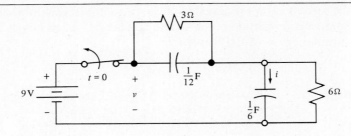

5.6.   For the circuit given in Problem 5.5, change the value of the $\frac{1}{12}$-F capacitor to $\frac{1}{3}$ F and repeat the problem.

5.7.   For the circuit given in Problem 5.5, replace the $\frac{1}{6}$-F capacitor by a 12-H inductor and repeat the problem.

5.8.   For the circuit given in Problem 5.5, replace the $\frac{1}{6}$-F capacitor by a 1.5-H inductor and repeat the problem.

fig. P5.9

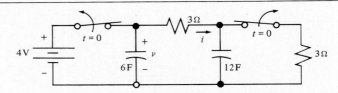

*5.9.   In the circuit shown in Fig. P5.9, both switches are opened when $t = 0$. Find $i(t)$ and $v(t)$ for all time. Sketch these functions.
Ans. $\frac{2}{3} + \frac{2}{3}(e^{-t/12} - 1)u(t)$, $4 + \frac{4}{3}(e^{-t/12} - 1)u(t)$

fig. P5.10

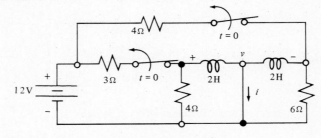

5.10.   Repeat Problem 5.9 for the circuit in Fig. P5.10.

fig. P5.11

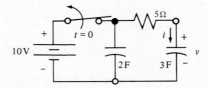

*5.11.   Repeat Problem 5.1 for the circuit in Fig. P5.11.
Ans.   10 V, 0A

fig. P5.12

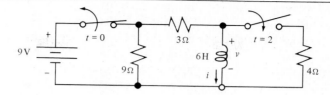

5.12. In the circuit given in Fig. P5.12, one switch is opened when $t = 0$ and one switch is closed when $t = 2$ s. Find $i(t)$ and $v(t)$ for all time. Sketch these functions.

5.13. Repeat Problem 5.10 for the case that the 4-Ω resistor is replaced by a 3-H inductor.

fig. P5.14

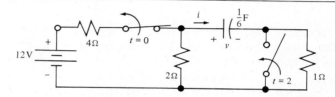

5.14. Repeat Problem 5.12 for the circuit shown in Fig. P5.14.

5.2   THE ZERO-STATE RESPONSE

Now that we have discussed the responses of first-order circuits that had no inputs (i.e., no independent sources) and nonzero initial (or boundary) conditions, let us next consider the case of responses to a nonzero input for a circuit in which all the initial conditions are equal to zero. Such a response is called a *zero-state response*.

We begin by determining the *step response* (i.e., the response that is due to a step function) for the simple series RC circuit shown in Fig. 5.21.

fig. 5.21

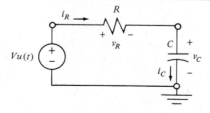

Before we determine the step response, say $v_C(t)$, mathematically, let us try to predict what will happen. Given that the capacitor is originally uncharged, for $t < 0$ the input voltage is zero. Thus, since there is no voltage to produce a current that will charge the capacitor, it should be obvious that $v_C(t) = 0$ for $t < 0$. At time $t = 0$ when the input voltage goes from 0 to V volts, since the voltage across the capacitor cannot change instantaneously (the capacitor acts as a short circuit to an instantaneous change of voltage), $v_C(0) = 0$. By KVL, $v_R(0) = V - 0 = V$ volts, and $i_R(0) = v_R(0)/R = V/R = i_C(0)$ amperes. Thus the voltage across the resistor changes instantaneously, as does the current through the resistor, which is the current through the capacitor. The nonzero current through the capacitor begins to charge it to some nonzero value. Since the input voltage remains constant for $t > 0$, by KVL, an increase in the capacitor voltage means a decrease in the resistor voltage. This in turn means a reduction in the resistor current, which is the capacitor current, and therefore the capacitor charges at a slower rate than before. The charging process continues until the capacitor is completely charged. After a long, long time, the input voltage seems more and more like a constant (dc), so as time goes to infinity, the capacitor behaves as an open circuit. Therefore, the final voltage across the capacitor is V volts. For the case of a zero-input response we have seen that a capacitor discharges exponentially. So we can take an educated guess that it will charge up exponentially as well.

To summarize, for $t < 0$ and for $t = 0$, the voltage across the capacitor is zero. For $t > 0$, the voltage across the capacitor increases at a slower and slower rate until it is completely charged to V volts. By KVL, the voltage across the resistor is zero for $t < 0$, it jumps to V volts instantaneously at $t = 0$, and decreases exponentially to zero as time increases. Furthermore, the current through the resistor—which equals the current through the capacitor—is proportional to the voltage across the resistor.

Let us now confirm our intuition with a mathematical analysis. By nodal analysis,

$$i_R = i_C$$

$$\frac{Vu(t) - v_C}{R} = C\frac{dv_C}{dt}$$

$$C\frac{dv_C}{dt} + \frac{v_C}{R} = \frac{Vu(t)}{R}$$

$$\frac{dv_C}{dt} + \frac{1}{RC}v_C = \frac{V}{RC}u(t)$$

Since

$$u(t) = \begin{cases} 0 & \text{for } t < 0 \\ 1 & \text{for } t \geq 0 \end{cases}$$

let us consider the above differential equation for the two time intervals $t < 0$ and $t \geq 0$.

First, for $t < 0$, the differential equation becomes

$$\frac{dv_C}{dt} + \frac{1}{RC}v_C = 0$$

Although it is a trivial solution, $v_C(t) = 0$ is clearly a solution of this equation. Since we want to find the response to the input under the circumstance that there is no initial voltage on the capacitor, and since the input is zero before time $t = 0$, this trivial solution is what we seek; that is,

$$v_C(t) = 0 \qquad \text{for } t < 0$$

Next, for $t \geq 0$, the differential equation becomes

$$\frac{dv_C}{dt} + \frac{1}{RC} v_C = \frac{V}{RC}$$

To solve this differential equation, we write

$$\frac{dv_C}{dt} = \frac{V}{RC} - \frac{1}{RC} v_C$$

$$= \frac{V - v_C}{RC}$$

and, dividing by $(V - v_C)/RC$, we get

$$\frac{RC}{V - v_C} \frac{dv_C}{dt} = 1$$

Integrating both sides with respect to t, we get

$$\int \frac{RC}{V - v_C} \frac{dv_C}{dt} \, dt = \int dt$$

By the chain rule, we have

$$RC \int \frac{dv_C}{V - v_C} = \int dt$$

or

$$-RC \int \frac{-dv_C}{V - v_C} = \int dt$$

which yields

$$-RC \ln(V - v_C) = t + K \tag{5.2}$$

where K is a constant of integration. Since $v_C(t) = 0$ for $t < 0$ and since the voltage across the capacitor cannot change instantaneously, $v_C(0) = 0$. Hence, substituting $t = 0$ in the above expression, we get

$$-RC \ln[V - v_C(0)] = 0 + K$$

or

$$-RC \ln V = K$$

which determines the value of K. Substituting this value into Equation (5.2), we obtain

$$-RC \ln(V - v_C) = t - RC \ln V$$

from which

$$RC \ln V - RC \ln(V - v_C) = t$$

$$RC[\ln V - \ln(V - v_C)] = t$$

$$RC \ln \frac{V}{V - v_C} = t$$

$$\ln \frac{V}{V - v_C} = \frac{t}{RC}$$

$$\frac{V}{V - v_C} = e^{t/RC}$$

$$\frac{V - v_C}{V} = e^{-t/RC}$$

$$V - v_C = Ve^{-t/RC}$$

$$v_C = V - Ve^{-t/RC}$$

Thus

$$v_C(t) = V(1 - e^{-t/RC}) \qquad \text{for } t \geq 0$$

which is the remainder of the solution. Hence, the complete solution is

$$v_C(t) = \begin{cases} 0 & \text{for } t < 0 \\ V(1 - e^{-t/RC}) & \text{for } t \geq 0 \end{cases}$$

which can be expressed by the single equation

$$v_C(t) = V(1 - e^{-t/RC})u(t) \qquad \text{for all } t$$

For the circuit above (Fig. 5.21), having determined the capacitor voltage we can now find the capacitor current (which is equal to the resistor current in this case) from

$$i_C(t) = C \frac{dv_C(t)}{dt}$$

For $t < 0$,

$$i_C(t) = C \frac{d(0)}{dt} = 0$$

for $t \geq 0$,

$$i_C(t) = C \frac{d}{dt} [V(1 - e^{-t/RC})]$$

$$= CV(-e^{-t/RC})\left(\frac{-1}{RC}\right)$$

$$= \frac{V}{R} e^{-t/RC}$$

Thus we can write

$$i_C(t) = \begin{cases} 0 & \text{for } t < 0 \\ \dfrac{V}{R} e^{-t/RC} & \text{for } t \geq 0 \end{cases}$$

or

$$i_C(t) = \frac{V}{R} e^{-t/RC} u(t) \qquad \text{for all } t$$

Another way of determining $i_C(t)$ is to take the derivative of $v_C(t)$ for all t as follows:

$$i_C(t) = C \frac{d}{dt}[V(1 - e^{-t/RC})u(t)]$$

Since

$$\frac{d}{dt}[f(t)g(t)] = f(t) \frac{dg(t)}{dt} + \frac{df(t)}{dt} g(t)$$

Then

$$i_C(t) = CV[(1 - e^{-t/RC})\delta(t) + (-e^{-t/RC})\left(\frac{-1}{RC}\right)u(t)]$$

$$= CV(1 - e^{-t/RC})\delta(t) + \frac{V}{R} e^{-t/RC} u(t)$$

By the sampling property of the impulse function,

$$CV(1 - e^{-t/RC})\delta(t) = CV(1 - e^0)\delta(t)$$

$$= 0\delta(t) = 0$$

Hence we have

$$i_C(t) = \frac{V}{R} e^{-t/RC} u(t) \qquad \text{for all } t$$

An alternative for finding i_C is to determine i_R since $i_C = i_R$. To do this, we can use KVL to write

$$v_R(t) = Vu(t) - v_C(t)$$

$$= Vu(t) - V(1 - e^{-t/RC})u(t)$$

$$= Vu(t) - Vu(t) + Ve^{-t/RC}u(t)$$

$$= Ve^{-t/RC}u(t)$$

from which

$$i_C(t) = i_R(t)$$

$$= \frac{V}{R} e^{-t/RC} u(t)$$

fig. 5.22

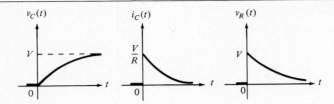

Sketches of $v_C(t)$, $i_C(t)$, and $v_R(t)$ are shown in Fig. 5.22. These results confirm the predictions that were made earlier.

By using the concept of duality, it is now a simple matter to analyze the simple parallel RL circuit shown in Fig. 5.23.

Since this circuit is the dual of the simple RC circuit given above (Fig. 5.21), we have

$$i_L(t) = I(1 - e^{-Rt/L})u(t)$$

$$v(t) = IRe^{-Rt/L}u(t)$$

$$i_R(t) = Ie^{-Rt/L}u(t)$$

Sketches of these functions are shown in Fig. 5.24.

fig. 5.23

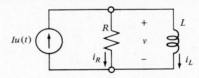

fig. 5.24

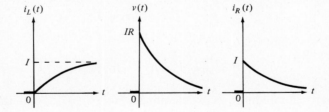

5.3   *TIME INVARIANCE*

Suppose that for the simple RC series circuit shown in Fig. 5.21 the input is a step of voltage in which the instantaneous change occurs at time $t = t_0$ rather than at time $t = 0$. In other words, suppose that we wish to find the zero-state step response to the input $Vu(t - t_0)$ volts. Then the differential equation describing the circuit becomes

$$\frac{dv_C}{dt} + \frac{1}{RC} v_C = \frac{V}{RC} u(t - t_0)$$

We now would like to find the solution of this equation. Again, we consider two intervals of time.

For $t < t_0$, the equation becomes

$$\frac{dv_C}{dt} + \frac{1}{RC}\, v_C = 0$$

and since there is no initial voltage across the capacitor, we have that

$$v_C(t) = 0 \qquad \text{for } t < t_0$$

For $t \geq t_0$, the equation becomes

$$\frac{dv_C}{dt} + \frac{1}{RC}\, v_C = \frac{V}{RC}$$

The solution to this equation subject to the constraint that $v_C(t_0) = 0$ is obtained as follows:

Proceeding as before, we obtain the expression

$$-RC \ln(V - v_C) = t + K \tag{5.2}$$

Since $v_C(t_0) = 0$, by substituting $t = t_0$ into this equation we get

$$-RC \ln[V - v_C(t_0)] = t_0 + K$$

or

$$K = -t_0 - RC \ln V$$

Substituting this value of K into Equation (5.2) yields

$$-RC \ln(V - v_C) = t + (-t_0 - RC \ln V)$$

from which

$$t = t_0 + RC \ln V - RC \ln(V - v_C)$$

$$= t_0 + RC \ln \frac{V}{V - v_C}$$

or

$$t - t_0 = RC \ln \frac{V}{V - v_C}$$

$$\frac{t - t_0}{RC} = \ln \frac{V}{V - v_C}$$

$$e^{(t - t_0)/RC} = \frac{V}{V - v_C}$$

$$e^{-(t - t_0)/RC} = \frac{V - v_C}{V}$$

$$Ve^{-(t - t_0)/RC} = V - v_C$$

$$v_C = V - Ve^{-(t - t_0)/RC}$$

Thus

$$v_C(t) = V(1 - e^{-(t-t_0)/RC}) \qquad \text{for } t \geq t_0$$

Combining this expression with the fact that $v_C(t) = 0$ for $t < t_0$, we get the single expression

$$v_C(t) = V(1 - e^{-(t-t_0)/RC})u(t - t_0) \qquad \text{for all } t$$

After a bit of manipulation, as we performed earlier, we can get the following formulas:

$$i_C(t) = \frac{V}{R}\, e^{-(t-t_0)/RC}u(t - t_0) \qquad \text{for all } t$$

$$v_R(t) = Ve^{-(t-t_0)/RC}u(t - t_0) \qquad \text{for all } t$$

Sketches of these functions are shown in Fig. 5.25. In these sketches t_0 is depicted as a positive quantity, but t_0 may be nonpositive as well.

fig. 5.25

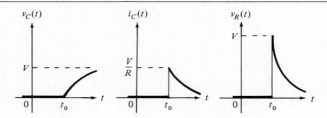

Notice that the only difference between these step responses and those due to the step input $Vu(t)$ volts is the "shift" of the response by t_0 seconds. In other words, the response is the same; only the time at which it starts is different—and this is because the input becomes nonzero at time $t = t_0$ rather than at time $t = 0$. A circuit such as this (that is, a circuit whose response is the same regardless of when the input is applied) is called a *time-invariant circuit*. For a time-invariant circuit, therefore, knowing the zero-state response $x(t)$ to an input $f(t)$, it is a simple matter to obtain the zero-state response to the input $f(t - t_0)$: It is simply $x(t - t_0)$. This fact is demonstrated for a first-order time-invariant circuit as follows.

Suppose that $y(t)$ is the solution of the first-order linear differential equation with constant coefficients

$$\frac{dx(t)}{dt} + ax(t) = f(t)$$

subject to the constraint that $x(0) = 0$. Then what is the solution of the differential equation

$$\frac{dx(t)}{dt} + ax(t) = f(t - t_0)$$

subject to the constraint that $x(t_0) = 0$? Since $y(t)$ is the solution of the original differential equation, we have the equality

$$\frac{dy(t)}{dt} + ay(t) = f(t)$$

Substituting $t - t_0$ for t in this equation, we get the equality

$$\frac{dy(t - t_0)}{d(t - t_0)} + ay(t - t_0) = f(t - t_0)$$

However, by the chain rule,

$$\frac{dy(t - t_0)}{dt} = \frac{dy(t - t_0)}{d(t - t_0)} \frac{d(t - t_0)}{dt}$$

But, since t_0 is a constant,

$$\frac{d(t - t_0)}{dt} = 1$$

and thus

$$\frac{dy(t - t_0)}{dt} = \frac{dy(t - t_0)}{d(t - t_0)}$$

Substituting this into the above equality, we obtain the equality

$$\frac{dy(t - t_0)}{dt} + ay(t - t_0) = f(t - t_0)$$

Thus, $y(t - t_0)$ is the solution of the differential equation in question.

The time-invariance property just discussed results when differential equations have constant coefficients—and such equations, in turn, result from circuits composed of resistors, capacitors, and inductors whose values are constant.

5.4 FORCED AND NATURAL RESPONSES

For the simple series RC circuit given previously (Fig. 5.21), the differential equation describing the voltage across the capacitor was

$$\frac{dv_C}{dt} + \frac{1}{RC} v_C = \frac{V}{RC} u(t)$$

and the solution of this equation is*

$$v_C(t) = V(1 - e^{-t/RC})u(t)$$

$$= Vu(t) - Ve^{-t/RC}u(t)$$

---

* From now on, when it is apparent that an expression is valid for all time, the phrase "for all t" will be omitted.

Note that this expression is the sum of two terms, one having the form of the input and the other having the form of the zero-input response. Is this always the case or is this the case only for this circuit? To answer this question, let us consider the general first-order linear differential equation with constant coefficients:

$$\frac{dx(t)}{dt} + ax(t) = f(t)$$

In this case $f(t)$ arises from the input and $x(t)$ represents the response. Let us multiply both sides of this equation by e^{at}. We therefore have

$$e^{at}\frac{dx(t)}{dt} + e^{at}ax(t) = e^{at}f(t)$$

Since

$$\frac{d}{dt}\left[e^{at}x(t)\right] = e^{at}\frac{dx(t)}{dt} + e^{at}ax(t)$$

then

$$\frac{d}{dt}\left[e^{at}x(t)\right] = e^{at}f(t)$$

Integrating both sides of this equation with respect to t, we get

$$\int \frac{d[e^{at}x(t)]}{dt}\,dt = \int e^{at}f(t)\,dt$$

By the chain rule,

$$\int d\left[e^{at}x(t)\right] = \int e^{at}f(t)\,dt$$

or

$$e^{at}x(t) = \int e^{at}f(t)\,dt + A$$

where A is a constant of integration. Multiplying both sides of this equation by e^{-at}, we obtain the solution

$$x(t) = e^{-at}\int e^{at}f(t)\,dt + Ae^{-at}$$

where the constant A is determined from an initial (boundary) condition. Thus, we see that in general the solution of the differential equation

$$\frac{dx(t)}{dt} + ax(t) = f(t)$$

is called the *complete response*, and it consists of the sum of two parts. The first part is

$$e^{-at}\int e^{at}f(t)\,dt$$

which basically is determined by the function $f(t)$. Since $f(t)$ results from the input of a circuit, we call this part of the solution the *forced response* (also known as the *steady-state response*), and we refer to $f(t)$ as the *forcing function*. The other part of the solution is

$$Ae^{-at}$$

which we recognize has the form of the *zero-input response* or *natural response*. (It is also called the *transient response*.)

Now let us be specific and consider the case that the forcing function is a constant, say b. Then we have the differential equation

$$\frac{dx(t)}{dt} + ax(t) = b$$

The forced response $x_f(t)$ is

$$x_f(t) = e^{-at} \int e^{at} b \, dt$$

$$= e^{-at}\left(\frac{1}{a} e^{at}\right)b = \frac{b}{a}$$

which is a constant, and the natural response $x_n(t)$ is

$$x_n(t) = Ae^{-at}$$

Thus the complete response is

$$\boxed{\begin{aligned} x(t) &= x_f(t) + x_n(t) \\ &= \frac{b}{a} + Ae^{-at} \end{aligned}}$$

where the constant A is determined from a boundary condition.

$\bullet \quad \bullet \quad \bullet$

Example

Let us now use the above result to find the zero-state step response for the simple series RL circuit shown in Fig. 5.26. Before analyzing this circuit, let us predict its behavior.

fig. 5.26

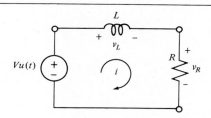

For $t < 0$ the input voltage is zero, as are all the inductor and resistor voltages and currents. At time $t = 0$ the input voltage becomes V volts. Since the current through an inductor cannot change instantaneously, the mesh current $i(0) = 0$. Thus, the voltage across the resistor is also zero. By KVL the voltage across the inductor at $t = 0$ is V volts. (The inductor acts as an open circuit to instantaneous current changes.) Since $v_L = L \, di_L/dt$, a voltage across the inductor means a positive rate of change of the current. Thus, the current starts to increase. This results in an increasing voltage across the resistor and, since the input voltage remains constant, a decreasing voltage across the inductor. The consequence of this is a positive rate of change of current that is less than before; that is, the current is still increasing but at a slower rate. This process continues until, after a long time, the input acts as a dc source. At that point, the inductor behaves as a short circuit, so $v_L = 0$, $i = V/R$ amperes, and $v_R = V$ volts. Because of the exponential nature of the RL zero-input circuit studied previously, we can anticipate this type of behavior.

Let us now proceed with a formal analysis. By KVL,

$$v_L + v_R = Vu(t)$$

and

$$L \frac{di}{dt} + Ri = Vu(t)$$

or

$$\frac{di}{dt} + \frac{R}{L} i = \frac{V}{L} u(t)$$

For $t < 0$, this equation becomes

$$\frac{di}{dt} + \frac{R}{L} i = 0$$

Since the input is zero for $t < 0$, and since the circuit is in the zero state initially, the solution to this equation which satisfies the boundary condition is the trivial solution $i = 0$. Thus,

$$i(t) = 0 \qquad \text{for } t < 0$$

For $t \geq 0$, the differential equation becomes

$$\frac{di}{dt} + \frac{R}{L} i = \frac{V}{L}$$

which is of the form

$$\frac{dx}{dt} + ax = b$$

where $x(t) = i(t)$, $a = R/L$, and $b = V/L$. Thus, from the discussion above, we find that the forced response is

$$i_f(t) = \frac{b}{a} = \frac{V/L}{R/L} = \frac{V}{R}$$

and the natural response is

$$i_n(t) = Ae^{-at} = Ae^{-Rt/L}$$

Hence the complete response is

$$i(t) = i_f(t) + i_n(t)$$

$$= \frac{V}{R} + Ae^{-Rt/L}$$

Since $i(t) = 0$ for $t < 0$, and since the current through an inductor cannot change instantaneously, $i(0) = 0$. Thus, substituting $t = 0$ into the previous equation we get

$$i(0) = \frac{V}{R} + Ae^0$$

from which

$$A = i(0) - \frac{V}{R}$$

$$= -\frac{V}{R}$$

Therefore,

$$i(t) = \frac{V}{R} - \frac{V}{R} e^{-Rt/L} \qquad \text{for } t \geq 0$$

Consequently, the final expression for the current is

$$i(t) = \left(\frac{V}{R} - \frac{V}{R} e^{-Rt/L} \right) u(t)$$

$$= \frac{V}{R} (1 - e^{-Rt/L}) u(t)$$

Furthermore,

$$v_R(t) = Ri(t)$$

$$= V(1 - e^{-Rt/L}) u(t)$$

In addition, by KVL,

$$v_L(t) = Vu(t) - v_R(t)$$

$$= Vu(t) - V(1 - e^{-Rt/L}) u(t)$$

$$= Ve^{-Rt/L} u(t)$$

Alternatively, we could have determined $v_L(t)$ as follows:

$$v_L(t) = L \frac{di(t)}{dt}$$

$$= L \frac{d}{dt} \left[\frac{V}{R} (1 - e^{-Rt/L}) u(t) \right]$$

$$= \frac{LV}{R} \left[(1 - e^{-Rt/L}) \delta(t) + (-e^{-Rt/L}) \left(-\frac{R}{L} \right) u(t) \right]$$

and since

$$(1 - e^{-Rt/L}) \delta(t) = 0$$

we have

$$v_L(t) = \frac{LV}{R} \left(\frac{R}{L} e^{-Rt/L} \right) u(t)$$

$$= Ve^{-Rt/L} u(t)$$

Sketches of these step responses are shown in Fig. 5.27.

fig. 5.27

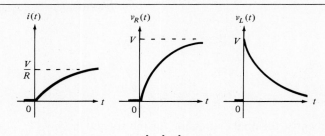

To analyze the simple parallel RC circuit shown in Fig. 5.28, with reference to the previous simple series RL circuit (Fig. 5.26), we merely apply the concept of duality to write

$$v(t) = IR(1 - e^{-t/RC}) u(t)$$

$$i_R(t) = I(1 - e^{-t/RC}) u(t)$$

$$i_C(t) = Ie^{-t/RC} u(t)$$

fig. 5.28

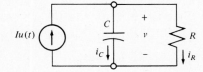

. . .

Example

Let us now consider the zero-state response $v_C(t)$ for the circuit shown in Fig. 5.29, where the input is the voltage pulse $v_s(t) = Vu(t) - Vu(t - t_0)$. A sketch of the input is given in Fig. 5.30.

fig. 5.29

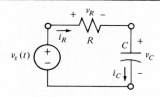

fig. 5.30

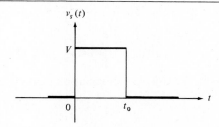

Just as before when we found the step response for this circuit, we get

$$\frac{dv_C}{dt} + \frac{1}{RC} v_C = \frac{V}{RC} [u(t) - u(t - t_0)]$$

For $t < 0$, this equation becomes

$$\frac{dv_C}{dt} + \frac{1}{RC} v_C = 0$$

and since the voltage across the capacitor is initially zero, we have

$$v_C(t) = 0 \qquad \text{for } t < 0$$

For $0 \le t < t_0$, the differential equation becomes

$$\frac{dv_C}{dt} + \frac{1}{RC} v_C = \frac{V}{RC}$$

This equation is of the form

$$\frac{dx(t)}{dt} + ax(t) = b$$

the solution to which is

$$x(t) = \frac{b}{a} + Ae^{-at}$$

Noting that $a = 1/RC$ and $b = V/RC$, we get

$$v_C(t) = V + Ae^{-t/RC}$$

Setting $t = 0$ in this equation, we obtain

$$v_C(0) = V + A$$

or

$$A = v_C(0) - V$$

Since the voltage across the capacitor cannot change instantaneously, and since $v_C(t) = 0$ for $t < 0$, then $v_C(0) = 0$. Hence

$$A = -V$$

Thus

$$v_C(t) = V - Ve^{-t/RC}$$

$$= V(1 - e^{-t/RC}) \qquad \text{for } 0 \le t < t_0$$

Now, for $t \ge t_0$, the differential equation again becomes

$$\frac{dv_C}{dt} + \frac{1}{RC} v_C = 0$$

But, since the voltage across the capacitor cannot change instantaneously,

$$v_C(t_0) = V(1 - e^{-t_0/RC})$$

and the solution to the differential equation (as we have seen in the section on zero-input responses) is, for $t \ge t_0$,

$$v_C(t) = v_C(t_0)e^{-(t-t_0)/RC}$$

$$= V(1 - e^{-t_0/RC})e^{-(t-t_0)/RC}$$

$$= V[e^{-(t-t_0)/RC} - e^{-t/RC}]$$

Therefore,

$$v_C(t) = \begin{cases} 0 & \text{for } t < 0 \\ V(1 - e^{-t/RC}) & \text{for } 0 \le t < t_0 \\ V[e^{-(t-t_0)/RC} - e^{-t/RC}] & \text{for } t \ge t_0 \end{cases}$$

or, in terms of a single expression,

$$v_C(t) = V(1 - e^{-t/RC})[u(t) - u(t - t_0)] + V[e^{-(t-t_0)/RC} - e^{-t/RC}]u(t - t_0)$$

$$= V(1 - e^{-t/RC})u(t) - Vu(t - t_0) + Ve^{-(t-t_0)/RC}u(t - t_0)$$

$$= V(1 - e^{-t/RC})u(t) - V(1 - e^{-(t-t_0)/RC})u(t - t_0)$$

A sketch of $v_C(t)$ is obtained as shown in Fig. 5.31.

fig. 5.31

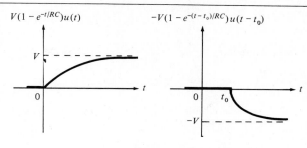

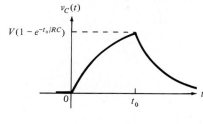

In the preceding example, the capacitor is initially uncharged. For $0 \le t < t_0$, the input voltage is V volts, and the capacitor begins to charge up as was the case for a step input studied previously. However, for $t \ge t_0$, the input voltage is again zero. Since the voltage source then acts as a short circuit, a zero-input circuit results, and the capacitor discharges through the resistor.

PROBLEMS *5.15. Find the zero-state response $v(t)$ for the circuit in Fig. P5.15 when $v_s(t) = u(t)$.
$Ans.$ $\frac{1}{2}(1 - e^{-2t/RC})u(t)$

fig. P5.15

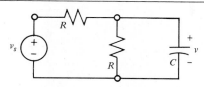

fig. P5.16

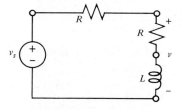

5.16. Repeat Problem 5.15 for the circuit shown in Fig. P5.16.

fig. P5.17

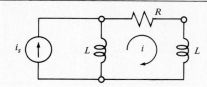

*5.17.   For the circuit in Fig. P5.17, find the zero-state response $i(t)$ when $i_s(t) = 2r(t)$.

Ans. $\dfrac{2L}{R}(1 - e^{-Rt/2L})u(t)$

fig. P5.18

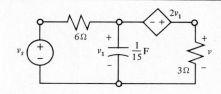

5.18.   Repeat Problem 5.15 for the circuit shown in Fig. P5.18.

fig. P5.19

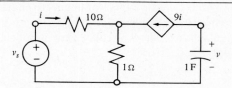

5.19.   Repeat Problem 5.15 for the circuit in Fig. P5.19.

fig. P5.20

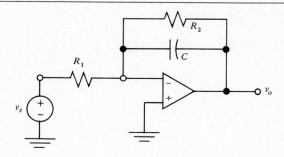

*5.20.   Find the zero-state step response $v_o(t)$ for the op amp circuit in Fig. P5.20.

Ans. $-\dfrac{R_2}{R_1}(1 - e^{-t/R_2C})u(t)$

5.21.   For the circuit shown in Fig. 5.29, find the voltage across the resistor and sketch this function.

fig. P5.22

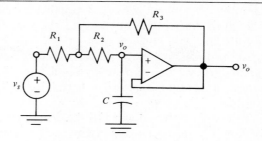

5.22.   For the op amp circuit in Fig. P5.22, suppose that $R_1 = R_2 = R_3 = R$. Find the zero-state step response $v_o(t)$.

5.23.   For the circuit given in Problem 5.22, interchange capacitor C and resistor R_3 and repeat the problem.

5.5   LINEARITY AND SUPERPOSITION

There is an alternative to the approach taken in the preceding simple RC circuit example (Figs. 5.29 and 5.30). What we can do instead is consider the input to be a sum of functions and then use the property of linearity, which we shall now justify.
    Consider the first-order differential equation

$$\frac{dx(t)}{dt} + ax(t) = f(t)$$

subject to the initial condition $x(0) = 0$. Suppose that $y_1(t)$ is the solution of this equation for the case that $f(t) = f_1(t)$. Then we have the equality

$$\frac{dy_1(t)}{dt} + ay_1(t) = f_1(t)$$

where $y_1(0) = 0$. Furthermore, suppose that $y_2(t)$ is the solution of the above differential equation for the case that $f(t) = f_2(t)$. Then we also have the equality

$$\frac{dy_2(t)}{dt} + ay_2(t) = f_2(t)$$

where $y_2(0) = 0$. Adding these two equalities, we get the following equalities:

$$\frac{dy_1(t)}{dt} + ay_1(t) + \frac{dy_2(t)}{dt} + ay_2(t) = f_1(t) + f_2(t)$$

$$\frac{dy_1(t)}{dt} + \frac{dy_2(t)}{dt} + ay_1(t) + ay_2(t) = f_1(t) + f_2(t)$$

$$\frac{d}{dt}[y_1(t) + y_2(t)] + a[y_1(t) + y_2(t)] = f_1(t) + f_2(t)$$

where $y_1(0) + y_2(0) = 0 + 0 = 0$. This last equality means that $y_1(t) + y_2(t)$ is the solution of the given differential equation for the case that $f(t) = f_1(t) + f_2(t)$ and subject to the constraint that $x(0) = 0$.

Based upon the discussion above, we can now conclude that if we know the zero-state response to $f_1(t)$ and the zero-state response to $f_2(t)$, it is a trivial matter to determine the zero-state response to $f_1(t) + f_2(t)$; it is simply the sum of the individual zero-state responses.

Clearly, this result can be extended to the case of inputs of the form $f_1(t) + f_2(t) + f_3(t) + \cdots + f_n(t)$. The zero-state response to such an input is simply the sum of the individual zero-state responses.

Furthermore, it should be clear that if $y(t)$ is the zero-state response to $f(t)$, then $Ky(t)$ is the zero-state response to $Kf(t)$.

$\bullet \quad \bullet \quad \bullet$

Example

Let us again find the zero-state response $v_C(t)$ for the simple series RC circuit given in the previous example (see Figs. 5.29 and 5.30).

Since we can write

$$v_s(t) = v_1(t) + v_2(t)$$

where $v_1(t) = Vu(t)$ and $v_2(t) = -Vu(t - t_0)$, suppose that $v_a(t)$ is the zero-state response to $v_1(t)$ and $v_b(t)$ is the zero-state response to $v_2(t)$. Then the zero-state response to $v_s(t)$ is

$$v_C(t) = v_a(t) + v_b(t)$$

From previous discussions we know that

$$v_a(t) = V(1 - e^{-t/RC})u(t)$$

and

$$v_b(t) = -V(1 - e^{-(t-t_0)/RC})u(t - t_0)$$

Thus

$$v_C(t) = V(1 - e^{-t/RC})u(t) - V(1 - e^{-(t-t_0)/RC})u(t - t_0)$$

which agrees with our previous analysis (albeit more complicated) of this circuit.

The expression for the voltage across the resistor can be obtained from KVL as follows:

$$v_R(t) = v_s(t) - v_C(t)$$
$$= [Vu(t) - Vu(t - t_0)] - [V(1 - e^{-t/RC})u(t) - V(1 - e^{-(t-t_0)/RC})u(t - t_0)]$$
$$= Ve^{-t/RC}u(t) - Ve^{-(t-t_0)/RC}u(t - t_0)$$

A sketch of $v_R(t)$ is shown in Fig. 5.32; a sketch of $v_C(t)$ was given previously in Fig. 5.31. Of course, we can obtain $i_C(t) = i_R(t)$ by using

$$i_C(t) = C\,\frac{dv_C(t)}{dt}$$

or

$$i_R(t) = \frac{v_R(t)}{R}$$

fig. 5.32

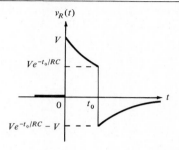

In either case,

$$i_C(t) = i_R(t) = \frac{V}{R}\,e^{-t/RC}u(t) - \frac{V}{R}\,e^{-(t-t_0)/RC}u(t - t_0)$$

The two sketches of $v_C(t)$ and $v_R(t)$ in Figs. 5.31 and 5.32 were typical. Let us now look at some extreme cases. First, suppose that the time constant $\tau = RC$ is small relative to t_0. In this case the capacitor has enough time essentially to completely charge before it discharges. The resulting plots are shown in Fig. 5.33.

fig. 5.33

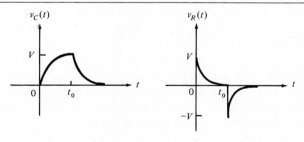

fig. 5.34

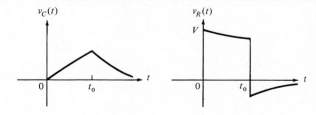

Next suppose that $\tau = RC$ is large compared to t_0. In this case, the capacitor barely charges up before the input becomes zero. Although charging occurs at an exponential rate, the very beginning of the exponential curve approximates a straight line. Typical plots are shown in Fig. 5.34.

· · ·

Even though the series RC circuit in the preceding example is very simple in structure, it is quite useful. Specifically, consider the case where the output is the resistor voltage as shown in Fig. 5.35. Suppose that the input voltage is a "square-wave" function as that given in Fig. 5.36.

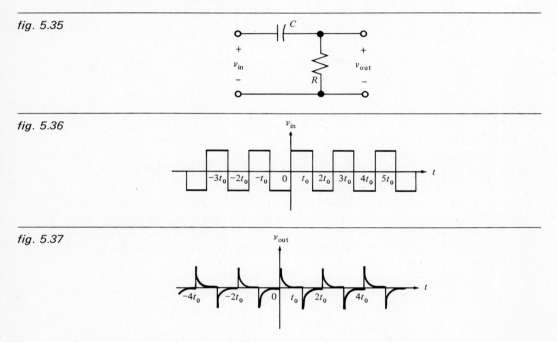

fig. 5.35

fig. 5.36

fig. 5.37

If $\tau = RC$ is much smaller than t_0 (written $\tau \ll t_0$), the resulting output voltage is as shown in Fig. 5.37. The "spikes" in the output voltage waveform are by no means impulses, but they do suggest them. In other words, the output voltage approximates the derivative of the input voltage. For this reason we call this circuit a (practical) *differentiator*.

Now consider the case where the output is the capacitor voltage as shown in Fig. 5.38. If $\tau = RC$ is much larger than t_0 (written $\tau \gg t_0$), for the same input voltage as above, the resulting output voltage is as shown in Fig. 5.39. Because of the large time constant, the line segments are approximately straight. Since the output voltage has the rough form of the integral of the input voltage, we call the circuit under consideration a (practical) *integrator*.*

* Even though such an integrator is useful, the op amp integrator considered previously is a much more accurate device.

fig. 5.38

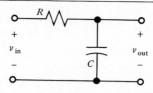

fig. 5.39

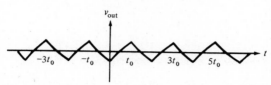

Both of the above simple circuits are used in the standard television set. Part of a TV video signal is a sequence or "train" of pulses, which provides the information used for horizontal and vertical synchronization. The video signal goes to a circuit called the *sync separator*, which removes the pulse train, and this is then applied to both a differentiator and an integrator. The output of the differentiator "triggers" the horizontal synchronization, while the output of the integrator is applied to the vertical oscillator for vertical synchronization purposes.

In the following example we have the situation in which both the input and initial condition (at $t = 0$) are nonzero. Even in this case, we can take advantage of the property of linearity.

· · ·

Example

Let us determine the voltage $v(t)$ across the capacitor for the circuit shown in Fig. 5.40, where the input current

$$i_s(t) = \begin{cases} 1 & \text{for } -\infty < t < 0 \\ -2 & \text{for } 0 \leq t < \infty \end{cases}$$

is shown in Fig. 5.41. By KCL,

$$i_C + i_R = i_s$$

so

$$3\frac{dv}{dt} + \frac{v}{2} = i_s$$

or

$$\frac{dv}{dt} + \frac{1}{6}v = \frac{1}{3}i_s$$

fig. 5.40

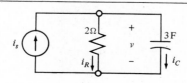

fig. 5.41

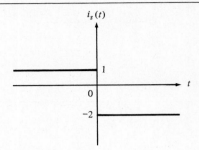

For $t < 0$, this differential equation becomes

$$\frac{dv}{dt} + \frac{1}{6}v = \frac{1}{3}$$

Since the circuit is a dc circuit for $t < 0$, the solution to this equation is a constant, K. Substituting K into the equation, we get

$$\frac{dK}{dt} + \frac{1}{6}K = \frac{1}{3}$$

$$K = \frac{1/3}{1/6} = 2$$

Hence

$$v(t) = 2 \qquad \text{for } t < 0$$

For $t \geq 0$, the differential equation becomes

$$\frac{dv}{dt} + \frac{1}{6}v = -\frac{2}{3}$$

and since the voltage across the capacitor cannot change instantaneously, $v(0) = 2$ V. Letting $a = \frac{1}{6}$ and $b = -\frac{2}{3}$, the solution to the differential equation is

$$v(t) = \frac{-2/3}{1/6} + Ae^{-t/6}$$

$$= -4 + Ae^{-t/6}$$

Setting $t = 0$ in this equation, we have

$$v(0) = -4 + Ae^0$$

$$2 = -4 + A$$

or

$$A = 6$$

Thus

$$v(t) = -4 + 6e^{-t/6} \qquad \text{for } t \geq 0$$

Hence we can write

$$v(t) = \begin{cases} 2 & \text{for } t < 0 \\ -4 + 6e^{-t/6} & \text{for } t \geq 0 \end{cases}$$

or the single expression

$$v(t) = 2 - 2u(t) + [-4 + 6e^{-t/6}]u(t)$$

$$= 2 - 6u(t) + 6e^{-t/6}u(t)$$

$$= 2 - 6(1 - e^{-t/6})u(t)$$

A sketch of $v(t)$ is shown in Fig. 5.42.

fig. 5.42

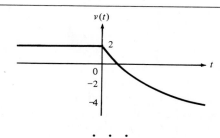

An alternative to finding $v(t)$ by the direct method discussed above is to use the property of linearity. The key to this approach is to note that we can write the input current as $i_s(t) = 1 - 3u(t) = i_1(t) + i_2(t)$.

If $v_1(t)$ is the response to $i_1(t) = 1$, then $v_1(t)$ is the solution to the equation

$$\frac{dv_1}{dt} + \frac{1}{6} v_1 = \frac{1}{3} i_1(t) = \frac{1}{3} \qquad \text{for all } t$$

and, as we have seen above,

$$v_1(t) = 2 \qquad \text{for all } t$$

If $v_2(t)$ is the response to $i_2(t) = -3u(t)$, then $v_2(t)$ is the solution to

$$\frac{dv_2}{dt} + \frac{1}{6} v_2 = \frac{1}{3} i_2(t) = -u(t)$$

For $t < 0$,

$$v_2(t) = 0$$

For $t \geq 0$, $v_2(t)$ satisfies

$$\frac{dv_2}{dt} + \frac{1}{6} v_2 = -1$$

Thus

$$v_2(t) = \frac{-1}{1/6} + Ae^{-t/6} = -6 + Ae^{-t/6}$$

from which

$$v_2(0) = -6 + A$$

$$0 = -6 + A$$

$$6 = A$$

Hence

$$v_2(t) = (-6 + 6e^{-t/6})u(t) = -6(1 - e^{-t/6})u(t)$$

By linearity, we have

$$v(t) = v_1(t) + v_2(t)$$
$$= 2 - 6(1 - e^{-t/6})u(t)$$

as was obtained previously.

   Because we are dealing with linear circuits, we can employ the principle of superposition. In essence, we did this in the above and a previous example when we expressed a forcing function as a sum of two other functions. For the case of a voltage source $v_s(t) = v_1(t) + v_2(t)$, a single voltage source whose value is $v_s(t)$ is equivalent to the series connection of two voltage sources—one whose value is $v_1(t)$ and the other whose value is $v_2(t)$. To apply the principle of superposition, set one voltage source to zero (replace it by a short circuit) and find the response due to the other source—then vice versa. The sum of these two responses is equal to the response due to $v_s(t)$. A similar argument can be used for a current-forcing function that can be expressed as a sum of (usually simpler) functions.

   Referring to the previous example, we know that for $t < 0$, $v(t) = 2$, and $v(0) = 2$ V. Thus, as was discussed in Chapter 4, for $t \geq 0$ we can replace the charged capacitor with an uncharged capacitor in series with a voltage source having value $v(0)u(t) = 2u(t)$. Furthermore, since $i_s(t) = -2$ for $t \geq 0$, to determine $v(t)$ for $t \geq 0$ we can analyze the circuit shown in Fig. 5.43.

   By the principle of superposition, we first determine the voltage $v_a(t)$ due to the current source. This situation is depicted in Fig. 5.44. This circuit has already been analyzed (see Fig. 5.28). Using that result, we have

$$v_a(t) = -2(2)(1 - e^{-t/(2)(3)}) = -4(1 - e^{-t/6}) \qquad \text{for } t \geq 0$$

fig. 5.43

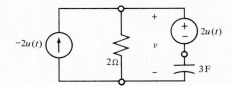

fig. 5.44

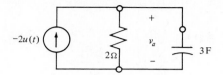

The voltage $v_b(t)$ due to the voltage source is found from Fig. 5.45. This circuit has also been analyzed previously (see Fig. 5.21). From that analysis,

$$v_b(t) = 2e^{-t/(2)(3)} = 2e^{-t/6} \qquad \text{for } t \geq 0$$

fig. 5.45

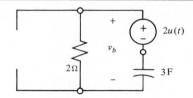

Hence

$$v(t) = v_a(t) + v_b(t)$$
$$= -4(1 - e^{-t/6}) + 2e^{-t/6}$$
$$= -4 + 6e^{-t/6} \qquad \text{for } t \geq 0$$

Together with the fact that $v(t) = 2$ for $t < 0$, we get

$$v(t) = 2 - 2u(t) + (-4 + 6e^{-t/6})u(t)$$
$$= 2 - 6u(t) + 6e^{-t/6}u(t)$$
$$= 2 - 6(1 - e^{-t/6})u(t)$$

5.6   SOME OTHER FORCING FUNCTIONS

Suppose now that we would like to determine zero-state responses that are due to inputs other than step functions. If the inputs are ramps or impulses, the responses can be determined by a simple operation on the step response. To see this, suppose that $y(t)$ is the solution to the differential equation

$$\frac{dx(t)}{dt} + ax(t) = f(t)$$

Then we have the following equality:

$$\frac{dy(t)}{dt} + ay(t) = f(t)$$

Differentiating both sides of this equation, we get the equality

$$\frac{d\left[\dfrac{dy(t)}{dt}\right]}{dt} + a\,\frac{dy(t)}{dt} = \frac{df(t)}{dt}$$

Hence, we see that for the differential equation

$$\frac{dx(t)}{dt} + ax(t) = \frac{df(t)}{dt}$$

the solution is $dy(t)/dt$. In other words, if $y(t)$ is the zero-state response to $f(t)$, then $dy(t)/dt$ is the zero-state response to $df(t)/dt$. Since an impulse is the derivative of a step, that is, since

$$\delta(t) = \frac{d}{dt}\,[u(t)]$$

the derivative of the zero-state response to $u(t)$ is equal to the zero-state response to $\delta(t)$.

Repeating the same argument as above and using integration instead of differentiation, since the integral of a step is a ramp, that is, since

$$r(t) = \int_{-\infty}^{t} u(t)\,dt$$

we conclude that the integral of the zero-state response to $u(t)$ is equal to the zero-state response to $r(t)$.

. . .

Example
Let us find the impulse responses for the simple RC circuit shown in Fig. 5.46.

---

fig. 5.46

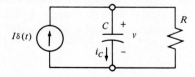

From our previous discussion (see Fig. 5.28), we know that the voltage across the capacitor due to an input of $Iu(t)$ is

$$v_s(t) = IR(1 - e^{-t/RC})u(t)$$

Therefore, the zero-state response due to the impulse is

$$v(t) = \frac{dv_s(t)}{dt}$$

$$= \frac{d}{dt} \left[IR(1 - e^{-t/RC})u(t) \right]$$

$$= IR(1 - e^{-t/RC})\delta(t) + IR(-e^{-t/RC})\left(\frac{-1}{RC}\right)u(t)$$

and since

$$(1 - e^{-t/RC})\delta(t) = 0$$

we find that the impulse response is

$$v(t) = \frac{I}{C} e^{-t/RC}u(t)$$

A sketch of $v(t)$ is shown in Fig. 5.47. Note that in this circuit the voltage across the capacitor does change instantaneously (at time $t = 0$). This is because there is an impulse of current produced by the source.

fig. 5.47

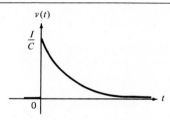

The current through the capacitor for this circuit is

$$i_C(t) = C \frac{dv(t)}{dt}$$

$$= C \frac{d}{dt} \left[\frac{I}{C} e^{-t/RC}u(t) \right]$$

$$= Ie^{-t/RC}\delta(t) + Ie^{-t/RC}\left(\frac{-1}{RC}\right)u(t)$$

By the sampling property,

$$e^{-t/RC}\delta(t) = \delta(t)$$

Hence

$$i_C(t) = I\delta(t) - \frac{I}{RC}\, e^{-t/RC}u(t)$$

A sketch of $i_C(t)$ is shown in Fig. 5.48.

fig. 5.48

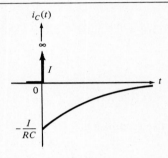

$$\bullet\ \ \bullet\ \ \bullet$$

Example
Consider the simple parallel *RL* circuit shown in Fig. 5.49.

fig. 5.49

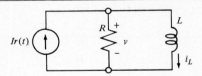

We mentioned previously (see Fig. 5.28) that the zero-state response of the inductor current for an input of $Iu(t)$ is

$$i_s(t) = I(1 - e^{-Rt/L})u(t)$$

Therefore the zero-state ramp response is

$$i_L(t) = \int_{-\infty}^{t} i_s(t)\, dt$$

$$= \int_{-\infty}^{t} I(1 - e^{-Rt/L})u(t)\, dt$$

$$= \int_{-\infty}^{t} Iu(t)\, dt - \int_{-\infty}^{t} Ie^{-Rt/L}u(t)\, dt$$

We already know that the integral of a step is a ramp. Thus

$$\int_{-\infty}^{t} Iu(t)\, dt = Ir(t)$$

In addition, for $t < 0$,

$$\int_{-\infty}^{t} Ie^{-Rt/L}u(t)\ dt = \int_{-\infty}^{t} 0\ dt = 0$$

and, for $t \geq 0$,

$$\int_{-\infty}^{t} Ie^{-Rt/L}u(t)\ dt = \int_{-\infty}^{0} Ie^{-Rt/L}(0)\ dt + \int_{0}^{t} Ie^{-Rt/L}(1)\ dt$$

$$= 0 + I \int_{0}^{t} e^{-Rt/L}\ dt$$

$$= -\frac{L}{R}\ Ie^{-Rt/L}\ \Big|_{0}^{t}$$

$$= -\frac{L}{R}\ I(e^{-Rt/L} - e^{0})$$

$$= \frac{LI}{R}\ (1 - e^{-Rt/L})$$

Thus, for all t,

$$\int_{-\infty}^{t} Ie^{-Rt/L}u(t)\ dt = \frac{LI}{R}\ (1 - e^{-Rt/L})u(t)$$

Hence

$$i_L(t) = Ir(t) - \frac{LI}{R}(1 - e^{-Rt/L})u(t)$$

Furthermore,

$$v_L(t) = L\ \frac{di_L(t)}{dt}$$

$$= L\ \frac{d}{dt}\left[\int_{-\infty}^{t} i_s(t)\ dt\right]$$

$$= Li_s(t)$$

$$= LI(1 - e^{-Rt/L})u(t)$$

$$\bullet \quad \bullet \quad \bullet$$

We have seen how the exponential functional form Ke^{st} can occur. Thus let us analyze a circuit in which the forcing function is exponential.

· · ·

Example

The circuit shown in Fig. 5.50 has an exponential forcing function. Let us determine the zero-state response $i_L(t)$ of this circuit.

fig. 5.50

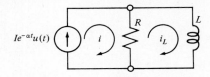

By mesh analysis, we have

$$i = Ie^{-\alpha t}u(t)$$

and

$$L\frac{di_L}{dt} + R(i_L - i) = 0$$

from which

$$\frac{di_L}{dt} + \frac{R}{L}i_L = \frac{R}{L}i$$

and substituting the above expression for i into this equation, we obtain

$$\frac{di_L}{dt} + \frac{R}{L}i_L = \frac{RI}{L}e^{-\alpha t}u(t)$$

For $t < 0$, the solution of this equation is $i_L(t) = 0$; for $t \geq 0$, this equation becomes

$$\frac{di_L}{dt} + \frac{R}{L}i_L = \frac{RI}{L}e^{-\alpha t}$$

and, since the current in an inductor cannot change instantaneously, we also have the constraint that $i_L(0) = 0$.

Since the differential equation is of the form

$$\frac{dx(t)}{dt} + ax(t) = f(t)$$

its solution, as we have seen, is

$$i_L(t) = e^{-at}\int e^{at}f(t)\,dt + Ae^{-at}$$

where $a = R/L$. Thus

$$i_L(t) = e^{-at} \int e^{at} \frac{RI}{L} e^{-\alpha t} \, dt + Ae^{-at}$$

$$= \frac{RI}{L} e^{-at} \int e^{(a-\alpha)t} \, dt + Ae^{-at}$$

$$= \frac{RI}{L} e^{-at} \frac{1}{a-\alpha} e^{(a-\alpha)t} + Ae^{-at}$$

$$= \frac{RI}{L(a-\alpha)} e^{-at+at-\alpha t} + Ae^{-at}$$

$$= \frac{RI}{L\left(\dfrac{R}{L} - \alpha\right)} e^{-\alpha t} + Ae^{-Rt/L}$$

$$= \frac{RI}{R - \alpha L} e^{-\alpha t} + Ae^{-Rt/L}$$

Setting $t = 0$, since $i_L(0) = 0$, we obtain

$$i_L(0) = \frac{RI}{R - \alpha L} e^0 + Ae^0$$

$$0 = \frac{RI}{R - \alpha L} + A$$

or

$$A = -\frac{RI}{R - \alpha L}$$

Thus

$$i_L(t) = \frac{RI}{R - \alpha L} (e^{-\alpha t} - e^{-Rt/L}) \qquad \text{for } t \geq 0$$

Hence

$$i_L(t) = \frac{RI}{R - \alpha L} (e^{-\alpha t} - e^{-Rt/L})u(t)$$

Now consider the special case $\alpha = a = R/L$. From the equation above, it is tempting to conclude that $i_L(t) = 0$ under this circumstance. However, note that in this case the coefficient $RI/(R - \alpha L)$ is undefined. Thus, to determine what the zero-state response is, we simply return to the above equation

$$i_L(t) = \frac{RI}{L} e^{-at} \int e^{(a - \alpha)t} \, dt + Ae^{-at}$$

and for $\alpha = a$, we have

$$i_L(t) = \frac{RI}{L} e^{-at} \int dt + Ae^{-at}$$

$$= \frac{RI}{L} te^{-at} + Ae^{-at}$$

Setting $t = 0$, we have

$$i_L(0) = \frac{RI}{L} (0)e^0 + Ae^0 \qquad \Rightarrow \qquad 0 = A$$

and thus

$$i_L(t) = \frac{RI}{L} te^{-Rt/L} \qquad \text{for } t \geq 0$$

Hence, the zero-state response is

$$i_L(t) = \frac{RI}{L} te^{-Rt/L} u(t)$$

. . .

5.7   SOME OTHER FIRST-ORDER CIRCUITS

Up to now the first-order circuits analyzed have contained a single resistance and either a single inductor or capacitor. In this section we shall investigate circuits that consist of a number of elements; included will be a dependent source and an op amp.

. . .

Example
The circuit shown in Fig. 5.51 contains a dependent current source. Let us determine the zero-state response $v_3(t)$ by the use of nodal analysis.

fig. 5.51

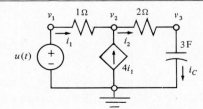

At node v_1,

$$v_1(t) = u(t)$$

At node v_2,

$$i_1 + 4i_1 - i_2 = 0$$

$$5i_1 - i_2 = 0$$

$$5\left(\frac{v_1 - v_2}{1}\right) - \left(\frac{v_2 - v_3}{2}\right) = 0$$

$$5(u(t) - v_2) - \frac{1}{2}(v_2 - v_3) = 0$$

$$10u(t) - 10v_2 - v_2 + v_3 = 0$$

$$11v_2 - v_3 = 10u(t)$$

At node v_3,

$$i_2 = i_C$$

$$\frac{v_2 - v_3}{2} = 3\frac{dv_3}{dt}$$

$$v_2 = 6\frac{dv_3}{dt} + v_3$$

Substituting this into the equation obtained at node v_2, we get

$$11\left(6\frac{dv_3}{dt} + v_3\right) - v_3 = 10u(t)$$

$$66\frac{dv_3}{dt} + 10v_3 = 10u(t)$$

$$\frac{dv_3}{dt} + \frac{5}{33}v_3 = \frac{5}{33}u(t)$$

For $t < 0$, this equation becomes

$$\frac{dv_3}{dt} + \frac{5}{33}v_3 = 0$$

and since the voltage across the capacitor is initially zero, we have

$$v_3(t) = 0 \qquad \text{for } t < 0$$

For $t \geq 0$, the differential equation becomes

$$\frac{dv_3}{dt} + \frac{5}{33}v_3 = \frac{5}{33}$$

Since this equation is of the form

$$\frac{dx(t)}{dt} + ax(t) = b$$

its solution is of the form

$$x(t) = \frac{b}{a} + Ae^{-at}$$

Thus, by setting $a = 5/33$ and $b = 5/33$, we get

$$v_3(t) = 1 + Ae^{-5t/33}$$

Setting $t = 0$ in this equation, we obtain

$$v_3(0) = 1 + Ae^0$$

or

$$A = v_3(0) - 1$$

Since the voltage across the capacitor cannot change instantaneously, and since $v_3(t) = 0$ for $t < 0$, then $v_3(0) = 0$. Thus

$$A = -1$$

Hence

$$v_3(t) = 1 - e^{-5t/33} \qquad \text{for } t \geq 0$$

Therefore, the zero-state response is

$$v_3(t) = (1 - e^{-5t/33})u(t)$$

An alternative way of determining the step response is with the use of Thévenin's theorem.

To find $v_{oc}(t)$, remove the capacitor. The resulting circuit is shown in Fig. 5.52.

fig. 5.52

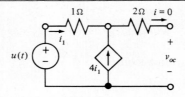

By KCL,

$$i_1 + 4i_1 = i = 0$$

$$5i_1 = 0$$

$$i_1 = 0$$

By KVL,

$$v_{oc} = -2i - 1i_1 + u(t) = u(t)$$

fig. 5.53

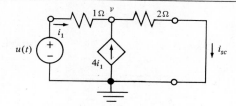

We can find i_{sc} from the circuit of Fig. 5.53. By KCL,

$$i_1 + 4i_1 = i_{sc}$$

$$5i_1 = i_{sc}$$

from which

$$5\left(\frac{u(t) - v}{1}\right) = \frac{v}{2}$$

Then

$$5u(t) = 5v + \frac{v}{2} = \frac{11}{2}v$$

so

$$v = \frac{10}{11}u(t)$$

and

$$i_{sc} = \frac{v}{2} = \frac{5}{11}u(t)$$

Thus

$$R_0 = \frac{v_{oc}}{i_{sc}} = \frac{u(t)}{\frac{5}{11}u(t)} = \tfrac{11}{5}\ \Omega$$

Replacing the circuit in Fig. 5.52 by its Thévenin equivalent results in the circuit shown in Fig. 5.54. This circuit was analyzed previously (see Fig. 5.21), at which time we obtained

$$v_3(t) = (1 - e^{-t/R_0C})u(t)$$

$$= (1 - e^{-5t/33})u(t)$$

and this, naturally, agrees with our previous result.

fig. 5.54

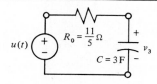

· · ·

Example

As a final example, let us find the step response $v_o(t)$ for the op amp circuit shown in Fig. 5.55.

fig. 5.55

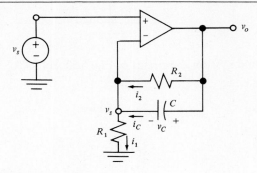

By nodal analysis, we have

$$i_2 + i_C = i_1$$

$$C \frac{d(v_o - v_s)}{dt} + \frac{v_o - v_s}{R_2} = \frac{v_s}{R_1}$$

from which

$$\frac{dv_o}{dt} + \frac{1}{R_2C} v_o = \frac{dv_s}{dt} + \left(\frac{1}{R_1C} + \frac{1}{R_2C} \right) v_s$$

when $v_s(t) = u(t)$, we have

$$\frac{dv_o}{dt} + \frac{1}{R_2C} v_o = \delta(t) + \left(\frac{1}{R_1C} + \frac{1}{R_2C} \right) u(t)$$

Of course we can use the property of linearity to write the right-hand side of this differential equation as $v_1(t) + v_2(t)$, where

$$v_1(t) = \delta(t) \qquad \text{and} \qquad v_2(t) = \left(\frac{1}{R_1C} + \frac{1}{R_2C} \right) u(t)$$

and proceed from there. However, a much simpler approach is to solve for $v_C(t)$ and then use the fact that $v_o(t) = v_C(t) + v_s(t)$. Doing this, we have by KCL

$$C \frac{dv_C}{dt} + \frac{v_C}{R_2} = \frac{v_s}{R_1}$$

from which

$$\frac{dv_C}{dt} + \frac{1}{R_2C} v_C = \frac{1}{R_1C} v_s = \frac{1}{R_1C} u(t)$$

For $t < 0$,

$$v_C(t) = 0$$

For $t \geq 0$,

$$v_C(t) = \frac{(1/R_1 C)}{(1/R_2 C)} + Ae^{-t/R_2 C}$$

$$= \frac{R_2}{R_1} + Ae^{-t/R_2 C}$$

and, since $v_C(0) = 0$, then $A = -R_2/R_1$. Thus

$$v_C(t) = \frac{R_2}{R_1}(1 - e^{-t/R_2 C})u(t)$$

and so the step response is

$$v_o(t) = v_C(t) + v_s(t) = \frac{1}{R_1}(R_1 + R_2 - R_2 e^{-t/R_2 C})u(t)$$

$$\cdot \quad \cdot \quad \cdot$$

Up to now we have investigated circuits having nonzero inputs and containing a single inductor or capacitor. The describing circuit equations that resulted were first-order differential equations. However, some circuits containing more than a single inductor or a single capacitor may also be described by first-order differential equations. Of course, a series or parallel (or both) connection of inductors can be combined into an equivalent inductance. The same statement can be made about capacitors. Yet there are connections, neither series nor parallel, for which the above-mentioned phenomenon occurs. As an example, consider the RC circuit shown in Fig. 5.56.

---

fig. 5.56

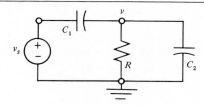

By nodal analysis,

$$C_1 \frac{d(v - v_s)}{dt} + C_2 \frac{dv}{dt} + \frac{v}{R} = 0$$

from which

$$(C_1 + C_2)\frac{dv}{dt} + \frac{v}{R} = C_1 \frac{dv_s}{dt}$$

or

$$\frac{dv}{dt} + \frac{1}{R(C_1 + C_2)}v = \frac{C_1}{C_1 + C_2}\frac{dv_s}{dt}$$

and this is a first-order differential equation. This result is a consequence of the fact that the capacitors and the voltage source form a loop. A similar situation occurs when a circuit contains inductors and current sources which form something called a "cut-set." These situations, as well as the concept of a "cut-set," are discussed further in Chapter 7.

Note that for the circuit of Fig. 5.56 it is the derivative of the input $v_s(t)$ that appears in the describing differential equation. For the case that $v_s(t) = r(t)$, the equation becomes

$$\frac{dv}{dt} + \frac{1}{R(C_1 + C_2)} v = \frac{C_1}{C_1 + C_2} u(t)$$

To find the zero-state ramp response, we need only solve this equation subject to the initial condition $v(0) = 0$ as described previously. Of course the step response is just the derivative of the ramp response, and the impulse response is the derivative of the step response.

PROBLEMS *5.24. Find the zero-state impulse response $v(t)$ for the circuit given in Problem 5.15.

Ans. $\dfrac{1}{RC} e^{-2t/RC} u(t)$

5.25. For the circuit given in Problem 5.16, find the zero-state response $v(t)$ for the case that $v_s(t) = r(t)$.

---

fig. P5.26

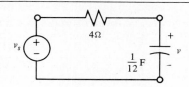

5.26. For the circuit in Fig. P5.26, find the zero-state response $v(t)$ for the case that $v_s(t) = (1 - e^{-3t}) u(t)$.

5.27. Repeat Problem 5.26 for $v_s(t) = e^{-t} u(t) - e^{-(t-1)} u(t-1)$.

---

fig. P5.28

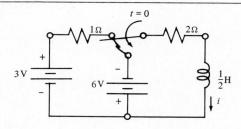

*5.28. In the circuit given in Fig. P5.28, find a single expression for $i(t)$ for all t.

Ans. $1 + 4(e^{-4t} - 1) u(t)$

5.29. For the circuit given in Problem 5.28, find $i(t)$ for $t \geq 0$ by replacing the inductor with a parallel connection of an inductor having no initial current and an appropriate current source and then applying the principle of superposition.

*5.30. The zero-state impulse response of a circuit is $i(t) = 10e^{-5t}u(t)$. Find the zero-state step response.

Ans. $2(1 - e^{-5t})u(t)$

*5.31. Find the zero-state response $v(t)$ of a circuit described by

$$\frac{dv}{dt} + 6v = 8e^{-2t}u(t)$$

given that $v(t) = 0$ for $t \le 0$.

Ans. $2(e^{-2t} - e^{-6t})u(t)$

*5.32. A circuit is described by

$$\frac{di}{dt} + 4i = u(t) + 2u(t - 1)$$

Find the zero-state response $i(t)$ given that $i(t) = 0$ for $t \le 0$.

Ans. $\frac{1}{4}(1 - e^{-4t})u(t) + \frac{1}{2}(1 - e^{-4(t-1)})u(t - 1)$

fig. P5.33

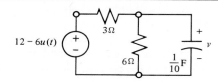

$12 - 6u(t)$, 3Ω, 6Ω, $\frac{1}{10}$F, $+ v -$

5.33. For the circuit given in Fig. P5.33, find $v(t)$.

5.34. For the circuit given in Problem 5.33, find $v(t)$ for $t \ge 0$ by replacing the capacitor with a series connection of a capacitor having no initial condition and an appropriate voltage source and then applying the principle of superposition.

5.35. For circuit given in Problem 5.20, let $R_1 = 1\ \Omega$, $R_2 = 2\ \Omega$, and $C = \frac{1}{4}$ F. Find the zero-state response $v_o(t)$ when (a) $v_s(t) = e^{-t}u(t)$ and (b) $v_s(t) = e^{-2t}u(t)$.

fig. P5.36

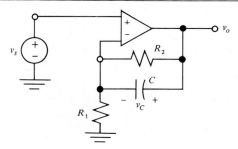

v_s, R_2, C, v_C, R_1, v_o

5.36. For the op amp circuit in Fig. P5.36, suppose that $R_1 = 1\ \Omega$, $R_2 = 2\ \Omega$, and $C = \frac{1}{4}$ F. Find the zero-state response $v_C(t)$ when (a) $v_s(t) = e^{-t}u(t)$ and (b) $v_s(t) = e^{-2t}u(t)$.

5.37. For the circuit given in Problem 5.36, write a differential equation in the variable $v_o(t)$. Given that $R_1 = R_2 = 2\ \Omega$ and $C = 1$ F, find the zero-state response $v_o(t)$ when $v_s(t) = (1 - e^{-t})u(t)$.

5.38. For the circuit given in Problem 5.22, interchange capacitor C and resistor R_2. Given that $R_1 = R_2 = R_3 = R$, find the zero-state response $v_o(t)$ when $v_s(t) = r(t)$.

Summary

1. A circuit containing either a single inductor or a single capacitor is described by a first-order differential equation.

2. If no independent source is present, the equation is homogeneous and the solution is the zero-input or natural response.

3. If the initial condition is zero and the equation nonhomogeneous, the solution is the zero-state response.

4. The complete response is the sum of two terms: One is the forced response and the other has the form of the natural response.

5. A time-invariant circuit is one in which the zero-state response to a delayed input can be obtained by delaying the zero-state response to the corresponding undelayed input.

6. In a linear circuit the zero-state response to a sum of inputs equals the sum of the individual zero-state responses.

7. In a linear circuit if the input is scaled, then the zero-state response is scaled by the same factor.

8. In a linear circuit if the input is differentiated, then the zero-state output is differentiated; if the input is integrated, so is the output.

9. Some zero-state responses for simple *RL* and *RC* series and parallel circuits are summarized in Table 5.1.

TABLE 5.1   Summary of Series and Parallel RC and RL Circuit Zero-State
Responses

	$v(t) = u(t)$	$v(t) = \delta(t)$
	$i(t) = \dfrac{1}{R}\, e^{-t/RC} u(t)$	$i(t) = \dfrac{1}{R}\left[\delta(t) - \dfrac{1}{RC}\, e^{-t/RC} u(t)\right]$
	$v_R(t) = e^{-t/RC} u(t)$	$v_R(t) = \delta(t) - \dfrac{1}{RC}\, e^{-t/RC} u(t)$
	$v_C(t) = (1 - e^{-t/RC}) u(t)$	$v_C(t) = \dfrac{1}{RC}\, e^{-t/RC} u(t)$
	$v(t) = u(t)$	$v(t) = \delta(t)$
	$i(t) = \dfrac{1}{R}\, (1 - e^{-Rt/L}) u(t)$	$i(t) = \dfrac{1}{L}\, e^{-Rt/L} u(t)$
	$v_R(t) = (1 - e^{-Rt/L}) u(t)$	$v_R(t) = \dfrac{R}{L}\, e^{-Rt/L} u(t)$
	$v_L(t) = e^{-Rt/L} u(t)$	$v_L(t) = \delta(t) - \dfrac{R}{L}\, e^{-Rt/L} u(t)$
	$i(t) = u(t)$	$i(t) = \delta(t)$
	$v(t) = R(1 - e^{-t/RC}) u(t)$	$v(t) = \dfrac{1}{C}\, e^{-t/RC} u(t)$
	$i_R(t) = (1 - e^{-t/RC}) u(t)$	$i_R(t) = \dfrac{1}{RC}\, e^{-t/RC} u(t)$
	$i_C(t) = e^{-t/RC} u(t)$	$i_C(t) = \delta(t) - \dfrac{1}{RC}\, e^{-t/RC} u(t)$
	$i(t) = u(t)$	$i(t) = \delta(t)$
	$v(t) = R e^{-Rt/L} u(t)$	$v(t) = R\left[\delta(t) - \dfrac{R}{L}\, e^{-Rt/L} u(t)\right]$
	$i_R(t) = e^{-Rt/L} u(t)$	$i_R(t) = \delta(t) - \dfrac{R}{L}\, e^{-Rt/L} u(t)$
	$i_L(t) = (1 - e^{-Rt/L}) u(t)$	$i_L(t) = \dfrac{R}{L}\, e^{-Rt/L} u(t)$

SECOND-ORDER CIRCUITS

6

Having dealt with circuits containing either a single inductor or capacitor, we shall now consider the case of two energy storage elements in a circuit. Most such networks, called *second-order circuits*, are described by second-order linear differential equations—the solution of which takes three forms.

As with first-order circuits, we start by considering the zero-input case, follow it with the situation in which there is an input present and the initial conditions (states) are zero (the zero-state case), and then proceed to circuits having both nonzero inputs and initial conditions.

We shall see that the complexity of analysis increases significantly when compared to that encountered with first-order circuits. It is because of this fact that we do not extend such an approach to higher-order circuits but, instead, take different points of view in the subsequent chapters.

6.1 THE SERIES RLC CIRCUIT

Suppose that for the zero-input series *RLC* circuit shown in Fig. 6.1, the initial current in the inductor is $i(0)$ and the initial voltage across the capacitor is $v(0)$. Writing a mesh equation, we get the integrodifferential equation

$$L \frac{di}{dt} + Ri + \frac{1}{C} \int_{-\infty}^{t} i \, dt = 0$$

fig. 6.1

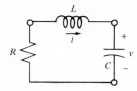

Taking the derivative of this equation yields the second-order, homogeneous, linear differential equation with constant coefficients

$$L \frac{d^2 i}{dt^2} + R \frac{di}{dt} + \frac{1}{C} i = 0$$

from which

$$\frac{d^2 i}{dt^2} + \frac{R}{L} \frac{di}{dt} + \frac{1}{LC} i = 0 \tag{6.1}$$

Again, from the circuit,

$$L \frac{di}{dt} + Ri + v = 0$$

Using the fact that

$$i = C \frac{dv}{dt}$$

we also get the second-order differential equation

$$LC \frac{d^2 v}{dt^2} + RC \frac{dv}{dt} + v = 0$$

from which

$$\frac{d^2 v}{dt^2} + \frac{R}{L} \frac{dv}{dt} + \frac{1}{LC} v = 0 \tag{6.2}$$

Both Equations (6.1) and (6.2) have the form

$$\frac{d^2 y(t)}{dt^2} + K_1 \frac{dy(t)}{dt} + K_2 y(t) = 0$$

so let us concentrate on solving this equation. Experience has shown that it is more convenient to use the constant 2α in place of K_1, and the constant $\omega_n{}^2$ in place of K_2. In other words, consider the equation

$$\frac{d^2 y(t)}{dt^2} + 2\alpha \frac{dy(t)}{dt} + \omega_n{}^2 y(t) = 0 \tag{6.3}$$

We have seen that the solution to a first-order, homogeneous, differential equation is

$$y(t) = Ae^{st}$$

where A and s are constants. However, is this a solution to the second-order homogeneous equation (6.3)? In order to find out, substitute this expression into the equation. The result is

$$\frac{d^2}{dt^2} (Ae^{st}) + 2\alpha \frac{d}{dt} (Ae^{st}) + \omega_n{}^2 (Ae^{st}) = 0 \tag{6.4}$$

from which

$$s^2 A e^{st} + 2\alpha s A e^{st} + \omega_n{}^2 A e^{st} = 0$$

Dividing both sides by $A e^{st}$ yields

$$s^2 + 2\alpha s + \omega_n{}^2 = 0 \qquad (6.5)$$

If this equality holds, then so does the one given in Equation (6.4). By the quadratic formula, this equality holds when

$$s = \frac{-2\alpha \pm \sqrt{4\alpha^2 - 4\omega_n{}^2}}{2}$$

or

$$s = -\alpha \pm \sqrt{\alpha^2 - \omega_n{}^2}$$

In other words, the two values of s that satisfy Equation (6.5) are

$$s_1 = -\alpha - \sqrt{\alpha^2 - \omega_n{}^2} \qquad \text{and} \qquad s_2 = -\alpha + \sqrt{\alpha^2 - \omega_n{}^2}$$

Thus, $A_1 e^{s_1 t}$ and $A_2 e^{s_2 t}$ satisfy Equation (6.3). That is,

$$\frac{d^2}{dt^2}(A_1 e^{s_1 t}) + 2\alpha \frac{d}{dt}(A_1 e^{s_1 t}) + \omega_n{}^2(A_1 e^{s_1 t}) = 0$$

and

$$\frac{d^2}{dt^2}(A_2 e^{s_2 t}) + 2\alpha \frac{d}{dt}(A_2 e^{s_2 t}) + \omega_n{}^2(A_2 e^{s_2 t}) = 0$$

Adding these two equations results in the equality

$$\frac{d^2}{dt^2}(A_1 e^{s_1 t} + A_2 e^{s_2 t}) + 2\alpha \frac{d}{dt}(A_1 e^{s_1 t} + A_2 e^{s_2 t}) + \omega_n{}^2(A_1 e^{s_1 t} + A_2 e^{s_2 t}) = 0$$

Therefore we see that, in addition to $A_1 e^{s_1 t}$ and $A_2 e^{s_2 t}$, their sum

$$y(t) = A_1 e^{s_1 t} + A_2 e^{s_2 t}$$

satisfies Equation (6.3). Since this is the most general expression (it contains both $A_1 e^{s_1 t}$ and $A_2 e^{s_2 t}$) that satisfies Equation (6.3), it is the solution.

In this solution, A_1 and A_2 are arbitrary constants. However, if we are given the initial conditions

$$y(0) \qquad \text{and} \qquad \frac{dy(t)}{dt}\bigg|_{t=0} \left(\text{more simply denoted as } \frac{dy(0)}{dt} \right)$$

using the fact that

$$y(t) = A_1 e^{s_1 t} + A_2 e^{s_2 t}$$

and

$$\frac{dy(t)}{dt} = s_1 A_1 e^{s_1 t} + s_2 A_2 e^{s_2 t}$$

we can determine the constants A_1 and A_2 from the two simultaneous equations

$$y(0) = A_1 + A_2 \qquad \text{and} \qquad \frac{dy(0)}{dt} = s_1 A_1 + s_2 A_2$$

Note that depending on the relative values of α and ω_n, the values of s_1 and s_2 can be either real or complex numbers. If $\alpha > \omega_n$, then $\alpha^2 - \omega_n{}^2 > 0$ so s_1 and s_2 are real numbers. This condition is referred to as the *overdamped* case. If $\alpha < \omega_n$, then $\alpha^2 - \omega_n{}^2 < 0$ (equivalently $\omega_n{}^2 - \alpha^2 > 0$), and therefore s_1 and s_2 are complex numbers. In particular,

$$s_1 = -\alpha - \sqrt{\alpha^2 - \omega_n{}^2} = -\alpha - \sqrt{-(\omega_n{}^2 - \alpha^2)} = -\alpha - j\omega_d$$
$$s_2 = -\alpha + \sqrt{\alpha^2 - \omega_n{}^2} = -\alpha + \sqrt{-(\omega_n{}^2 - \alpha^2)} = -\alpha + j\omega_d$$

where $j = \sqrt{-1}$ and $\omega_d = \sqrt{\omega_n{}^2 - \alpha^2}$. This condition is referred to as the *under-damped* case. Finally, the special case that $\alpha = \omega_n$, called the *critically damped* case, results in $s_1 = s_2 = -\alpha$. For this condition,

$$\begin{aligned} y(t) &= A_1 e^{s_1 t} + A_2 e^{s_2 t} \\ &= A_1 e^{-\alpha t} + A_2 e^{-\alpha t} \\ &= (A_1 + A_2) e^{-\alpha t} \\ &= A e^{-\alpha t} \end{aligned}$$

which cannot be the solution to Equation (6.3) since in general it is impossible to satisfy the two initial conditions $y(0)$ and $dy(0)/dt$ with the single constant A. The solution of Equation (6.3) for this special case will be discussed shortly. Let us now discuss these three cases in detail.

Returning to the series *RLC* circuit, we have for the second-order current differential equation (6.1), as well as the voltage equation (6.2),

$$2\alpha = \frac{R}{L} \quad \Rightarrow \quad \alpha = \frac{R}{2L}$$

and

$$\omega_n{}^2 = \frac{1}{LC} \quad \Rightarrow \quad \omega_n = \frac{1}{\sqrt{LC}}$$

As an example, suppose that $R = 5\ \Omega$, $L = \frac{1}{2}\ H$, and $C = \frac{1}{8}\ F$. Also, suppose that $i(0) = 1\ A$ and $v(0) = 2\ V$. Then

$$\alpha = \frac{R}{2L} = 5$$

$$\omega_n = \frac{1}{\sqrt{LC}} = 4$$

so $\alpha > \omega_n$, and the circuit is overdamped. Since

$$s_1 = -\alpha - \sqrt{\alpha^2 - \omega_n^2} = -8$$

$$s_2 = -\alpha + \sqrt{\alpha^2 - \omega_n^2} = -2$$

then the expression for the current $i(t)$ has the form

$$i(t) = A_1 e^{-8t} + A_2 e^{-2t}$$

Using the initial condition $i(0) = 1$, we get the equation

$$1 = A_1 + A_2 \tag{6.6}$$

Taking the derivative of $i(t)$ yields

$$\frac{di(t)}{dt} = -8A_1 e^{-8t} - 2A_2 e^{-2t} \tag{6.7}$$

The second initial condition given was $v(0) = 2$, not $di(0)/dt$. However, referring to the circuit, by KVL we have that

$$L\frac{di(t)}{dt} = -Ri(t) - v(t)$$

so

$$\frac{di(t)}{dt} = -\frac{R}{L}i(t) - \frac{1}{L}v(t)$$

$$= -10i(t) - 2v(t)$$

Setting $t = 0$ yields

$$\frac{di(0)}{dt} = -10(1) - 2(2) = -14$$

Hence, setting $t = 0$ in Equation (6.7) results in

$$-14 = -8A_1 - 2A_2 \tag{6.8}$$

The solution of the simultaneous Equations (6.6) and (6.8) is

$$A_1 = 2 \quad \text{and} \quad A_2 = -1$$

Thus, the solution of Equation (6.1) is

$$i(t) = 2e^{-8t} - e^{-2t} \qquad \text{for } t \geq 0$$

Since it is the solution to Equation (6.2), the expression for the voltage $v(t)$ has the form

$$v(t) = B_1 e^{-8t} + B_2 e^{-2t}$$

Setting $t = 0$ yields

$$2 = B_1 + B_2 \qquad\qquad (6.9)$$

Taking the derivative of $v(t)$ results in

$$\frac{dv(t)}{dt} = -8B_1 e^{-8t} - 2B_2 e^{-2t} \qquad\qquad (6.10)$$

From the circuit,

$$i(t) = C \frac{dv(t)}{dt}$$

or

$$\frac{dv(t)}{dt} = \frac{1}{C} i(t)$$

so

$$\frac{dv(0)}{dt} = \frac{1}{C} i(0) = 8$$

Setting $t = 0$ in Equation (6.10) yields

$$8 = -8B_1 - 2B_2 \qquad\qquad (6.11)$$

The solution of the simultaneous equations (6.9) and (6.11) is found to be

$$B_1 = -2 \qquad \text{and} \qquad B_2 = 4$$

Thus, the solution of Equation (6.2) is

$$v(t) = -2e^{-8t} + 4e^{-2t} \qquad \text{for } t \geq 0$$

Of course, once $i(t)$ has been determined, it is not necessary to solve Equation (6.2) since from the circuit we have that

$$v(t) = -L \frac{di(t)}{dt} - Ri(t)$$

$$= -\frac{1}{2}(-16e^{-8t} + 2e^{-2t}) - 5(2e^{-8t} - e^{-2t})$$

$$= -2e^{-8t} + 4e^{-2t}$$

which confirms the above result.

From our expressions for $i(t)$ and $v(t)$, setting $t = 0$, we get

$$i(0) = 2e^{-0} - e^{-0} = 2 - 1 = 1 \text{ A}$$

and

$$v(0) = -2e^{-0} + 4e^{-0} = -2 + 4 = 2 \text{ V}$$

as required. Plots of $v(t)$ and $i(t)$ for $t \geq 0$ are shown in Figs. 6.2 and 6.3, respectively.

fig. 6.2

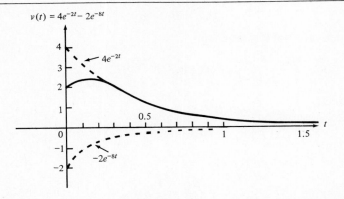

fig. 6.3

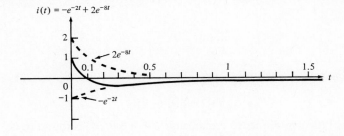

The initial current in the inductor, and hence through each element, is a clockwise current. The initial voltage across the capacitor is a positive value with the polarity indicated. The capacitor, therefore, tends to discharge by producing a counterclockwise current. The net effect is that, due to the inductor current, the capacitor will first charge to a greater voltage before it begins to discharge. The current through the inductor goes to zero when $t \approx 0.116$ s, at which time the capacitor is charged to its maximum value of approximately 2.38 V. After this time the current becomes counterclockwise (negative in value) and the energy stored is eventually dissipated in the resistor as heat. Note that if the direction of the initial current had the opposite direction (or if the polarity of the initial voltage was opposite) the capacitor would not have charged to a greater value before it discharged.

For this example, the energy stored in the inductor at $t = 0$ is

$$w_L(0) = \frac{1}{2}Li^2(0) = \frac{1}{2}\left(\frac{1}{2}\right)(1)^2 = \tfrac{1}{4} \text{ J}$$

and the energy stored in the capacitor at $t = 0$ is

$$w_C(0) = \frac{1}{2}Cv^2(0) = \frac{1}{2}\left(\frac{1}{8}\right)(2)^2 = \tfrac{1}{4} \text{ J}$$

so that the total energy stored at $t = 0$ is $\tfrac{1}{2}$ J. The power absorbed by the resistor is

$$p_R(t) = Ri^2(t) = 5(-e^{-2t} + 2e^{-8t})^2 = 5(e^{-4t} - 4e^{-10t} + 4e^{-16t})$$

Thus the energy absorbed by the resistor is

$$w_R = \int_0^\infty p_R(t)\, dt = \int_0^\infty 5(e^{-4t} - 4e^{-10t} + 4e^{-16t})\, dt$$

$$= 5\left(-\frac{1}{4}e^{-4t} + \frac{4}{10}e^{-10t} - \frac{4}{16}e^{-16t}\right)\Bigg|_0^\infty$$

$$= 5\left(\frac{1}{4} - \frac{4}{10} + \frac{4}{16}\right) = 5\left(\frac{1}{10}\right) = \tfrac{1}{2} \text{ J}$$

as it should be.

In the second of the three cases (underdamping) to be considered, we shall see quite different waveforms. Of course, we could change the values of R, L, and C in the previous example to get the underdamped case ($\omega_n > \alpha$); the initial conditions $v(0)$ and $i(0)$ have no effect on the damping of a circuit. However, let us discuss the underdamped case with a different example.

6.2 THE PARALLEL RLC CIRCUIT

For the zero-input parallel RLC circuit shown in Fig. 6.4, the initial conditions (state) of the circuit are $i(0)$ and $v(0)$. Since

$$v = L\frac{di}{dt} \qquad \text{and} \qquad \frac{v}{R} + i + C\frac{dv}{dt} = 0$$

fig. 6.4

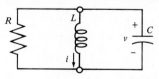

then

$$\frac{L}{R}\frac{di}{dt} + i + LC\frac{d^2i}{dt^2} = 0$$

so

$$\frac{d^2i}{dt^2} + \frac{1}{RC}\frac{di}{dt} + \frac{1}{LC}i = 0$$

which has the form of Equation (6.3), where

$$\alpha = \frac{1}{2RC} \quad \text{and} \quad \omega_n = \frac{1}{\sqrt{LC}}$$

(Note that these results could have been obtained from the series RLC zero-input circuit by duality.)

Suppose that for this circuit $R = \frac{1}{12} \Omega$, $L = \frac{1}{50}$ H, and $C = 2$ F, and the initial conditions are $i(0) = 1$ A and $v(0) = -0.14$ V. Then $\alpha = 3$ and $\omega_n = 5$, so $\omega_n > \alpha$ and the circuit is underdamped. Since $\omega_d = \sqrt{\omega_n{}^2 - \alpha^2} = \sqrt{25 - 9} = 4$, then

$$s_1 = -\alpha - j\omega_d = -3 - j4$$

$$s_2 = -\alpha + j\omega_d = -3 + j4$$

and the expression for the current $i(t)$ has the form

$$i(t) = A_1 e^{(-3 - j4)t} + A_2 e^{(-3 + j4)t}$$

$$= A_1 e^{-3t} e^{-j4t} + A_2 e^{-3t} e^{j4t}$$

$$= e^{-3t}(A_1 e^{-j4t} + A_2 e^{j4t})$$

Using Euler's formula

$$\boxed{e^{j\theta} = \cos \theta + j \sin \theta}$$

which will be derived later, we get

$$i(t) = e^{-3t}(A_1 \cos 4t - jA_1 \sin 4t + A_2 \cos 4t + jA_2 \sin 4t)$$

$$= e^{-3t}[(A_1 + A_2)\cos 4t + j(-A_1 + A_2)\sin 4t]$$

$$= e^{-3t}(B_1 \cos 4t + B_2 \sin 4t)$$

Letting $t = 0$, we get

$$i(0) = e^{-0}(B_1 \cos 0 + B_2 \sin 0)$$

$$1 = B_1$$

Taking the derivative of $i(t)$ results in

$$\frac{di(t)}{dt} = e^{-3t}(4B_2 \cos 4t - 4B_1 \sin 4t) - 3e^{-3t}(B_1 \cos 4t + B_2 \sin 4t)$$

and, letting $t = 0$,

$$\frac{di(0)}{dt} = 4B_2 - 3B_1 = 4B_2 - 3$$

so

$$B_2 = \frac{1}{4}\left(\frac{di(0)}{dt} + 3\right)$$

For the circuit, though, we have

$$L \frac{di(t)}{dt} = v(t) \qquad \text{or} \qquad \frac{di(t)}{dt} = \frac{1}{L} \, v(t)$$

so

$$\frac{di(0)}{dt} = 50v(0)$$

$$= 50(-0.14) = -7$$

Thus

$$B_2 = \frac{1}{4}(-7 + 3) = -1$$

and the expression for the current is

$$i(t) = e^{-3t}(\cos 4t - \sin 4t) \qquad \text{for } t \geq 0$$

Before attempting to sketch this function, let us derive an alternative form for this type of expression.

In the present situation—the underdamped case—the solution to Equation (6.3) has the form

$$y(t) = A_1 e^{s_1 t} + A_2 e^{s_2 t}$$

$$= A_1 e^{-\alpha t} e^{-j\omega_d t} + A_2 e^{-\alpha t} e^{j\omega_d t}$$

$$= e^{-\alpha t}(A_1 e^{-j\omega_d t} + A_2 e^{j\omega_d t})$$

where $\omega_d = \sqrt{\omega_n{}^2 - \alpha^2}$. Using Euler's formula and combining constants, we have just seen an example of the fact that for the underdamped case we can rewrite $y(t)$ in the following form:

$$y(t) = e^{-\alpha t}(B_1 \cos \omega_d t + B_2 \sin \omega_d t) \tag{6.12}$$

Even so, it is possible to further combine the two sinusoids having frequency ω_d into a single sinusoid of frequency ω_d. One possibility is $B \cos(\omega_d t - \phi)$, where we call B the *amplitude* of the sinusoid and ϕ is the *phase angle*.*

Now we want to determine under what conditions the following equality holds:

$$B_1 \cos \omega_d t + B_2 \sin \omega_d t = B \cos(\omega_d t - \phi) \tag{6.13}$$

Using the trigonometric identity for the cosine of the difference of two angles, we get

$$B_1 \cos \omega_d t + B_2 \sin \omega_d t = B(\cos \omega_d t \cos \phi + \sin \omega_d t \sin \phi)$$

$$= B \cos \phi \cos \omega_d t + B \sin \phi \sin \omega_d t$$

*Among other possibilities are $D \sin(\omega_d t - \theta)$ and $E \cos(\omega_d t + \varphi)$.

and equality holds when

$$B_1 = B \cos \phi \qquad \text{and} \qquad B_2 = B \sin \phi$$

from which

$$\frac{B_2}{B_1} = \frac{B \sin \phi}{B \cos \phi} = \tan \phi$$

or

$$\phi = \tan^{-1}\left(\frac{B_2}{B_1}\right)$$

The angle ϕ can be depicted in the right triangle shown in Fig. 6.5.

fig. 6.5

But

$$B_1 = B \cos \phi \Rightarrow B = \frac{B_1}{\cos \phi} = \frac{B_1}{B_1/\sqrt{B_1{}^2 + B_2{}^2}}$$

So

$$B = \sqrt{B_1{}^2 + B_2{}^2}$$

In summary, the conditions for which Equation (6.13) holds are

$$B = \sqrt{B_1{}^2 + B_2{}^2} \qquad \text{and} \qquad \phi = \tan^{-1}\left(\frac{B_2}{B_1}\right)$$

and the expression given by Equation (6.12) can be rewritten as

$$y(t) = Be^{-\alpha t} \cos(\omega_d t - \phi)$$

where $\omega_d = \sqrt{\omega_n{}^2 - \alpha^2}$. A typical sketch of $y(t)$ for $t \geq 0$ is shown in Fig. 6.6.

As a consequence of this discussion, we call α the *damping factor*, since its value determines how fast the sinusoid's amplitude diminishes. We call ω_d the *damped natural frequency*—or *damped frequency*, for short. For the case that the damping factor $\alpha = 0$,

fig. 6.6

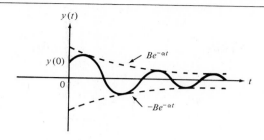

we have that $\omega_d = \sqrt{\omega_n{}^2 - \alpha^2} = \omega_n$, and we refer to ω_n as the *undamped natural frequency*—or *natural frequency*,* for short.

In the present example, we have that

$$i(t) = e^{-3t}(\cos 4t - \sin 4t) = e^{-3t}(B_1 \cos 4t + B_2 \sin 4t)$$

so

$$B = \sqrt{B_1{}^2 + B_2{}^2} = \sqrt{1 + 1} = \sqrt{2}$$

and

$$\phi = \tan^{-1} \frac{B_2}{B_1} = \tan^{-1}\left(\frac{-1}{1}\right) = -\frac{\pi}{4} \text{ rad}$$

Thus, we can also write the current as

$$i(t) = \sqrt{2}e^{-3t} \cos\left(4t + \frac{\pi}{4}\right) \qquad \text{for } t \geq 0$$

To find $v(t) = L \, di(t)/dt$ from this expression for $i(t)$ involves manipulations to be discussed in Chapter 8. So instead, let us use

$$v(t) = \frac{1}{50} \frac{d}{dt}[e^{-3t}(\cos 4t - \sin 4t)]$$

$$= \frac{1}{50} \left[e^{-3t}(-4 \sin 4t - 4 \cos 4t) - 3e^{-3t}(\cos 4t - \sin 4t)\right]$$

$$= \frac{1}{50} e^{-3t}[-7 \cos 4t - \sin 4t]$$

In this case

$$B = \sqrt{(-7)^2 + (-1)^2} = \sqrt{50} = 5\sqrt{2}$$

*Although some texts refer to ω_n as the "resonance frequency," except for very simple circuits like the series and parallel *RLC* circuits, the resonance frequency of the circuit (to be discussed in Chapter 10) is generally different from the natural frequency.

and

$$\phi = \tan^{-1}\left(\frac{-1}{-7}\right) = 3.28 = -3.0 \text{ rad}$$

so

$$v(t) = \frac{\sqrt{2}}{10} e^{-3t} \cos(4t + 3.0) \qquad \text{for } t \geq 0$$

What's that? Your calculator says that $\tan^{-1}(-1/-7)$ is equal to 0.142 rad (8 degrees)? If so, be careful. Your calculator cannot distinguish between $\tan^{-1}(-1/-7)$ and $\tan^{-1}(1/7)$—these are different quantities! And it is the former we require in this example. Plots of $i(t)$ and $v(t)$ for $t \geq 0$ are shown in Figs. 6.7 and 6.8, respectively.

 fig. 6.7

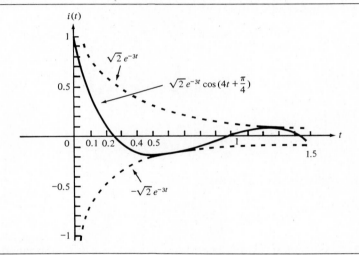

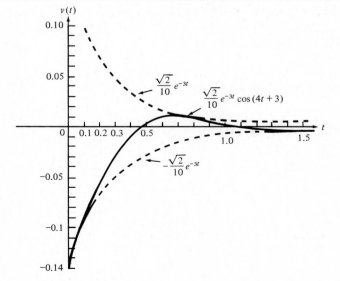 fig. 6.8

What happens in the underdamped case is that the energy that is stored in the inductor and capacitor eventually gets dissipated by the resistor, of course. However, the energy does not simply go from the L and C directly to the R. Instead, the energy is transferred back and forth between the two energy storage elements—the resistance taking its toll during the process. The result of all of this is a response that is oscillatory (changes sign more than once). In particular, the response is the product of a real exponential and a sinusoid—called a *damped sinusoid.*

In the example above, although oscillatory, the responses (current and voltage) went to zero before too many oscillations occurred. This is due to the fact that ω_n is not much larger than α (that is, slight underdamping). By reducing the damping (making α smaller), the circuit becomes more underdamped, and more oscillations occur before the responses get close to zero. A sketch of an underdamped response where α is small is shown in Fig. 6.9.

fig. 6.9

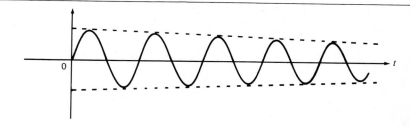

For the parallel RLC circuit when $R = \infty$ or for the series RLC circuit when $R = 0$, we have the underdamped case with no damping ($\alpha = 0$). A consequence of this is a response that will be perfectly sinusoidal, and thus will never decrease in amplitude. Needless to say, such a situation cannot be physically constructed with ordinary capacitors and inductors. There are, however, electronic ways for producing such a result.

Having considered the cases of overdamping ($\alpha > \omega_n$) and underdamping ($\omega_n > \alpha$), it is now time to investigate the critical damping condition ($\alpha = \omega_n$). We shall do this while we are analyzing a new circuit.

6.3 A SERIES-PARALLEL RLC CIRCUIT

We have seen for the series RLC circuit and the parallel RLC circuit that the natural frequency in both cases is $\omega_n = 1/\sqrt{LC}$. As a consequence, it is frequently assumed that this is a general result for any circuit containing one inductor and one capacitor. The circuit in Fig. 6.10 will show that such an assumption is unwarranted.

fig. 6.10

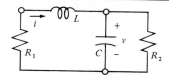

Applying KVL to the mesh on the left, we get

$$R_1 i + L \frac{di}{dt} + v = 0 \tag{6.14}$$

while applying KCL at the node common to the inductor and the capacitor, we obtain

$$i = C \frac{dv}{dt} + \frac{v}{R_2} \tag{6.15}$$

Substituting Equation (6.15) into Equation (6.14) yields

$$R_1 \left(C \frac{dv}{dt} + \frac{v}{R_2} \right) + L \frac{d}{dt} \left(C \frac{dv}{dt} + \frac{v}{R_2} \right) + v = 0$$

from which we obtain

$$LC \frac{d^2 v}{dt^2} + \left(R_1 C + \frac{L}{R_2} \right) \frac{dv}{dt} + \left(\frac{R_1}{R_2} + 1 \right) v = 0$$

or

$$\frac{d^2 v}{dt^2} + \left(\frac{R_1}{L} + \frac{1}{R_2 C} \right) \frac{dv}{dt} + \frac{R_1 + R_2}{R_2 LC} v = 0 \tag{6.16}$$

Thus, for this circuit,

$$\alpha = \frac{R_1}{2L} + \frac{1}{2R_2 C} \quad \text{and} \quad \omega_n = \sqrt{\frac{R_1 + R_2}{R_2 LC}}$$

Suppose that $R_1 = 3\ \Omega$, $R_2 = 1\ \Omega$, $L = 1$ H, and $C = 1$ F. Then

$$\alpha = \frac{3}{2} + \frac{1}{2} = 2 \quad \text{and} \quad \omega_n = \sqrt{\frac{4}{1}} = 2$$

and, since $\alpha = \omega_n$, the circuit is critically damped. As mentioned before,

$$y(t) = A_1 e^{s_1 t} + A_2 e^{s_2 t}$$

is not the solution to Equation (6.3) for this case. To determine what is, let us return to the equation. Using the critically damped condition $\alpha = \omega_n$, Equation (6.3) becomes

$$\frac{d^2 y(t)}{dt^2} + 2\alpha \frac{dy(t)}{dt} + \alpha^2 y(t) = 0$$

or

$$\frac{d^2 y(t)}{dt^2} + \alpha \frac{dy(t)}{dt} + \alpha \frac{dy(t)}{dt} + \alpha^2 y(t) = 0$$

$$\frac{d}{dt} \left[\frac{dy(t)}{dt} + \alpha y(t) \right] + \alpha \left[\frac{dy(t)}{dt} + \alpha y(t) \right] = 0 \tag{6.17}$$

If we define the function $f(t)$ by

$$f(t) = \frac{dy(t)}{dt} + \alpha y(t)$$

then Equation (6.17) becomes

$$\frac{df(t)}{dt} + \alpha f(t) = 0$$

which is a first-order differential equation whose solution is

$$f(t) = A_1 e^{-\alpha t}$$

Thus

$$\frac{dy(t)}{dt} + \alpha y(t) = A_1 e^{-\alpha t}$$

Multiplying both sides by $e^{\alpha t}$ yields

$$e^{\alpha t} \frac{dy(t)}{dt} + e^{\alpha t} \alpha y(t) = A_1$$

or

$$\frac{d}{dt} [e^{\alpha t} y(t)] = A_1$$

Integrating both sides with respect to t results in

$$\int \frac{d[e^{\alpha t} y(t)]}{dt} \, dt = \int A_1 \, dt$$

or

$$e^{\alpha t} y(t) = A_1 t + A_2$$

where A_2 is a constant of integration. Multiplying both sides by $e^{-\alpha t}$ gives

$$\boxed{y(t) = A_1 t e^{-\alpha t} + A_2 e^{-\alpha t} = (A_1 t + A_2) e^{-\alpha t}}$$

and this is the solution of Equation (6.3) for the case of critical damping.

Returning now to the zero-input series-parallel *RLC* circuit in Fig. 6.10, the expression for $v(t)$—that is, the solution of Equation (6.16) for the case of critical damping—is

$$v(t) = A_1 t e^{-2t} + A_2 e^{-2t}$$

Given the initial conditions $i(0) = 1$ A and $v(0) = 2$ V, then setting $t = 0$ yields

$$v(0) = 0 + A_2$$

$$2 = A_2$$

Next,

$$\frac{dv(t)}{dt} = -2A_1 t e^{-2t} + A_1 e^{-2t} - 2A_2 e^{-2t}$$

so

$$\frac{dv(0)}{dt} = A_1 - 2A_2 = A_1 - 4$$

From Equation (6.15)

$$i(t) = C\frac{dv(t)}{dt} + \frac{v(t)}{R_2} \qquad\qquad (6.18)$$

from which

$$\frac{dv(t)}{dt} = -\frac{v(t)}{R_2 C} + \frac{i(t)}{C} = -v(t) + i(t)$$

Setting $t = 0$,

$$\frac{dv(0)}{dt} = -v(0) + i(0) = -2 + 1 = -1$$

so

$$A_1 - 4 = -1 \Rightarrow A_1 = 3$$

Therefore, the expression for $v(t)$ is

$$v(t) = 3te^{-2t} + 2e^{-2t} \qquad \text{for } t \geq 0$$
$$= (3t + 2)e^{-2t} \qquad \text{for } t \geq 0$$

In addition, from Equation (6.18),

$$i(t) = 1\frac{d}{dt}\left[(3t + 2)e^{-2t}\right] + (3t + 2)e^{-2t}$$

$$= -2(3t + 2)e^{-2t} + 3e^{-2t} + (3t + 2)e^{-2t}$$

$$= -(3t + 2)e^{-2t} + 3e^{-2t} = (-3t + 1)e^{-2t} \qquad \text{for } t \geq 0$$

Sketches of $v(t)$ and $i(t)$ for $t \geq 0$ are shown in Figs. 6.11 and 6.12, respectively.

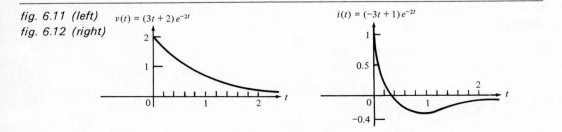

fig. 6.11 (left)
fig. 6.12 (right)

   An important fact may become evident when we look at the results of the previous examples. The natural (zero-input) response of a given circuit has the same form regardless of the variable (voltage or current) that is selected as the output. In the last example, both the capacitor voltage and inductor current have the form $(A_1t + A_2)e^{-\alpha t}$—the critically damped case. Any other variable (e.g., capacitor current, inductor voltage, resistor currents and voltages) will also have this form—the value of α being the same. This result is due to the fact that the voltage and current for a resistor differ by a constant, while inductors and capacitors are elements with differential and integral relationships. Performing such operations as scaling, differentiating, and integrating will not change the form or the value of α for a critically damped expression. Furthermore, applying KVL and KCL (adding such expressions) also does not change the form or value of α. Of course, the same conclusion can be deduced for the overdamped and underdamped cases.

PROBLEMS *6.1.   For the circuit shown in Fig. P6.1, find $v(t)$ and $i(t)$ for all t.

$Ans.$ $1 - [1 + \frac{5}{8}e^{-3t} - \frac{13}{8}e^{-t}]u(t)$, $1 - [1 - \frac{15}{2}e^{-3t} + \frac{13}{2}e^{-t}]u(t)$

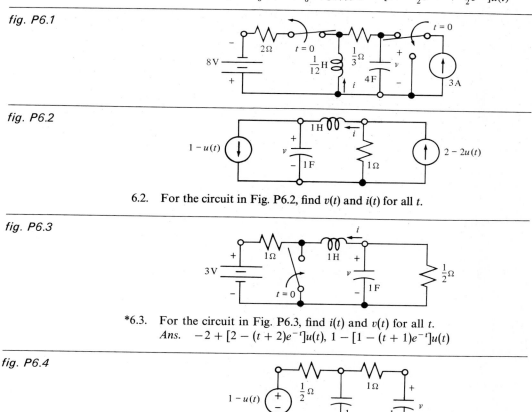

fig. P6.1

fig. P6.2

6.2.   For the circuit in Fig. P6.2, find $v(t)$ and $i(t)$ for all t.

fig. P6.3

*6.3.   For the circuit in Fig. P6.3, find $i(t)$ and $v(t)$ for all t.

$Ans.$ $-2 + [2 - (t + 2)e^{-t}]u(t)$, $1 - [1 - (t + 1)e^{-t}]u(t)$

fig. P6.4

6.4.   For the circuit in Fig. P6.4, find $v(t)$ for all t.

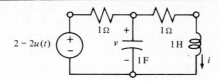

6.5. For the circuit in Fig. P6.5, find $v(t)$ and $i(t)$ for all t.

Ans. $1 - \left[1 - \sqrt{2}e^{-t}\cos\left(t + \frac{\pi}{4}\right)\right]u(t), 1 - \left[1 - \sqrt{2}e^{-t}\cos\left(t - \frac{\pi}{4}\right)\right]u(t)$

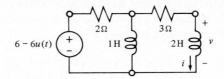

6.6. For the circuit shown in Fig. P6.6, find $i(t)$ for all t.

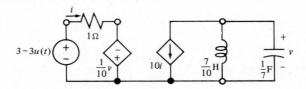

6.7. For the circuit in Fig. P6.7, find $v(t)$ for all t.

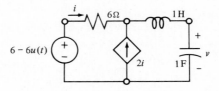

6.8. For the circuit shown in Fig. P6.8, find $v(t)$ for all t.

6.4 CIRCUITS WITH NONZERO INPUTS

Having considered some second-order circuits with zero inputs and nonzero initial conditions, let us now look at second-order circuits with nonzero inputs. We shall limit our discussion to forcing functions that are ramps, steps, impulses, or various combinations of these functions. Other types of forcing functions will be dealt with in subsequent chapters when additional concepts are introduced.

Consider the series *RLC* circuit shown in Fig. 6.13, whose input is a voltage step function.

fig. 6.13

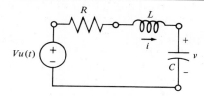

We find the zero-state response, by KVL,

$$Vu(t) = Ri + L\frac{di}{dt} + v$$

Since

$$i = C\frac{dv}{dt}$$

then

$$LC\frac{d^2v}{dt^2} + RC\frac{dv}{dt} + v = Vu(t)$$

or

$$\frac{d^2v}{dt^2} + \frac{R}{L}\frac{dv}{dt} + \frac{1}{LC}v = \frac{V}{LC}u(t)$$

For $t < 0$, this equation becomes

$$\frac{d^2v}{dt^2} + \frac{R}{L}\frac{dv}{dt} + \frac{1}{LC}v = 0$$

and the solution is $v(t) = 0$. However, for $t \geq 0$,

$$\frac{d^2v}{dt^2} + \frac{R}{L}\frac{dv}{dt} + \frac{1}{LC}v = \frac{V}{LC} \qquad (6.19)$$

As in the case of first-order equations, the solution $v(t)$ consists of two parts—the forced response $v_f(t)$ and the natural response $v_n(t)$. That is, the solution is of the form

$$v(t) = v_f(t) + v_n(t)$$

where $v_n(t)$ is the solution to

$$\frac{d^2v}{dt^2} + 2\alpha\frac{dv}{dt} + \omega_n^2 v = 0$$

and $\alpha = R/2L$, $\omega_n = 1/\sqrt{LC}$.

In this case, the forcing function is a constant. Thus, the forced response is a constant, say K. If the natural response $v_n(t)$ is substituted into the left-hand side of

Equation (6.19), the result is zero. Thus the forced response $v_f(t) = K$ must satisfy Equation (6.19); that is,

$$\frac{d^2 K}{dt^2} + \frac{R}{L}\frac{dK}{dt} + \frac{1}{LC}K = \frac{V}{LC}$$

or

$$\frac{1}{LC}K = \frac{V}{LC}$$

so that

$$K = V$$

Hence the solution to Equation (6.19) for the overdamped case is

$$v(t) = V + A_1 e^{s_1 t} + A_2 e^{s_2 t}$$

For the underdamped case, it is

$$v(t) = V + A_1 e^{-\alpha t}\cos \omega_d t + A_2 e^{-\alpha t}\sin \omega_d t$$

For the critically damped case, it is

$$v(t) = V + A_1 t e^{-\alpha t} + A_2 e^{-\alpha t}$$

The constants A_1 and A_2 are determined from the circuit topology and the initial conditions $i(0) = v(0) = 0$.

Since there is no energy initially stored in the inductor and capacitor in this circuit, when the input voltage is zero, the voltage $v(t)$ and the current $i(t)$ will be zero. At $t = 0$, when the input voltage becomes V volts, since the voltage across a capacitor and the current through an inductor cannot change instantaneously, $v(0) = 0$ and $i(0) = 0$. After a long time, the input acts as a constant and the inductor behaves as a short circuit and the capacitor behaves as an open circuit. Thus, eventually, the voltage across the capacitor will be V volts and the current through the inductor will be zero amperes. The shape of the voltage and current waveforms between initial and final values will depend upon whether the circuit is overdamped, underdamped, or critically damped.

· · ·

Example

For the circuit given in Fig. 6.13, the underdamped case results when $R = 12\ \Omega$, $L = 2$ H, and $C = \frac{1}{50}$ F, since $\alpha = 3$ and $\omega_n = 5$. Also $\omega_d = 4$. If $V = \frac{2}{5}$ V, then the solution of Equation (6.18) is

$$v(t) = \frac{2}{5} + A_1 e^{-3t}\cos 4t + A_2 e^{-3t}\sin 4t$$

Setting $t = 0$ yields

$$v(0) = \frac{2}{5} + A_1$$

$$0 = \frac{2}{5} + A_1 \Rightarrow A_1 = -\frac{2}{5}$$

Taking the derivative of $v(t)$, we get

$$\frac{dv(t)}{dt} = -3A_1e^{-3t}\cos 4t - 4A_1e^{-3t}\sin 4t - 3A_2e^{-3t}\sin 4t + 4A_2e^{-3t}\cos 4t$$

so

$$\frac{dv(0)}{dt} = -3A_1 + 4A_2$$

$$= \frac{6}{5} + 4A_2$$

From the circuit,

$$i(t) = C\frac{dv(t)}{dt}$$

so

$$\frac{dv(t)}{dt} = \frac{1}{C}i(t)$$

and

$$\frac{dv(0)}{dt} = \frac{1}{C}i(0) = 0$$

Thus

$$0 = \frac{6}{5} + 4A_2$$

from which

$$A_2 = -\frac{3}{10}$$

Therefore, for $t \geq 0$,

$$v(t) = \frac{2}{5} - \frac{2}{5}e^{-3t}\cos 4t - \frac{3}{10}e^{-3t}\sin 4t$$

$$= \frac{2}{5} - \frac{1}{10}e^{-3t}(4\cos 4t + 3\sin 4t)$$

$$= \frac{2}{5} - \frac{1}{2}e^{-3t}\cos(4t - 0.64)$$

Combining this with the fact that $v(t) = 0$ for $t < 0$ results in the zero-state voltage step response

$$v(t) = \left[\frac{2}{5} - \frac{1}{2}e^{-3t}\cos(4t - 0.64)\right]u(t)$$

and therefore the zero-state current step response is

$$i(t) = C \frac{dv}{dt}$$

$$= \frac{1}{50} \frac{d}{dt} \left[\left(\frac{2}{5} - \frac{2}{5} e^{-3t} \cos 4t - \frac{3}{10} e^{-3t} \sin 4t \right) u(t) \right]$$

$$= \frac{1}{50} \left[\left(\frac{2}{5} - \frac{2}{5} e^{-3t} \cos 4t - \frac{3}{10} e^{-3t} \sin 4t \right) \delta(t) \right.$$

$$+ \left(\frac{6}{5} e^{-3t} \cos 4t + \frac{8}{5} e^{-3t} \sin 4t \right.$$

$$\left. + \frac{9}{10} e^{-3t} \sin 4t - \frac{12}{10} e^{-3t} \cos 4t \right) u(t) \right]$$

$$= \frac{1}{50} \left[\left(\frac{2}{5} - \frac{2}{5} \right) \delta(t) + \left(\frac{25}{10} e^{-3t} \sin 4t \right) u(t) \right]$$

$$= \left(\frac{1}{20} e^{-3t} \sin 4t \right) u(t)$$

Sketches of $v(t)$ and $i(t)$ are shown in Figs. 6.14 and 6.15, respectively.

Suppose now that the input is changed from the step function $\frac{2}{5} u(t)$ volts to the impulse function $\frac{2}{5} \delta(t)$. Then using the reasoning discussed in the chapter on first-

---

fig. 6.14

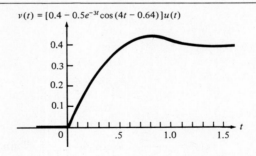

$$v(t) = [0.4 - 0.5 e^{-3t} \cos(4t - 0.64)] u(t)$$

---

fig. 6.15

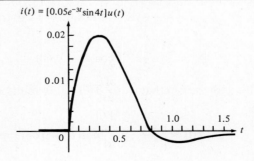

$$i(t) = [0.05 e^{-3t} \sin 4t] u(t)$$

order circuits, the (zero-state) impulse response is just the derivative of the (zero-state) step response. Hence the resulting voltage across the $\frac{1}{50}$-F capacitor is

$$v_\delta(t) = \frac{dv(t)}{dt} = \left(\frac{5}{2} e^{-3t} \sin 4t\right) u(t)$$

and the zero-state current impulse response is

$$i_\delta(t) = \frac{di(t)}{dt} = \frac{d}{dt}\left[\left(\frac{1}{20}e^{-3t} \sin 4t\right) u(t)\right]$$

$$= \left(\frac{1}{20}e^{-3t} \sin 4t\right)\delta(t) + \frac{1}{20}(-3e^{-3t} \sin 4t + 4e^{-3t} \cos 4t)u(t)$$

$$= \left[\frac{1}{4}e^{-3t} \cos(4t + 0.64)\right] u(t)$$

As a consequence of the fact that R, L, and C have constant values, the resulting differential equation describing the circuit has constant coefficients. As was the case for first-order circuits, this means that the time-invariance property holds for such second-order circuits—and higher-order circuits, too. In other words, if $y(t)$ is the response to the input $x(t)$ subject to the initial condition $y(0)$, then the response to the input $x(t - t_0)$ subject to the boundary condition $y(t_0) = y(0)$ is $y(t - t_0)$. For example, the zero-state current response to the input $(2/5)u(t - 2)$ is

$$i(t - 2) = \left[\frac{1}{20}e^{-3(t-2)} \sin 4(t - 2)\right] u(t - 2)$$

In addition, since the circuit is linear, scaling the input by the constant K also scales the zero-state response by the same constant K. For example, the current zero-state (unit) step response to the unit step function $u(t) = (5/2)[(2/5)u(t)]$ is

$$i_u(t) = \frac{5}{2} i(t) = \left(\frac{1}{8}e^{-3t} \sin 4t\right) u(t)$$

· · ·

Let us now look at a practical series *RLC* circuit in which not only is the input nonzero, but an initial condition is nonzero as well.

· · ·

Example
The circuit shown in Fig. 6.16 is in essence an old-fashioned (nonelectronic) automobile ignition system. The input of this circuit is a 12-V automobile battery. The series resistance consists of a ballast resistor, the resistance of the ignition switch, and the series resistance of the inductor (known as the "ignition coil"). The capacitor (called a "condenser" by mechanics) is in parallel with a switch (called the "points") that is closed for $t < 0$ and open for $t \geq 0$. (In actuality, the points

fig. 6.16

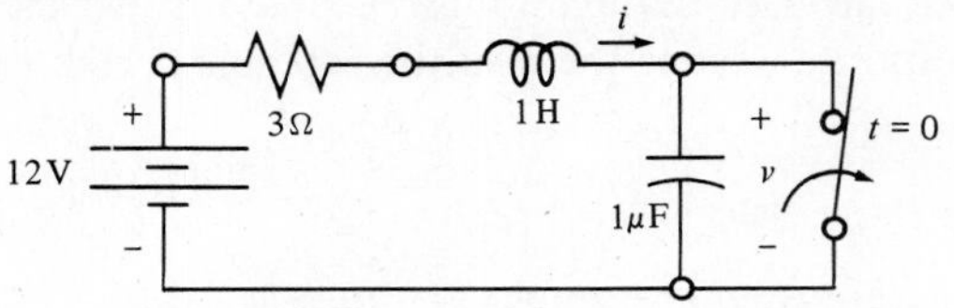

open and close periodically—the rate depending on the engine rpm.) It is the voltage produced across the coil that is applied to the spark plugs. This in turn produces a spark that ignites the fuel mixture.

For the given circuit, for $t < 0$, clearly $v(t) = 0$ and $i(t) = 12/3 = 4$ A. Thus, $v(0) = 0$ and $i(0) = 4$ A. For $t \geq 0$, by KVL,

$$Ri + L\frac{di}{dt} + \frac{1}{C}\int_0^t i\,dt + v(0) = 12$$

Thus

$$3i + \frac{di}{dt} + 10^6 \int_0^t i\,dt + 0 = 12$$

and taking the derivative of this equation we get

$$\frac{d^2i}{dt^2} + 3\frac{di}{dt} + 10^6\,i = 0$$

Hence

$$\alpha = \frac{3}{2} < \omega_n = 10^3$$

and this circuit is (very) underdamped. We then have that

$$\omega_d = \sqrt{\omega_n{}^2 - \alpha^2} \approx \omega_n = 10^3$$

and since the forced response is zero, the approximate current has the form

$$i(t) \approx A_1 e^{-3t/2} \cos 10^3 t + A_2 e^{-3t/2} \sin 10^3 t$$

Setting $t = 0$, we have that

$$i(0) = A_1 + 0 \Rightarrow A_1 = 4$$

Next

$$\frac{di(t)}{dt} = -\frac{3}{2}(4)e^{-3t/2} \cos 10^3 t - 4(10^3)e^{-3t/2} \sin 10^3 t$$

$$-\frac{3}{2}A_2 e^{-3t/2} \sin 10^3 t + 10^3 A_2 e^{-3t/2} \cos 10^3 t$$

from which

$$\frac{di(0)}{dt} = -6 - 0 - 0 + 10^3 A_2$$

However, from the circuit (for $t \geq 0$) by KVL

$$12 = 3i + 1\frac{di}{dt} + v$$

so

$$\frac{di}{dt} = 12 - 3i - v$$

and

$$\frac{di(0)}{dt} = 12 - 3i(0) - v(0) = 12 - 12 - 0 = 0$$

Thus

$$-6 + 10^3 A_2 = 0 \Rightarrow A_2 = 6(10^{-3})$$

The current expression is then

$$i(t) \approx 4e^{-3t/2} \cos 10^3 t + 6(10^{-3})e^{-3t/2} \sin 10^3 t$$

and, since the second term is considerably smaller than the first, a good approximation is

$$i(t) \approx 4e^{-3t/2} \cos 10^3 t$$

and the voltage across the inductor is approximately

$$v_L(t) = L\frac{di(t)}{dt} \approx 1\frac{d}{dt}(4e^{-3t/2} \cos 10^3 t)$$

$$\approx -\frac{3}{2}(4)e^{-3t/2} \cos 10^3 t - 4(10^3)e^{-3t/2} \sin 10^3 t$$

In this expression the first term is considerably smaller than the second, so a good approximation is

$$v_L(t) \approx -4(10^3)e^{-3t/2} \sin 10^3 t$$

Let's evaluate this formula the first time the sine term equals unity—that is, for $t = (\pi/2)$ ms. Then we get

$$v_L\left[\frac{\pi}{2}(10^{-3})\right] \approx -4(10^3)e^{-(3/2)(\pi/2)10^{-3}}$$

$$\approx -4000 \text{ V}$$

Although the values of L and C given in this example are practical in nature, they were still chosen to yield computational convenience. In a typical automobile, the voltage required to produce a spark across the gap of a spark plug is between 6000 and 10,000 V.

Having analyzed a series *RLC* circuit prior to this example, we could have used the previously obtained differential equation (6.19) to determine $v(t)$. In that case the solution would contain a nonzero forced response. However, we instead wrote a differential equation in the variable $i(t)$.

. . .

Let us now consider the parallel *RLC* circuit shown in Fig. 6.17.

fig. 6.17

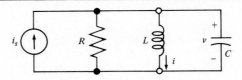

Since

$$v = L\frac{di}{dt}$$

and

$$\frac{v}{R} + i + C\frac{dv}{dt} = i_s$$

then

$$\frac{L}{R}\frac{di}{dt} + i + LC\frac{d^2i}{dt^2} = i_s$$

or

$$\frac{d^2i}{dt^2} + \frac{1}{RC}\frac{di}{dt} + \frac{1}{LC}i = \frac{1}{LC}i_s \qquad (6.20)$$

Again, the solution of this equation has the form

$$i(t) = i_f(t) + i_n(t)$$

where $i_f(t)$ is the forced response and $i_n(t)$ is the natural response. In determining the natural response—whether the circuit is overdamped, underdamped, or critically damped—we use the fact that

$$\alpha = \frac{1}{2RC} \qquad \text{and} \qquad \omega_n = \frac{1}{\sqrt{LC}}$$

Suppose that $R = 6\ \Omega$, $L = 7$ H, $C = \frac{1}{42}$ F, $i_s(t) = 6u(t)$, $v(0) = 0$ and $i(0) = -4$ A. Then Equation (6.20) becomes

$$\frac{d^2i}{dt^2} + 7\frac{di}{dt} + 6i = 36u(t)$$

and

$$\alpha = \frac{7}{2} > \omega_n = \sqrt{6}$$

so the circuit is overdamped. Since

$$s = -\alpha \pm \sqrt{\alpha^2 - \omega_n^2} = -\frac{7}{2} \pm \sqrt{\frac{49}{4} - 6} = -\frac{7}{2} \pm \frac{5}{2}$$

then

$$s_1 = -6 \qquad \text{and} \qquad s_2 = -1$$

and therefore the natural response has the form

$$i_n(t) = A_1 e^{-6t} + A_2 e^{-t}$$

For $t \geq 0$, the forcing function is a constant, so the forced response is a constant K that can be obtained by substituting K into the differential equation. Also, from the circuit, for the dc case we have $i_f(t) = 6$ A. Hence the complete response has the form

$$i(t) = 6 + A_1 e^{-6t} + A_2 e^{-t}$$

Setting $t = 0$, we get

$$i(0) = 6 + A_1 + A_2 = -4 \Rightarrow A_1 + A_2 = -10 \qquad (6.21)$$

Also

$$\frac{di(t)}{dt} = -6A_1 e^{-6t} - A_2 e^{-t}$$

so

$$\frac{di(0)}{dt} = -6A_1 - A_2$$

But from the circuit,

$$v(t) = L \frac{di(t)}{dt} \Rightarrow \frac{di(t)}{dt} = \frac{1}{L} v(t)$$

and

$$\frac{di(0)}{dt} = \frac{1}{7} v(0) = 0 = -6A_1 - A_2 \qquad (6.22)$$

From Equations (6.21) and (6.22) we get

$$A_1 = 2 \qquad \text{and} \qquad A_2 = -12$$

Therefore the complete response is

$$i(t) = 6 + 2e^{-6t} - 12e^{-t} \qquad \text{for } t \geq 0$$

Needless to say, a zero-state second-order circuit with more than one independent source can be analyzed with the use of the principle of superposition. Even for the case that an L or C has a nonzero initial condition, we can model it as an appropriate connection of an independent source and that element with a zero initial condition. The principle of superposition can then be applied.

For the circuit in Fig. 6.17, the equivalent zero-state circuit is shown in Fig. 6.18. In this circuit $i_L(0) = 0$ and $v(0) = 0$. To find $i_1(t)$, which is that portion of $i_L(t)$ due solely

fig. 6.18

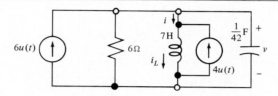

to the $6u(t)$-ampere current source, set the other independent current source to zero. The resulting differential equation is

$$\frac{d^2 i_L}{dt^2} + 7\frac{di_L}{dt} + 6i_L = 36u(t)$$

and, as described above, the solution for $t \geq 0$ is

$$i_1(t) = 6 + A_1 e^{-6t} + A_2 e^{-t}$$

Then

$$0 = i_1(0) = 6 + A_1 + A_2 \Rightarrow A_1 + A_2 = -6$$

and

$$\frac{di_1(t)}{dt} = -6A_1 e^{-6t} - A_2 e^{-t} = \frac{1}{L}v(t)$$

so

$$\frac{di_1(0)}{dt} = -6A_1 - A_2 = \frac{1}{L}v(0) = 0$$

Solving for A_1 and A_2 we get

$$A_1 = \frac{6}{5} \quad \text{and} \quad A_2 = -\frac{36}{5}$$

and therefore

$$i_1(t) = 6 + \frac{6}{5}e^{-6t} - \frac{36}{5}e^{-t} \qquad \text{for } t \geq 0$$

To find $i_2(t)$, which is that portion of $i_L(t)$ due solely to the $4u(t)$-ampere current source, set the other independent current source to zero. The resulting differential equation is

$$\frac{d^2 i_L}{dt^2} + 7\frac{di_L}{dt} + 6i_L = 24u(t)$$

and the solution for $t \geq 0$ is

$$i_2(t) = 4 + B_1 e^{-6t} + B_2 e^{-t}$$

Then

$$0 = i_2(0) = 4 + B_1 + B_2 \Rightarrow B_1 + B_2 = -4$$

and

$$\frac{di_2(t)}{dt} = -6B_1 e^{-6t} - B_2 e^{-t} = \frac{1}{L}v(t)$$

so

$$\frac{di_2(0)}{dt} = -6B_1 - B_2 = \frac{1}{L}v(0) = 0$$

Solving for B_1 and B_2 we get

$$B_1 = \frac{4}{5} \quad \text{and} \quad B_2 = -\frac{24}{5}$$

Thus

$$i_2(t) = 4 + \frac{4}{5}e^{-6t} - \frac{24}{5}e^{-t} \quad \text{for } t \geq 0$$

Hence

$$i_L(t) = i_1(t) + i_2(t) = 10 + 2e^{-6t} - 12e^{-t} \quad \text{for } t \geq 0$$

and finally

$$i(t) = i_L(t) - 4u(t)$$

so

$$i(t) = 6 + 2e^{-6t} - 12e^{-t} \quad \text{for } t \geq 0$$

Even though this approach required many more steps to obtain the answer than was used by the preceding method, it was presented to demonstrate the principle of superposition.

Let us write the differential equation for one more second-order circuit having a nonzero input. The detailed analysis for this particular circuit is part of the subsequent problem set (Problems 6.11 and 6.20).

Consider the series-parallel circuit shown in Fig. 6.19. Since by KVL

$$v = R_1 i + L\frac{di}{dt} \tag{6.23}$$

fig. 6.19

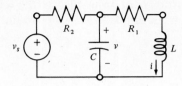

then

$$\frac{dv}{dt} = R_1 \frac{di}{dt} + L \frac{d^2i}{dt^2} \tag{6.24}$$

Also, by KCL

$$C \frac{dv}{dt} = \frac{v_s - v}{R_2} - i \tag{6.25}$$

Thus, substituting Equations (6.23) and (6.24) into (6.25) yields

$$C \left(R_1 \frac{di}{dt} + L \frac{d^2i}{dt^2} \right) = \frac{v_s}{R_2} - \frac{1}{R_2} \left(R_1 i + L \frac{di}{dt} \right) - i$$

from which

$$\frac{d^2i}{dt^2} + \left(\frac{R_1}{L} + \frac{1}{R_2 C} \right) \frac{di}{dt} + \left(\frac{R_1 + R_2}{R_2 LC} \right) i = \frac{v_s}{R_2 LC} \tag{6.26}$$

For this equation

$$\alpha = \frac{R_1}{2L} + \frac{1}{2R_2 C} \quad \text{and} \quad \omega_n = \sqrt{\frac{R_1 + R_2}{R_2 LC}}$$

and these determine whether the circuit is overdamped, underdamped, or critically damped.

In determining differential equation (6.26) as well as the other second-order differential equations in this chapter, we did not proceed by exclusively using either nodal or mesh analysis. Although such an approach is valid, it may not be the easiest algebraically. For the circuit in Fig. 6.19, Equation (6.26) was obtained from Equations (6.23) and (6.25). The latter two equations are known as the "state equations" of the circuit, and this subject is the topic of the next chapter.

PROBLEMS *6.9.   For the series *RLC* circuit shown in Fig. P6.9, find the zero-state response $v(t)$ when (a) $v_s(t) = u(t)$ and (b) $v_s(t) = \delta(t)$.
Ans.   (a) $(1 - 2te^{-2t} - e^{-2t})u(t)$, (b) $4te^{-2t}u(t)$

fig. P6.9

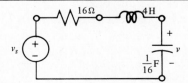

6.10. For the circuit given in Problem 6.4 of the previous problem set, change the independent voltage source to $u(t)$ volts and find the zero-state step response $v(t)$.

*6.11. For the circuit given in Problem 6.5, change the independent voltage source to $u(t)$ volts and find the zero-state step response $v(t)$.

$Ans.$ $\dfrac{1}{2}\left[1 - \sqrt{2}e^{-t}\cos\left(t + \dfrac{\pi}{4}\right)\right]u(t)$

6.12. For the circuit given in Problem 6.7, change the independent voltage source to $u(t)$ volts and find the zero-state step response $v(t)$.

6.13. For the circuit given in Problem 6.8, change the independent voltage source to $u(t)$ volts and find the zero-state step response $v(t)$.

6.14. For the circuit given in Problem 6.2, change the independent current sources to $u(t)$ amperes and find the zero-state step response $v(t)$.

fig. P6.15

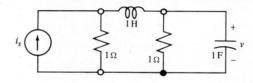

6.15. For the circuit shown in Fig. P6.15, find the zero-state response $v(t)$ when (a) $i_s(t) = u(t)$ and (b) $i_s(t) = \delta(t)$.

fig. P6.16

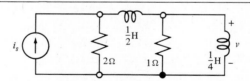

6.16. Repeat Problem 6.15 for the circuit in Fig. P6.16.

fig. P6.17

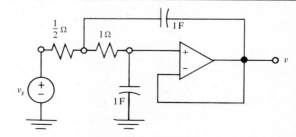

*6.17. Repeat Problem 6.9 for the op amp circuit in Fig. P6.17.

$Ans.$ $(1 + e^{-2t} - 2e^{-t})u(t)$

6.18. For the op amp circuit in Fig. P6.18, let $v_s(t) = u(t)$ and find the zero-state step response when (a) $C = \frac{1}{4}$ F and (b) $C = \frac{1}{16}$ F.

fig. P6.18

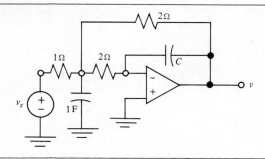

fig. P6.19

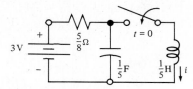

6.19.   For the circuit in Fig. P6.19, find $i(t)$ for all t.

fig. P6.20

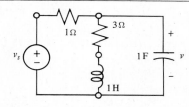

6.20.   For the circuit shown in Fig. P6.20, find the zero-state response $v(t)$ given that $v_s(t) = \frac{1}{3}(1 - e^{-3t})u(t)$.

fig. P6.21

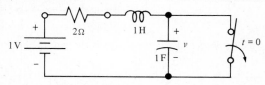

*6.21.   Find $v(t)$ for all t for the circuit in Fig. P6.21.
         Ans. $(1 - \frac{1}{2}te^{-t} - e^{-t})u(t)$

6.22.   For the circuit given in Problem 6.4 of the previous problem set, change the independent voltage source to $1 - \frac{1}{8}u(t)$ and find the resulting response $v(t)$ for all t.

Summary

1.   The natural response of a second-order circuit is either overdamped, underdamped, or critically damped.
2.   An overdamped response is the sum of two (real) exponentials.

3. An underdamped response is a damped sinusoid.
4. A critically damped response is the sum of an exponential and another exponential (with the same time constant) multiplied by time.
5. As with the case of first-order circuits, the complete response is the sum of two terms—one is the forced response and the other has the form of the natural response.
6. The comments about the linearity and time-invariance properties made for first-order circuits hold for second-order circuits (and higher-order circuits, too). (See items 5 through 8 in the summary for Chapter 5.)

STATE-VARIABLE ANALYSIS

7

Introduction

We have studied first- and second-order circuits in the two preceding chapters. What is done for a third- or higher-order circuit? Although a higher-order differential equation can be written, we shall not attempt to solve it directly—solving a second-order differential equation is involved enough. Instead, for an nth-order circuit, we shall write a set of n simultaneous first-order differential equations, which by means of a matrix formulation can be expressed as a single first-order matrix differential equation. Although the solution of such a matrix equation is analogous to the case of an ordinary differential equation, we can find a numerical solution instead of the closed-form solutions that we have been obtaining.

7.1 ZERO-INPUT CIRCUITS

Let us return to the series RLC circuit shown in Fig. 7.1. Suppose that for this circuit, which has a zero input, the inductor current at time $t = 0$ is $i_L(0)$ and the capacitor voltage at time $t = 0$ is $v_C(0)$. Thus we can say that the "condition" or "state" of the circuit at time $t = 0$ is specified by the inductor current and the capacitor voltage. For this reason we call the pair of numbers $[i_L(0), v_C(0)]$ the *initial state* of the circuit.

fig. 7.1

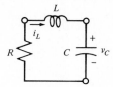

Extending this concept, we can refer to the pair $[i_L(t), v_C(t)]$ as the *state* of the circuit at time t. Furthermore, the variables i_L and v_C are called the *state variables* of the circuit.

330

Applying KVL to this circuit, we obtain

$$L \frac{di_L}{dt} + v_C + Ri_L = 0 \tag{7.1}$$

Furthermore,

$$i_L = C \frac{dv_C}{dt} \tag{7.2}$$

Substituting this equation into Equation (7.1), after division by LC, we get

$$\frac{d^2v_C}{dt^2} + \frac{R}{L} \frac{dv_C}{dt} + \frac{1}{LC} v_C = 0$$

This, of course, can be written as

$$\frac{d^2v_C}{dt^2} + 2\alpha \frac{dv_C}{dt} + \omega_n{}^2 v_C = 0$$

where $\alpha = R/2L$ and $\omega_n = 1/\sqrt{LC}$. We may then proceed, as we have discussed earlier, to solve this differential equation for $v_C(t)$.

In Equations (7.1) and (7.2), the only variables present are state variables. Let us rewrite these equations such that the coefficients of the derivatives are unity, and the derivatives alone are on one side of the equation. In other words, let us write Equations (7.1) and (7.2) in the form

$$\frac{di_L}{dt} = -\frac{R}{L} i_L - \frac{1}{L} v_C \quad \text{and} \quad \frac{dv_C}{dt} = \frac{1}{C} i_L$$

We call equations in this form, where all the variables present are state variables, *state equations.*

We may now write the above two state equations as a single matrix state equation as follows:

$$\begin{bmatrix} \dfrac{di_L}{dt} \\ \dfrac{dv_C}{dt} \end{bmatrix} = \begin{bmatrix} -\dfrac{R}{L} & -\dfrac{1}{L} \\ \dfrac{1}{C} & 0 \end{bmatrix} \begin{bmatrix} i_L \\ v_C \end{bmatrix}$$

By definition, the derivative of a matrix is

$$\frac{d}{dt} \begin{bmatrix} a_{11} & a_{12} & \cdots & a_{1m} \\ a_{21} & a_{22} & \cdots & a_{2m} \\ \vdots & \vdots & \ddots & \vdots \\ a_{n1} & a_{n2} & \cdots & a_{nm} \end{bmatrix} = \begin{bmatrix} \dfrac{da_{11}}{dt} & \dfrac{da_{12}}{dt} & \cdots & \dfrac{da_{1m}}{dt} \\ \dfrac{da_{21}}{dt} & \dfrac{da_{22}}{dt} & \cdots & \dfrac{da_{2m}}{dt} \\ \vdots & \vdots & \ddots & \vdots \\ \dfrac{da_{n1}}{dt} & \dfrac{da_{n2}}{dt} & \cdots & \dfrac{da_{nm}}{dt} \end{bmatrix}$$

Therefore, we can write the above matrix state equation as

$$\frac{d}{dt}\begin{bmatrix} i_L \\ v_C \end{bmatrix} = \begin{bmatrix} -\dfrac{R}{L} & -\dfrac{1}{L} \\ \dfrac{1}{C} & 0 \end{bmatrix}\begin{bmatrix} i_L \\ v_C \end{bmatrix}$$

If we define the matrices $\mathbf{X}(t)$ and $\mathbf{A}$ by

$$\mathbf{X}(t) = \begin{bmatrix} i_L \\ v_C \end{bmatrix} \quad \text{and} \quad \mathbf{A} = \begin{bmatrix} -\dfrac{R}{L} & -\dfrac{1}{L} \\ \dfrac{1}{C} & 0 \end{bmatrix}$$

then the matrix state equation can be written as

$$\frac{d\mathbf{X}(t)}{dt} = \mathbf{A}\mathbf{X}(t)$$

Since the components of the column matrix $\mathbf{X}(t)$ comprise the state of the circuit, we shall call $\mathbf{X}(t)$ the *state vector*. Consequently, $\mathbf{X}(0)$ will be called the *initial-state vector*. The process of finding the solution to the state equation is known as *state-variable analysis*.

We may now ask, "What is the solution to the state equation?" To answer this, consider the first-order ordinary (scalar) differential equation

$$\frac{dx(t)}{dt} = ax(t)$$

This equation can be rewritten as

$$\frac{dx(t)}{dt} - ax(t) = 0$$

We have seen that the solution to this differential equation is

$$x(t) = x(0)e^{at} \qquad \text{for } t \geq 0$$

In the same vein as this, it can be shown that the solution to the matrix state equation

$$\frac{d\mathbf{X}(t)}{dt} = \mathbf{A}\mathbf{X}(t)$$

is

$$\mathbf{X}(t) = e^{\mathbf{A}t}\mathbf{X}(0) \qquad \text{for } t \geq 0$$

For any specific value of t and any real number a, e^{at} is also a real number. However, since $\mathbf{A}$ is a matrix, $e^{\mathbf{A}t}$ is also a matrix (called the *state-transition matrix*). It is for this reason that we write the solution of the state equation as we do—with $\mathbf{X}(0)$ multiplying

$e^{\mathbf{A}t}$ on the right—so that we have the proper ordering of the matrices for matrix multiplication.

One technique for evaluating $e^{\mathbf{A}t}$ is by using a power series expansion. In freshman calculus we learned that the Taylor series expansion of e^{at} is

$$e^{at} = 1 + at + \frac{a^2 t^2}{2!} + \frac{a^3 t^3}{3!} + \cdots$$

The analogous result for $e^{\mathbf{A}t}$ is

$$e^{\mathbf{A}t} = \mathbf{I} + \mathbf{A}t + \frac{\mathbf{A}^2 t^2}{2!} + \frac{\mathbf{A}^3 t^3}{3!} + \cdots$$

where $\mathbf{I}$ is the identity matrix.

Calculating the matrix $e^{\mathbf{A}t}$ this way by hand is simply out of the question. The amount of time and effort required is just too much. Writing a computer program, or having one available as a subroutine, that calculates $e^{\mathbf{A}t}$ is one approach for the computer solution of the state equation. However, we shall shortly discuss a simple numerical technique for solving state equations that can be readily implemented with a digital computer. Before we do, though, let us get some practice writing state equations for circuits with zero inputs.

Although the next few examples use inspection to obtain state equations, this process can be used often. Unfortunately, such a hit-and-miss approach will not be productive for some types of circuits. Toward the end of this chapter, after the reader has received some feel for writing state equations, we shall formally present the rules to follow in order to obtain state equations.

In the following examples, the rule of thumb is that we choose the inductor currents and the capacitor voltages as the state variables of a circuit.

· · ·

Example

Consider the zero-input parallel *RLC* shown in Fig. 7.2. The state variables for this circuit are i_L and v_C. We now wish, therefore, to express di_L/dt in terms of i_L and v_C.

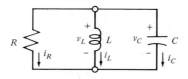

We also wish to express dv_C/dt in terms of i_L and v_C. To obtain the former, we use the fact that

$$L \frac{di_L}{dt} = v_C$$

from which

$$\frac{di_L}{dt} = \frac{1}{L} \, v_C$$

This is the first of the two state equations.

To obtain the latter, we use the fact that

$$C \frac{dv_C}{dt} = i_C$$

However, since i_C is not a state variable, we can apply KCL and write

$$C \frac{dv_C}{dt} = -i_L - i_R$$

Here i_L is a state variable, but i_R is not. But, by Ohm's law,

$$C \frac{dv_C}{dt} = -i_L - \frac{v_C}{R}$$

whence we get the second state equation

$$\frac{dv_C}{dt} = -\frac{1}{C} \, i_L - \frac{1}{RC} \, v_C$$

Expressing the two state equations as a single matrix state equation, we have

$$\frac{d}{dt}\begin{bmatrix} i_L \\ v_C \end{bmatrix} = \begin{bmatrix} \dfrac{di_L}{dt} \\ \dfrac{dv_C}{dt} \end{bmatrix} = \begin{bmatrix} 0 & \dfrac{1}{L} \\ -\dfrac{1}{C} & -\dfrac{1}{RC} \end{bmatrix}\begin{bmatrix} i_L \\ v_C \end{bmatrix}$$

which has the form

$$\frac{d\mathbf{X}(t)}{dt} = \mathbf{A}\mathbf{X}(t)$$

$$\bullet \quad \bullet \quad \bullet$$

Example
Another zero-input *RLC* circuit is shown in Fig. 7.3. For this circuit, we again have two state variables, i_L and v_C.

fig. 7.3

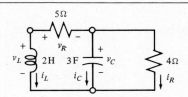

We may write

$$2\frac{di_L}{dt} = v_L$$

$$= v_R + v_C$$

$$= -5i_L + v_C$$

and thus

$$\frac{di_L}{dt} = -\frac{5}{2}i_L + \frac{1}{2}v_C$$

is our first state equation. We can obtain the second state equation as follows:

$$3\frac{dv_C}{dt} = i_C$$

$$= -i_L - i_R$$

$$= -i_L - \frac{v_C}{4}$$

and therefore

$$\frac{dv_C}{dt} = -\frac{1}{3}i_L - \frac{1}{12}v_C$$

Writing the state equations in matrix form, we have

$$\frac{d}{dt}\begin{bmatrix} i_L \\ v_C \end{bmatrix} = \begin{bmatrix} -\frac{5}{2} & \frac{1}{2} \\ -\frac{1}{3} & -\frac{1}{12} \end{bmatrix}\begin{bmatrix} i_L \\ v_C \end{bmatrix}$$

which, of course, has the form

$$\frac{d\mathbf{X}(t)}{dt} = \mathbf{AX}(t)$$

· · ·

Example

Consider the zero-input circuit in Fig. 7.4, which has three state variables: i_L, v_1 and v_2. We may now write the first state equation immediately: that is,

$$4\frac{di_L}{dt} = v_2$$

from which

$$\frac{di_L}{dt} = \frac{1}{4}v_2$$

fig. 7.4

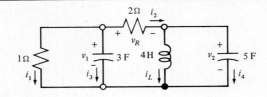

Next we have

$$3\frac{dv_1}{dt} = i_3 = -i_1 - i_2$$

$$= -\frac{v_1}{1} - \frac{v_R}{2}$$

$$= -v_1 - \frac{(v_1 - v_2)}{2}$$

$$= -\frac{3}{2}v_1 + \frac{1}{2}v_2$$

and therefore

$$\frac{dv_1}{dt} = -\frac{1}{2}v_1 + \frac{1}{6}v_2$$

which is the second state equation.

Finally,

$$5\frac{dv_2}{dt} = i_4$$

$$= -i_L + i_2$$

$$= -i_L + \frac{v_R}{2}$$

$$= -i_L + \frac{v_1 - v_2}{2}$$

and thus the third state equation is

$$\frac{dv_2}{dt} = -\frac{1}{5}i_L + \frac{1}{10}v_1 - \frac{1}{10}v_2$$

Expressing the state equations in the matrix form

$$\frac{d\mathbf{X}(t)}{dt} = \mathbf{A}\mathbf{X}(t)$$

we have

$$\frac{d}{dt}\begin{bmatrix} i_L \\ v_1 \\ v_2 \end{bmatrix} = \begin{bmatrix} 0 & 0 & \frac{1}{4} \\ 0 & -\frac{1}{2} & \frac{1}{6} \\ -\frac{1}{5} & \frac{1}{10} & -\frac{1}{10} \end{bmatrix} \begin{bmatrix} i_L \\ v_1 \\ v_2 \end{bmatrix}$$

. . .

PROBLEMS *7.1.   Write the state equation for the zero-input circuit shown in Fig. P7.1.

Ans. $\dfrac{dv}{dt} = \dfrac{-1}{(R_1/3 + R_2)C} v$

fig. P7.1

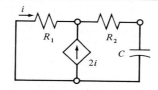

fig. P7.2

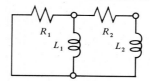

7.2.   Write the matrix state equation for the zero-input circuit in Fig. P7.2.

fig. P7.3

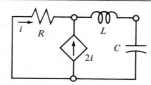

7.3.   Repeat Problem 7.2 for the circuit in Fig. P7.3.

fig. P7.4

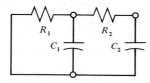

*7.4.   Repeat Problem 7.2 for the circuit in Fig. P7.4.

Ans. $\dfrac{d}{dt}\begin{bmatrix} i_L \\ v \end{bmatrix} = \begin{bmatrix} -\dfrac{R}{3L} & -\dfrac{1}{L} \\ \dfrac{1}{C} & 0 \end{bmatrix} \begin{bmatrix} i_L \\ v \end{bmatrix}$

fig. P7.5

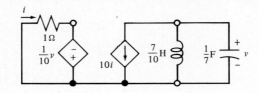

*7.5.   Repeat Problem 7.2 for the circuit in Fig. P7.5.

$$
Ans. \quad \frac{d}{dt}\begin{bmatrix} i_L \\ v \end{bmatrix} = \begin{bmatrix} 0 & \dfrac{10}{7} \\ -7 & -7 \end{bmatrix}\begin{bmatrix} i_L \\ v \end{bmatrix}
$$

fig. P7.6

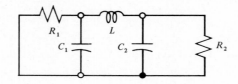

7.6.   Repeat Problem 7.2 for the circuit in Fig. P7.6.

fig. P7.7

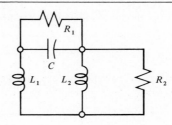

7.7.   Repeat Problem 7.2 for the circuit in Fig. P7.7.

fig. P7.8

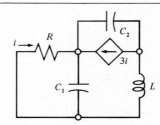

7.8.   Repeat Problem 7.2 for the circuit in Fig. P7.8.

fig. P7.9

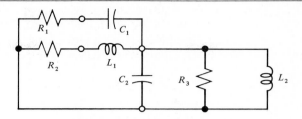

7.9.   Repeat Problem 7.2 for the circuit in Fig. P7.9.

fig. P7.10

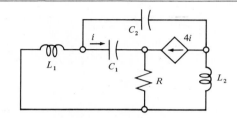

7.10.   Repeat Problem 7.2 for the circuit in Fig. P7.10.

7.2.   NUMERICAL SOLUTION OF THE STATE EQUATION

As we have mentioned before, the solution of the state equation

$$\frac{d\mathbf{X}(t)}{dt} = \mathbf{A}\mathbf{X}(t)$$

is

$$\mathbf{X}(t) = e^{\mathbf{A}t}\mathbf{X}(0) \qquad \text{for } t \geq 0$$

Clearly, if we have the capability of determining the matrix $e^{\mathbf{A}t}$ for any value of t, then it is merely a matter of multiplication of this matrix by the zero-state vector $\mathbf{X}(0)$ to obtain the corresponding state $\mathbf{X}(t)$.

Alternatively, we shall now present a numerical technique (known as *Euler's method*) that will allow us to determine the state of the circuit at any time, with any degree of accuracy, without having to evaluate the matrix $e^{\mathbf{A}t}$.

We begin by first considering the case of a single state variable $x(t)$ described by the first-order differential state equation

$$\frac{dx(t)}{dt} = Ax(t)$$

subject to the initial state $x(0)$. The extension of the procedure to be described to the case of two, three, four, or more variables is immediate and straightforward.

Whether the variable $x(t)$ describes voltage or current (or anything else, for that matter) is immaterial. Suppose that $x(t)$ describes the (one-dimensional) position of a particle. Then $dx(t)/dt$ is the velocity of the particle. In particular, the initial position $(t = 0)$ is $x(0)$ and the initial velocity is $dx(0)/dt$. From elementary physics we know that if velocity is constant, then distance is the product of velocity and time. Thus, for the case that the initial velocity is a constant, the position of the particle at time Δt is

$$x(\Delta t) = x(0) + \frac{dx(0)}{dt}\,\Delta t$$

However, in general, velocity (initial or otherwise) is not a constant. But, if we make the time interval Δt sufficiently small, the velocity during this time interval will be approximately constant.* Thus we can write the approximate formula

$$x(\Delta t) \approx x(0) + \frac{dx(0)}{dt}\,\Delta t$$

and the smaller the interval Δt, the better the approximation. From the state equation, we have

$$x(\Delta t) \approx x(0) + Ax(0)\Delta t$$

Thus we see that if we choose Δt sufficiently small, we can approximately determine the position at time $t = \Delta t$ from the initial position $x(0)$, the constant A, and the number Δt.

Once we have determined the approximate position at time $t = \Delta t$, using reasoning as above, we can determine the approximate position at time $t = 2\Delta t$ from

$$x(2\Delta t) \approx x(\Delta t) + \frac{dx(\Delta t)}{dt}\,\Delta t$$

Using the state equation, we have

$$x(2\Delta t) \approx x(\Delta t) + Ax(\Delta t)\Delta t$$

Repeating this process, we get

$$x(3\Delta t) \approx x(2\Delta t) + Ax(2\Delta t)\Delta t$$

and so on, and thus a general expression is

$$x([k + 1]\,\Delta t) \approx x(k\Delta t) + Ax(k\Delta t)\Delta t$$
$$\approx x(k\Delta t)[1 + A\Delta t]$$

for $k = 0, 1, 2, 3, \ldots.$

· · ·

Example

For a parallel RC zero-input circuit, we choose the voltage $v(t)$ across the capacitor as the state variable. We have already seen that

$$\frac{dv}{dt} + \frac{1}{RC}\,v = 0$$

---

* An exception is when $x(t)$ is discontinuous somewhere in the interval Δt. Fortunately, though, the voltage across a capacitor and the current through an inductor do not change instantaneously.

from which we obtain the state equation

$$\frac{dv}{dt} = -\frac{1}{RC}\,v = Av$$

where $A = -1/RC$. Even though we already know that the solution of this differential equation is $v(t) = v(0)e^{-t/RC}$ for $t \geq 0$, let us determine the approximate numerical solution as described above for the case that $R = 1\ \Omega$, $C = 1$ F, and $v(0) = 1$ V. We arbitrarily select $\Delta t = 0.1$ s. (In the calculations below for comparison purposes, the numbers on the right in the parentheses are the actual values of the voltage obtained from formula $v(t) = e^{-t}$.) We have the following results:

$$v(0.1) \approx v(0)[1 + A\Delta t] = 1[1 + (-1)(0.1)] = 0.9 \quad (0.905)$$

$$v(0.2) \approx v(0.1)[1 + A\Delta t] = 0.9[0.9] = 0.81 \qquad (0.819)$$

$$v(0.3) \approx v(0.2)[1 + A\Delta t] = 0.81[0.9] = 0.729 \qquad (0.741)$$

$$v(0.4) \approx v(0.3)[1 + A\Delta t] = 0.729[0.9] = 0.656 \qquad (0.670)$$

$$v(0.5) \approx v(0.4)[1 + A\Delta t] = 0.656[0.9] = 0.5904 \qquad (0.607)$$

and so on. In this example we see that the choice of $\Delta t = 0.1$ s results in values of $v(t)$ that are progressively farther from the actual value. This indicates that the choice of Δt is too large, and that the selection of $\Delta t = 0.01$ s would be better, and $\Delta t = 0.001$ s even better than that. Of course, the choice of a smaller value of Δt requires more calculations for a given interval of time—the price of greater accuracy. However, this is not a problem for a digital computer. Just remember, though, even if $\Delta t = 0.001$ s is used, you may want the computer to print out values of the voltage only every 0.1 s.

· · ·

In the above example we were able to conclude that $\Delta t = 0.1$ s was too large by comparing the numerical results with the answer! Needless to say, when applying this technique to a situation where we don't yet know the actual answer, this is what we do. Arbitrarily pick a value for Δt that seems to be small, and then let the computer print out the results. Next, pick a new value of Δt—say, 10 percent of the original value. Repeat the process (remember, the computer is doing all the work) and compare the results with those of the first run. If the values obtained from the second run do not differ significantly (less than 1 percent) from those calculated previously, then Δt is small enough. If the difference is significant, again decrease Δt by a factor of ten, and calculate a new set of numbers. Continue this routine until there is an insignificant difference between the latest two runs. At such a time, a small enough value of Δt has been obtained.

This numerical technique—*Euler's method*— is rather elementary, and consequently, not as accurate as some of the more sophisticated procedures that are available. However, such techniques are beyond the scope of this book and are left for more advanced courses in engineering or mathematics.

Although the discussion above specifically considered the case of a single state variable, a similar approach can be taken for two or more state variables.

Let us investigate the case of two state variables, $x_1(t)$ and $x_2(t)$, given that the initial-state vector is

$$\mathbf{X}(0) = \begin{bmatrix} x_1(0) \\ x_2(0) \end{bmatrix}$$

The state equation has the form

$$\frac{d}{dt}\begin{bmatrix} x_1(t) \\ x_2(t) \end{bmatrix} = \begin{bmatrix} \dfrac{dx_1(t)}{dt} \\ \dfrac{dx_2(t)}{dt} \end{bmatrix} = \begin{bmatrix} a_{11} & a_{12} \\ a_{21} & a_{22} \end{bmatrix}\begin{bmatrix} x_1(t) \\ x_2(t) \end{bmatrix}$$

If we make the time interval Δt sufficiently small, we can write the approximate formulas

$$x_1(\Delta t) \approx x_1(0) + \frac{dx_1(0)}{dt}\,\Delta t$$

$$x_2(\Delta t) \approx x_2(0) + \frac{dx_2(0)}{dt}\,\Delta t$$

Using the state equations

$$\frac{dx_1(t)}{dt} = a_{11}x_1(t) + a_{12}x_2(t)$$

$$\frac{dx_2(t)}{dt} = a_{21}x_1(t) + a_{22}x_2(t)$$

we get the approximations

$$x_1(\Delta t) \approx x_1(0) + [a_{11}x_1(0) + a_{12}x_2(0)]\,\Delta t$$
$$x_2(\Delta t) \approx x_2(0) + [a_{12}x_1(t) + a_{22}x_2(0)]\,\Delta t$$

Writing these formulas in matrix form, we have

$$\mathbf{X}(\Delta t) \approx \mathbf{X}(0) + \frac{d\mathbf{X}(0)}{dt}\,\Delta t$$

However, using the state equation

$$\frac{d\mathbf{X}(0)}{dt} = \mathbf{A}\mathbf{X}(0)$$

substituting this into the previous approximation, we obtain

$$\mathbf{X}(\Delta t) \approx \mathbf{X}(0) + \mathbf{A}\mathbf{X}(0)\Delta t$$

Hence we see that if we choose Δt sufficiently small, we can accurately determine the state at time $t = \Delta t$ from the initial state $\mathbf{X}(0)$, the matrix $\mathbf{A}$, and the number Δt.

Once we have determined the approximate state at time $t = \Delta t$, we can determine the approximate state at time $t = 2\Delta t$ and so forth. Assuming that Δt is sufficiently small and using reasoning as above, we eventually obtain the general approximations

$$x_1([k + 1]\, \Delta t) \approx x_1(k\Delta t) + \frac{dx_1(k\Delta t)}{dt}\, \Delta t$$

$$x_2([k + 1]\, \Delta t) \approx x_2(k\Delta t) + \frac{dx_2(k\Delta t)}{dt}\, \Delta t$$

Using the state equations, we get

$$x_1([k + 1]\, \Delta t) \approx x_1(k\Delta t) + [a_{11}x_1(k\Delta t) + a_{12}x_2(k\Delta t)]\, \Delta t$$
$$x_2([k + 1]\, \Delta t) \approx x_2(k\Delta t) + [a_{21}x_1(k\Delta t) + a_{22}x_2(k\Delta t)]\, \Delta t$$

Writing these formulas in matrix form, we have

$$\mathbf{X}([k + 1]\, \Delta t) \approx \mathbf{X}(k\Delta t) + \frac{d\mathbf{X}(k\Delta t)}{dt}\, \Delta t$$

which, after using the state equation, can be expressed as

$$\mathbf{X}([k + 1]\, \Delta t) \approx \mathbf{X}(k\Delta t) + \mathbf{A}\mathbf{X}(k\Delta t)\Delta t$$

for $k = 0, 1, 2, 3, \ldots$.

Although the discussion above specifically considered the case of two state variables, it should be clear that the same approach is valid for any number of state variables.

$\cdot \quad \cdot \quad \cdot$

Example
Suppose that for the circuit shown in Fig. 7.5, the initial state is $i_L(0) = 1$ A and $v_C(0) = 1$ V.

fig. 7.5

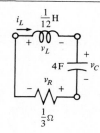

We can write the state equations as follows:

$$\frac{1}{12} \frac{di_L}{dt} = v_L$$

$$\frac{1}{12} \frac{di_L}{dt} = -v_R - v_C$$

$$\frac{1}{12} \frac{di_L}{dt} = -\frac{1}{3} i_L - v_C$$

or

$$\frac{di_L}{dt} = -4i_L - 12v_C$$

and

$$4 \frac{dv_C}{dt} = i_L$$

or

$$\frac{dv_C}{dt} = \frac{1}{4} i_L$$

The state equation in the form

$$\frac{d\mathbf{X}(t)}{dt} = \mathbf{A}\mathbf{X}(t)$$

is

$$\begin{bmatrix} \dfrac{di_L}{dt} \\[2ex] \dfrac{dv_C}{dt} \end{bmatrix} = \frac{d}{dt} \begin{bmatrix} i_L \\[2ex] v_C \end{bmatrix} = \begin{bmatrix} -4 & -12 \\[2ex] \frac{1}{4} & 0 \end{bmatrix} \begin{bmatrix} i_L \\[2ex] v_C \end{bmatrix}$$

Suppose we select $\Delta t = 0.1$ s. From the initial state, we can now calculate the approximate state at time $t = 0.1$ s. We may either use the following two formulas

$$i_L(\Delta t) \approx i_L(0) + \frac{di_L(0)}{dt} \Delta t$$

and

$$v_C(\Delta t) \approx v_C(0) + \frac{dv_C(0)}{dt} \Delta t$$

from which

$$i_L(\Delta t) \approx i_L(0) + [-4i_L(0) - 12v_C(0)]\, \Delta t$$

$$i_L(0.1) \approx 1 + [-4 - 12](0.1) = -0.6$$

and

$$v_C(\Delta t) \approx v_C(0) + \left[\frac{1}{4} i_L(0)\right]\Delta t$$

$$v_C(0.1) \approx 1 + \frac{1}{4}(0.1) = 1.025$$

or the corresponding matrix form

$$\begin{bmatrix} i_L(\Delta t) \\ v_C(\Delta t) \end{bmatrix} \approx \begin{bmatrix} i_L(0) \\ v_C(0) \end{bmatrix} + \begin{bmatrix} -4 & -12 \\ \frac{1}{4} & 0 \end{bmatrix} \begin{bmatrix} i_L(0) \\ v_C(0) \end{bmatrix} \Delta t$$

from which

$$\begin{bmatrix} i_L(0.1) \\ v_C(0.1) \end{bmatrix} \approx \begin{bmatrix} 1 \\ 1 \end{bmatrix} + \begin{bmatrix} -4 & -12 \\ \frac{1}{4} & 0 \end{bmatrix} \begin{bmatrix} 1 \\ 1 \end{bmatrix} (0.1)$$

$$\approx \begin{bmatrix} 1 \\ 1 \end{bmatrix} + \begin{bmatrix} -16 \\ \frac{1}{4} \end{bmatrix} (0.1) = \begin{bmatrix} 1 \\ 1 \end{bmatrix} + \begin{bmatrix} -1.6 \\ 0.025 \end{bmatrix}$$

$$\approx \begin{bmatrix} -0.6 \\ 1.025 \end{bmatrix}$$

In a similar manner we can determine the approximate state at time $t = 0.2$ s by using

$$i_L(2\Delta t) \approx i_L(\Delta t) + \frac{di_L(\Delta t)}{dt} \Delta t$$

and

$$v_C(2\Delta t) \approx v_C(\Delta t) + \frac{dv_C(\Delta t)}{dt} \Delta t$$

from which

$$i_L(2\Delta t) \approx i_L(\Delta t) + [-4i_L(\Delta t) - 12v_C(\Delta t)]\, \Delta t$$

$$i_L(0.2) \approx -0.6 + [-4(-0.6) - 12(1.025)](0.1) = -1.59$$

and

$$v_C(2\Delta t) \approx v_C(\Delta t) + \left[\frac{1}{4} i_L(\Delta t)\right]\Delta t$$

$$v_C(0.2) \approx 1.025 + \left[\frac{1}{4}(-0.6)\right](0.1) = 1.01$$

or by using

$$\begin{bmatrix} i_L(2\Delta t) \\ v_C(2\Delta t) \end{bmatrix} \approx \begin{bmatrix} i_L(\Delta t) \\ v_C(\Delta t) \end{bmatrix} + \begin{bmatrix} -4 & -12 \\ \frac{1}{4} & 0 \end{bmatrix} \begin{bmatrix} i_L(\Delta t) \\ v_C(\Delta t) \end{bmatrix} \Delta t$$

$$\begin{bmatrix} i_L(0.2) \\ v_C(0.2) \end{bmatrix} \approx \begin{bmatrix} -0.6 \\ 1.025 \end{bmatrix} + \begin{bmatrix} -4 & -12 \\ \frac{1}{4} & 0 \end{bmatrix} \begin{bmatrix} -0.6 \\ 1.025 \end{bmatrix} (0.1)$$

$$\approx \begin{bmatrix} -0.6 \\ 1.025 \end{bmatrix} + \begin{bmatrix} -9.9 \\ -0.15 \end{bmatrix} (0.1) = \begin{bmatrix} -0.6 \\ 1.025 \end{bmatrix} + \begin{bmatrix} -0.99 \\ -0.015 \end{bmatrix}$$

$$\approx \begin{bmatrix} -1.59 \\ 1.01 \end{bmatrix}$$

Continuing in this manner, we can calculate approximate values for

$$\begin{bmatrix} i_L(0.3) \\ v_C(0.3) \end{bmatrix} \qquad \begin{bmatrix} i_L(0.4) \\ v_C(0.4) \end{bmatrix}$$

and so forth.

· · ·

Example
The zero-input circuit shown in Fig. 7.6 has three state variables. It is a simple matter to obtain the state equations

$$\frac{di_1}{dt} = -i_1 - v$$

$$\frac{di_2}{dt} = -i_2 + v$$

$$\frac{dv}{dt} = \frac{1}{2}(i_1 - i_2)$$

fig. 7.6

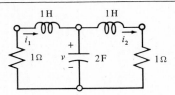

Figure 7.7 shows a FORTRAN computer program in which the current $i_2(t)$ is printed out every 0.1 s from $t = 0$ to $t = 10$ s. The initial conditions were chosen arbitrarily to be $i_1(0) = 0$ A, $i_2(0) = 1$ A, and $v(0) = -2$ V. Lines 14, 15 and 16 are the approximations used to obtain the states at time $(k + 1)\Delta t$ from the states at time $k\Delta t$. The value used for the time increment is $\Delta t = 0.001$ s. (Using

fig. 7.7

```
00001 PROGRAM SVAP(INPUT,OUTPUT,TAPE6=OUTPUT)
00002 REAL I1,I2,I1K,I2K
00003 WRITE(6,30)
00004 30 FORMAT(6X,*TIME*,4X,*CURRENT*)
00005 X=0.
00006 V=-2.
00007 I1=0.
00008 I2=1.
00009 T=0.
00010 DELT=0.001
00011 40 WRITE(6,50) T,I2
00012 50 FORMAT(3X,F7.2,5X,F5.2)
00013 DO 60 J=1,100
00014    VK=V+0.5*(I1-I2)*DELT
00015    I1K=I1+(-I1-V)*DELT
00016    I2K=I2+(-I2+V)*DELT
00017    V=VK
00018    I1=I1K
00019    I2=I2K
00020 60 T=T+DELT
00021 X=X+1.
00022 IF(100.-X)70,40,40
00023 70 CONTINUE
00024 END
```

$\Delta t = 0.0001$ s yields the same results.) A computer printout and plot of $i_2(t)$ versus t are given in Fig. 7.8. (See page 348). The computer plot was done by a subroutine that is not a part of the program shown.

· · ·

PROBLEMS   7.11.   For the simple RC circuit shown in Fig. P7.11, use Euler's method, with $\Delta t = 0.001$ s, to determine a sufficient number of values to plot $v(t)$ in the interval

(a)   0–1 s, given that $R = 4\ \Omega$ and $v(0) = 8$ V.

(b)   1–3 s, given that $R = 2.4\ \Omega$ and $v(1) = 2.943$ V.

fig. P7.11

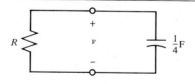

fig. P7.12

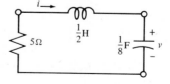

7.12.   For the series RLC circuit in Fig. P7.12 given the initial conditions $i(0) = 1$ A and $v(0) = 2$ V, use Euler's method, with $\Delta t = 0.001$ s, to plot $i(t)$ and $v(t)$ in the interval 0–2 s.

fig. 7.8

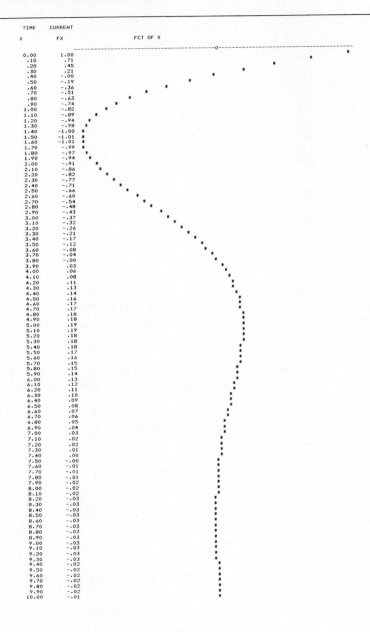

fig. P7.13

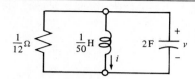

7.13. Repeat Problem 7.12 for the parallel *RLC* circuit in Fig. P7.13 subject to the initial conditions $i(0) = 1$ A and $v(0) = -0.14$ V.

fig. P7.14

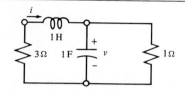

7.14. For the series-parallel *RLC* circuit in Fig. P7.14, given the initial conditions $i(0) = 1$ A and $v(0) = 2$ V, use Euler's method, with $\Delta t = 0.001$ s, to plot $i(t)$ and $v(t)$ in the interval 0–2.5 s.

fig. P7.15

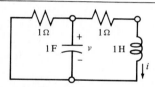

7.15. For the circuit shown in Fig. P7.15, repeat Problem 7.14 for the interval 0–4 s, and subject to the initial conditions $i(0) = 1$ A and $v(0) = 1$ V.

7.3 CIRCUITS WITH NONZERO INPUTS

Having already considered circuits with no input, our next step is to consider circuits having nonzero inputs.

Let us again look at the series *RLC* circuit, which is shown in Fig. 7.9. We have already analyzed this circuit for the case when $v_s(t)$ is a step function by writing a second-order differential equation in the variable v_C and solving it. Now let us see how to extend the ideas of state-variable analysis to deal with circuits of this type—that is, circuits with nonzero inputs.

fig. 7.9

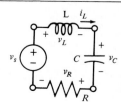

Again we choose our state variables to be the inductor current i_L and the capacitor voltage v_C. We can write

$$L \frac{di_L}{dt} = v_L$$

and, by KVL,

$$L \frac{di_L}{dt} = -v_R - v_C + v_s$$

$$= -Ri_L - v_C + v_s$$

so that

$$\frac{di_L}{dt} = -\frac{R}{L} i_L - \frac{1}{L} v_C + \frac{1}{L} v_s$$

In this case we have expressed the derivative of one of the state variables in terms of both state variables and the input. (The input is in general a given time-varying quantity, not a circuit variable.)

We also have that

$$C \frac{dv_C}{dt} = i_L$$

from which

$$\frac{dv_C}{dt} = \frac{1}{C} i_L$$

We have seen that when $v_s = 0$, we can write the above state equations in the form

$$\frac{d\mathbf{X}(t)}{dt} = \mathbf{A}\mathbf{X}(t)$$

However, since $v_s \neq 0$ in general, in order to write the state equations in matrix form, we must include another term in the matrix expression.

For the state equations given above, we can write

$$\frac{d}{dt} \begin{bmatrix} i_L \\ v_C \end{bmatrix} = \begin{bmatrix} \dfrac{di_L}{dt} \\ \dfrac{dv_C}{dt} \end{bmatrix} = \begin{bmatrix} -\dfrac{R}{L} & -\dfrac{1}{L} \\ \dfrac{1}{C} & 0 \end{bmatrix} \begin{bmatrix} i_L \\ v_C \end{bmatrix} + \begin{bmatrix} \dfrac{1}{L} \\ 0 \end{bmatrix} v_s$$

In this case we added the product of a matrix (whose entries are scalars) and the input such that the resulting additional terms yield the proper state equations.

The form of the above matrix state equation is

$$\frac{d\mathbf{X}(t)}{dt} = \mathbf{A}\mathbf{X}(t) + \mathbf{B}w(t)$$

where $\mathbf{X}(t)$ is the state vector and $w(t)$ results from the input. This form is for the case of a single input; we shall discuss the case of multiple inputs shortly.

The above matrix state equation is a linear, nonhomogeneous, first-order matrix differential equation. Although its closed-form solution is more complicated than that for the homogeneous case, the numerical solution is essentially the same. We again use the fact that for sufficiently small Δt, we get the approximation

$$\mathbf{X}(\Delta t) \approx \mathbf{X}(0) + \frac{d\mathbf{X}(0)}{dt} \Delta t$$

Using the fact that

$$\frac{d\mathbf{X}(t)}{dt} = \mathbf{A}\mathbf{X}(t) + \mathbf{B}w(t)$$

the approximate formula becomes

$$\mathbf{X}(\Delta t) \approx \mathbf{X}(0) + [\mathbf{A}\mathbf{X}(0) + \mathbf{B}w(0)] \Delta t$$

or

$$\mathbf{X}(\Delta t) \approx \mathbf{X}(0) + \mathbf{A}\mathbf{X}(0)\Delta t + \mathbf{B}w(0)\Delta t$$

and proceeding in the manner as we did for the zero-input response, we get the general form

$$\mathbf{X}([k + 1] \Delta t) \approx \mathbf{X}(k\Delta t) + \mathbf{A}\mathbf{X}(k\Delta t)\Delta t + \mathbf{B}w(k\Delta t)\Delta t$$

for $k = 0, 1, 2, \ldots.$

Note that the first two terms in the sum on the right-hand side of the approximation, that is,

$$\mathbf{X}(k\Delta t) + \mathbf{A}\mathbf{X}(k\Delta t)\Delta t$$

constitute the expression which yields the zero-input response of the circuit. It is the third term,

$$\mathbf{B}w(k\Delta t)\Delta t$$

that contributes the part of the response due to the input.

Furthermore, note that when the initial-state vector is zero, the use of the above approximation results in the zero-state response. Otherwise the complete response is obtained.

For the case when the input is a unit step function [i.e., when $w(t) = u(t)$], since $u(t) = 1$ for $t \geq 0$, the above approximation becomes

$$\mathbf{X}([k + 1] \Delta t) \approx \mathbf{X}(k\Delta t) + \mathbf{A}\mathbf{X}(k\Delta t)\Delta t + \mathbf{B}\Delta t$$

$$\bullet \quad \bullet \quad \bullet$$

Example

Let us write a matrix state equation of the form

$$\frac{d\mathbf{X}}{dt} = \mathbf{A}\mathbf{X} + \mathbf{B}w$$

for the circuit in Fig. 7.10.

fig. 7.10

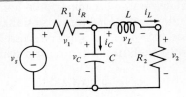

For this circuit we have that

$$L \frac{di_L}{dt} = v_L$$

$$= -v_2 + v_C$$

$$= -R_2 i_L + v_C$$

so that

$$\frac{di_L}{dt} = -\frac{R_2}{L} i_L + \frac{1}{L} v_C$$

Furthermore,

$$C \frac{dv_C}{dt} = i_C$$

$$= -i_L + i_R$$

$$= -i_L + \frac{v_1}{R_1}$$

$$= -i_L + \frac{v_s - v_C}{R_1}$$

$$= -i_L - \frac{v_C}{R_1} + \frac{v_s}{R_1}$$

from which

$$\frac{dv_C}{dt} = -\frac{1}{C} i_L - \frac{1}{R_1 C} v_C + \frac{1}{R_1 C} v_s$$

We can write these state equations in matrix form as

$$\frac{d}{dt} \begin{bmatrix} i_L \\ \\ v_C \end{bmatrix} = \begin{bmatrix} \dfrac{di_L}{dt} \\ \\ \dfrac{dv_C}{dt} \end{bmatrix} = \begin{bmatrix} -\dfrac{R_2}{L} & \dfrac{1}{L} \\ \\ -\dfrac{1}{C} & -\dfrac{1}{R_1 C} \end{bmatrix} \begin{bmatrix} i_L \\ \\ v_C \end{bmatrix} + \begin{bmatrix} 0 \\ \\ \dfrac{1}{R_1 C} \end{bmatrix} v_s$$

For this circuit, we have the approximations

$$i_L([k + 1]\,\Delta t) \approx i_L(k\Delta t) + \frac{di_L(k\Delta t)}{dt}\,\Delta t$$

$$\approx i_L(k\Delta t) + \left[-\frac{R_2}{L}\,i_L(k\Delta t) + \frac{1}{L}\,v_C(k\Delta t)\right]\Delta t$$

and

$$v_C([k + 1]\,\Delta t) \approx v_C(k\Delta t) + \frac{dv_C(k\Delta t)}{dt}\,\Delta t$$

$$\approx v_C(k\Delta t) + \left[-\frac{1}{C}\,i_L(k\Delta t) - \frac{1}{R_1 C}\,v_C(k\Delta t) + \frac{1}{R_1 C}\,v_s(k\Delta t)\right]\Delta t$$

or, in matrix form,

$$\begin{bmatrix} i_L([k + 1]\,\Delta t) \\ v_C([k + 1]\,\Delta t) \end{bmatrix} \approx \begin{bmatrix} i_L(k\Delta t) \\ v_C(k\Delta t) \end{bmatrix} + \frac{d}{dt}\begin{bmatrix} i_L(k\Delta t) \\ v_C(k\Delta t) \end{bmatrix}\Delta t$$

$$\approx \begin{bmatrix} i_L(k\Delta t) \\ \\ v_C(k\Delta t) \end{bmatrix} + \begin{bmatrix} -\dfrac{R_2}{L} & \dfrac{1}{L} \\ \\ -\dfrac{1}{C} & -\dfrac{1}{R_1 C} \end{bmatrix}\begin{bmatrix} i_L(k\Delta t) \\ \\ v_C(k\Delta t) \end{bmatrix}\Delta t$$

$$+ \begin{bmatrix} 0 \\ \\ \dfrac{1}{R_1 C} \end{bmatrix} v_s(k\Delta t)\Delta t$$

· · ·

Example

Let us now consider a circuit in which there is a dependent source as well as an independent source, as shown in Fig. 7.11.

We may now write

$$L\,\frac{di_L}{dt} = v_L$$

$$= -v_R - v_C + v_s$$

$$= -Ri - v_C + v_s$$

fig. 7.11

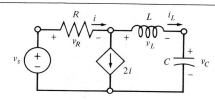

However, by KCL,

$$i = 2i + i_L \qquad \text{or} \qquad -i = i_L$$

Thus

$$L \frac{di_L}{dt} = Ri_L - v_C + v_s$$

from which

$$\frac{di_L}{dt} = \frac{R}{L} i_L - \frac{1}{L} v_C + \frac{1}{L} v_s$$

In addition,

$$C \frac{dv_C}{dt} = i_L$$

so that

$$\frac{dv_C}{dt} = \frac{1}{C} i_L$$

Expressing the state equations in matrix form, we have

$$\begin{bmatrix} \dfrac{di_L}{dt} \\[2ex] \dfrac{dv_C}{dt} \end{bmatrix} = \begin{bmatrix} \dfrac{R}{L} & -\dfrac{1}{L} \\[2ex] \dfrac{1}{C} & 0 \end{bmatrix} \begin{bmatrix} i_L \\[2ex] v_C \end{bmatrix} + \begin{bmatrix} \dfrac{1}{L} \\[2ex] 0 \end{bmatrix} v_s$$

$$\cdot \quad \cdot \quad \cdot$$

Example
The circuit in Fig. 7.12 has a single input and three state variables: i_L, v_1 and v_2.

fig. 7.12

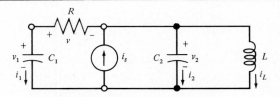

For this circuit we can write

$$L \frac{di_L}{dt} = v_2 \qquad \text{or} \qquad \frac{di_L}{dt} = \frac{1}{L} v_2$$

Also,

$$C_1 \frac{dv_1}{dt} = i_1 = -\frac{v}{R} = -\frac{1}{R}(v_1 - v_2)$$

so that

$$\frac{dv_1}{dt} = -\frac{1}{RC_1} v_1 + \frac{1}{RC_1} v_2$$

Furthermore,

$$C_2 \frac{dv_2}{dt} = i_2$$

$$= -i_L - i_1 + i_s$$

$$= -i_L + \frac{v}{R} + i_s$$

$$= -i_L + \frac{v_1 - v_2}{R} + i_s$$

and thus

$$\frac{dv_2}{dt} = -\frac{1}{C_2} i_L + \frac{1}{RC_2} v_1 - \frac{1}{RC_2} v_2 + \frac{1}{C_2} i_s$$

Writing these equations in matrix form, we have

$$\begin{bmatrix} \dfrac{di_L}{dt} \\[2mm] \dfrac{dv_1}{dt} \\[2mm] \dfrac{dv_2}{dt} \end{bmatrix} = \begin{bmatrix} 0 & 0 & \dfrac{1}{L} \\[2mm] 0 & -\dfrac{1}{RC_1} & \dfrac{1}{RC_1} \\[2mm] -\dfrac{1}{C_2} & \dfrac{1}{RC_2} & -\dfrac{1}{RC_2} \end{bmatrix} \begin{bmatrix} i_L \\[2mm] v_1 \\[2mm] v_2 \end{bmatrix} + \begin{bmatrix} 0 \\[2mm] 0 \\[2mm] \dfrac{1}{C_2} \end{bmatrix} i_s$$

$$\cdot \ \cdot \ \cdot$$

Now let us consider the situation in which a circuit has more than one input. We have seen that when we have more than one state variable, we utilize a state vector. Similarly, when we have more than one independent source, we may employ a vector due to the inputs. We then can write a matrix state equation of the form

$$\frac{d\mathbf{X}}{dt} = \mathbf{AX} + \mathbf{BW}$$

where $\mathbf{W}$ is a vector due to the inputs and $\mathbf{B}$ is a matrix whose dimension conforms to the equation. Specifically, we know that if there are n state variables, then $\mathbf{A}$ is an $n \times n$ matrix. Now, if there are m inputs, then $\mathbf{B}$ is an $n \times m$ matrix.

$\cdot \quad \cdot \quad \cdot$

Example

The circuit in Fig. 7.13 has two inputs and two state variables.

fig. 7.13

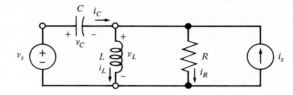

To begin with,

$$L \frac{di_L}{dt} = v_L = -v_C + v_s \qquad \text{or} \qquad \frac{di_L}{dt} = -\frac{1}{L} v_C + \frac{1}{L} v_s$$

Also,

$$C \frac{dv_C}{dt} = i_C$$

$$= i_L + \frac{v_L}{R} - i_s$$

$$= i_L + \frac{-v_C + v_s}{R} - i_s$$

and hence

$$\frac{dv_C}{dt} = \frac{1}{C} i_L - \frac{1}{RC} v_C + \frac{1}{RC} v_s - \frac{1}{C} i_s$$

The state equations can be written in the form

$$\frac{d\mathbf{X}}{dt} = \mathbf{AX} + \mathbf{BW}$$

as follows:

$$\frac{d}{dt} \begin{bmatrix} i_L \\ v_C \end{bmatrix} = \begin{bmatrix} \dfrac{di_L}{dt} \\ \dfrac{dv_C}{dt} \end{bmatrix} = \begin{bmatrix} 0 & -\dfrac{1}{L} \\ \dfrac{1}{C} & -\dfrac{1}{RC} \end{bmatrix} \begin{bmatrix} i_L \\ v_C \end{bmatrix} + \begin{bmatrix} \dfrac{1}{L} & 0 \\ \dfrac{1}{RC} & -\dfrac{1}{C} \end{bmatrix} \begin{bmatrix} v_s \\ i_s \end{bmatrix}$$

$\cdot \quad \cdot \quad \cdot$

We can obtain a numerical solution to the state equation

$$\frac{d\mathbf{X}}{dt} = \mathbf{AX} + \mathbf{BW}$$

just as we did for the case of a single input by using the approximation

$$\mathbf{X}([k + 1]\,\Delta t) \approx \mathbf{X}(k\Delta t) + \mathbf{A}\mathbf{X}(k\Delta t)\Delta t + \mathbf{B}\mathbf{W}(k\Delta t)\Delta t$$

for $k = 0, 1, 2, \dots$.

• • •

Example

The zero-state circuit shown in Fig. 7.14 has three state variables.

fig. 7.14

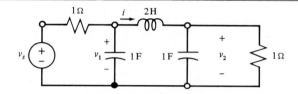

The following state equations can easily be obtained (see Problem 7.28):

$$\frac{di(t)}{dt} = \frac{1}{2}(v_1 - v_2)$$

$$\frac{dv_1(t)}{dt} = -i - v_1 + v_s$$

$$\frac{dv_2(t)}{dt} = i - v_2$$

A FORTRAN program for the case $v_s(t) = u(t)$ is given in Fig. 7.15, in which the voltage $v_2(t)$ is printed out every 0.1 s from $t = 0$ to $t = 10$ s. Lines 15, 16 and

fig. 7.15

```
00001 PROGRAM SVAPN(INPUT,OUTPUT,TAPE6=OUTPUT)
00002 REAL I,IK
00003 WRITE(6,30)
00004 30 FORMAT(6X,*TIME*,4X,*VOLTAGE*)
00005 X=0.
00006 I=0.
00007 V1=0.
00008 V2=0.
00009 T=0.
00010 DELT=0.001
00011 40 WRITE(6,50) T,V2
00012 50 FORMAT(3X,F7.2,5X,F5.2)
00013 DO 60 J=1,100
00014    VS=1.
00015    IK=I+0.5*(V1-V2)*DELT
00016    V1K=V1+(VS-I-V1)*DELT
00017    V2K=V2+(I-V2)*DELT
00018    I=IK
00019    V1=V1K
00020    V2=V2K
00021 60 T=T+DELT
00022 X=X+1.
00023 IF(100.-X)70,40,40
00024 70 CONTINUE
00025 END
```

17 are the approximations used to obtain the states at time $(k + 1)\Delta t$ from the states at time $k\Delta t$. The value used for the time increment is $\Delta t = 0.001$ s. (Using $\Delta t = 0.0001$ s yields the same results.) A printout and computer plot of $v_2(t)$ versus t are shown in Fig. 7.16.

fig. 7.16

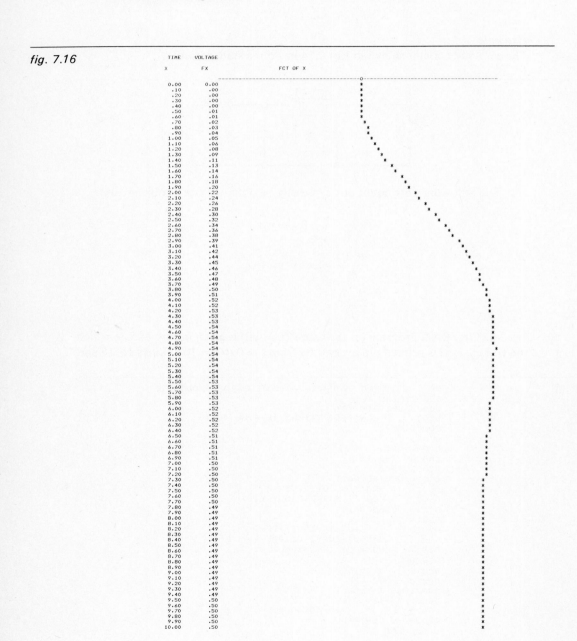

fig. 7.17

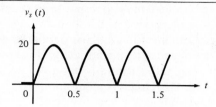

fig. 7.18

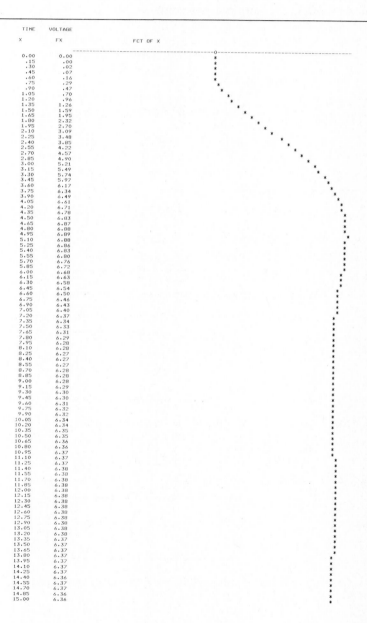

TIME	VOLTAGE	
X	FX	FCT OF X
0.00	0.00	
.15	.00	
.30	.02	
.45	.07	
.60	.16	
.75	.29	
.90	.47	
1.05	.70	
1.20	.96	
1.35	1.26	
1.50	1.59	
1.65	1.95	
1.80	2.32	
1.95	2.70	
2.10	3.09	
2.25	3.48	
2.40	3.85	
2.55	4.22	
2.70	4.57	
2.85	4.90	
3.00	5.21	
3.15	5.49	
3.30	5.74	
3.45	5.97	
3.60	6.17	
3.75	6.34	
3.90	6.49	
4.05	6.61	
4.20	6.71	
4.35	6.78	
4.50	6.83	
4.65	6.87	
4.80	6.88	
4.95	6.89	
5.10	6.88	
5.25	6.86	
5.40	6.83	
5.55	6.80	
5.70	6.76	
5.85	6.72	
6.00	6.68	
6.15	6.63	
6.30	6.58	
6.45	6.54	
6.60	6.50	
6.75	6.46	
6.90	6.43	
7.05	6.40	
7.20	6.37	
7.35	6.34	
7.50	6.33	
7.65	6.31	
7.80	6.29	
7.95	6.28	
8.10	6.28	
8.25	6.27	
8.40	6.27	
8.55	6.27	
8.70	6.28	
8.85	6.28	
9.00	6.28	
9.15	6.29	
9.30	6.30	
9.45	6.30	
9.60	6.31	
9.75	6.32	
9.90	6.32	
10.05	6.34	
10.20	6.34	
10.35	6.35	
10.50	6.35	
10.65	6.36	
10.80	6.36	
10.95	6.37	
11.10	6.37	
11.25	6.37	
11.40	6.38	
11.55	6.38	
11.70	6.38	
11.85	6.38	
12.00	6.38	
12.15	6.38	
12.30	6.38	
12.45	6.38	
12.60	6.38	
12.75	6.38	
12.90	6.38	
13.05	6.38	
13.20	6.38	
13.35	6.37	
13.50	6.37	
13.65	6.37	
13.80	6.37	
13.95	6.37	
14.10	6.37	
14.25	6.37	
14.40	6.36	
14.55	6.37	
14.70	6.37	
14.85	6.36	
15.00	6.36	

Having numerically determined the step response for the circuit given in Fig. 7.14, let us determine the zero-state response to the input voltage $v_s(t) = |20 \sin 2\pi t|u(t)$ shown in Fig. 7.17. The only modification to the program needed is to change the line corresponding to the input voltage $v_s(t)$ appropriately. However, for aesthetic reasons, the voltage is printed out every 0.15 s from $t = 0$ to $t = 15$ s. The computer printout and plot are given in Fig. 7.18. This circuit configuration is often employed as a "power supply filter," a network whose input is a "rectified" sine wave specified by $v_s(t)$ in Fig. 7.17 and whose output $v_2(t)$ is very nearly a constant.

PROBLEMS *7.16. Write the matrix state equation for the circuit shown in Fig. P7.16.

Ans.
$$\frac{d}{dt}\begin{bmatrix} i \\ v \end{bmatrix} = \begin{bmatrix} 0 & -\tfrac{1}{2} \\ 4 & -1 \end{bmatrix}\begin{bmatrix} i \\ v \end{bmatrix} + \begin{bmatrix} \tfrac{1}{2} \\ 0 \end{bmatrix}u(t)$$

fig. P7.16

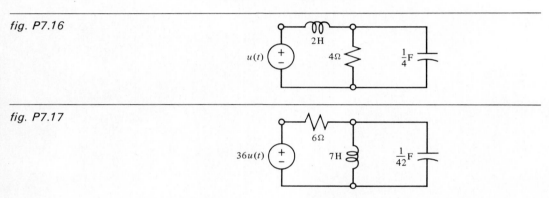

fig. P7.17

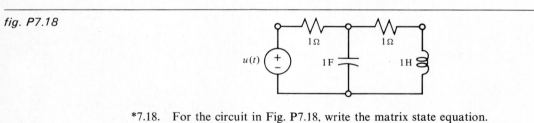

7.17. Write the matrix state equation for the circuit shown in Fig. P7.17.

fig. P7.18

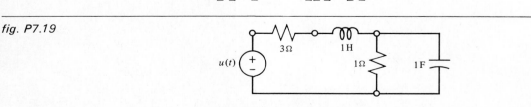

*7.18. For the circuit in Fig. P7.18, write the matrix state equation.

Ans.
$$\frac{d}{dt}\begin{bmatrix} i \\ v \end{bmatrix} = \begin{bmatrix} -1 & 1 \\ -1 & -1 \end{bmatrix}\begin{bmatrix} i \\ v \end{bmatrix} + \begin{bmatrix} 0 \\ 1 \end{bmatrix}u(t)$$

fig. P7.19

7.19. For the circuit in Fig. P7.19, write the matrix state equation.

fig. P7.20

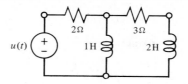

7.20.   Write the matrix state equation for the circuit in Fig. P7.20.

fig. P7.21

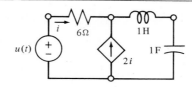

*7.21.   Write the matrix state equation for the circuit shown in Fig. P7.21.

$$Ans. \quad \frac{d}{dt}\begin{bmatrix} i_L \\ v \end{bmatrix} = \begin{bmatrix} -2 & -1 \\ 1 & 0 \end{bmatrix}\begin{bmatrix} i_L \\ v \end{bmatrix} + \begin{bmatrix} 1 \\ 0 \end{bmatrix}u(t)$$

fig. P7.22

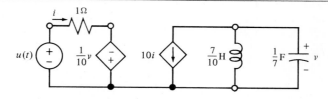

7.22.   For the circuit in Fig. P7.22, write the state equation.

7.23.   For the circuit given in Problem 6.15 at the end of Chapter 6, write the matrix state equation.

7.24.   For the circuit given in Problem 6.16, write the matrix state equation.

*7.25.   For the circuit given in Problem 6.17, write the matrix state equation.

$$Ans. \quad \frac{d}{dt}\begin{bmatrix} v_1 \\ v_2 \end{bmatrix} = \begin{bmatrix} 0 & 1 \\ -2 & -3 \end{bmatrix}\begin{bmatrix} v_1 \\ v_2 \end{bmatrix} + \begin{bmatrix} 0 \\ 2 \end{bmatrix}v_s$$

7.26.   For the circuit given in Problem 6.18, write the matrix state equation.

7.27.   For the circuit given in Problem 6.20, write the matrix state equation.

7.28.   Write the matrix state equation for the circuit shown in Fig. 7.14.

7.29.   Use Euler's method to find the zero-state response of a series RLC circuit ($R = 12\ \Omega$, $L = 2$ H, $C = \frac{1}{50}$ F) whose input voltage is $0.4u(t)$. Use $\Delta t = 0.001$ s and plot $i_L(t)$ and $v_C(t)$ up to $t = 2$ s.

7.30.   For the circuit given in Problem 7.16, find the zero-state step responses $i_L(t)$ and $v_C(t)$ using Euler's method. Use $\Delta t = 0.001$ s and plot $i_L(t)$ and $v_C(t)$ up to $t = 10$ s.

7.31.   For the circuit given in Problem 7.17, use Euler's method to find the inductor current $i_L(t)$ (directed downward) and the capacitor voltage $v(t)$ (plus on the top plate) given that $i_L(0) = -4$ A and $v(0) = 0$. Use $\Delta t = 0.001$ s and plot $i_L(t)$ and $v_C(t)$ up to $t = 5$ s.

7.32.   Use Euler's method to find the zero-state step responses $i_L(t)$ and $v_C(t)$ for the circuit given in Problem 7.18. Use $\Delta t = 0.001$ s and plot $i_L(t)$ and $v_C(t)$ up to $t = 4$ s.

7.33.   Repeat Problem 7.32 for the circuit given in Problem 7.21.

7.34.   For the circuit given in Problem 7.22, find the zero-state step response $v(t)$ using Euler's method.

7.35.   Repeat Problem 7.34 for the case where the input voltage is $|10 \sin 2\pi t| u(t)$.

7.4   RULES FOR WRITING STATE EQUATIONS

Having seen how state equations were written for numerous examples, we should now have a little feeling for this procedure. However, the judicious choice of circuits was responsible for the relative ease in which the results were obtained. There are some situations, though, that are more subtle. One difficulty arises when we cannot express the voltage across a resistor as a sum of capacitor and independent source voltages, and we also cannot express the current through that resistor as a sum of inductor and independent source currents. Thus, it is now time—as was promised earlier—to give a formal set of rules for writing state equations.

Before we do, however, we need to define another graph-theoretical term. Given a connected graph, a set of edges whose removal disconnects some of the nodes is called a *cut-set* provided that no portion of this set (called a *subset*) of edges will also disconnect some of the nodes.

· · ·

Example

For the connected graph shown in Fig. 7.19, the set of edges $\{e_2, e_3, e_5\}$ is a cut-set since the removal of these edges disconnects node b from the other three nodes, whereas removal of any two of the three edges will not. Moreover, $\{e_1, e_3, e_4, e_5\}$ is also a cut-set since the removal of these edges disconnects nodes a and b from nodes c and d. The set of edges $\{e_1, e_2, e_4, e_5\}$ is not a cut-set since a subset of it, specifically $\{e_1, e_2, e_4\}$, disconnects node a from the other three nodes.

fig. 7.19

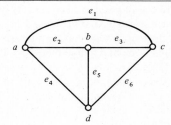

For each cut-set there corresponds a node or a supernode that is obtained by passing a closed curve through the edges comprising the cut-set. Shown in Fig. 7.20 are the supernodes (indicated by dashed lines) associated with the cut-sets

fig. 7.20

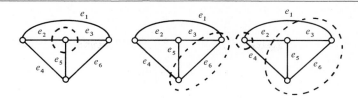

$\{e_2, e_3, e_5\}$, $\{e_1, e_3, e_4, e_5\}$, and $\{e_1, e_2, e_4\}$. In the last case, two apparently different situations are depicted. However, as far as KCL is concerned, they are equivalent.

$\cdots$

Given a connected graph and a spanning tree, for each branch of the tree there is a unique cut-set consisting of that tree branch and nonbranch edges. Such a cut-set is called a *fundamental cut-set* (with respect to that tree).

$\cdots$

Example
In the graph in Fig. 7.21 the spanning tree is indicated by bold edges. The fundamental cut-set corresponding to branch e_3 is $\{e_1, e_3, e_6\}$. For branch e_4 it is $\{e_1, e_2, e_4\}$, and for branch e_5 it is $\{e_1, e_2, e_5, e_6\}$.

fig. 7.21

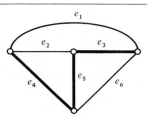

$\cdots$

With these new definitions, it is now possible to give a formal procedure for obtaining state equations:

Step 1. Pick a spanning tree such that voltage sources and capacitors correspond to branches, whereas current sources and inductors correspond to nonbranch edges. Furthermore, if possible, an element whose voltage controls a dependent source should correspond to a branch, and one whose current controls a dependent source should correspond to a nonbranch edge. More than one such tree may exist—or none at all. (An example of the latter will be given later.)

Step 2. Arbitrarily assign a voltage to each tree branch capacitor and a current to each nonbranch inductor; these are the state variables. If possible, express the voltage across each element corresponding to a branch and the current through each element corresponding to a nonbranch in terms of voltage sources, current sources, and state variables. If it is not possible, assign a new voltage variable to a resistor corresponding to a branch and new current variable to a resistor corresponding to a nonbranch.

Step 3. Apply KVL to the fundamental loop determined by each nonbranch inductor.

Step 4. Apply KCL to the node or supernode corresponding to the fundamental cut-set determined by each branch capacitor.

Step 5. Apply KVL to the fundamental loop determined by each resistor with a new current variable assigned in step 2.

Step 6. Apply KCL to the node or supernode corresponding to the fundamental cut-set determined by each resistor with a new voltage variable assigned in step 2.

Step 7. Solve the simultaneous equations obtained from steps 5 and 6 for the new variables in terms of the voltage sources, current sources, and state variables.

Step 8. Substitute the expressions obtained in step 7 into the equations determined in steps 3 and 4.

· · ·

Example

In the circuit shown in Fig. 7.22 the tree selected is indicated with bold lines. According to step 1, the only option available is to choose between R_1 and R_2 as a tree element. The former choice was made arbitrarily, as was the polarity of the voltage across the capacitor and the direction of the current through the inductor.

fig. 7.22

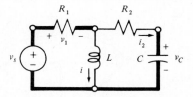

Since the voltage across R_1 cannot be expressed readily in terms of v_s, v_C, and i, according to step 2, we assign the new variable v_1. Similarly, we assign the current i_2.

From step 3,

$$L\frac{di}{dt} = -v_1 + v_s \quad \text{or} \quad \frac{di}{dt} = -\frac{1}{L}v_1 + \frac{1}{L}v_s$$

and from step 4 we have

$$-i_2 = C\frac{dv_C}{dt} \Rightarrow \frac{dv_C}{dt} = \frac{1}{C}i_2$$

From step 5, we have

$$-i_2 R_2 = v_C - v_s + v_1$$

and, from step 6,

$$\frac{v_1}{R_1} = i_2 + i$$

Solving these two equations for v_1 and i_2 in terms of v_s, v_C, and i, we get

$$i_2 = \frac{-R_1}{R_1 + R_2} i - \frac{1}{R_1 + R_2} v_C + \frac{1}{R_1 + R_2} v_s$$

and

$$v_1 = \frac{R_1 R_2}{R_1 + R_2} i - \frac{R_1}{R_1 + R_2} v_C + \frac{R_1}{R_1 + R_2} v_s$$

Substituting these expressions for i_2 and v_1 into the equations obtained for steps 3 and 4 results in the state equations

$$\frac{di}{dt} = \frac{-R_1 R_2}{L(R_1 + R_2)} i + \frac{R_1}{L(R_1 + R_2)} v_C + \frac{R_2}{L(R_1 + R_2)} v_s$$

$$\frac{dv_C}{dt} = \frac{-R_1}{C(R_1 + R_2)} i - \frac{1}{C(R_1 + R_2)} v_C + \frac{1}{C(R_1 + R_2)} v_s$$

• • •

Example

The circuit shown in Fig. 7.23 contains a current-dependent current source and four state variables: i_1, i_2, v_1, and v_2.

fig. 7.23

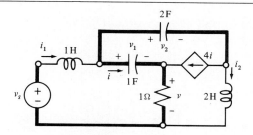

According to step 1, the two capacitors and voltage source correspond to tree branches, whereas the inductors and current source do not. Consequently, the 1-Ω resistor must correspond to a tree branch. Thus it is not possible to have the element (the 1-F capacitor) whose current controls the dependent current source correspond to a nonbranch.

According to step 2, we assign capacitor voltages and inductor currents as shown. Since an expression, in terms of the voltage source and the state variables, for the voltage across the 1-Ω resistor may not be immediately recognizable, we assign the new variable, v.

From step 3, for the 1-H inductor,

$$1 \frac{di_1}{dt} = v_s - v - v_1 \tag{7.3}$$

and, for the 2-H inductor,

$$2\frac{di_2}{dt} = v_1 - v_2 + v \Rightarrow \frac{di_2}{dt} = \frac{1}{2}v_1 - \frac{1}{2}v_2 + \frac{1}{2}v \tag{7.4}$$

For step 4, the fundamental cut-set determined by the 1-F capacitor consists of the capacitor, the two inductors and the dependent current source. Applying KCL to the corresponding supernode, we get

$$i + 4i = i_1 - i_2$$

$$5i = i_1 - i_2$$

$$5\left(1\frac{dv_1}{dt}\right) = i_1 - i_2 \Rightarrow \frac{dv_1}{dt} = \frac{1}{5}i_1 - \frac{1}{5}i_2 \tag{7.5}$$

For the fundamental cut-set determined by the 2-F capacitor, we obtain

$$2\frac{dv_2}{dt} = 4i + i_2 \Rightarrow \frac{dv_2}{dt} = 2i + \frac{1}{2}i_2 \tag{7.6}$$

From step 6, the fundamental cut-set determined by the 1-Ω resistor consists of the resistor and the two inductors. Applying KCL to the corresponding supernode, we get

$$\frac{v}{1} = i_1 - i_2$$

Substituting this result into Equations (7.3) and (7.4) yields the state equations

$$\frac{di_1}{dt} = -i_1 + i_2 - v_1 + v_s$$

and

$$\frac{di_2}{dt} = \frac{1}{2}i_1 - \frac{1}{2}i_2 + \frac{1}{2}v_1 - \frac{1}{2}v_2$$

Equation (7.5) is already a state equation. To convert Equation (7.6) into a state equation, we use the fact that, by KCL,

$$i + 4i = \frac{v}{1}$$

or

$$5i = v \Rightarrow i = \frac{1}{5}v = \frac{1}{5}i_1 - \frac{1}{5}i_2$$

Substituting this into Equation (7.6), we get the final state equation

$$\frac{dv_2}{dt} = \frac{2}{5}i_1 + \frac{1}{10}i_2$$

Therefore, the matrix state equation is

$$\frac{d}{dt}\begin{bmatrix} i_1 \\ i_2 \\ v_1 \\ v_2 \end{bmatrix} = \begin{bmatrix} -1 & 1 & -1 & 0 \\ \frac{1}{2} & -\frac{1}{2} & \frac{1}{2} & -\frac{1}{2} \\ \frac{1}{5} & -\frac{1}{5} & 0 & 0 \\ \frac{2}{5} & \frac{1}{10} & 0 & 0 \end{bmatrix}\begin{bmatrix} i_1 \\ i_2 \\ v_1 \\ v_2 \end{bmatrix} + \begin{bmatrix} 1 \\ 0 \\ 0 \\ 0 \end{bmatrix} v_s$$

· · ·

Finally, let us consider the case that we cannot select a spanning tree as specified in step 1. Such a situation arises when the circuit contains a loop consisting of capacitors and voltage sources or a cut-set consisting of inductors and current sources. For the former case, it is not possible to have a tree containing all the capacitors and voltage sources, so we shall leave a capacitor out of the tree. Consequently, this capacitor's voltage is not a state variable. For the latter case, every tree must contain an inductor or current source, so we shall place an inductor in the tree. The result is that this inductor's current is not a state variable.

· · ·

Example
In the circuit in Fig. 7.24 the two capacitors and the voltage source form a loop, so in selecting a tree we do not include both capacitors as branches—we arbitrarily leave out C_2. The resulting spanning tree is indicated with bold lines. After

fig. 7.24

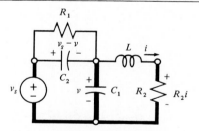

assigning the state variables i and v, we see that the voltage across C_2 can be expressed in terms of a state variable (the voltage v) and the voltage source. The voltage across R_2 can also be expressed in terms of a state variable (the current i).

From step 3 of our procedure, we get

$$L\frac{di}{dt} = v - R_2 i \Rightarrow \frac{di}{dt} = -\frac{R_2}{L}i + \frac{1}{L}v$$

which is the first state equation. The fundamental cut-set determined by C_1 consists of C_1, C_2, L, and R_1. Thus, from step 4, applying KCL to the corresponding node (the node common to both capacitors), we obtain

$$C_1\frac{dv}{dt} = C_2\frac{d}{dt}(v_s - v) + \frac{1}{R_1}(v_s - v) - i$$

from which

$$(C_1 + C_2) \frac{dv}{dt} = -i - \frac{1}{R_1} v + \frac{1}{R_1} v_s + C_2 \frac{dv_s}{dt}$$

and thus the second state equation is

$$\frac{dv}{dt} = \frac{-1}{C_1 + C_2} i - \frac{1}{R_1(C_1 + C_2)} v + \frac{1}{R_1(C_1 + C_2)} v_s + \frac{C_2}{C_1 + C_2} \frac{dv_s}{dt}$$

Hence we see that, in addition to the function $v_s(t)$, we also have the appearance of $dv_s(t)/dt$, the derivative of $v_s(t)$. If it makes you feel more comfortable, we can give the latter a different name—say, $f_s(t)$. That is, $f_s(t) = dv_s(t)/dt$. Doing so, we can write the matrix state equation

$$\frac{d}{dt} \begin{bmatrix} i \\ v \end{bmatrix} = \begin{bmatrix} -\dfrac{R_2}{L} & \dfrac{1}{L} \\ \dfrac{-1}{C_1 + C_2} & \dfrac{-1}{R_1(C_1 + C_2)} \end{bmatrix} \begin{bmatrix} i \\ v \end{bmatrix} + \begin{bmatrix} 0 & 0 \\ \dfrac{1}{R_1(C_1 + C_2)} & \dfrac{C_2}{C_1 + C_2} \end{bmatrix} \begin{bmatrix} v_s \\ f_s \end{bmatrix}$$

$$\cdots$$

PROBLEMS *7.36.   Write the matrix state equation for the circuit shown in Fig. P7.36.

Ans. $\dfrac{d}{dt} \begin{bmatrix} i \\ v \end{bmatrix} = \begin{bmatrix} -4 & -2 \\ \tfrac{4}{3} & -1 \end{bmatrix} \begin{bmatrix} i \\ v \end{bmatrix} + \begin{bmatrix} -1 \\ \tfrac{1}{3} \end{bmatrix} v_s$

fig. P7.36

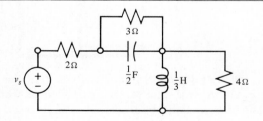

fig. P7.37

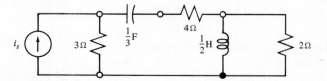

7.37.   For the circuit in Fig. P7.37, write the matrix state equation.

7.38.   For the circuit in Fig. P7.38, write the matrix state equation.

fig. P7.38

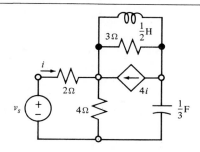

fig. P7.39

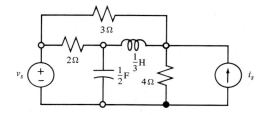

*7.39.   Write the matrix state equation for the circuit shown in Fig. P7.39.

$$\text{Ans.} \quad \frac{d}{dt}\begin{bmatrix} i \\ v \end{bmatrix} = \begin{bmatrix} -\frac{36}{7} & 3 \\ -2 & -1 \end{bmatrix}\begin{bmatrix} i \\ v \end{bmatrix} + \begin{bmatrix} -\frac{36}{7} & -\frac{12}{7} \\ 0 & 1 \end{bmatrix}\begin{bmatrix} i_s \\ v_s \end{bmatrix}$$

fig. P7.40

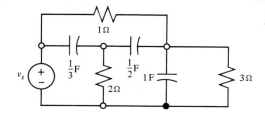

7.40.   For the circuit in Fig. P7.40, write the matrix state equation for the case that the state variables are the voltages across the following:
(a)   The $\frac{1}{3}$-F and $\frac{1}{2}$-F capacitors.
(b)   The $\frac{1}{3}$-F and 1-F capacitors.
(c)   The $\frac{1}{2}$-F and 1-F capacitors.

fig. P7.41

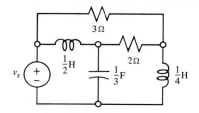

7.41.   Write the matrix state equation for the circuit in Fig. P7.41.

fig. P7.42

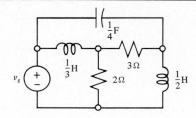

7.42.   For the circuit in Fig. P7.42, write the matrix state equation.

fig. P7.43

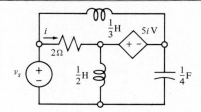

7.43.   Write the matrix state equation for the circuit shown in Fig. P7.43.

fig. P7.44

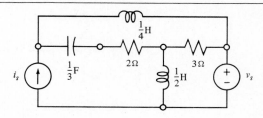

7.44.   For the circuit in Fig. P7.44, write the matrix state equation.

fig. P7.45

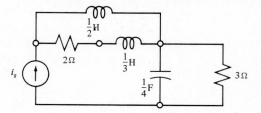

7.45.   In the circuit shown in Fig. P7.45, write the matrix state equation using as the state variables, the capacitor voltage and the current through (a) the $\frac{1}{3}$-H inductor and (b) $\frac{1}{2}$-H inductor.

Summary

1. The currents through the inductors and the voltages across the capacitors describe the state of a circuit. These voltages and currents are the state variables.
2. An nth-order circuit can be described by a set of n first-order differential equations whose variables are the state variables.
3. These simultaneous equations can be written as a single matrix state equation.
4. The zero-input or zero-state responses of a circuit can be determined by numerical techniques.
5. The numerical solution approach allows for the analysis of circuits with nonzero inputs and initial conditions as well.

SINUSOIDAL ANALYSIS 8

Introduction

Step and impulse functions are useful in determining the responses of circuits when they are first turned on or when sudden or irregular changes occur in the input; this is called transient analysis. However, to see how a circuit responds to a regular or repetitive input—the steady-state analysis—the function that is by far the most useful is the sinusoid.

The sinusoid is an extremely important and ubiquitous function. To begin with, the shape of ordinary household voltage is sinusoidal. Consumer radio transmissions are either amplitude modulation (AM), in which the amplitude of a sinusoid is changed or modulated according to some information signal, or frequency modulation (FM), in which the frequency of a sinusoid is modulated. Consumer television uses AM for the picture (video) and FM for the sound (audio). Sinusoids even occur in subtle ways, for as we shall see in a subsequent chapter (Fourier series), a nonsinusoidal waveform (like a sawtooth or a pulse train) is in essence just a sum of sinusoids! For these reasons and more, sinusoidal analysis is a fundamental topic in the study of electrical networks.

In this chapter we shall see that although we can analyze sinusoidal circuits with the techniques discussed previously, by generalizing the sinusoid we can perform analysis in a more simplified manner that avoids the direct solution of differential equations.

8.1 TIME DOMAIN ANALYSIS

Of all the functions encountered in electrical engineering, perhaps the single most important one is the sinusoid.

From trigonometry recall the plot of $\cos x$ versus the angle x (whose units are

372

radians), as shown in Fig. 8.1. Changing the angle from x to ωt, we get the plot shown in Fig. 8.2, where t is time in seconds and thus ω must have radians per second as units. We say that ω is the *radian* or *angular frequency* of $\cos \omega t$. In order to plot $\cos \omega t$ versus

fig. 8.1

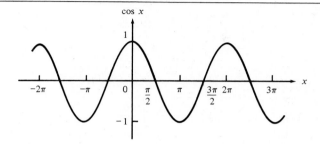

fig. 8.2

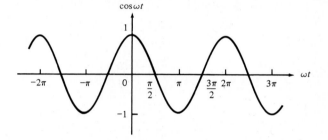

fig. 8.3

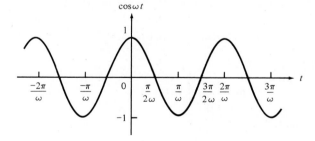

time t, divide the numbers on the horizontal axis (abscissa) by ω. The result is shown in Fig. 8.3. From this plot it is easy to see that this function goes through a complete cycle in $2\pi/\omega$ seconds. We call the time to complete one cycle the *period* of the sinusoid and denote the period by T; that is, the number of seconds per cycle is

$$T = \frac{2\pi}{\omega} \quad \text{seconds}$$

The inverse of this quantity is the number of cycles that occur in 1 s. That is, the number of cycles per second is

$$\frac{1}{T} = \frac{\omega}{2\pi} \quad \frac{1}{\text{seconds}}$$

The number of cycles per second is designated by f and is called the *(real) frequency* of the sinusoid. The term "cycles per second" has been replaced by "hertz" (abbreviated Hz).* Thus, the relationships between real and radian frequencies are

$$f = \frac{\omega}{2\pi} \qquad \text{hertz (Hz)}$$

and

$$\omega = 2\pi f \qquad \text{radians per second (rad/s)}†$$

In order to obtain a plot of the general sinusoid $\cos(\omega t - \phi)$ versus t, reconsider the plot of $\cos \omega t$ versus ωt in Fig. 8.2. Replacing ωt by $\omega t - \phi$, where ϕ is a positive quantity, corresponds to translating (or shifting) the sinusoid to the right by the amount ϕ, and we obtain the plot shown in Fig. 8.4. Then, dividing by ω, we get Fig. 8.5. Of course, if ϕ is a negative quantity the shift is to the left. We call ϕ the *phase angle* or simply *angle* of the sinusoid. When $\phi = \pi/2$ rad (90°), note that

$$\cos\left(\omega t - \frac{\pi}{2}\right) = \sin \omega t$$

fig. 8.4

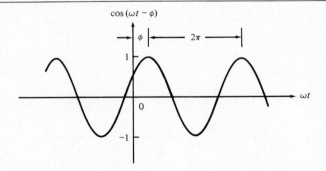

fig. 8.5

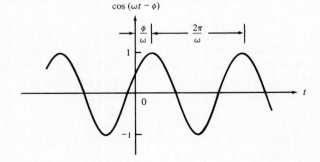

* Named for the German physicist Heinrich Hertz (1857–1894).
† It has been suggested that the term "radians per second" be replaced by "metz"—abbreviated Mz—in honor of the American electrical engineer Charles P. Steinmetz (1865–1923).

Similarly,

$$\sin\left(\omega t + \frac{\pi}{2}\right) = \cos \omega t$$

Consider now a circuit with a sinusoidal input. In general, the response consists of the natural response and the forced response. In particular, let us concentrate on the forced (or steady-state) response. We know that the integral or derivative of a sinusoid having a given frequency is again a sinusoid of that frequency. For a resistor the voltage and the current differ by a constant, whereas for an inductor and a capacitor the relationships between voltage and current are given in terms of integrals and derivatives. Thus, for each of the elements, a sinusoidal input produces a sinusoidal response. Since combining sinusoids of frequency ω results in a single sinusoid of frequency ω, we can deduce that the forced response of an RLC circuit whose input is a sinusoid of frequency ω is also a sinusoid of frequency ω.

· · ·

Example
Consider the sinusoidal circuit shown in Fig. 8.6 in which $v_s(t)$ is the input and $v(t)$ is the output.

fig. 8.6

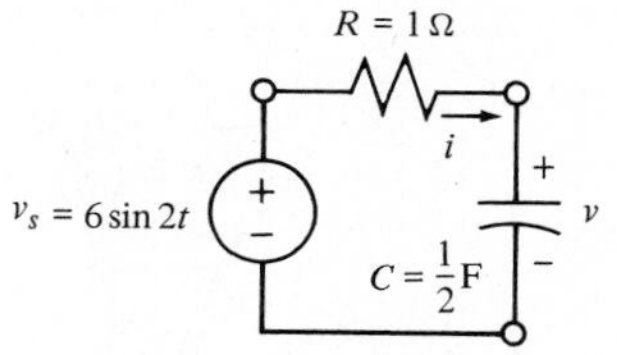

Since

$$v_s = Ri + v$$

and

$$i = C\frac{dv}{dt}$$

then

$$v_s = RC\frac{dv}{dt} + v$$

or

$$\frac{dv}{dt} + \frac{1}{RC}v = \frac{v_s}{RC}$$

Thus

$$\frac{dv}{dt} + 2v = 12 \sin 2t \qquad (8.1)$$

Since the forced response is a sinusoid of frequency $\omega = 2$ rad/s, it has the general form $A \cos(2t - \phi)$. However, in order to avoid using the trigonometric identity for the cosine of the sum of two angles, let us use the alternative form

$$v(t) = A_1 \cos 2t + A_2 \sin 2t$$

Substituting this into Equation (8.1) results in

$$(-2A_1 \sin 2t + 2A_2 \cos 2t) + 2(A_1 \cos 2t + A_2 \sin 2t) = 12 \sin 2t$$

or

$$(-2A_1 + 2A_2)\sin 2t + (2A_1 + 2A_2)\cos 2t = 12 \sin 2t$$

Equating coefficients yields the simultaneous equations

$$-2A_1 + 2A_2 = 12$$
$$2A_1 + 2A_2 = 0$$

the solution of which is $A_1 = -3$ and $A_2 = 3$. Hence, the forced response is

$$v(t) = -3 \cos 2t + 3 \sin 2t$$

or [see Equation (6.13) on page 305]

$$v(t) = 3\sqrt{2} \cos\left(2t - \frac{3\pi}{4}\right)$$

$$= 3\sqrt{2} \cos\left(2t - \frac{\pi}{2} - \frac{\pi}{4}\right)$$

$$= 3\sqrt{2} \sin\left(2t - \frac{\pi}{4}\right)$$

A plot of the functions $v_s(t)$ and $v(t)$ versus ωt is shown in Fig. 8.7. Since $v_s(t)$ reaches its peak before $v(t)$, we can say that $v_s(t)$ *leads* $v(t)$ by $\pi/4$ radians (or 45°) or that $v(t)$ *lags* $v_s(t)$ by $\pi/4$ radians (or 45°). Since the output of the circuit lags the input, the given circuit is an example of what is called a *lag network*.

fig. 8.7

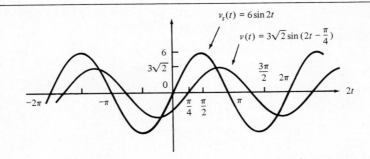

The current in this circuit is

$$i(t) = C \frac{dv(t)}{dt} = \frac{1}{2} \frac{dv(t)}{dt}$$

$$= \frac{1}{2} (6 \sin 2t + 6 \cos 2t)$$

$$= 3 \cos 2t + 3 \sin 2t$$

$$= 3\sqrt{2} \cos\left(2t - \frac{\pi}{4}\right)$$

$$= 3\sqrt{2} \cos\left(2t - \frac{\pi}{2} + \frac{\pi}{4}\right)$$

$$= 3\sqrt{2} \sin\left(2t + \frac{\pi}{4}\right)$$

Thus, the current $i(t)$ leads the voltage $v_s(t)$ by 45° and $i(t)$ leads the voltage $v(t)$ by 90°.

· · ·

If two sinusoids having the same frequency reach their peak (or maximum) values at exactly the same time, we say that they are *in phase*. For example, $3 \sin 2\pi t$ and $5 \sin 2\pi t$ are in phase. So are $12 \cos(3t - \pi/12)$ and $8 \cos(3t - \pi/12)$. Even $4 \cos(\omega t - \pi/3)$ and $7 \sin(\omega t + \pi/6)$ are in phase! That's because $\sin(\omega t + \pi/2)$ $= \cos \omega t$ implies that $4 \cos(\omega t - \pi/3) = 4 \sin(\omega t + \pi/2 - \pi/3) = 4 \sin(\omega t + \pi/6)$.

For the case that one sinusoid reaches its maximum value at the exact time when a second sinusoid (of the same frequency) reaches its minimum value, we say that they are *180° out of phase*. For instance, $3 \cos \omega t$ and $-4 \cos \omega t =$ $4 \cos (\omega t - \pi) = 4 \cos (\omega t + \pi)$ are 180° out of phase. In general, $A \cos(\omega t - \phi)$ and $-A \cos (\omega t - \phi)$ are 180° out of phase.

Just as a positive angle greater than 180° can be expressed as a negative angle between $-180°$ and 0°, when one sinusoid lags another by an amount greater than 180° (and less than 360°), we can say that the former leads the latter by an amount less than 180°. Of course a similar statement can be made for the case that one sinusoid leads another by more than 180°. When referring to two sinusoids of the same frequency, if they are not in phase or 180° out of phase, we shall say that one leads or lags the other by an amount between 0° and 180°. For example, if $v_1(t) = 3 \cos(\omega t + 70°)$ and $v_2(t)$ $= 4 \cos(\omega t - 150°)$, instead of saying that $v_1(t)$ leads $v_2(t)$ by 220°, we shall say that $v_1(t)$ lags $v_2(t)$ by 140° (or, equivalently, $v_2(t)$ leads $v_1(t)$ by 140°).

· · ·

Example

Now consider the series *RLC* sinusoidal circuit shown in Fig. 8.8.

We have already seen that

$$\frac{d^2v}{dt^2} + \frac{R}{L} \frac{dv}{dt} + \frac{1}{LC} v = \frac{v_s}{LC}$$

fig. 8.8

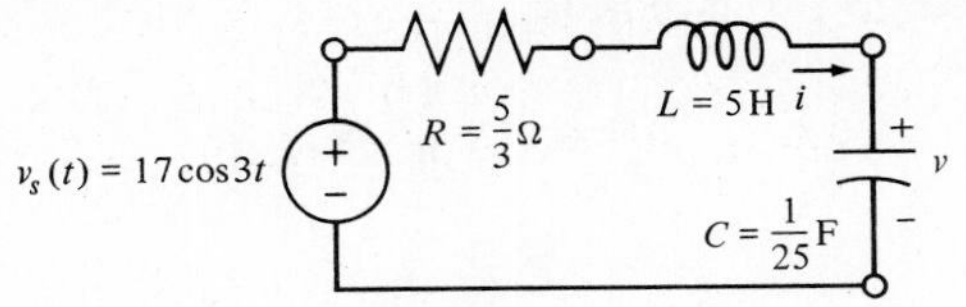

Thus

$$\frac{d^2v}{dt^2} + \frac{1}{3}\frac{dv}{dt} + 5v = 85 \cos 3t$$

and the forced response has the form

$$v(t) = A_1 \cos 3t + A_2 \sin 3t$$

Substituting this into the differential equation and collecting terms we get

$$(-4A_1 + A_2)\cos 3t + (-A_1 - 4A_2)\sin 3t = 85 \cos 3t$$

Equating coefficients yields the pair of simultaneous equations

$$-4A_1 + A_2 = 85$$
$$-A_1 - 4A_2 = 0$$

the solution of which is $A_1 = -20$ and $A_2 = 5$. Thus

$$v(t) = -20 \cos 3t + 5 \sin 3t$$
$$= 5\sqrt{17} \cos(3t - 2.9 \text{ rad})$$
$$= 5\sqrt{17} \cos(3t - 166°)$$

and the current is

$$i(t) = C\frac{dv}{dt} = \frac{1}{25}\frac{dv}{dt}$$

$$= \frac{1}{25}(60 \sin 3t + 15 \cos 3t)$$

$$= \frac{3\sqrt{17}}{5} \cos(3t - 1.33 \text{ rad})$$

$$= \frac{3\sqrt{17}}{5} \cos(3t - 76°)$$

$\bullet \quad \bullet \quad \bullet$

As indicated by the previous two examples, sinusoidal analysis is straightforward. However, the amount of arithmetic required can be extremely tedious. A lot of the drudgery involved in analyzing sinusoidal circuits can be eliminated if we utilize the concept of a "complex sinusoid" and the principle of superposition. Such an approach not only will be numerically advantageous, but also will give rise to concepts that will prove invaluable in the future. In order to proceed, though, it will be necessary to study complex numbers.

PROBLEMS 8.1. For the sinusoidal circuit shown in Fig. P8.1, find the forced response $v(t)$. Does $v(t)$ lead or lag $v_s(t)$?

fig. P8.1

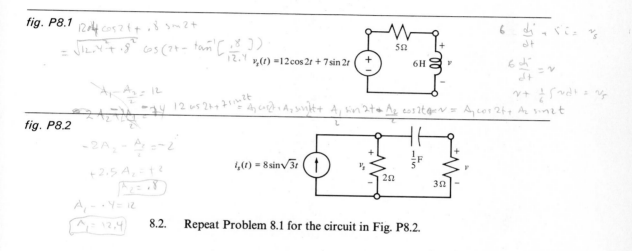

fig. P8.2

8.2. Repeat Problem 8.1 for the circuit in Fig. P8.2.

fig. P8.3

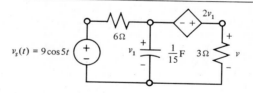

*8.3. Repeat Problem 8.1 for the circuit in Fig. P8.3.
Ans. $3.71 \cos(5t - 15.95°)$, lag

fig. P8.4

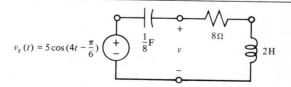

*8.4. Repeat Problem 8.1 for the circuit in Fig. P8.4.
Ans. $5.66 \cos(4t - 21.9°)$, lead
*8.5. Repeat Problem 8.1 for the circuit in Fig. P8.5.
Ans. $1.6 \sin 2t$, lag

fig. P8.5

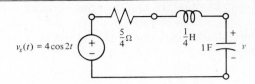

fig. P8.6

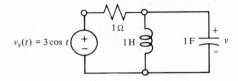

8.6. Repeat Problem 8.1 for the circuit in Fig. P8.6.

fig. P8.7

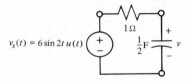

*8.7. Find the zero-state response $v(t)$ for the circuit in Fig. P8.7.
   Ans. $3[\sqrt{2} \cos (2t - 135°) + e^{-2t}]u(t)$

8.8. For the circuit given in Problem 8.5, find the zero-state response $v(t)$ for the case that the voltage source is $v_s(t) = 4 \cos 2t\, u(t)$.

8.9. For the circuit given in Problem 8.6, find the zero-state response $v(t)$ for the case that the voltage source is $v_s(t) = 3 \cos t\, u(t)$.

8.10. Repeat Problem 8.9 for the case that the resistor is replaced by a $\frac{1}{2}$-Ω resistor.

8.2 COMPLEX NUMBERS

As mentioned previously, we designate the constant $\sqrt{-1}$ by j; that is,

$$j = \sqrt{-1}$$

Raising j to powers, we get

$$j^2 = (\sqrt{-1})^2 = -1$$
$$j^3 = j^2 j = -j$$
$$j^4 = j^2 j^2 = (-1)(-1) = 1$$
$$j^5 = j^4 j = j$$

and so on.

A *complex number* $\mathbf{A}$ is a number of the form

$$\mathbf{A} = a + jb$$

where a and b are real numbers. We call a the *real part* of $\mathbf{A}$, denoted

$$\text{Re}(\mathbf{A}) = a$$

and we call b the *imaginary part* of $\mathbf{A}$, denoted

$$\text{Im}(\mathbf{A}) = b$$

If $a = 0$, we say that $\mathbf{A}$ is *purely imaginary*, and if $b = 0$, we say that $\mathbf{A}$ is *purely real*.

Two complex numbers $\mathbf{A}_1 = a_1 + jb_1$ and $\mathbf{A}_2 = a_2 + jb_2$ are *equal* if both

$$a_1 = a_2$$

and

$$b_1 = b_2$$

The *sum* of two complex numbers is

$$\mathbf{A}_1 + \mathbf{A}_2 = (a_1 + jb_1) + (a_2 + jb_2) = (a_1 + a_2) + j(b_1 + b_2)$$

and the *difference* is

$$\mathbf{A}_1 - \mathbf{A}_2 = (a_1 + jb_1) - (a_2 + jb_2) = (a_1 - a_2) + j(b_1 - b_2)$$

The *complex conjugate* of $\mathbf{A} = a + jb$ is denoted by $\mathbf{A}^*$ and is defined as

$$\mathbf{A}^* = a - jb$$

The *product* of two complex numbers is

$$\mathbf{A}_1\mathbf{A}_2 = (a_1 + jb_1)(a_2 + jb_2)$$

$$= a_1a_2 + ja_1b_2 + ja_2b_1 + j^2b_1b_2$$

$$= (a_1a_2 - b_1b_2) + j(a_1b_2 + a_2b_1)$$

The *quotient* is

$$\frac{\mathbf{A}_1}{\mathbf{A}_2} = \frac{a_1 + jb_1}{a_2 + jb_2} = \frac{(a_1 + jb_1)(a_2 - jb_2)}{(a_2 + jb_2)(a_2 - jb_2)}$$

$$= \frac{a_1a_2 + b_1b_2}{a_2{}^2 + b_2{}^2} + j\frac{b_1a_2 - a_1b_2}{a_2{}^2 + b_2{}^2}$$

Although addition and subtraction of complex numbers are simple operations, multiplication and division are a little more complicated. This is a consequence of the form of a complex number; that is, a complex number $\mathbf{A}$ can be written as $\mathbf{A} = a + jb$. We call this the *rectangular form* of the complex number $\mathbf{A}$. Fortunately, there is an alternative form for a complex number that lends itself very nicely to multiplication and division, although not to addition and subtraction. In order to develop this form, we shall first derive "Euler's* formula," which was mentioned earlier.

---

* Named for the Swiss mathematician, Leonhard Euler (1707–1783).

Let θ be a real variable. Define the function $f(\theta)$ by

$$f(\theta) = \cos\,\theta + j\,\sin\,\theta$$

Taking the derivative with respect to θ yields

$$\frac{df(\theta)}{d\theta} = -\sin\,\theta + j\,\cos\,\theta$$

$$= j^2\,\sin\,\theta + j\,\cos\,\theta$$

$$= j(\cos\,\theta + j\,\sin\,\theta)$$

$$= jf(\theta)$$

from which

$$\frac{1}{f(\theta)}\frac{df(\theta)}{d\theta} = j$$

Integrating both sides with respect to θ results in

$$\int \frac{1}{f(\theta)}\frac{df(\theta)}{d\theta}\,d\theta = \int j\,d\theta$$

and, by the chain rule,

$$\int \frac{df(\theta)}{f(\theta)} = j\int d\theta$$

Evaluating the integrals

$$\ln f(\theta) = j\theta + K$$

where K is a constant of integration. Thus

$$\ln(\cos\,\theta + j\,\sin\,\theta) = j\theta + K$$

Setting $\theta = 0$ gives

$$\ln 1 = K$$

$$0 = K$$

so

$$\ln(\cos\,\theta + j\,\sin\,\theta) = j\theta$$

Taking powers of e, we get

$$\boxed{\cos\,\theta + j\,\sin\,\theta = e^{j\theta}}$$

and this result is known as *Euler's formula*. Replacing θ by $-\theta$ we obtain another version

$$\cos\,\theta - j\,\sin\,\theta = e^{-j\theta}$$

Thus, we see that $e^{j\theta}$ is a complex number such that

$$\mathrm{Re}(e^{j\theta}) = \cos\theta$$

$$\mathrm{Im}(e^{j\theta}) = \sin\theta$$

If M is a real constant, then

$$Me^{j\theta} = M(\cos\theta + j\sin\theta)$$

$$= M\cos\theta + jM\sin\theta$$

is a complex number where

$$\mathrm{Re}(Me^{j\theta}) = M\cos\theta$$

$$\mathrm{Im}(Me^{j\theta}) = M\sin\theta$$

Given a complex number $\mathbf{A} = a + jb$, is there a complex number of the form $Me^{j\theta}$ (where $M > 0$) which equals $\mathbf{A}$? If there is, then

$$a + jb = M\cos\theta + jM\sin\theta$$

or

$$a = M\cos\theta \qquad \text{and} \qquad b = M\sin\theta$$

so

$$\cos\theta = \frac{a}{M} \qquad \text{and} \qquad \sin\theta = \frac{b}{M}$$

from which

$$\tan\theta = \frac{\sin\theta}{\cos\theta} = \frac{b/M}{a/M} = \frac{b}{a}$$

or

$$\theta = \tan^{-1}\left(\frac{b}{a}\right)$$

Pictorially, we have the right triangle shown in Fig. 8.9.

fig. 8.9

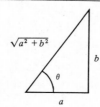

From

$$a = M \cos \theta$$

we get

$$M = \frac{a}{\cos \theta} = \frac{a}{a/\sqrt{a^2 + b^2}} = \sqrt{a^2 + b^2}$$

In summary,

$$\boxed{a + jb = Me^{j\theta}}$$

when

$$\boxed{\begin{array}{ll} M = \sqrt{a^2 + b^2} & \qquad a = M \cos \theta \\ \theta = \tan^{-1}\left(\dfrac{b}{a}\right) & \text{or} \qquad b = M \sin \theta \end{array}}$$

We say that $Me^{j\theta}$ is the *exponential* or *polar form* of a complex number, where M is called the *magnitude* and θ is called the *angle* or *argument* of the complex number. We denote the magnitude of $\mathbf{A}$ by $|\mathbf{A}|$ and the angle of $\mathbf{A}$ by $\text{ang}(\mathbf{A})$.

Note that by using the polar forms of the complex numbers $\mathbf{A}_1 = M_1 e^{j\theta_1}$ and $\mathbf{A}_2 = M_2 e^{j\theta_2}$, the product is

$$\mathbf{A}_1 \mathbf{A}_2 = (M_1 e^{j\theta_1})(M_2 e^{j\theta_2})$$

$$= M_1 M_2 e^{j\theta_1} e^{j\theta_2}$$

$$= M_1 M_2 e^{j(\theta_1 + \theta_2)}$$

Thus

$$|\mathbf{A}_1 \mathbf{A}_2| = M_1 M_2 = |\mathbf{A}_1| \, |\mathbf{A}_2|$$

and

$$\text{ang}(\mathbf{A}_1 \mathbf{A}_2) = \text{ang}(\mathbf{A}_1) + \text{ang}(\mathbf{A}_2)$$

The quotient is

$$\frac{\mathbf{A}_1}{\mathbf{A}_2} = \frac{M_1 e^{j\theta_1}}{M_2 e^{j\theta_2}}$$

$$= \frac{M_1}{M_2} e^{j\theta_1} e^{-j\theta_2}$$

$$= \frac{M_1}{M_2} e^{j(\theta_1 - \theta_2)}$$

Thus

$$\left|\frac{\mathbf{A}_1}{\mathbf{A}_2}\right| = \frac{M_1}{M_2} = \frac{|\mathbf{A}_1|}{|\mathbf{A}_2|}$$

and

$$\text{ang}\left(\frac{\mathbf{A}_1}{\mathbf{A}_2}\right) = \text{ang}(\mathbf{A}_1) - \text{ang}(\mathbf{A}_2)$$

Note that

$$(Me^{j\theta})^* = (M\cos\theta + jM\sin\theta)^*$$
$$= M\cos\theta - jM\sin\theta$$
$$= Me^{-j\theta}$$

$\cdot$ $\cdot$ $\cdot$

Example

If $\mathbf{A}_1 = -3 + j4 = M_1 e^{j\theta_1}$, then

$$|\mathbf{A}_1| = M_1 = \sqrt{(-3)^2 + (4)^2} = 5$$

and

$$\theta_1 = \tan^{-1}\left(\frac{4}{-3}\right) = 2.21 \text{ rad} = 126.9°$$

while if $\mathbf{A}_2 = -5 - j12 = M_2 e^{j\theta_2}$, then

$$|\mathbf{A}_2| = M_2 = \sqrt{(-5)^2 + (-12)^2} = 13$$

and

$$\theta_2 = \tan^{-1}\left(\frac{-12}{-5}\right) = -1.97 \text{ rad} = -112.6°$$

Thus

$$\mathbf{A}_1\mathbf{A}_2 = 5e^{j126.9°}13e^{-j112.6°}$$
$$= 5(13)e^{j(126.9-112.6)} = 65e^{j14.3°}$$

Also,

$$\mathbf{A}_1\mathbf{A}_2 = 65\cos(14.3°) + j65\sin(14.3°) = 63 + j16$$

$\cdot$ $\cdot$ $\cdot$

PROBLEMS *8.11. Find the exponential form of the following complex numbers given in rectangular form:

(a) $4 + j7$ (d) $-1 - j6$ (g) $j7$
(b) $3 - j5$ (e) 4 (h) $-j2$
(c) $-2 + j3$ (f) -5

Ans.   (a) $8.06e^{j60.3°}$; (c) $3.61e^{j123.7°}$; (e) 4; (h) $2e^{-j90°}$

*8.12. Find the rectangular form of the following complex numbers given in exponential form:

(a) $3e^{j70°}$ (d) $4e^{-j150°}$ (g) $2e^{j180°}$

(b) $2e^{j120°}$ (e) $6e^{j90°}$ (h) $2e^{-j180°}$

(c) $5e^{-j60°}$ (f) $e^{-j90°}$

Ans. (b) $-1 + j1.732$; (d) $-3.46 - j2$; (e) $j6$; (h) -2

*8.13. Find the rectangular form of the sum $\mathbf{A}_1 + \mathbf{A}_2$ given that

(a) $\mathbf{A}_1 = 3e^{j30°}$, $\mathbf{A}_2 = 4e^{j60°}$.

(b) $\mathbf{A}_1 = 3e^{j30°}$, $\mathbf{A}_2 = 4e^{-j30°}$.

(c) $\mathbf{A}_1 = 5e^{-j60°}$, $\mathbf{A}_2 = 2e^{j120°}$.

(d) $\mathbf{A}_1 = 4e^{j45°}$, $\mathbf{A}_2 = 2e^{-j90°}$.

Ans. (a) $4.6 + j4.96$; (c) $1.5 - j2.6$

8.3   FREQUENCY DOMAIN ANALYSIS

Given an *RLC* circuit whose input is the sinusoid $A \cos(\omega t + \theta)$, the forced response is a sinusoid of frequency ω, say $B \cos(\omega t + \phi)$. If the input is shifted 90°, that is, $A \sin(\omega t + \theta)$, then by the time-invariance property the corresponding response is $B \sin(\omega t + \phi)$. Furthermore, if the input is scaled by a constant K, then by the property of linearity so will the response; that is, the response to $KA \sin(\omega t + \theta)$ is $KB \sin(\omega t + \phi)$. By the principle of superposition, the response to

$$A \cos(\omega t + \theta) + KA \sin(\omega t + \theta)$$

is

$$B \cos(\omega t + \phi) + KB \sin(\omega t + \phi)$$

Consider now the special case that the constant $K = \sqrt{-1} = j$. Thus, the response to

$$Ae^{j(\omega t + \theta)}$$

is

$$Be^{j(\omega t + \phi)}$$

We therefore see that whether we wish to find the response to $A \cos(\omega t + \theta)$ or the response to $Ae^{j(\omega t + \theta)}$, we must determine the same two constants B and ϕ. We have seen examples of what it is like to find a response when the input is the sinusoid $A \cos(\omega t + \theta)$. Now let us look at an example of finding the response to the *complex sinusoid* $Ae^{j(\omega t + \theta)}$.

$\bullet \quad \bullet \quad \bullet$

Example

Reconsider the *RC* circuit shown in Fig. 8.10.

fig. 8.10

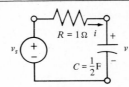

From

$$v_s = Ri + v \qquad \text{and} \qquad i = C\frac{dv}{dt}$$

we have that

$$\frac{dv}{dt} + \frac{1}{RC}v = \frac{v_s}{RC} \qquad \text{or} \qquad \frac{dv}{dt} + 2v = 2v_s$$

The forced response to

$$v_s = 6\sin 2t = 6\cos\left(2t - \frac{\pi}{2}\right) = A\cos(\omega t + \theta)$$

is

$$v = B\cos(2t + \phi)$$

Instead of solving directly for B and ϕ, let us use the fact that the response to

$$v_s = Ae^{j(\omega t + \theta)} = 6e^{j(2t - \pi/2)}$$

is

$$v = Be^{j(2t + \phi)}$$

Substitution into the differential equation yields

$$\frac{d}{dt}(Be^{j(2t + \phi)}) + 2(Be^{j(2t + \phi)}) = 12e^{j(2t - \pi/2)}$$

or

$$2jBe^{j2t}e^{j\phi} + 2Be^{j2t}e^{j\phi} = 12e^{j2t}e^{-j\pi/2}$$

Dividing both sides by $2e^{j2t}$ gives

$$(j + 1)Be^{j\phi} = 6e^{-j\pi/2}$$

from which

$$Be^{j\phi} = \frac{6}{1 + j}e^{-j\pi/2} = \frac{6(1 - j)}{(1 + j)(1 - j)}e^{-j\pi/2}$$
$$= 3(1 - j)e^{-j\pi/2}$$
$$= 3(\sqrt{2}e^{-j\pi/4})e^{-j\pi/2}$$
$$= 3\sqrt{2}e^{-j3\pi/4}$$

Thus $B = 3\sqrt{2}$ and $\phi = -3\pi/4$. We conclude therefore that the response to

$$v_s = 6e^{j(2t - \pi/2)}$$

is

$$v = 3\sqrt{2}e^{j(2t - 3\pi/4)}$$

and the response to

$$v_s = 6 \sin 2t$$

is

$$v = 3\sqrt{2} \cos\left(2t - \frac{3\pi}{4}\right) = 3\sqrt{2} \sin\left(2t - \frac{\pi}{4}\right)$$

$$\cdot \quad \cdot \quad \cdot$$

Although it may be somewhat easier computationally to find the response to a (real) sinusoid by considering instead the corresponding complex sinusoidal case, the determination of the appropriate differential equation is still required. By introducing the appropriate concepts, however, we are able to eliminate this requirement, and the result is the simplification of sinusoidal analysis.

Just as the sinusoid $A \cos(\omega t + \theta)$ has frequency ω, amplitude A, and phase angle θ, we can say that the complex sinusoid

$$A e^{j(\omega t + \theta)} = A e^{j\theta} e^{j\omega t}$$

also has frequency ω, amplitude A, and phase angle θ. Let us denote the complex number

$$\mathbf{A} = A e^{j\theta}$$

by

$$\mathbf{A} = A \underline{/\theta}$$

which is known as the *phasor* representation of the (real or complex) sinusoid.

Why do this? We begin to answer this question by considering a resistor (Fig. 8.11). By Ohm's law,

$$v = Ri$$

---

fig. 8.11

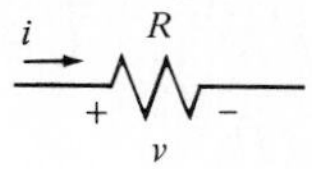

If the current is a complex sinusoid, say $i = I e^{j(\omega t + \theta)}$, then so is the voltage, say $v = V e^{j(\omega t + \phi)}$. Thus,

$$V e^{j(\omega t + \phi)} = R I e^{j(\omega t + \theta)}$$

$$V e^{j\omega t} e^{j\phi} = R I e^{j\omega t} e^{j\theta}$$

and dividing by $e^{j\omega t}$ yields

$$V e^{j\phi} = R I e^{j\theta}$$

From this equation we deduce that $V = RI$ and $\phi = \theta$. Using the phasor forms of the voltage and current, we can write

$$V \underline{/\phi} = RI \underline{/\theta}$$

Thus, we have the phasor equation

$$\boxed{\mathbf{V} = R\mathbf{I}}$$

where $\mathbf{V} = V \underline{/\phi}$ and $\mathbf{I} = I \underline{/\theta}$. Since $\phi = \theta$, the voltage and the current are in phase for a resistor.

For an inductor (Fig. 8.12), if $i = Ie^{j(\omega t + \theta)}$ and $v = Ve^{j(\omega t + \phi)}$, since

$$v = L \frac{di}{dt}$$

fig. 8.12

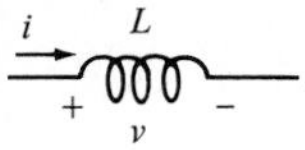

then

$$Ve^{j(\omega t + \phi)} = L \frac{d}{dt} \left(Ie^{j(\omega t + \theta)}\right)$$

$$Ve^{j\omega t}e^{j\phi} = j\omega LIe^{j\omega t}e^{j\theta}$$

so

$$Ve^{j\phi} = j\omega LIe^{j\theta}$$

and, using the phasor form of the voltage and current,

$$V \underline{/\phi} = j\omega LI \underline{/\theta}$$

or

$$\boxed{\mathbf{V} = j\omega L\mathbf{I}}$$

where $\mathbf{V} = V \underline{/\phi}$ and $\mathbf{I} = I \underline{/\theta}$. Note that since

$$\text{ang}(\mathbf{V}) = \text{ang}(j\omega L\mathbf{I}) = \text{ang}(j\omega L) + \text{ang}(\mathbf{I})$$

then

$$\phi = 90° + \theta$$

Thus, we see that for an inductor the voltage leads the current by 90° or, equivalently, the current lags the voltage by 90°.

For a capacitor (Fig. 8.13), if $i = Ie^{j(\omega t + \theta)}$ and $v = Ve^{j(\omega t + \phi)}$ since

$$i = C\frac{dv}{dt}$$

fig. 8.13

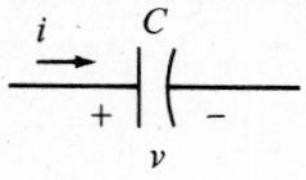

then

$$Ie^{j(\omega t + \theta)} = C\frac{d}{dt}(Ve^{j(\omega t + \phi)})$$

$$Ie^{j\omega t}e^{j\theta} = j\omega CVe^{j\omega t}e^{j\phi}$$

so

$$Ie^{j\theta} = j\omega CVe^{j\phi} \qquad \text{and} \qquad I\underline{/\theta} = j\omega CV\underline{/\phi}$$

or

$$\mathbf{I} = j\omega C\mathbf{V}$$

from which

$$\boxed{\mathbf{V} = \frac{1}{j\omega C}\mathbf{I}}$$

Using reasoning as was used for the inductor, we deduce that for a capacitor, the current leads the voltage (the voltage lags the current) by 90°.

If you would like an easy way to remember whether current leads or lags voltage for a capacitor or an inductor, just think of the word "capacitor" as the key. Let the first three letters stand for "*current and potential*"; doing so indicates that current comes before (leads) voltage for a capacitor. This, of course, implies that current lags voltage for an inductor.

We now see that for the resistor, inductor, and capacitor we get a phasor equation of the form

$$\boxed{\mathbf{V} = \mathbf{Z}\mathbf{I}}$$

and we call the quantity $\mathbf{Z}$ the *impedance* of the element. In other words, the impedance of an R-ohm resistor is

$$\boxed{Z_R = R}$$

the impedance of an *L*-henry inductor is

$$\mathbf{Z}_L = j\omega L$$

and the impedance of a *C*-farad capacitor is

$$\mathbf{Z}_C = \frac{1}{j\omega C}$$

The equation

$$\mathbf{V} = \mathbf{ZI}$$

is the phasor version of Ohm's law. From this equation we see that impedance is the ratio of voltage phasor to current phasor; that is,

$$\mathbf{Z} = \frac{\mathbf{V}}{\mathbf{I}}$$

Furthermore,

$$\mathbf{I} = \frac{\mathbf{V}}{\mathbf{Z}}$$

In this expression, the reciprocal of impedance appears. As this can occur frequently, it is useful to give it a name. We call the ratio of current phasor to voltage phasor, *admittance*, and denote it by the symbol **Y**; that is,

$$\mathbf{Y} = \frac{\mathbf{I}}{\mathbf{V}}$$

from which

$$\mathbf{Y}_R = \frac{1}{R} \qquad \mathbf{Y}_L = \frac{1}{j\omega L} \qquad \mathbf{Y}_C = j\omega C$$

and

$$\mathbf{I} = \mathbf{YV} \qquad \text{or} \qquad \mathbf{V} = \frac{\mathbf{I}}{\mathbf{Y}}$$

As was for the resistive case, the units of impedance are ohms and those of admittance are mhos.

• • •

Example
Consider a sinusoid whose frequency is $\omega = 100$ rad/s. For a 25-Ω resistor,

$$\mathbf{Z}_R = 25\ \Omega \quad \text{and} \quad \mathbf{Y}_R = \frac{1}{25} = 0.04\ \mho$$

for a $\frac{1}{20}$-H inductor,

$$\mathbf{Z}_L = j5\ \Omega \quad \text{and} \quad \mathbf{Y}_L = \frac{1}{j5} = -j(0.2)\ \mho$$

and, for a $\frac{1}{50}$-F capacitor,

$$\mathbf{Z}_C = \frac{1}{j2} = -j(0.5)\ \Omega \quad \text{and} \quad \mathbf{Y}_C = j2\ \mho$$

• • •

Suppose that for a sinusoidal *RLC* circuit, writing KVL around a loop results in the equation

$$v_1(t) + v_2(t) + v_3(t) + \cdots + v_n(t) = 0$$

For the case of a real sinusoid, the equation takes the form

$$V_1 \cos(\omega t + \theta_1) + V_2 \cos(\omega t + \theta_2) + \cdots + V_n \cos(\omega t + \theta_n) = 0$$

whereas, for the case of a complex sinusoid,

$$V_1 e^{j(\omega t + \theta_1)} + V_2 e^{j(\omega t + \theta_2)} + \cdots + V_n e^{j(\omega t + \theta_n)} = 0$$

Dividing both sides by $e^{j\omega t}$ yields

$$V_1 e^{j\theta_1} + V_2 e^{j\theta_2} + \cdots + V_n e^{j\theta_n} = 0$$

or, using phasor notation,

$$V_1 \underline{/\theta_1} + V_2 \underline{/\theta_2} + \cdots + V_n \underline{/\theta_n} = 0$$

so

$$\mathbf{V}_1 + \mathbf{V}_2 + \cdots + \mathbf{V}_n = 0$$

where $\mathbf{V}_i = V_i \underline{/\theta_i}$. Thus, we see that in addition to voltage functions of time, KVL holds for voltage phasors. We can similarly draw the conclusion that KCL holds for current phasors as well as time functions.

For inductors and capacitors, as well as resistors, the phasor relationships between voltage and current are

$$\mathbf{V} = \mathbf{ZI}$$

$$\mathbf{I} = \frac{\mathbf{V}}{\mathbf{Z}}$$

$$\mathbf{Z} = \frac{\mathbf{V}}{\mathbf{I}}$$

that is, these are the phasor versions of Ohm's law. By using KVL and KCL, we can analyze a sinusoidal circuit without resorting to differential equations. We simply look at the circuit from the phasor point of view and then proceed just as was done for resistive circuits using nodal, mesh, or loop analysis. For resistive circuits, we manipulate real numbers, whereas for sinusoidal circuits involving capacitors and/or inductors, complex number arithmetic is required.

If we analyze a circuit by using time functions and the various differential and integral relationships between voltage and current (and hence are required to solve differential equations), we say that we are working in the *time domain*. However, if we have a sinusoidal circuit, we can use the phasor representations of the time functions and the impedances (or admittances) of the various (nonsource) elements or combinations of elements. This is said to be the *frequency domain* approach.

Making the transformation from the time to the frequency domain is a simple matter. Just express a function of time by its phasor representation and resistors, inductors, and capacitors (or combinations of them) by their impedances or admittances. Proceed as for resistive circuits to find the phasor representations of the desired responses. The transformation back to the time domain is then done by inspection.

· · ·

Example

Reconsider the series *RLC* sinusoidal circuit shown in Fig. 8.14. Converting to phasors (i.e., in the frequency domain), we obtain Fig. 8.15.

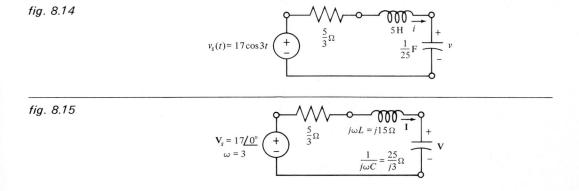

fig. 8.14

fig. 8.15

By KVL,

$$V_s = RI + j\omega LI + \frac{1}{j\omega C} I$$

$$17\underline{/0°} = \frac{5}{3} I + j15I + \frac{25}{j3} I$$

$$= \left(\frac{5}{3} + j15 - j\frac{25}{3}\right)I$$

$$= \left(\frac{5}{3} + j\frac{20}{3}\right)I = \frac{5}{3}(1 + j4)I$$

$$I = \frac{3}{5} \frac{17\underline{/0°}}{1 + j4} = \frac{3}{5} \frac{17\underline{/0°}}{\sqrt{17}\underline{/76°}}$$

Thus

$$I = \frac{3}{5} \sqrt{17}\underline{/-76°}$$

so the corresponding sinusoidal time function is

$$i(t) = \frac{3}{5} \sqrt{17} \cos(3t - 76°)$$

Since

$$V = \frac{1}{j\omega C} I$$

then

$$V = \frac{25}{j3}\left(\frac{3}{5} \sqrt{17}\underline{/-76°}\right) = \frac{5\sqrt{17}\underline{/-76°}}{1\underline{/90°}}$$

$$V = 5\sqrt{17}\underline{/-166°}$$

and the corresponding sinusoidal time function is

$$v(t) = 5\sqrt{17} \cos(3t - 166°)$$

Note that we can also obtain V by voltage division as follows:

$$V = \frac{Z_C}{Z_C + Z_R + Z_L} V_s$$

$$= \frac{\dfrac{25}{j3}}{\dfrac{25}{j3} + \dfrac{5}{3} + j15} (17\underline{/0°})$$

$$V = \frac{\dfrac{25}{j3}}{\dfrac{5}{3} + j\dfrac{20}{3}} (17 \underline{/0°}) = \frac{\dfrac{5}{j}}{1 + j4} (17 \underline{/0°})$$

$$= \frac{5 \underline{/-90°}}{\sqrt{17} \; \underline{/76°}} (17 \underline{/0°})$$

$$= 5\sqrt{17} \; \underline{/-166°}$$

. . .

The voltage divider formula used above can be derived as was the case for resistive circuits by considering the circuit in Fig. 8.16.

fig. 8.16

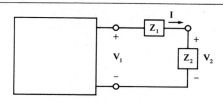

By KVL,

$$V_1 = Z_1 I + Z_2 I$$
$$= (Z_1 + Z_2)I$$

from which

$$I = \frac{V_1}{Z_1 + Z_2} \tag{8.2}$$

and since

$$V_2 = Z_2 I$$

we have that

$$\boxed{V_2 = \frac{Z_2 V_1}{Z_1 + Z_2}}$$

Note also that the current through the series connection of Z_1 and Z_2 is given by Equation (8.2). That is, the series combination behaves as a single impedance Z, where $Z = Z_1 + Z_2$ (see Fig. 8.17).

fig. 8.17

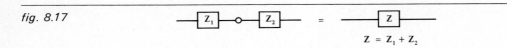

$$Z = Z_1 + Z_2$$

fig. 8.18

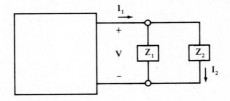

Consider now the case of current division depicted by Fig. 8.18. Since

$$\mathbf{I}_1 = \frac{\mathbf{V}}{\mathbf{Z}_1} + \frac{\mathbf{V}}{\mathbf{Z}_2} \qquad \text{or} \qquad \mathbf{I}_1 = \left(\frac{1}{\mathbf{Z}_1} + \frac{1}{\mathbf{Z}_2}\right)\mathbf{V}$$

then

$$\mathbf{V} = \frac{\mathbf{I}_1}{\dfrac{1}{\mathbf{Z}_1} + \dfrac{1}{\mathbf{Z}_2}} = \frac{\mathbf{Z}_1\mathbf{Z}_2\mathbf{I}_1}{\mathbf{Z}_1 + \mathbf{Z}_2} \qquad (8.3)$$

But

$$\mathbf{I}_2 = \frac{\mathbf{V}}{\mathbf{Z}_2}$$

so

$$\boxed{\mathbf{I}_2 = \frac{\mathbf{Z}_1\mathbf{I}_1}{\mathbf{Z}_1 + \mathbf{Z}_2}}$$

is the current division formula. From Equation (8.3) we find that the impedance of the parallel connection of $\mathbf{Z}_1$ and $\mathbf{Z}_2$ is

$$\boxed{\mathbf{Z} = \frac{\mathbf{V}}{\mathbf{I}_1} = \frac{1}{\dfrac{1}{\mathbf{Z}_1} + \dfrac{1}{\mathbf{Z}_2}} = \frac{\mathbf{Z}_1\mathbf{Z}_2}{\mathbf{Z}_1 + \mathbf{Z}_2}}$$

Thus two impedances in parallel behave as a single impedance $\mathbf{Z}$, where $1/\mathbf{Z} = 1/\mathbf{Z}_1 + 1/\mathbf{Z}_2$ (see Fig. 8.19).

fig. 8.19

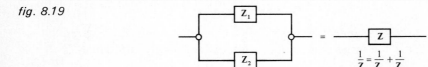

$$\frac{1}{Z} = \frac{1}{Z_1} + \frac{1}{Z_2}$$

We now conclude that impedances in series and parallel combine as do resistances. Using dual arguments, we may deduce that admittances in parallel and series combine, as do conductances (see Fig. 8.20).

fig. 8.20

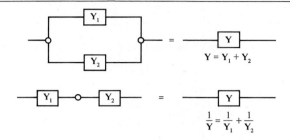

fig. 8.21

Recall the lag network encountered previously (see Fig. 8.21). By voltage division,

$$\mathbf{V}_C = \frac{\mathbf{Z}_C \mathbf{V}_1}{\mathbf{Z}_C + \mathbf{Z}_R}$$

$$= \frac{\dfrac{1}{j\omega C} \mathbf{V}_1}{\dfrac{1}{j\omega C} + R}$$

$$= \frac{\mathbf{V}_1}{1 + j\omega RC} = \frac{|\mathbf{V}_1|\underline{/\phi}}{\sqrt{1 + (\omega RC)^2}\underline{/\tan^{-1}(\omega RC)}}$$

$$= \frac{|\mathbf{V}_1|}{\sqrt{1 + (\omega RC)^2}}\underline{/\phi - \tan^{-1}(\omega RC)}$$

Thus, when ω, R, and C have positive values, the phase angle of the capacitor (output) voltage is less than the source (input) voltage. This means that the output lags the input.

As far as the capacitor is concerned, and hence the voltage across it, the effect of a series connection of a voltage source and a resistor is the same as that of a parallel connection of a current source and a resistor. Therefore, an alternative form of a lag network is as shown in Fig. 8.22, where the output voltage lags the input current.

fig. 8.22

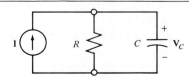

fig. 8.23

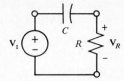

Next consider the network shown in Fig. 8.23. Again, by voltage division,

$$\mathbf{V}_R = \frac{\mathbf{Z}_R \mathbf{V}_1}{\mathbf{Z}_R + \mathbf{Z}_C}$$

$$= \frac{R\mathbf{V}_1}{R + \dfrac{1}{j\omega C}} = \frac{j\omega RC\mathbf{V}_1}{1 + j\omega RC}$$

$$= \frac{(\omega RC\,\underline{/90°})(|\mathbf{V}_1|\,\underline{/\phi})}{\sqrt{1 + (\omega RC)^2}\,\underline{/\tan^{-1}(\omega RC)}}$$

$$= \frac{\omega RC|\mathbf{V}_1|}{\sqrt{1 + (\omega RC)^2}}\,\underline{/\phi + 90° - \tan^{-1}(\omega RC)}$$

However, $\tan^{-1}(\omega RC)$ varies from 0° to 90° for positive values of ω, R, and C. We conclude that the phase angle of the resistor output voltage is greater than the input source voltage's. In other words, the output voltage leads the input voltage. For this reason we say that this circuit is a *lead network*.

Suppose that in the previous circuit (Fig. 8.23), the voltage source is replaced by a nonideal source as shown in Fig. 8.24. By voltage division,

$$V_o = \frac{R\mathbf{V}_1}{R + \dfrac{1}{j\omega C} + R_s} = \frac{j\omega RC\mathbf{V}_1}{1 + j\omega(R + R_s)C}$$

$$= \frac{\omega RC|\mathbf{V}_1|}{\sqrt{1 + [\omega(R + R_s)C]^2}}\,\underline{/\phi + 90° - \tan^{-1}[\omega(R + R_s)C]}$$

fig. 8.24

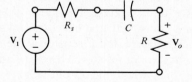

Thus we again conclude that this circuit is a lead network, and for the reason mentioned above, so is the circuit in Fig. 8.25.

These are a few of the different types of lag and lead networks that are used to change or "shift" the phase angles of sinusoids in some way. The result of such an action can be the compensation or stabilization of certain electronic circuits, such as

fig. 8.25

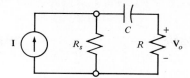

amplifiers and various types of control systems. In some applications it may be desirable to get destabilized results as can be accomplished by using three lead networks to obtain a "phase-shift oscillator," an electronic device that produces sinusoids.

Since *RLC* circuits are linear, regardless of whether or not the inputs are sinusoidal, we can apply the principle of superposition if we so desire.

· · ·

Example
The circuit shown in Fig. 8.26 is a single-stage transistor amplifier (the simplified transistor model is in the dashed box), which includes two batteries that are used to "bias" the transistor appropriately. Let us determine the voltage $v(t)$ by using the principle of superposition.

fig. 8.26

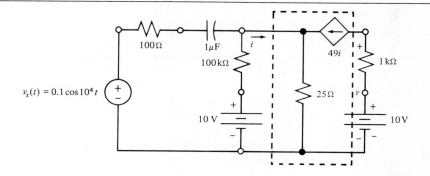

We begin by finding that part of $v(t)$, call it $v_{ac}(t)$, due to the input $v_s(t)$. Setting the two batteries to zero, we obtain the frequency domain circuit shown in Fig. 8.27. In order to write equations at two nodes instead of three, we can

fig. 8.27

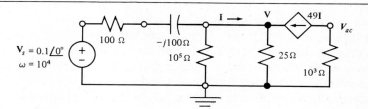

combine the 100-Ω resistor and the 1-μF capacitor into the single impedance $100 - j100 = 100(1 - j)\ \Omega$. By nodal analysis, at the node labeled **V**, we get

$$\frac{\mathbf{V}_s - \mathbf{V}}{100(1 - j)} + 49\mathbf{I} = \frac{\mathbf{V}}{10^5} + \frac{\mathbf{V}}{25}$$

Examining the right-hand side of the equality, we see that $\mathbf{V}/25$ is 4000 times as great as $\mathbf{V}/10^5$. Therefore, for all intents and purposes, we can drop the term $\mathbf{V}/10^5$. Furthermore, by KCL we have that

$$\mathbf{I} + 49\mathbf{I} = \frac{\mathbf{V}}{25} \Rightarrow \mathbf{I} = \frac{\mathbf{V}}{1250}$$

Substituting this fact into the previous equation (after $\mathbf{V}/10^5$ is dropped), we get

$$\frac{\mathbf{V}_s - \mathbf{V}}{100(1 - j)} + 49\,\frac{\mathbf{V}}{1250} = \frac{\mathbf{V}}{25}$$

Multiplying both sides of this equation by $12{,}500\,(1 - j)$ yields

$$125(\mathbf{V}_s - \mathbf{V}) + 490(1 - j)\mathbf{V} = 500(1 - j)\mathbf{V}$$

and from this expression we obtain

$$\mathbf{V} = \frac{125\mathbf{V}_s}{135 - j10}$$

Thus,

$$\mathbf{V}_{ac} = -49\mathbf{I}(10^3) = -49\left(\frac{\mathbf{V}}{1250}\right)(10^3)$$

$$= \frac{-49}{1.25}\,\mathbf{V} = \frac{-49}{1.25}\left(\frac{125\mathbf{V}_s}{135 - j10}\right)$$

$$= \frac{-4900}{135 - j10}\,\mathbf{V}_s = \frac{4900\,\underline{/-180°}}{135.4\,\underline{/-4.24°}}\,\mathbf{V}_s$$

$$= (36.2\,\underline{/-175.8°})\,\mathbf{V}_s = 3.62\,\underline{/-175.8°}$$

and, therefore, $v_{ac}(t) = 3.62\cos(10^4 t - 175.8°)$.

In the process of finding that portion of $v(t)$ due to each battery, we set $v_s(t)$ to zero. The result is a dc circuit. But a capacitor acts as an open circuit to dc. How does this conform to the fact that the impedance of a capacitor is $1/j\omega C$? Remembering that the ω in $1/j\omega C$ is the frequency of the sinusoid $A\cos(\omega t + \theta)$, for the special case that $\omega = 0$ the sinusoid becomes $A\cos\theta$, which is a constant. In other words, a dc circuit is a degenerate case of a sinusoidal circuit—the case that $\omega = 0$. In such a situation, the impedance of a capacitor is infinite; that is, the capacitor behaves as an open circuit. (By duality, the impedance of an inductor is zero for the dc case, and an inductor behaves as a short circuit.)

As a consequence of this discussion, in order to find that portion of $v(t)$ due to the batteries, we need only analyze the dc circuit of Fig. 8.28 to find $v_{dc}(t)$. Since this is a resistive circuit that has been given in previous problem sets (Problems 2.17, 2.29, and 3.28), we shall simply state that $v_{dc}(t) = 5.16$ V.

fig. 8.28

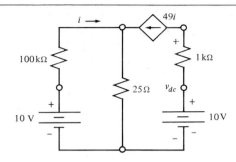

Hence, by the principle of superposition, we have that

$$v(t) = v_{dc}(t) + v_{ac}(t)$$

$$= 5.16 + 3.62 \cos(10^4 t - 175.8°) \text{ V}$$

· · ·

Other resistive network concepts apply equally well to sinusoidal circuits if such networks are described in the frequency domain—that is, in terms of phasors and impedances (or admittances). One very important example is Thévenin's theorem, which says in essence that (with a few exceptions, of course) a circuit is equivalent to an appropriate voltage source in series with an appropriate impedance. This situation is depicted in Fig. 8.29.

fig. 8.29

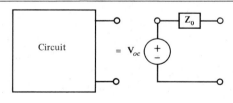

· · ·

Example
For the circuit shown in Fig. 8.30, we can determine the effect of the circuit on the load impedance $\mathbf{Z}_L$ by employing Thévenin's theorem.

We first determine $\mathbf{V}_{oc}$ from Fig. 8.31. By nodal analysis, at the node labeled $\mathbf{V}_{oc}$,

$$\frac{(5\angle -30°) - \mathbf{V}_{oc}}{-j2} = \frac{\mathbf{V}_{oc} - \mathbf{V}}{j8}$$

fig. 8.30

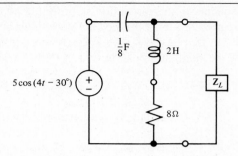

fig. 8.31

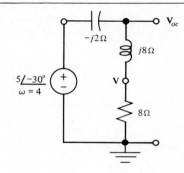

or

$$-4(5\underline{/-30°}) + 4\mathbf{V}_{oc} = \mathbf{V}_{oc} - \mathbf{V}$$

so

$$3\mathbf{V}_{oc} + \mathbf{V} = 20\underline{/-30°}\qquad\qquad(8.4)$$

At the node labeled **V**,

$$\frac{\mathbf{V}_{oc} - \mathbf{V}}{j8} = \frac{\mathbf{V}}{8}$$

or

$$\mathbf{V}_{oc} - \mathbf{V} = j\mathbf{V}$$

so

$$\mathbf{V}_{oc} = \mathbf{V} + j\mathbf{V} = (1 + j)\mathbf{V}$$

From this equation,

$$\mathbf{V} = \frac{1}{1 + j}\,\mathbf{V}_{oc}$$

and substituting this into Equation (8.4) yields

$$3\mathbf{V}_{oc} + \frac{1}{1 + j}\,\mathbf{V}_{oc} = 20\underline{/-30°}$$

or

$$\frac{4 + 3j}{1 + j} \, \mathbf{V}_{oc} = 20 \underline{/-30°}$$

from which

$$\mathbf{V}_{oc} = \frac{1 + j}{4 + 3j} \, (20 \underline{/-30°})$$

$$= \frac{\sqrt{2} \underline{/45°}}{5 \underline{/36.9°}} \, (20 \underline{/-30°})$$

$$= 4\sqrt{2} \underline{/-21.9°} = 5.66 \underline{/-21.9°}$$

and

$$v_{oc}(t) = 4\sqrt{2} \, \cos(4t - 21.9°) = 5.66 \, \cos(4t - 21.9°)$$

To find the Thévenin equivalent impedance $\mathbf{Z}_0$, set the independent source to zero. The resulting circuit is shown in Fig. 8.32.

fig. 8.32

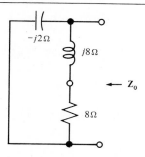

We can find the expression for $\mathbf{Z}_0$ by combining impedances in series and parallel. In particular, (the symbol $\|$ means "in parallel with"),

$$\mathbf{Z}_0 = (\mathbf{Z}_L + \mathbf{Z}_R) \| \mathbf{Z}_C$$

$$= \frac{(j8 + 8)(-j2)}{j8 + 8 - j2} = \frac{16(1 - j)}{2(4 + j3)} = \frac{8}{25} - j \frac{56}{25} = 2.26 \underline{/-81.9°} \, \Omega$$

In the frequency domain, therefore, we have the circuit in Fig. 8.33. To represent this circuit in the time domain we can use the *RLC* combination from which we determined $\mathbf{Z}_0$, thus yielding the circuit of Fig. 8.34.

fig. 8.33

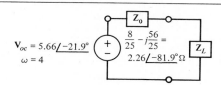

fig. 8.34

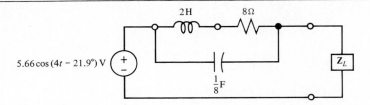

Alternatively, we can use the fact that $\mathbf{Z}_0 = \frac{8}{25} - j\frac{56}{25}$ Ω has the form of a resistor in series with a capacitor; that is,

$$\mathbf{Z}_0 = R + \frac{1}{j\omega C} = R - \frac{j}{\omega C}$$

Thus,

$$R = \frac{8}{25}\ \Omega$$

$$\frac{1}{\omega C} = \frac{56}{25} \Rightarrow C = \frac{25}{224}\ \text{F}$$

and the equivalent circuit is shown in Fig. 8.35.

fig. 8.35

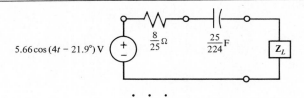

$\cdot\ \cdot\ \cdot$

For the impedance $\mathbf{Z} = a + jb$, we call a the *resistance* and b the *reactance*. For the case that b is negative (as in the above example) we say that the reactance is *capacitive* and denote it by X_C. If b is positive, we say that the reactance is *inductive* and denote it by X_L. In other words, an impedance has the form $\mathbf{Z} = R + jX$.

$\cdot\ \cdot\ \cdot$

Example
Now consider the sinusoidal circuit in Fig. 8.36, which contains a dependent source.

fig. 8.36

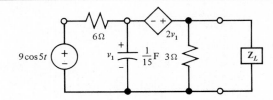

fig. 8.37

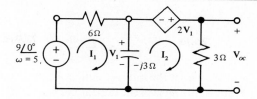

We find $\mathbf{V}_{oc}$ from the circuit with the load removed (see Fig. 8.37). By mesh analysis, for the first mesh,

$$6\mathbf{I}_1 - j3(\mathbf{I}_1 - \mathbf{I}_2) = 9$$

from which

$$(2 - j)\mathbf{I}_1 + j\mathbf{I}_2 = 3 \tag{8.5}$$

For the second mesh,

$$-\mathbf{V}_1 - 2\mathbf{V}_1 + 3\mathbf{I}_2 = 0 \qquad \text{or} \qquad -3\mathbf{V}_1 + 3\mathbf{I}_2 = 0$$

from which

$$j3(\mathbf{I}_1 - \mathbf{I}_2) + \mathbf{I}_2 = 0$$

or

$$j3\mathbf{I}_1 + (1 - j3)\mathbf{I}_2 = 0 \tag{8.6}$$

Writing Equations (8.5) and (8.6) in matrix form, we get

$$\begin{bmatrix} 2 - j & j \\ j3 & 1 - j3 \end{bmatrix} \begin{bmatrix} \mathbf{I}_1 \\ \mathbf{I}_2 \end{bmatrix} = \begin{bmatrix} 3 \\ 0 \end{bmatrix}$$

Calculating determinants

$$\Delta = \begin{vmatrix} 2 - j & j \\ j3 & 1 - j3 \end{vmatrix} = (2 - j)(1 - j3) - (j3)(j) = 2 - j7$$

$$\Delta_1 = \begin{vmatrix} 3 & j \\ 0 & 1 - j3 \end{vmatrix} = 3(1 - j3)$$

$$\Delta_2 = \begin{vmatrix} 2 - j & 3 \\ j3 & 0 \end{vmatrix} = -j9$$

By Cramer's rule,

$$\mathbf{I}_1 = \frac{\Delta_1}{\Delta} = \frac{3(1 - j3)}{2 - j7} = \frac{3(23 + j)}{53}$$

$$\mathbf{I}_2 = \frac{\Delta_2}{\Delta} = \frac{-j9}{2 - j7} = \frac{-j9(2 + j7)}{(2 - j7)(2 + j7)} = \frac{9(7 - j2)}{53}$$

Thus

$$\mathbf{V}_{oc} = 3\mathbf{I}_2 = \frac{27}{53}(7 - j2) = \frac{27}{\sqrt{53}} \underline{/\tan^{-1}(-2/7)} = 3.71 \underline{/-16°}$$

To determine $\mathbf{Z}_0$ we set the independent voltage source to zero, but leave the dependent source as is. Because of the dependent source, $\mathbf{Z}_0$ cannot be obtained by combining elements in series and parallel. Instead, in the circuit shown in Fig. 8.38, $\mathbf{Z}_0 = \mathbf{V}_0/\mathbf{I}_0$.

fig. 8.38

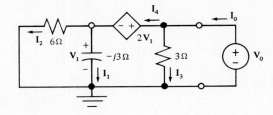

By KVL,

$$\mathbf{V}_1 = -2\mathbf{V}_1 + \mathbf{V}_0$$

or

$$3\mathbf{V}_1 = \mathbf{V}_0$$

$$\mathbf{V}_1 = \frac{\mathbf{V}_0}{3}$$

Also,

$$\mathbf{I}_0 = \mathbf{I}_3 + \mathbf{I}_4 = \mathbf{I}_3 + \mathbf{I}_2 + \mathbf{I}_1$$

$$= \frac{\mathbf{V}_0}{3} + \frac{\mathbf{V}_1}{6} + \frac{\mathbf{V}_1}{-j3}$$

$$= \frac{\mathbf{V}_0}{3} + \frac{\mathbf{V}_0}{18} + \frac{\mathbf{V}_0}{-j9} = \mathbf{V}_0\left(\frac{7 + j2}{18}\right)$$

so

$$\mathbf{Z}_0 = \frac{\mathbf{V}_0}{\mathbf{I}_0} = \frac{18}{7 + j2} = \frac{18(7 - j2)}{53}$$

$$= \frac{126}{53} - j\frac{36}{53} = 2.47 \underline{/-16°}$$

Thus we have the situation depicted in Fig. 8.39.

fig. 8.39

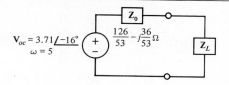

Since $\mathbf{Z}_0$ has the form

$$R + \frac{1}{j\omega C} = R - \frac{j}{\omega C}$$

then

$$R = \frac{126}{53}\,\Omega$$

and

$$\frac{1}{\omega C} = \frac{36}{53} \Rightarrow C = \frac{53}{36\omega} = \frac{53}{180}\,F$$

so that in the time domain we obtain the circuit shown in Fig. 8.40.

fig. 8.40

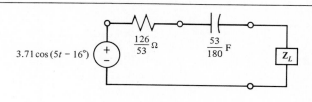

. . .

For the last two examples, what is the load impedance $\mathbf{Z}_L$ that will absorb the maximum power? We shall proceed to answer this question in the next chapter.

PROBLEMS   8.14.   Using frequency domain analysis, repeat Problem 8.1, which was given in the first problem set of this chapter.
8.15.   Using frequency domain analysis, repeat Problem 8.2.
*8.16.   Using frequency domain analysis, repeat Problem 8.3 by using nodal analysis.
Ans.   3.71 cos(5t − 15.95°), lag
*8.17.   Using frequency domain analysis, repeat Problem 8.4 by using voltage division.
Ans.   5.66 cos(4t − 21.9°), lead
*8.18.   Using frequency domain analysis, repeat Problem 8.5.
Ans.   1.6 cos(2t − 90°), lag
8.19.   Using frequency domain analysis, repeat Problem 8.6.

fig. P8.20

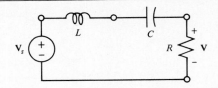

8.20. For the series *RLC* sinusoidal circuit shown in Fig. P8.20, find an expression for V in terms of R, L, C, ω, and V_s.

8.21. For the circuit given in Problem 8.20, suppose that $R = 4\,\Omega$, $L = 2\,H$, and $\omega = 3\,\text{rad/s}$. Determine whether $v(t)$ leads or lags $v_s(t)$ when C is (a) $\frac{1}{8}$ F, (b) $\frac{1}{18}$ F, and (c) $\frac{1}{32}$ F.

fig. P8.22

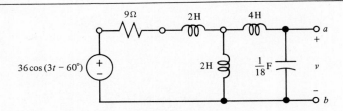

8.22. For the circuit in Fig. P8.22, find $v(t)$.

8.23. For the circuit given in Problem 8.22, find the impedance seen by the independent voltage source.

fig. P8.24

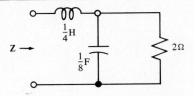

*8.24. For the *RLC* connection shown in Fig. P8.24, find the impedance Z when (a) $\omega = 1$, (b) $\omega = 4$, (c) $\omega = 8\,\text{rad/s}$.

Ans. (a) $\frac{32}{17} - j\frac{15}{68}\,\Omega$

fig. P8.25

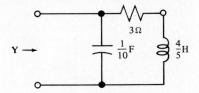

8.25. For the *RLC* connection in Fig. P8.25, find the admittance Y when (a) $\omega = 1$, (b) $\omega = 5$, and (c) $\omega = 10\,\text{rad/s}$.

8.26. For the circuit given in Problem 8.22, find the Thévenin equivalent circuit with respect to terminals a and b.

fig. P8.27

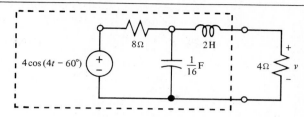

*8.27. For the circuit shown in Fig. P8.27, find $v(t)$ by first replacing that portion of the circuit in the dashed box by its Thévenin equivalent.

Ans. $\dfrac{4}{\sqrt{17}}\cos(4t - 164°)$

fig. P8.28

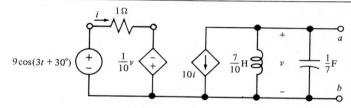

8.28. Find the Thévenin equivalent of the circuit in Fig. P8.28.

fig. P8.29

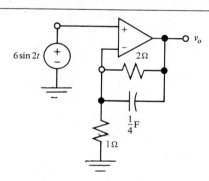

*8.29. For the op amp circuit in Fig. P8.29, find $v_o(t)$.
Ans. $6\sqrt{5}\cos(2t - 116.6°)$

fig. P8.30

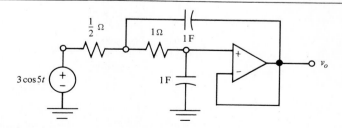

8.30. Find $v_o(t)$ for the op amp circuit in Fig. P8.30.

8.31. For the circuit given in Problem 8.30, find the impedance seen by the independent voltage source.

fig. P8.32

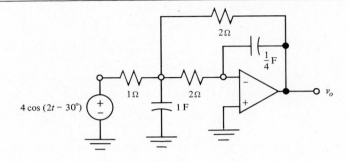

8.32. For the op amp circuit in Fig. P8.32, find $v_o(t)$.

8.33. For the circuit given in Problem 8.32, find the impedance seen by the independent voltage source.

fig. P8.34

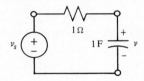

8.34. For the simple series RC circuit shown in Fig. P8.34, find $v(t)$ given that

$$v_s(t) = \frac{1}{\pi} - \frac{1}{2} \sin \pi t - \frac{2}{3\pi} \cos 2\pi t$$

8.35. Repeat Problem 8.34 for the case that $v_s(t) = 10 \cos t + 10 \cos 10t$.

fig. P8.36

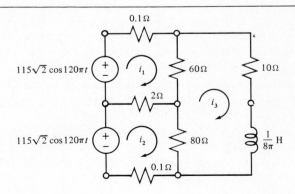

8.36. Find the mesh currents $i_1(t)$, $i_2(t)$, and $i_3(t)$, for the circuit shown in Fig. P8.36.

fig. P8.37

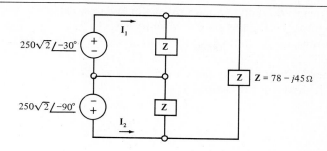

8.37. For the circuit in Fig. P8.37, find the currents I_1 and I_2.

fig. P8.38

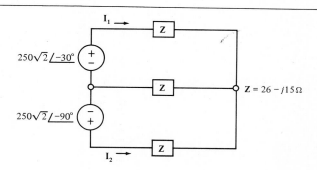

*8.38.   Repeat Problem 8.37 for the circuit shown in Fig. P8.38.
Ans.   6.8 $\underline{/30°}$, 6.8 $\underline{/-90°}$

Summary

1. If the input of a linear, time-invariant circuit is a sinusoid, the response is a sinusoid of the same frequency.
2. A complex number **c** can be written in rectangular form as $\mathbf{c} = a + jb$, where a is the real part of **c** and b is the imaginary part of **c**. It can also be written in polar form as $\mathbf{c} = Me^{j\theta}$, where M (also denoted $|\mathbf{c}|$) is a positive number called the magnitude of **c** and θ is the angle of **c**.
3. Finding the magnitude and phase angle of a sinusoidal steady-state response can be accomplished with either real or complex sinusoids.
4. If the output of a sinusoidal circuit reaches its peak before the input, the circuit is a lead network. Conversely, it is a lag network.
5. Using the concepts of phasors and impedance (or admittance), sinusoidal circuits can be analyzed in the

frequency domain in a manner analogous to resistive
circuits by using the phasor versions of KCL, KVL, nodal
analysis, mesh analysis, and loop analysis.

6. Such important circuit concepts as the principle of
   superposition and Thévenin's theorem are also applicable
   in the frequency domain.

POWER 9

Introduction

In our previous analyses our major concern has been with determining voltages and currents. In many applications (e.g., electric utilities), energy or power supplied and absorbed are extremely important parameters. Knowing the instantaneous power is useful, since many electrical and electronic devices have maximum instantaneous or "peak" power ratings that, for satisfactory operation, should not be exceeded. By averaging instantaneous power we get "average power," which is the average rate at which energy is supplied or absorbed. Power usages vary from a fraction of a watt for small electronic circuits to millions of watts for large electric utilities.

We have previously defined power to be the product of voltage and current. For the case that voltage and current are constants (the dc case), the instantaneous power is equal to the average value of the power. This result is a consequence of the fact that the instantaneous and average values of a constant are equal. A sinusoid, however, has an average value of zero. This does not mean that power due to a sinusoid is zero. We shall see what effect sinusoids have in terms of power.

Just as we were able to determine under what condition maximum power transfer occurred for resistive circuits, in this chapter we deal with *RLC* sinusoidal circuits.

Although our prime interest in this chapter will be with the sinusoid, we shall also be able to discuss average power for the case of repetitive or periodic nonsinusoidal waveforms by using the notion of effective values. Yet it is the sinusoid that gets the greatest attention, since it is in this form that electric power is produced and distributed.

9.1 AVERAGE POWER

Recall that for the arbitrary linear element shown in Fig. 9.1, the instantaneous power absorbed by the element is

$$p(t) = v(t)i(t)$$

For the case that the voltage is a sinusoid, say

$$v(t) = V \cos(\omega t + \theta_1)$$

fig. 9.1

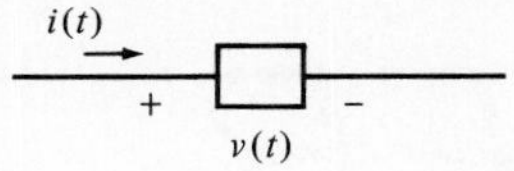

the current will also be sinusoidal, say

$$i(t) = I \cos(\omega t + \theta_2)$$

Then the instantaneous power is

$$p(t) = [V \cos(\omega t + \theta_1)][I \cos(\omega t + \theta_2)]$$

Using the trigonometric identity

$$(\cos \alpha)(\cos \beta) = \frac{1}{2} \cos(\alpha - \beta) + \frac{1}{2} \cos(\alpha + \beta)$$

we get

$$p(t) = VI\left[\frac{1}{2} \cos(\theta_1 - \theta_2) + \frac{1}{2} \cos(2\omega t + \theta_1 + \theta_2)\right]$$

$$= \frac{VI}{2} \cos(\theta_1 - \theta_2) + \frac{VI}{2} \cos(2\omega t + \theta_1 + \theta_2)$$

If this is the instantaneous power, what is the average power P? The second term on the right in the previous equation is a sinusoid of frequency 2ω (radians per second). We know that the average value of a sinusoid is zero; this can be verified with calculus. Thus, the *average power absorbed* is

$$P = \frac{1}{2} VI \cos(\theta_1 - \theta_2)$$

Since $\cos(-x) = \cos x$, we can write

$$\boxed{P = \frac{1}{2} VI \cos \theta}$$

where θ is the difference in phase angles between voltage and current—that is, either $\theta_1 - \theta_2$ or $\theta_2 - \theta_1$.

For the special case of a resistor (see Fig. 9.2), since the voltage and current are in phase, the average power absorbed is

$$P_R = \frac{1}{2} VI \cos 0 = \frac{1}{2} VI$$

fig. 9.2

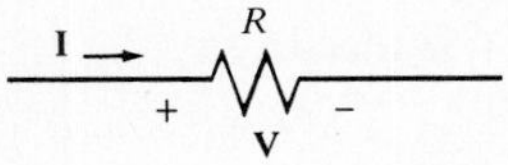

From the fact that

$$V = RI$$

we get

$$P_R = \frac{1}{2}(RI)I$$

or

$$\boxed{P_R = \frac{1}{2}RI^2}$$

Also,

$$P_R = \frac{1}{2}V\left(\frac{V}{R}\right)$$

or

$$\boxed{P_R = \frac{1}{2}\frac{V^2}{R}}$$

Let us now find the average power absorbed by an impedance that is purely imaginary (reactive)—that is, $\mathbf{Z} = jX$, where X is a positive or a negative number, as depicted by Fig. 9.3.

fig. 9.3

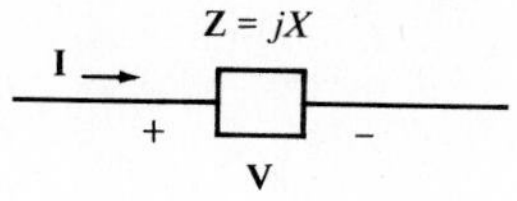

Since

$$\mathbf{Z} = \frac{\mathbf{V}}{\mathbf{I}} = \frac{V\underline{/\phi_1}}{I\underline{/\phi_2}} = \frac{V}{I}\underline{/(\phi_1 - \phi_2)}$$

for positive X,

$$jX = X\underline{/90°}$$

whereas, for negative X,

$$jX = -j(-X) = -X\underline{/-90°}$$

we have

$$\mathbf{Z} = |X| \underline{/ \pm 90^\circ} = \frac{V}{I} \underline{/ (\phi_1 - \phi_2)}$$

Thus the average power P_X absorbed by this impedance is

$$P_X = \tfrac{1}{2} VI \cos(\phi_1 - \phi_2) = \tfrac{1}{2} VI \cos(\pm 90^\circ)$$
$$= 0$$

That is, purely reactive impedances, like (ideal) inductors and capacitors or combinations of the two, absorb zero average power, and are referred to as *lossless elements*.

· · ·

Example
Consider the *RC* circuit shown in Fig. 9.4, which has the phasor form shown in Fig. 9.5.

fig. 9.4 (left)
fig. 9.5 (right)

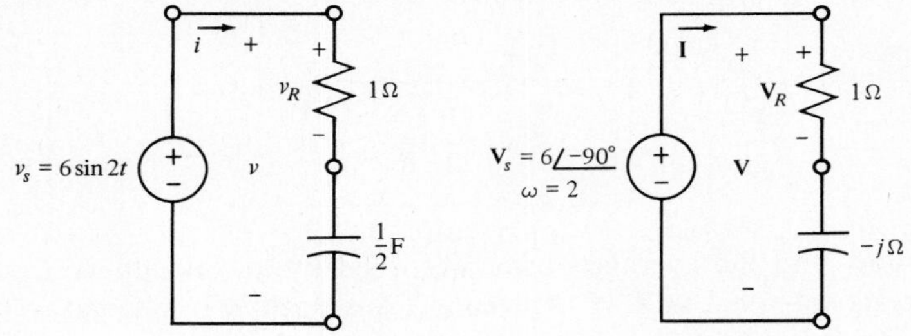

Since

$$\mathbf{I} = \frac{\mathbf{V}_s}{\mathbf{Z}} = \frac{6 \underline{/-90^\circ}}{1 - j} = \frac{6 \underline{/-90^\circ}}{\sqrt{2} \underline{/-45^\circ}}$$

$$\mathbf{I} = 3\sqrt{2} \underline{/-45^\circ}$$

then the average power absorbed by the series *RC* impedance (supplied by the voltage source) is

$$P = \frac{1}{2} VI \cos \theta$$

$$= \frac{1}{2} (6)(3\sqrt{2}) \cos(-90^\circ + 45^\circ)$$

$$= 9\sqrt{2} \cos(-45^\circ)$$

$$= 9 \text{ W}$$

By voltage division, the voltage across the resistor is

$$\mathbf{V}_R = \frac{\mathbf{Z}_R}{\mathbf{Z}_R + \mathbf{Z}_C}\mathbf{V}_s$$

$$= \frac{1}{1-j}(6\,\underline{/-90^\circ}) = \frac{6\,\underline{/-90^\circ}}{\sqrt{2}\,\underline{/-45^\circ}}$$

$$= \frac{6}{\sqrt{2}}\,\underline{/-45^\circ} = V_R\,\underline{/\,\phi_R}$$

and the power absorbed by the resistor is

$$P_R = \frac{1}{2}\frac{V_R^2}{R}$$

$$= \frac{1}{2}\left(\frac{18}{1}\right) = 9\,\text{W}$$

Of course, this conclusion could have been obtained using the fact that the impedance of a capacitor is purely imaginary, so the power absorbed by the capacitor is $P_c = 0$, and all of the power supplied by the source must, therefore, be absorbed by the resistor.

. . .

In the preceding example, we specifically considered a resistor connected in series with a capacitor. However, since an arbitrary load $\mathbf{Z}_L$ generally has a real and an imaginary part, i.e., $\mathbf{Z}_L = R_L + jX_L$, we can model $\mathbf{Z}_L$ as a resistance R_L connected in series with a reactance X_L (remember, X_L can be positive or negative) as shown in Fig. 9.6. Since a pure reactance absorbs zero average power, if the current through $\mathbf{Z}_L$ is $\mathbf{I}_L$, then the average power absorbed by $\mathbf{Z}_L$ is

$$P = \frac{1}{2}|\mathbf{I}_L|^2R_L$$

fig. 9.6

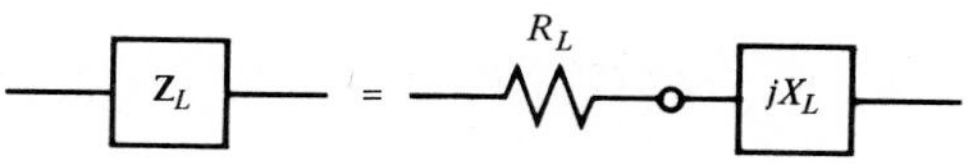

Now let us answer a question posed earlier. Given a circuit having a load impedance $\mathbf{Z}_L = R_L + jX_L$, as shown in Fig. 9.7, what impedance $\mathbf{Z}_L$ will result in the maximum average power being absorbed by the load?

fig. 9.7 (left)
fig. 9.8 (right)

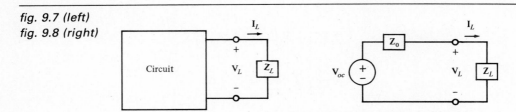

The effect on the load is the same if the circuit is replaced by its Thévenin equivalent, as shown in Fig. 9.8, where $\mathbf{Z}_0 = R_0 + jX_0$. We have already seen that if $\mathbf{Z}_0 = R_0$ and $\mathbf{Z}_L = R_L$, then the maximum power is absorbed by the load (i.e., maximum power is transferred to the load) when $R_L = R_0$. However, for the general case, we have

$$\mathbf{I}_L = \frac{\mathbf{V}_{oc}}{\mathbf{Z}_L + \mathbf{Z}_0} = I_L \underline{/\phi_1}$$

Let

$$\mathbf{V}_{oc} = V_{oc} \underline{/\phi_{oc}}$$

Since

$$|\mathbf{I}_L| = \frac{|\mathbf{V}_{oc}|}{|\mathbf{Z}_L + \mathbf{Z}_0|}$$

and

$$\mathbf{Z}_L + \mathbf{Z}_0 = R_L + jX_L + R_0 + jX_0 = (R_L + R_0) + j(X_L + X_0)$$

then

$$I_L = \frac{V_{oc}}{\sqrt{(R_L + R_0)^2 + (X_L + X_0)^2}}$$

and the average power absorbed by the load is

$$P_L = \frac{1}{2} I_L^2 R_L$$

$$= \frac{1}{2} \frac{V_{oc}^2 R_L}{(R_L + R_0)^2 + (X_L + X_0)^2}$$

The quantity V_{oc} is a constant which depends upon the given circuit. Once the values of R_0 and R_L have been selected, the term $(X_L + X_0)^2$ tends only to reduce P_L. To make P_L as large as possible, make $(X_L + X_0)^2$ zero by choosing $X_L = -X_0$. Then

$$P_L = \frac{1}{2} \frac{R_L V_{oc}^2}{(R_L + R_0)^2}$$

and just as in the case of maximum power transfer when $\mathbf{Z}_0 = R_0$ and $\mathbf{Z}_L = R_L$, the quantity P_L is maximum when $R_L = R_0$. Thus, in order to get maximum power to $\mathbf{Z}_L$, we select

$$\boxed{\mathbf{Z}_L = R_L + jX_L = R_0 - jX_0 = \mathbf{Z}_0{}^*}$$

Under this condition, the power absorbed by $\mathbf{Z}_L$ is

$$P_{L(max)} = \frac{1}{2} \frac{R_0 V_{oc}^2}{(R_0 + R_0)^2} = \frac{V_{oc}^2}{8R_0}$$

Suppose now that $\mathbf{Z}_L$ is restricted to be purely real (resistive). What value of R_L would result in maximum power transfer to the load? For the circuit in Fig. 9.9, we have

$$\mathbf{I}_L = \frac{\mathbf{V}_{oc}}{R_L + \mathbf{Z}_0} = I_L \underline{/\phi_1}$$

fig. 9.9

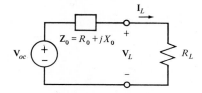

where

$$\mathbf{V}_{oc} = V_{oc} \underline{/\phi_{oc}}$$

Since

$$P_L = \frac{1}{2} I_L{}^2 R_L$$

and

$$I_L = \frac{V_{oc}}{\sqrt{(R_L + R_0)^2 + X_0{}^2}}$$

then

$$P_L = \frac{1}{2} \frac{V_{oc}^2 R_L}{(R_L + R_0)^2 + X_0{}^2}$$

To determine for what value of R_L this is maximum, set $dP_L/dR_L = 0$. Thus

$$\frac{dP_L}{dR_L} = \frac{1}{2}\left(\frac{[(R_L + R_0)^2 + X_0{}^2]V_{oc}^2 - 2(R_L + R_0)V_{oc}^2 R_L}{[(R_L + R_0)^2 + X_0{}^2]^2}\right) = 0$$

or equality holds when

$$[(R_L + R_0)^2 + X_0{}^2]V_{oc}^2 = 2(R_L + R_0)V_{oc}^2 R_L$$

from which

$$R_L{}^2 + 2R_0 R_L + R_0{}^2 + X_0{}^2 = 2R_L{}^2 + 2R_0 R_L$$

or

$$R_L{}^2 = R_0{}^2 + X_0{}^2$$

and hence

$$\boxed{R_L = \sqrt{R_0{}^2 + X_0{}^2} = |\mathbf{Z}_0|}$$

This is the condition for either maximum P_L or minimum P_L. Since it is obvious that minimum load power occurs when $R_L = 0$, this is the condition for maximum power transfer to a purely resistive load R_L.

$$\cdot \quad \cdot \quad \cdot$$

Example

The circuit shown in Fig. 9.10 was considered previously (see Fig. 8.36). It was determined that

$$\mathbf{Z}_0 = \frac{126}{53} - j\frac{36}{53} = 2.47\underline{/-16°}$$

---

fig. 9.10

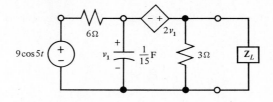

Thus maximum power is delivered to the load $\mathbf{Z}_L$ when

$$\mathbf{Z}_L = \mathbf{Z}_0{}^* = \frac{126}{53} + j\frac{36}{53} = 2.47\underline{/16°}\ \Omega$$

Since $\mathbf{Z}_L$ has the form $\mathbf{Z}_L = R_L + j\omega L$, we have that

$$R_L = \frac{126}{53}\ \Omega$$

and

$$\omega L = \frac{36}{53} \Rightarrow L = \frac{36}{53\omega} = \frac{36}{265} \text{ H}$$

That is, $\mathbf{Z}_L$ is given by the circuit shown in Fig. 9.11 and the power absorbed by $\mathbf{Z}_L$ is

$$P_{L_{(max)}} = \frac{V_{oc}^2}{8R_0}$$

$$= \frac{(27/\sqrt{53})^2}{8(126/53)} = 0.723 \text{ W}$$

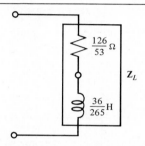

fig. 9.11

The purely resistive load R_L that absorbs the maximum amount of power is

$$R_L = \sqrt{R_0^2 + X_0^2}$$

$$= \sqrt{\left(\frac{126}{53}\right)^2 + \left(\frac{36}{53}\right)^2} = \frac{131}{53} \Omega$$

and the power delivered to R_L is

$$P_L = \frac{1}{2} \frac{V_{oc}^2 R_L}{(R_L + R_0)^2 + X_0^2}$$

$$= \frac{1}{2} \frac{(27/\sqrt{53})^2(131/53)}{\left(\frac{131}{53} + \frac{126}{53}\right)^2 + \left(\frac{36}{53}\right)^2} = 0.709 \text{ W}$$

$$\bullet \quad \bullet \quad \bullet$$

PROBLEMS   ***9.1.**   For the *RLC* circuit shown in Fig. P9.1, find the average power absorbed by the 4-Ω load resistor for the case that (a) $C = \frac{1}{8}$ F, (b) $C = \frac{1}{18}$ F, and (c) $C = \frac{1}{32}$ F.
Ans.   (a) 7.38 W, (b) 12.5 W, (c) 5.29 W

fig. P9.1

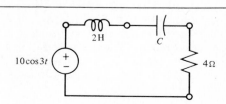

$10\cos 3t$

fig. P9.2

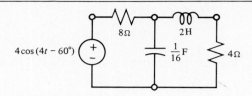

9.2.   Consider the circuit in Fig. P9.2.
(a) Find the average power dissipated by each resistor.
(b) Replace the 4-Ω load resistor by an impedance that absorbs the maximum average power, and determine this maximum power.
(c) Replace the 4-Ω load resistor by a pure resistance that absorbs the maximum power for resistive loads, and determine this power.

fig. P9.3

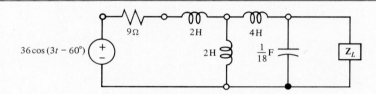

*9.3.   Consider the circuit shown in Fig. P9.3.
(a) Find the load impedance $\mathbf{Z}_L$ that absorbs the maximum average power, and determine this power.
(b) Find the load resistance that absorbs the maximum power for resistive loads, and determine this power.
Ans.   (a) $0.5 + j9.5$ Ω, 18 W; (b) 9.51 Ω, 1.8 W

fig. P9.4

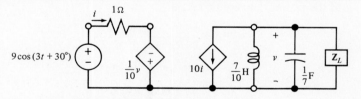

9.4.   Repeat Problem 9.3 for the circuit in Fig. P9.4.
*9.5.   For the op amp circuit given in Problem 8.29, in the last problem set of Chapter 8, find the average power absorbed by each resistor.
Ans.   18 W
9.6.   For the op amp circuit given in Problem 8.30, find the average power absorbed by each resistor.
*9.7.   For the op amp circuit given in Problem 8.32, find the average power absorbed by each resistor.
Ans.   4.15 W, 0.64 W, 1.23 W
9.8.   For the circuit given in Problem 8.37, (a) find the average power absorbed by each impedance, (b) find the average power supplied by each source, (c) find the instantaneous power supplied by each source, and (d) find the total instantaneous power supplied by both sources.

9.9.   Repeat Problem 9.8 for the circuit given in Problem 8.38.

fig. P9.10

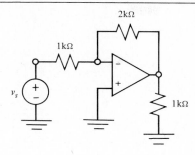

*9.10.   Consider the op amp circuit shown in Fig. P9.10. For the case that
$v_s(t) = \cos(\omega t + \phi)$,
(a)   Find the average power absorbed by each resistor.
(b)   Find the average power supplied by the independent source.
(c)   Why doesn't the power supplied in (b) equal the power absorbed in
(a)?

Ans.   (a) 0.5 mW, 1 mW, 2 mW; (b) 0.5 mW

fig. P9.11

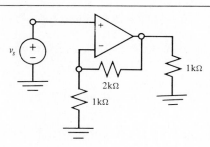

9.11.   Repeat Problem 9.10 for the op amp circuit in Fig. P9.11.
9.12.   What is the average power absorbed by the op amp (a) in Problem
9.10 and (b) in Problem 9.11?

9.2   EFFECTIVE VALUES

We have seen that for a resistor (see Fig. 9.12) with a sinusoidal current through it, say
$i(t) = I \cos(\omega t + \phi)$, the average power dissipated by R is $\frac{1}{2}I^2R$. What direct current I_e
would yield the same power dissipation by R?

fig. 9.12

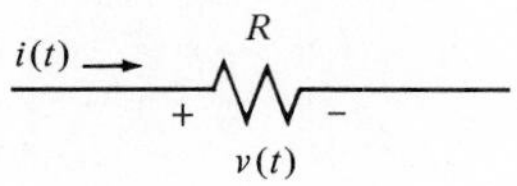

$i(t) \longrightarrow$ $\overset{R}{\underset{+ \quad -}{\wedge\wedge\wedge}}$
$v(t)$

Since the power dissipated by R due to a direct current I_e is $I_e{}^2R$, we get the equality

$$I_e{}^2R = \frac{1}{2}I^2R$$

when

$$I_e{}^2 = \frac{I^2}{2}$$

or

$$\boxed{I_e = \frac{I}{\sqrt{2}} \approx 0.707I}$$

Thus, the constant current $I_e = I/\sqrt{2}$ and the sinusoidal current $i(t) = I\cos(\omega t + \phi)$ result in the same average power dissipation in R. For this reason, we say that $I/\sqrt{2}$ is the *effective value* of the sinusoid whose amplitude is I.

Since the voltage across R is

$$v(t) = Ri(t) = RI\cos(\omega t + \phi) = V\cos(\omega t + \phi)$$

then the effective value of $v(t)$ is $V_e = V/\sqrt{2}$. Furthermore, the average power absorbed by R is

$$P_R = \frac{1}{2}VI\cos\theta$$

where $\theta = 0$, since the voltage and current are in phase. Thus,

$$P_R = \frac{VI}{2}$$

$$= \frac{V}{\sqrt{2}}\frac{I}{\sqrt{2}} = V_eI_e$$

Alternatively, since $V = RI$, then

$$P_R = \frac{RI^2}{2} = R\frac{I}{\sqrt{2}}\frac{I}{\sqrt{2}} = RI_e{}^2$$

and

$$P_R = \frac{V}{2}\left(\frac{V}{R}\right) = \frac{1}{R}\frac{V}{\sqrt{2}}\frac{V}{\sqrt{2}} = \frac{V_e{}^2}{R}$$

In summary, given a resistor R having current $i(t) = I \cos(\omega t + \phi)$ through it and voltage $v(t) = V \cos(\omega t + \phi)$ across it, then the power absorbed by the resistor is

$$P_R = V_e I_e = I_e^2 R = \frac{V_e^2}{R}$$

where $V_e = V/\sqrt{2}$ and $I_e = I/\sqrt{2}$ are the effective values of the voltage and the current, respectively.

For an arbitrary element,

$$P = \frac{1}{2} VI \cos \theta$$

or

$$P = V_e I_e \cos \theta$$

where θ is the difference in phase between the voltage and the current.

· · ·

Example
Ordinary household voltage is designated typically as 115 V ac, 60 Hz. The term "ac," which stands for *alternating current*, indicates a sinusoid, and 60 Hz is the frequency of the sinusoid. However, the 115 V does not refer to the amplitude of the sinusoid but rather its effective value. Thus, the amplitude of the sinusoid is $115\sqrt{2} \approx 163$ V.

· · ·

Again, for the resistor (Fig. 9.12), consider the case that the current (or the voltage) is nonsinusoidal. Since the instantaneous power absorbed or dissipated by the resistor is

$$p(t) = v(t)i(t)$$

by Ohm's law,

$$v(t) = Ri(t)$$

we get

$$p(t) = Ri^2(t)$$

Summing the instantaneous power from time t_1 to t_2 and dividing by the interval $\Delta t = t_2 - t_1$, yields the average power dissipated in the interval; that is,

$$P = \frac{1}{t_2 - t_1} \int_{t_1}^{t_2} p(t)\, dt$$

$$= \frac{1}{t_2 - t_1} \int_{t_1}^{t_2} Ri^2(t)\, dt$$

Suppose that the current (and hence, the voltage) is "repetitive" or "periodic," and the period is T. Then the average power dissipated in one period is

$$P = \frac{1}{T} \int_{t_0}^{t_0+T} p(t) \, dt$$

$$= \frac{1}{T} \int_{t_0}^{t_0+T} Ri^2(t) \, dt$$

where t_0 is an arbitrary value. Again, let I_e be the direct current that results in the same amount of average power dissipation. Thus

$$RI_e^2 = \frac{1}{T} \int_{t_0}^{t_0+T} Ri^2(t) \, dt$$

from which the expression for I_e, the effective value of $i(t)$, is

$$I_e = \sqrt{\frac{1}{T} \int_{t_0}^{t_0+T} i^2(t) \, dt}$$

$$\underbrace{\text{square } root \quad \text{mean} \quad square}_{\substack{\text{average} \\ \text{or} \\ \text{mean}}}$$

and for this reason the effective value is also known as the *root-mean-square* or *rms* value. Similarly, the effective or rms value of $v(t)$ is

$$V_e = \sqrt{\frac{1}{T} \int_{t_0}^{t_0+T} v^2(t) \, dt}$$

It is a routine exercise in integral calculus to confirm that the rms value of the sinusoid $A \cos(\omega t + \phi)$, whose period is $T = 2\pi/\omega$, is $A/\sqrt{2}$.

· · ·

Example
The "sawtooth" waveform shown in Fig. 9.13 is periodic, where the period is $T = 1$ second. Although the average or mean value of this function is $1/2$ (why?), the rms

fig. 9.13

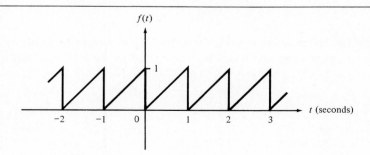

value of $f(t)$ is

$$\sqrt{\frac{1}{T}\int_0^T f^2(t)\,dt} = \sqrt{\frac{1}{1}\int_0^1 t^2\,dt}$$

$$= \sqrt{\left.\frac{t^3}{3}\right|_0^1} = \frac{1}{\sqrt{3}}$$

In other words, if the current through a resistor R is $i(t) = f(t)$, then the average power dissipated by R is

$$P = RI_e^2$$

$$= R\left(\frac{1}{\sqrt{3}}\right)^2 = \frac{R}{3} \qquad \text{watts}$$

. . .

Returning now to the sinusoidal case, consider the general load shown in Fig. 9.14, where $v(t) = V\cos(\omega t + \phi_1)$ and $i(t) = I\cos(\omega t + \phi_2)$. Then the average power absorbed by the load is

$$P = \frac{1}{2}VI\cos\theta$$

fig. 9.14

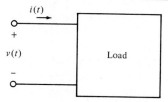

where $\theta = \phi_1 - \phi_2$. In terms of effective values,

$$P = \frac{V}{\sqrt{2}}\frac{I}{\sqrt{2}}\cos\theta = V_e I_e \cos\theta \tag{9.1}$$

If the network is purely resistive, then the average power it absorbs is

$$P_R = V_e I_e$$

However, if the network is not purely resistive, even though it might "appear" at first glance that $V_e I_e$ is the average power absorbed, this quantity is given by Equation (9.1). We call $V_e I_e$ the *apparent power* and use the units *volt-amperes* (VA) to distinguish this quantity from the actual power (the average power) that is absorbed by the load.

The quantity $\cos \theta$ is called the *power factor* (abbreviated pf) and is the ratio of average power to apparent power; that is,

$$pf = \frac{\text{average power}}{\text{apparent power}} = \cos \theta = \frac{P}{V_e I_e}$$

The angle θ is referred to as the *power factor angle.*

The impedance of the load is

$$\mathbf{Z} = \frac{\mathbf{V}}{\mathbf{I}} = \frac{V \underline{/\phi_1}}{I \underline{/\phi_2}} = \frac{V}{I} \underline{/(\phi_1 - \phi_2)} = \frac{V}{I} \underline{/\theta}$$

so that the angle of the load is the pf angle. The general form of a load impedance is $\mathbf{Z} = R + jX$. If $X > 0$, we say that the load is *inductive*; if $X < 0$, we say it is *capacitive.* Assuming a positive resistance R, then for an inductive load,

$$\theta = \text{ang } \mathbf{Z} = \tan^{-1} \frac{X}{R} > 0$$

that is, the pf is positive; for a capacitive load,

$$\theta = \text{ang } \mathbf{Z} = \tan^{-1} \frac{X}{R} < 0$$

that is, the pf is negative. The former case ($\phi_1 > \phi_2$) implies that the current lags the voltage, while the latter case ($\phi_1 < \phi_2$) implies that the current leads the voltage. In the vernacular of the power industry, we refer to the pf as being *lagging* or *leading*, respectively. In other words, if the current lags the voltage for a load (the inductive case), the result is a lagging pf, while if the current leads the voltage (the capacitive case), the result is a leading pf.

You may ask, "Why talk about lagging and leading pf at all? Why not just talk about the impedance of the load?" One reason is that in some applications (e.g., power systems) a load may not consist simply of ordinary inductors, capacitors, and resistors. In the following example, the load is an electric motor.

· · ·

Example
A 1000-W electric motor is connected to a source of 200 V rms, 60 Hz, and the result is a lagging pf of 0.8. Since the pf is lagging, the angle of the current is less than the voltage angle by $\cos^{-1}(0.8) = 36.9°$. Arbitrarily selecting the voltage angle to be zero, we depict the given situation in Fig. 9.15.

fig. 9.15

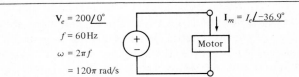

$V_e = 200\underline{/0°}$

$f = 60\,\text{Hz}$

$\omega = 2\pi f$

$= 120\pi \text{ rad/s}$

$\mathbf{I}_m = I_e\underline{/-36.9°}$

Motor

Since

$$P = V_e I_e \cos \theta$$

then

$$I_e = \frac{P}{V_e \cos \theta} = \frac{1000}{200(0.8)} = \frac{25}{4} = 6.25$$

and therefore

$$\mathbf{I}_m = 6.25 \underline{/-36.9°}$$

Of course, since the motor absorbs 1000 W of average power, then the source supplies this power. But suppose now that a 28-μF capacitor is placed in parallel with the motor. Since the current through the motor does not change, then we have the situation shown in Fig. 9.16.

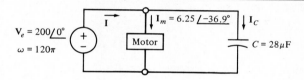

fig. 9.16

Since

$$\mathbf{I}_C = \frac{\mathbf{V}_e}{\mathbf{Z}_C} = j\omega C \mathbf{V}_e$$

$$= (1 \underline{/90°})(120\pi)(28)(10^{-6})(200 \underline{/0°})$$

$$= 2.11 \underline{/90°} = j2.11$$

But

$$\mathbf{I}_m = 6.25 \underline{/-36.9°} = 5 - j3.75$$

Thus

$$\mathbf{I} = \mathbf{I}_m + \mathbf{I}_C = 5 - j1.64 = 5.26 \underline{/-18.1°}$$

and the pf for the new load (the motor in parallel with the capacitor) is

$$\cos(18.1°) = 0.95$$

and, since the pf angle is positive, the pf is lagging.

We now see that placing a capacitor in parallel with the motor increases the lagging pf from 0.8 to 0.95. Since capacitors absorb no average power, and since the motor absorbs 1000 W in either case, the voltage source supplies 1000 W in both cases. Note, however, that originally a current of 6.25 A rms is drawn from the voltage source, while after the capacitor is connected in parallel, the current is 5.26 A rms—a decrease of about 16 percent.

· · ·

There are situations in which these current changes have significant implications. One important case occurs when the voltage source is an electric utility company generator and the load is some demand by a consumer. Between the two is the power transmission line, which is not ideal and has some associated nonzero resistance.

· · ·

Example
Suppose that a large consumer of electricity requires 10 kW of power by using 230 V rms at a pf angle of 60° lagging. We conclude that the current drawn by the load has an effective value of

$$I_e = \frac{P}{V_e \cos \theta} = \frac{10,000}{230(0.5)} \approx 87 \text{ A}$$

If the transmission line resistance is $R = 0.1 \ \Omega$ (see Fig. 9.17), by KVL,

$$\mathbf{V}_e = 0.1\mathbf{I}_L + \mathbf{V}_L$$

$$\approx (8.7 \underline{/-60°}) + (230 \underline{/0°}) = 234.5 \underline{/-1.8°}$$

fig. 9.17

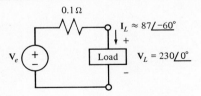

Thus, the voltage which must be generated by the power company must have an effective value of $V_e \approx 234.5$ V. Also, the line loss is $I_e^2 R = 756$ W. Although the utility gets paid for 10 kW of power, it has to produce 10,756 W.

If the consumer can increase the pf from 0.5 to 0.9 lagging, the load current will become $I_e \approx 48$ A rms, so the line loss will be approximately 233 W, which is about 30 percent of the loss at the lower pf. It's for this reason that an electric utility takes power factor into consideration when dealing with large consumers.

· · ·

An average household uses relatively small amounts of power and typically has a reasonably high pf. Thus, power factor is not a consideration in home electricity usage.

9.3 COMPLEX POWER

In the chapter on sinusoidal analysis, we found it computationally convenient to introduce the abstract concept of a complex sinusoid—a real sinusoid was just a special case. So, too, we now discuss and apply the notion of "complex power."

Again, consider the arbitrary load depicted in Fig. 9.18, where $\mathbf{V} = V\underline{/\phi_1}$, $\mathbf{I} = I\underline{/\phi_2}$, $V_e = V/\sqrt{2}$, $I_e = I/\sqrt{2}$.

fig. 3.18

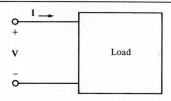

Since

$$P = V_e I_e \cos \theta$$

and

$$\cos \theta = \text{Re}[e^{j\theta}]$$

then

$$
\begin{aligned}
P &= V_e I_e(\text{Re}[e^{j\theta}]) \\
&= V_e I_e(\text{Re}[e^{j(\phi_1 - \phi_2)}]) \\
&= V_e I_e(\text{Re}[e^{j\phi_1} e^{-j\phi_2}]) \\
&= \text{Re}[V_e I_e e^{j\phi_1} e^{-j\phi_2}] \\
&= \text{Re}[(V_e e^{j\phi_1})(I_e e^{-j\phi_2})]
\end{aligned}
$$

But

$$(Ke^{j\phi})* = (Ke^{-j\phi})$$

so

$$P = \text{Re}[(V_e e^{j\phi_1})(I_e e^{j\phi_2})*]$$

That is, the average power absorbed by the load is equal to the real part of some complex quantity. Defining $\mathbf{V}_{\text{rms}}$ and $\mathbf{I}_{\text{rms}}$ by

$$\mathbf{V}_{\text{rms}} = V_e e^{j\phi_1}$$

$$\mathbf{I}_{\text{rms}} = I_e e^{j\phi_2}$$

then

$$P = \text{Re}[\mathbf{V}_{\text{rms}}\mathbf{I}^*_{\text{rms}}]$$

That is, the average power absorbed is the real part of the quantity

$$\mathbf{P} = \mathbf{V}_{rms}\mathbf{I}^*_{rms}$$

which is called the *complex power* absorbed by the load; that is,

$$\mathbf{P} = (V_e e^{j\phi_1})(I_e e^{-j\phi_2})$$
$$= V_e I_e e^{j(\phi_1 - \phi_2)}$$
$$= V_e I_e e^{j\theta}$$

Thus we see that the magnitude of the complex power is the apparent power

$$|\mathbf{P}| = V_e I_e$$

and the angle of the complex power is the pf angle. By Euler's formula, we can write the complex power in rectangular form as follows:

$$\mathbf{P} = V_e I_e(\cos\theta + j\sin\theta)$$
$$= V_e I_e \cos\theta + jV_e I_e \sin\theta$$
$$= P + jQ$$

where $P = V_e I_e \cos\theta$ is the average power absorbed by the load—also called the *real power*—and $Q = V_e I_e \sin\theta$ is called the *reactive power*. As mentioned before, the unit of P is the watt. The unit of Q is the *var*—short for *volt-ampere* reactive. The unit of complex power $\mathbf{P}$, like its magnitude $|\mathbf{P}|$ (apparent power) is the volt-ampere. We thus have

$$V_e I_e = |\mathbf{P}| = \sqrt{P^2 + Q^2}$$

and the angle of $\mathbf{P}$ is the pf angle θ, so

$$\theta = \tan^{-1}\left(\frac{Q}{P}\right)$$

From Fig. 9.19 we deduce that the power factor is

$$\cos\theta = \frac{P}{\sqrt{P^2 + Q^2}} = \frac{P}{|\mathbf{P}|}$$

fig. 9.19

Example
In Fig. 9.20 the source is given in terms of the ordinary phasor representation.

fig. 9.20

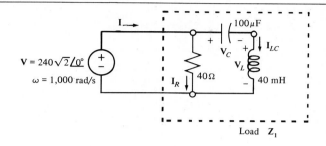

Load $\mathbf{Z}_1$

The impedance loading the voltage source is

$$\mathbf{Z}_1 = \frac{\mathbf{V}}{\mathbf{I}} = \frac{R\left(\dfrac{1}{j\omega C} + j\omega L\right)}{R + \dfrac{1}{j\omega C} + j\omega L}$$

$$= \frac{R(1 - \omega^2 LC)}{1 - \omega^2 LC + j\omega RC}$$

$$= \frac{40(1 - 4)}{(1 - 4) + j4} = \frac{-120}{-3 + j4} = \frac{120}{3 - j4} = \frac{120}{5\big/-53°}$$

$$= 24\big/53°\ \Omega$$

Thus

$$\mathbf{I} = \frac{\mathbf{V}}{\mathbf{Z}_1} = \frac{240\sqrt{2}\big/0°}{24\big/53°} = 10\sqrt{2}\big/-53°$$

so that the complex power supplied by the source is

$$\mathbf{P}_C = \mathbf{V}_{\text{rms}} \mathbf{I}^*_{\text{rms}} = (240\big/0°)(10\big/53°)$$

$$= 2400\big/53° = 1440 + j1920$$

Hence, the apparent power is $|\mathbf{P}| = 2400$ VA, the pf angle is ang$(\mathbf{P}) = 53°$, and thus the pf is $\cos 53° = 0.6$ lagging. The real power is $P = 1440$ W and the reactive power is $Q = 1920$ vars.

Because a too small pf may cost money, let us try to increase the pf. Since

$$\text{pf} = \cos\theta = \frac{P}{|\mathbf{P}|}$$

one way to increase the pf is to increase the real power P. But, real power costs money. So instead, let's place another load in parallel with the given one. This will not affect the voltage and current for the original load, but it will have an effect on the current through the voltage source. The resulting situation is depicted in Fig. 9.21, where $\mathbf{Z_1} = 24\,\underline{/53°}$, $\mathbf{I_1} = 10\sqrt{2}\,\underline{/-53°}$, and $\mathbf{P_1} = 2400\,\underline{/53°} = 1440 + j1920$. Suppose we wish to raise the lagging pf to 0.9. Since $\cos^{-1}(0.9) = 25.84°$,

fig. 9.21

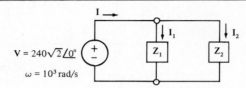

then let us call the current drain on the source $\mathbf{I} = \sqrt{2}I\,\underline{/-25.84°}$. The resulting complex power supplied by the source is

$$\mathbf{P} = \mathbf{V_{rms}}\,\mathbf{I_{rms}^*} = (240\,\underline{/0°})(I\,\underline{/25.84°})$$

$$= 240I\,\underline{/25.84°}$$

As we wish the real power to remain the same as before, we have

$$P = |\mathbf{P}|\cos\theta = (240I)(0.9) = 1440$$

from which

$$I = \frac{1440}{(240)(0.9)} = 6.67 \Rightarrow \mathbf{I} = 6.67\sqrt{2}\,\underline{/-25.84°}$$

By KCL,

$$\mathbf{I_2} = \mathbf{I} - \mathbf{I_1} = (6.67\sqrt{2}\,\underline{/-25.84°}) - (10\sqrt{2}\,\underline{/-53°})$$

$$= \sqrt{2}[6 - j2.9 - (6 - j8)] = j5.1\sqrt{2} = 5.1\sqrt{2}\,\underline{/90°}$$

Thus

$$\mathbf{Z_2} = \frac{\mathbf{V}}{\mathbf{I_2}} = \frac{240\,\underline{/0°}}{5.1\,\underline{/90°}} = 47\,\underline{/-90°} = -j47\ \Omega$$

and this is purely reactive, which might have been expected because of the constraint that there be no additional real power supplied by the source. This impedance can be realized with a capacitor by equating

$$\frac{1}{j\omega C} = \frac{-j}{\omega C} = -j47$$

from which

$$\frac{1}{\omega C} = 47 \Rightarrow C = \frac{1}{47\omega} = \frac{1}{47(10^3)} = 21.3 \ \mu F$$

Return now to the original circuit given in this example. The current through the resistor is

$$\mathbf{I}_R = \frac{\mathbf{V}}{40} = 6\sqrt{2} \ \underline{/0°}$$

and the complex power absorbed by the resistor is

$$\mathbf{P}_R = (240 \ \underline{/0°})(6 \ \underline{/0°}) = 1440 \ \text{W}$$

which, of course, is purely real. The current through the series LC combination is

$$\mathbf{I}_{LC} = \frac{\mathbf{V}}{\dfrac{1}{j\omega C} + j\omega L} = \frac{240\sqrt{2} \ \underline{/0°}}{\dfrac{10}{j} + j40}$$

$$= -j8\sqrt{2} = 8\sqrt{2} \ \underline{/-90°}$$

The voltage across the capacitor is

$$\mathbf{V}_C = \frac{1}{j\omega C} \mathbf{I}_{LC} = \frac{-j8\sqrt{2}}{j(10^{-1})} = -80\sqrt{2} = 80\sqrt{2} \ \underline{/180°}$$

and the voltage across the inductor is

$$\mathbf{V}_L = j\omega L \mathbf{I}_{LC} = j40(-j8\sqrt{2}) = 320\sqrt{2}$$

The complex power absorbed by the capacitor is

$$\mathbf{P}_C = \frac{1}{2} \mathbf{V}_C \mathbf{I}_{LC}^* = (80 \ \underline{/180°})(8 \ \underline{/90°}) = 640 \ \underline{/-90°} = -j640$$

and the complex power absorbed by the inductor is

$$\mathbf{P}_L = \frac{1}{2} \mathbf{V}_L \mathbf{I}_{LC}^* = (320 \ \underline{/0°})(8 \ \underline{/90°}) = 2560 \ \underline{/90°} = j2560$$

The sum of the complex powers absorbed by the resistor, capacitor, and inductor is

$$\mathbf{P}_R + \mathbf{P}_C + \mathbf{P}_L = 1440 - j640 + j2560 = 1440 + j1920$$

and this is exactly the complex power supplied by the source. This result is a consequence of the fact that, as with the case of real power, complex power is conserved.

· · ·

PROBLEMS   *9.13.   Repeat Problem 9.10(a) for the functions shown in Fig. P9.13.
Ans.   4.5 mW, 9 mW, 18 mW

fig. P9.13a,b,c,d

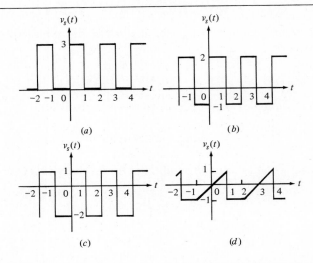

(a)   (b)

(c)   (d)

fig. P9.14

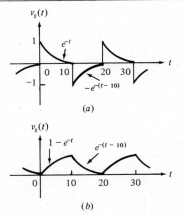

(a)

(b)

9.14.   Repeat Problem 9.11(a) for the exponential functions shown in Fig. P9.14.

fig. P9.15

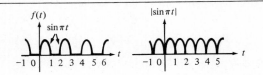

9.15.   Find the effective values of the "rectified" sine waves shown in Fig. P9.15.

[*Hint:* $\sin^2 x = \dfrac{1}{2}(1 - \cos 2x)$.]

9.16. A voltage source $v_s(t) = 6 \cos(2t - 90°)$, a resistor R, and a capacitor C are all connected in series. Find R and C given that the voltage across the capacitor is 3 V rms and the resistor dissipates 9 W.

fig. P9.17

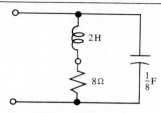

9.17. The load shown in Fig. P9.17 operates at 60 Hz.
(a) Find the pf and pf angle of this load.
(b) Is the pf leading or lagging?
(c) To what value should the capacitor be changed to get a lagging pf of 0.8?

*9.18. A 115-V-rms, 60-Hz electric hair dryer absorbs 500 W at a lagging power factor of 0.95. What is the rms value of the current drawn by this dryer?
Ans. 4.58 A

9.19. Two loads, which are connected in parallel, operate at 230 V rms. One load absorbs 500 W at a pf of 0.8 lagging, and the other absorbs 1000 W at a pf of 0.9 lagging.
(a) What is the current drawn by the combined load?
(b) Find the pf of the combined load.
(c) Is this pf leading or lagging?

9.20. The combined load given in Problem 9.19 is connected to a source with an internal resistance of 1 Ω. What is the rms value of the source?

9.21. To the combined load given in Problem 9.19, a third load that operates at 1500 W at a pf of 0.9 leading is connected in parallel.
(a) What is the current drawn by the resulting composite load?
(b) Find the pf of the composite load.
(c) Is this pf leading or lagging?

9.22. The parallel connection of two 115 V rms, 60 Hz loads operates at 2000 W with a lagging pf of 0.95. One load absorbs 1200 W at a pf of 0.8 lagging.
(a) What is the power absorbed and the pf angle of the other load?
(b) What reactive element should be placed in parallel with the load to result in an overall unity pf?

fig. P9.23

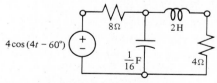

*9.23. Consider the circuit shown in Fig. P9.23.
(a) Find the complex power supplied by the source.
(b) Find the apparent power supplied by the source.
(c) Find the power factor and determine whether it is leading or lagging.
Ans. (a) $0.686 \, \underline{/-31°}$ VA, (b) 0.686 VA, (c) 0.86 leading

fig. P9.24

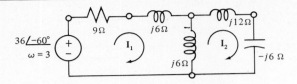

$36\underline{/-60°}$ $\omega = 3$

$9\,\Omega$ I_1 $j6\,\Omega$ $j6\,\Omega$ I_2 $j12\,\Omega$ $-j6\,\Omega$

9.24. For the circuit shown in Fig. P9.24, the mesh currents are

$$\mathbf{I_1} = 2\sqrt{2}\,\underline{/-105°} \quad \text{and} \quad \mathbf{I_2} = \sqrt{2}\,\underline{/-105°}$$

(a) What is the complex power absorbed by the capacitor?
(b) What is the real power absorbed by the capacitor?
(c) What is the complex power absorbed by the resistor?
(d) What is the real power absorbed by the resistor?

9.25. Consider the circuit given in Problem 9.24.
(a) Find the complex power absorbed by each inductor.
(b) Find the complex power supplied by the source.
(c) Is the complex power absorbed equal to the complex power supplied (is complex power conserved)?

9.26. Consider the circuit given in Problem 9.24.
(a) Find the apparent power supplied by the source.
(b) Find the apparent power absorbed by each of the remaining elements.
(c) Is the apparent power absorbed equal to the apparent power supplied (is apparent power conserved)?

*9.27. A load operating at 440 V rms draws 5 A rms at a leading pf of 0.95.
(a) Determine the complex power absorbed by the load.
(b) Determine the apparent power absorbed by the load.
(c) Determine the impedance of the load.

Ans. (a) 2200 $\underline{/-18.2°}$ VA, (b) 2200 VA, (c) 88 $\underline{/-18.2°}\ \Omega$

9.4 SINGLE-PHASE THREE-WIRE SYSTEMS

Previously, we mentioned that ordinary household voltage was sinusoidal, having an rms value of 115 V and a frequency of 60 Hz. We shall now discuss the subject further.

Coming from a power line into a home's service panel or "fuse box" are three wires. One wire is bare—it has no insulation. This is the *neutral* or *ground* wire, so called since it must be connected to a water pipe or a rod (or both) that goes into the ground. (The earth is the reference potential of zero volts.) Another wire is red and a third is black. The voltage between the red and neutral wire (plus at the red wire) is $v_{rn}(t) = 115\sqrt{2}$ $\cos(120\pi t + \phi)$—that is, 115 V (rms) ac, 60 Hz. The voltage $v_{bn}(t)$ between the black and the neutral wires (plus at the black wire) is 180° out of phase with $v_{rn}(t)$; that is, $v_{bn}(t) = -v_{rn}(t)$. This means that the voltage between the neutral and black wires (with the plus at the neutral wire) is $v_{nb}(t) = -v_{bn}(t)$. Thus $v_{nb}(t) = v_{rn}(t)$, and therefore these two voltages have the same phase. Since these voltages remain fairly constant for a wide variety of practical loads (appliances), we can reasonably approximate the *single-phase*

fig. 9.22

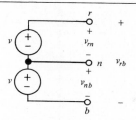

three-wire source just described as shown in Fig. 9.22, where $v_{rn}(t) = v_{nb}(t) = v(t)$ $= 115\sqrt{2}\cos(120\pi t + \phi)$. By KVL, we have $v_{rb}(t) = 230\sqrt{2}\cos(120\pi t + \phi)$. Thus, in addition to two 115-V, 60-Hz ac sources, there is a supply of 230 V (rms) ac, 60 Hz. This higher voltage is useful for higher-power appliances, since for a given higher power requirement, a higher voltage uses less current than a lower voltage. The result is a lower line loss.

The simple single-phase three-wire system of Fig. 9.23 (which is represented in the frequency domain), consists of a single-phase three-wire source and two identical loads.

fig. 9.23

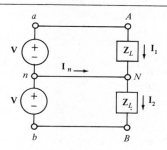

By Ohm's law,

$$\mathbf{I}_1 = \frac{\mathbf{V}}{\mathbf{Z}_L} = \mathbf{I}_2$$

so, by KCL,

$$\mathbf{I}_n = \mathbf{I}_2 - \mathbf{I}_1 = 0$$

and we see that there is no current in the "neutral wire" (the short circuit between nodes n and N). If this wire is removed (replaced by an open circuit), the voltage across each load $\mathbf{Z}_L$ is still $\mathbf{V}$ and the currents $\mathbf{I}_1$ and $\mathbf{I}_2$ remain $\mathbf{V}/\mathbf{Z}_L$.

Returning to the circuit shown in Fig. 9.23, the short circuits between nodes a and A and between nodes b and B are the *lines* of the system. Suppose that they both have the same impedance $\mathbf{Z}_g$. Since each such impedance is in series with an impedance $\mathbf{Z}_L$, we can treat the resulting circuit as the simple system above where the loads are $\mathbf{Z}_L + \mathbf{Z}_g$. We can then deduce that again $\mathbf{I}_n = 0$, and the neutral wire can be removed without affecting the voltages and currents. But what if the neutral wire has a nonzero impedance? (This also corresponds to a line impedance.) Furthermore, what happens if another load $\mathbf{Z}$ is connected between nodes A and B? To answer these questions, consider the circuit shown in Fig. 9.24.

fig. 9.24

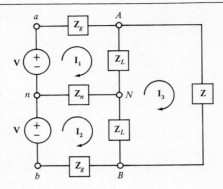

By mesh analysis, for mesh $\mathbf{I}_1$,

$$\mathbf{V} = (\mathbf{Z}_g + \mathbf{Z}_L + \mathbf{Z}_n)\mathbf{I}_1 - \mathbf{Z}_n\mathbf{I}_2 - \mathbf{Z}_L\mathbf{I}_3 \qquad (9.2)$$

and, for mesh $\mathbf{I}_2$,

$$\mathbf{V} = -\mathbf{Z}_n\mathbf{I}_1 + (\mathbf{Z}_g + \mathbf{Z}_L + \mathbf{Z}_n)\mathbf{I}_2 - \mathbf{Z}_L\mathbf{I}_3 \qquad (9.3)$$

Subtracting Equation (9.3) from Equation (9.2) we obtain

$$0 = (\mathbf{Z}_g + \mathbf{Z}_L + \mathbf{Z}_n)(\mathbf{I}_1 - \mathbf{I}_2) + \mathbf{Z}_n(\mathbf{I}_1 - \mathbf{I}_2)$$

or

$$0 = (\mathbf{Z}_g + \mathbf{Z}_L + 2\mathbf{Z}_n)(\mathbf{I}_1 - \mathbf{I}_2)$$

Thus

$$0 = \mathbf{I}_1 - \mathbf{I}_2 \Rightarrow \mathbf{I}_1 = \mathbf{I}_2$$

and

$$\mathbf{I}_n = \mathbf{I}_2 - \mathbf{I}_1 = 0$$

as before. Note that this result was obtained without writing the equation for mesh $\mathbf{I}_3$; that is, the value for $\mathbf{Z}$ does not influence this outcome. Specifically, for the special case that $\mathbf{Z}$ is infinite (an open circuit), then $\mathbf{I}_3 = 0$, and again subtracting Equation (9.3) from Equation (9.2) yields $\mathbf{I}_n = 0$.

The circuit in Fig. 9.24 has a symmetry with respect to the neutral connection between nodes n and N, and this symmetry (we call the load comprised of both $\mathbf{Z}_L$'s and $\mathbf{Z}$ *symmetrical* or *balanced*) results in a zero neutral current regardless of the neutral impedance $\mathbf{Z}_n$.

$$\cdot \quad \cdot \quad \cdot$$

Example

The load for a single-phase three-wire system is generally not balanced. For the circuit in Fig. 9.24, suppose that $\mathbf{Z}_g = 0.1\ \Omega$, $\mathbf{Z}_n = 2\ \Omega$, the load impedance between nodes A and N is 60 Ω, the load impedance between nodes N and B is

80 Ω, and $\mathbf{Z} = 10 + j15$ Ω. Suppose that the rms phasor for the sources is $\mathbf{V} = 115\underline{/0°}$. Then for mesh $\mathbf{I}_1$ we get

$$115 = 62.1\mathbf{I}_1 - 2\mathbf{I}_2 - 60\mathbf{I}_3$$

and, for mesh $\mathbf{I}_2$,

$$115 = -2\mathbf{I}_1 + 82.1\mathbf{I}_2 - 80\mathbf{I}_3$$

Finally, for mesh $\mathbf{I}_3$,

$$-60\mathbf{I}_1 - 80\mathbf{I}_2 + (150 + j15)\mathbf{I}_3 = 0$$

Solving these equations we get the rms mesh currents (see Problem 8.36)

$$\mathbf{I}_1 = 13.8\underline{/-49.26°}$$

$$\mathbf{I}_2 = 13.51\underline{/-50.71°}$$

$$\mathbf{I}_3 = 12.66\underline{/-55.8°}$$

Thus the power delivered to the loads is $P_{AN} + P_{NB} + P_{AB}$, where

$$P_{AN} = |\mathbf{I}_1 - \mathbf{I}_3|^2(60) = (1.88)^2(60) = 212 \text{ W}$$

$$P_{NB} = |\mathbf{I}_2 - \mathbf{I}_3|^2(80) = (1.44)^2(80) = 165 \text{ W}$$

$$P_{AB} = |\mathbf{I}_3|^2(10) = (12.66)^2(10) = 1603 \text{ W}$$

Thus

$$P_{AN} + P_{NB} + P_{AB} = 1980 \text{ W}$$

The line loss is $P_{aA} + P_{bB} + P_{nN}$, where

$$P_{aA} = |\mathbf{I}_1|^2(0.1) = (13.8)^2(0.1) = 19 \text{ W}$$

$$P_{bB} = |\mathbf{I}_2|^2(0.1) = (13.51)^2(0.1) = 18.3 \text{ W}$$

$$P_{nN} = |\mathbf{I}_1 - \mathbf{I}_2|^2(2) = (0.44)^2(2) = 0.4 \text{ W}$$

so

$$P_{aA} + P_{bB} + P_{nN} = 37.7 \text{ W}$$

and the total power supplied by the sources must be

$$P = 1980 + 37.7 \approx 2018 \text{ W}$$

As a check, the power supplied by the upper source is

$$|\mathbf{V}|\,|\mathbf{I}_1|\cos\theta_1 = (115)(13.8)\cos 49.26° = 1035 \text{ W}$$

and the power supplied by the lower source is

$$|\mathbf{V}|\,|\mathbf{I}_2|\cos\theta_2 = (115)(13.51)\cos 50.71° = 984 \text{ W}$$

for a total of $1035 + 984 = 2019$ W. The apparent discrepancy in power absorbed and power supplied is due to roundoff error (after all, 1 out of 2018 is an error of only about 0.05 percent).

. . .

9.5   THREE-PHASE CIRCUITS

A single-phase three-wire system contains a source that produces two sinusoidal voltages that have the same amplitude and the same phase. A system that contains a source that produces (sinusoidal) voltages with different phases is called a *polyphase system*. The importance of this concept lies in the fact that the vast majority of the generation and distribution of electric power in this country is accomplished with polyphase systems. The most common polyphase system is the balanced three-phase system, which has the property that it supplies constant instantaneous power. This results in less vibration of the rotating machinery used to generate electric power.

A *Y* (or *wye*)-*connected three-phase source* in the frequency domain is shown in Fig. 9.25. The terminals *a*, *b*, and *c* are called the *line terminals* and *n* is called the

fig. 9.25

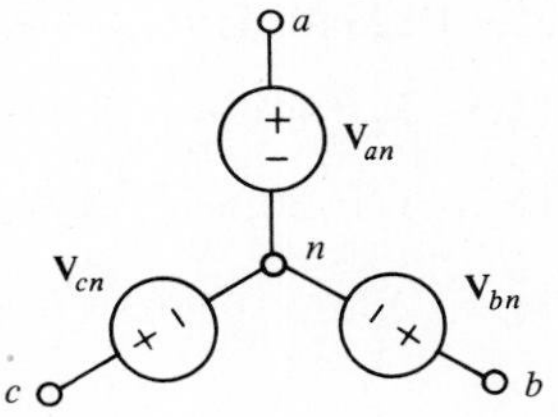

neutral terminal. The source is said to be *balanced* if the voltages $\mathbf{V}_{an}$, $\mathbf{V}_{bn}$, and $\mathbf{V}_{cn}$, called *phase voltages*, have the same amplitude and sum to zero—that is, if

$$|\mathbf{V}_{an}| = |\mathbf{V}_{bn}| = |\mathbf{V}_{cn}|$$

and

$$\mathbf{V}_{an} + \mathbf{V}_{bn} + \mathbf{V}_{cn} = 0$$

Suppose that the amplitude of the sinusoids is V. If we arbitrarily select the angle of $\mathbf{V}_{an}$ to be zero; that is, if $\mathbf{V}_{an} = V\underline{/0°}$, then the two situations that result in a balanced source are:

Case I	Case II
$\mathbf{V}_{an} = V\underline{/0°}$	$\mathbf{V}_{an} = V\underline{/0°}$
$\mathbf{V}_{bn} = V\underline{/-120°}$	$\mathbf{V}_{bn} = V\underline{/120°} = V\underline{/-240°}$
$\mathbf{V}_{cn} = V\underline{/-240°} = V\underline{/120°}$	$\mathbf{V}_{cn} = V\underline{/240°} = V\underline{/-120°}$

For the former case, $v_{an}(t)$ leads $v_{bn}(t)$ by 120° and $v_{bn}(t)$ leads $v_{cn}(t)$ by 120°. It is, therefore, called a *positive* or *abc phase sequence*. Similarly, the latter case is called a *negative* or *acb phase sequence*. Clearly, a negative phase sequence can be converted to a positive phase sequence simply by relabeling the terminals. Thus we need only consider positive phase sequences.

It is a simple matter to determine the voltages between the lines, called *line voltages*. By KVL,

$$\mathbf{V}_{ab} = \mathbf{V}_{an} - \mathbf{V}_{bn} = V\underline{/0°} - V\underline{/-120°}$$

$$= V \cos 0° + jV \sin 0° - [V \cos(-120°) + jV \sin(-120°)]$$

$$= V + \frac{1}{2} V + j\frac{\sqrt{3}}{2} V = \frac{3}{2} V + j\frac{\sqrt{3}}{2} V$$

$$= \frac{\sqrt{3}}{2} V(\sqrt{3} + j) = \frac{\sqrt{3}}{2} V(2\underline{/30°}) = \sqrt{3} V\underline{/30°}$$

Similarly,

$$\mathbf{V}_{bc} = \sqrt{3} V\underline{/-90°}$$

and

$$\mathbf{V}_{ca} = \sqrt{3} V\underline{/-210°} = \sqrt{3} V\underline{/150°}$$

Let us now connect our balanced source to a *balanced Y-connected three-phase load* as shown in Fig. 9.26. If we denote the current through $\mathbf{Z}$ from terminal x to terminal y by $\mathbf{I}_{xy}$, then the line currents are

$$\mathbf{I}_{AN} = \frac{\mathbf{V}_{an}}{\mathbf{Z}} = \frac{V\underline{/0°}}{\mathbf{Z}} = \frac{V}{\mathbf{Z}}$$

$$\mathbf{I}_{BN} = \frac{\mathbf{V}_{bn}}{\mathbf{Z}} = \frac{V\underline{/-120°}}{\mathbf{Z}} = \frac{V}{\mathbf{Z}}(1\underline{/-120°})$$

$$= \mathbf{I}_{AN}(1\underline{/-120°})$$

$$\mathbf{I}_{CN} = \frac{\mathbf{V}_{cn}}{\mathbf{Z}} = \frac{V\underline{/-240°}}{\mathbf{Z}} = \frac{V}{\mathbf{Z}}(1\underline{/-240°})$$

$$= \mathbf{I}_{AN}(1\underline{/-240°}) = \mathbf{I}_{AN}(1\underline{/120°})$$

fig. 9.26

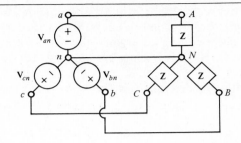

By KCL,

$$\mathbf{I}_{Nn} = \mathbf{I}_{AN} + \mathbf{I}_{BN} + \mathbf{I}_{CN}$$

$$= \mathbf{I}_{AN} + \mathbf{I}_{AN}(1\,/\!-120°) + \mathbf{I}_{AN}(1\,/120°)$$

$$= \mathbf{I}_{AN}\left[1 + \left(-\frac{1}{2} - j\frac{\sqrt{3}}{2}\right) + \left(-\frac{1}{2} + j\frac{\sqrt{3}}{2}\right)\right]$$

$$= 0$$

so we see that there is no current in the "neutral wire" (the short circuit between terminals n and N). We deduce that the neutral wire can be removed without affecting the voltages and currents.

If the lines (depicted as short circuits between a and A, b and B, and c and C) all have the same impedance, the effective load is still balanced, so the neutral current is zero and the neutral wire can again be removed. If the current in the neutral wire is zero when the neutral wire's impedance is both zero and infinity, we may deduce that the current is also zero for any value in between. For the case of a balanced source, balanced load and balanced (equal) line impedances (called a *balanced system*), whether or not there is a neutral wire connecting terminals n and N, it is convenient to think of these two points as being connected by a short circuit. This way we can analyze the system on a "per-phase" basis—that is, as three separate single-phase problems. This can greatly simplify the analysis of the system.

·   ·   ·

Example

For the previously given balanced three-phase system (Fig. 9.26), suppose that $\mathbf{V}_{an} = 120\,/0°$ is an rms phasor. Assuming a positive phase sequence, then $\mathbf{V}_{bn} = 120\,/\!-120°$ and $\mathbf{V}_{cn} = 120\,/120°$. Thus, the rms line voltage phasors are:

$$\mathbf{V}_{ab} = 120\sqrt{3}\,/30° \qquad \mathbf{V}_{bc} = 120\sqrt{3}\,/\!-90° \qquad \mathbf{V}_{ca} = 120\sqrt{3}\,/150°$$

Thus, the line voltage is $120\sqrt{3} = 208$ V rms.

If $\mathbf{Z} = 30 + j40 = 50\underline{/53°}$, then the rms line current $\mathbf{I}_{aA} = \mathbf{I}_{AN}$ is

$$\mathbf{I}_{aA} = \frac{\mathbf{V}_{an}}{\mathbf{Z}} = \frac{120\underline{/0°}}{50\underline{/53°}} = 2.4\underline{/-53°}$$

and the power absorbed by the single load $\mathbf{Z}$ between terminals A and N is

$$P_{AN} = |\mathbf{V}_{an}||\mathbf{I}_{AN}|\cos 53°$$
$$= 120(2.4)(0.6) = 172.8 \text{ W}$$

and thus the total power absorbed by the three-phase load is $3(172.8) = 518.4$ W.

The powers mentioned above are average powers. How does this compare with the instantaneous power? For the single load between nodes A and N, the instantaneous power absorbed is

$$p_{AN}(t) = v_{an}(t)i_{AN}(t)$$
$$= [120\sqrt{2}\cos \omega t][2.4\sqrt{2}\cos(\omega t - 53°)]$$
$$= 576\cos \omega t \cos(\omega t - 53°)$$

which is a time-varying quantity. However, the total instantaneous power absorbed by the three-phase load is

$$p(t) = p_{AN}(t) + p_{BN}(t) + p_{CN}(t)$$

where

$$p_{BN}(t) = v_{bn}(t)i_{BN}(t)$$
$$= [120\sqrt{2}\cos(\omega t - 120°)][2.4\sqrt{2}\cos(\omega t - 53° - 120°)]$$

and

$$p_{CN}(t) = v_{cn}(t)i_{CN}(t)$$
$$= [120\sqrt{2}\cos(\omega t - 240°)][2.4\sqrt{2}\cos(\omega t - 53° - 240°)]$$

Using the trigonometric identity

$$\cos \alpha \cos \beta = \frac{1}{2}\cos(\alpha + \beta) + \frac{1}{2}\cos(\alpha - \beta)$$

we get

$$p(t) = 288[\cos(2\omega t - 53°) + \cos 53°]$$
$$+ 288[\cos(2\omega t - 53° - 240°) + \cos 53°]$$
$$+ 288[\cos(2\omega t - 53° - 120°) + \cos 53°]$$

It's a routine complex arithmetic matter to show that since the three sinusoids are spaced 120° apart, they sum to zero. Thus,

$$p(t) = 288[3 \cos 53°] = 518.4 \text{ W}$$

so we see that the instantaneous power absorbed by the load (supplied by the source) is constant! This result holds for any balanced Y–Y-connected three-phase system, which is one reason for its importance.

$$\cdot\quad\cdot\quad\cdot$$

Example
Suppose now that the balanced three-phase system given in Fig. 9.26 has a line voltage of 250 V rms and the total power absorbed by the load is 600 W at a leading pf angle of 30°. Thus, the per-phase voltage (e.g., $\mathbf{V}_{an}$) has an rms magnitude given by

$$|\mathbf{V}_{an}| = \frac{|\mathbf{V}_{ab}|}{\sqrt{3}} = \frac{250}{\sqrt{3}}\,\text{V}$$

Since the per-phase power absorbed is 200 W, from the formula

$$P_{AN} = |\mathbf{V}_{an}|\,|\mathbf{I}_{AN}|\cos\theta$$

we have that the magnitude of the rms line current is

$$|\mathbf{I}_{AN}| = \frac{P_{AN}}{|\mathbf{V}_{an}|\cos\theta} = \frac{200}{\dfrac{250}{\sqrt{3}}\left(\dfrac{\sqrt{3}}{2}\right)} = 1.6\,\text{A}$$

The magnitude of the per-phase impedance is

$$|\mathbf{Z}| = \frac{|\mathbf{V}_{an}|}{|\mathbf{I}_{AN}|} = \frac{250/\sqrt{3}}{1.6} = 90\,\Omega$$

and since the pf is leading, the current leads the voltage, and therefore

$$\mathbf{Z} = 90\underline{/-30°} = 78 - j45\,\Omega$$

$$\cdot\quad\cdot\quad\cdot$$

More common than a Y-connected three-phase load is a Δ-*connected (delta-connected)* load. A Y-connected source with a balanced Δ-connected load is shown in Fig. 9.27. We see that the individual loads are connected directly across the lines, and

fig. 9.27

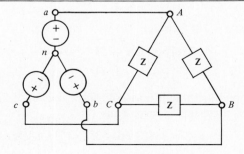

consequently it is relatively easier to add or remove loads on a single phase with a Δ-connected load than with a Y-connected load. As a matter of fact, it may not even be possible to do so for a Y-connected load since the neutral terminal may not be accessible.

Suppose that the voltage between terminals x and y (plus at x) is denoted by $\mathbf{V}_{xy}$. For a balanced source,

$$\mathbf{V}_{an} = V\underline{/0^\circ} \qquad \mathbf{V}_{bn} = V\underline{/-120^\circ} \qquad \mathbf{V}_{cn} = V\underline{/-240^\circ}$$

and, as was shown previously, the line voltages are

$$\mathbf{V}_{ab} = \sqrt{3}V\underline{/30^\circ} \qquad \mathbf{V}_{bc} = \sqrt{3}V\underline{/-90^\circ} \qquad \mathbf{V}_{ca} = \sqrt{3}V\underline{/-210^\circ}$$

which are the phase voltages for a Δ-connected load. The phase currents are

$$\mathbf{I}_{AB} = \frac{\mathbf{V}_{ab}}{\mathbf{Z}} \qquad \mathbf{I}_{BC} = \frac{\mathbf{V}_{bc}}{\mathbf{Z}} \qquad \mathbf{I}_{CA} = \frac{\mathbf{V}_{ca}}{\mathbf{Z}}$$

Thus, by KCL, the line current $\mathbf{I}_{aA}$ is

$$\mathbf{I}_{aA} = \mathbf{I}_{AB} - \mathbf{I}_{CA} = \frac{\mathbf{V}_{ab}}{\mathbf{Z}} - \frac{\mathbf{V}_{ca}}{\mathbf{Z}} = \frac{1}{\mathbf{Z}}(\mathbf{V}_{ab} - \mathbf{V}_{ca})$$

$$= \frac{1}{\mathbf{Z}}(\sqrt{3}\,V\underline{/30^\circ} - \sqrt{3}\,V\underline{/-210^\circ}) = \frac{\sqrt{3}\,V}{\mathbf{Z}}(1\underline{/30^\circ} - 1\underline{/150^\circ})$$

$$= \frac{\sqrt{3}V}{\mathbf{Z}}\left(\frac{\sqrt{3}}{2} + j\frac{1}{2} + \frac{\sqrt{3}}{2} - j\frac{1}{2}\right) = \frac{3V}{\mathbf{Z}} = \frac{3V\underline{/0^\circ}}{\mathbf{Z}}$$

Similarly,

$$\mathbf{I}_{bB} = \frac{3V\underline{/-120^\circ}}{\mathbf{Z}} \qquad \mathbf{I}_{cC} = \frac{3V\underline{/-240^\circ}}{\mathbf{Z}}$$

Since the magnitude of a phase current (e.g., $\mathbf{I}_{AB}$) is

$$|\mathbf{I}_{AB}| = \frac{|\mathbf{V}_{ab}|}{|\mathbf{Z}|} = \frac{\sqrt{3}\,V}{|\mathbf{Z}|}$$

and the magnitude of a line current (e.g., $\mathbf{I}_{aA}$) is

$$|\mathbf{I}_{aA}| = \frac{3V}{|\mathbf{Z}|}$$

then

$$|\mathbf{I}_{aA}| = \sqrt{3}\,|\mathbf{I}_{AB}|$$

That is, a line current is $\sqrt{3}$ times as great as a phase current. Furthermore, since the phase currents are 120° apart, they are a balanced set—as are the line currents.

. . .

Example

Suppose that, for the previously given balanced three-phase system with Y-connected source and Δ-connected load (Fig. 9.27), the line voltage is 250 V rms and the total power absorbed is 600 W at a leading pf angle of 30°. Since the per-phase load power absorbed is 200 W, from the formula

$$P_{AB} = |\mathbf{V}_{ab}| \, |\mathbf{I}_{AB}| \cos \theta$$

we find that the magnitude of the rms phase current (e.g., $\mathbf{I}_{AB}$) is

$$|\mathbf{I}_{AB}| = \frac{P_{AB}}{|\mathbf{V}_{ab}| \cos \theta} = \frac{200}{250(\sqrt{3}/2)} = 0.92 \text{ A}$$

and of the rms line current (e.g., $\mathbf{I}_{aA}$) is

$$|\mathbf{I}_{aA}| = \sqrt{3} \, |\mathbf{I}_{AB}| = 1.6 \text{ A}$$

The magnitude of the per-phase impedance is

$$|\mathbf{Z}| = \frac{|\mathbf{V}_{ab}|}{|\mathbf{I}_{AB}|} = \frac{250}{0.92} = 271 \ \Omega$$

and since the pf is leading, the current leads the voltage, and therefore

$$\mathbf{Z} = 271 \underline{/-30°} = 234 - j135 \ \Omega$$

. . .

Suppose now that a balanced three-phase load is Y-connected, the magnitude of the line voltage is V_L volts rms and the line current is I_L amperes rms. If the power P_1 absorbed per phase is due to rms voltage V and current I with pf angle θ, then

$$P_1 = VI \cos \theta = \frac{V_L}{\sqrt{3}} I_L \cos \theta$$

and therefore the total power absorbed is

$$P = 3P_1 = \sqrt{3} V_L I_L \cos \theta$$

For the case of a balanced Δ-connected load,

$$P_1 = VI \cos \theta$$

$$= V_L \left(\frac{I_L}{\sqrt{3}} \right) \cos \theta$$

and, again,

$$P = 3P_1 = \sqrt{3} V_L I_L \cos \theta$$

Thus we see that regardless of whether a balanced load is Y-connected or Δ-connected, in terms of the line voltage, line current, and load impedance phase angle, we can use the same formula for the total power absorbed by the load:

$$P = \sqrt{3}\ V_L I_L \cos \theta$$

In our discussion above, we have assumed that the source is balanced and Y-connected. Although the same conclusions can be obtained for a balanced, Δ-connected source, for practical reasons such sources are uncommon. If a Δ-connected source is not balanced exactly, a large current can circulate around the loop formed by the elements comprising the delta. The result is the reduction of the source's current-delivering capacity and an increase in the system's losses.

The suggestion has been made that a Y connection is equivalent to a Δ connection, and vice versa. But under what conditions is equivalence obtained? To answer this question, consider the Δ and Y connections shown in Fig. 9.28.

fig. 9.28

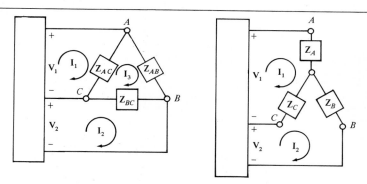

For the Δ connection the mesh equations are

$$\mathbf{V}_1 = \mathbf{Z}_{AC}\mathbf{I}_1 - \mathbf{Z}_{AC}\mathbf{I}_3$$

$$\mathbf{V}_2 = \mathbf{Z}_{BC}\mathbf{I}_2 - \mathbf{Z}_{BC}\mathbf{I}_3$$

$$0 = -\mathbf{Z}_{AC}\mathbf{I}_1 - \mathbf{Z}_{BC}\mathbf{I}_2 + (\mathbf{Z}_{AB} + \mathbf{Z}_{BC} + \mathbf{Z}_{AC})\mathbf{I}_3$$

From the third equation,

$$\mathbf{I}_3 = \frac{\mathbf{Z}_{AC}\mathbf{I}_1}{\mathbf{Z}_{AB} + \mathbf{Z}_{BC} + \mathbf{Z}_{AC}} + \frac{\mathbf{Z}_{BC}\mathbf{I}_2}{\mathbf{Z}_{AB} + \mathbf{Z}_{BC} + \mathbf{Z}_{AC}}$$

and substituting this expression for $\mathbf{I}_3$ into the first two equations we get

$$\mathbf{V}_1 = \frac{\mathbf{Z}_{AB}\mathbf{Z}_{AC} + \mathbf{Z}_{AC}\mathbf{Z}_{BC}}{\mathbf{Z}_{AB} + \mathbf{Z}_{BC} + \mathbf{Z}_{AC}}\ \mathbf{I}_1 - \frac{\mathbf{Z}_{AC}\mathbf{Z}_{BC}}{\mathbf{Z}_{AB} + \mathbf{Z}_{BC} + \mathbf{Z}_{AC}}\ \mathbf{I}_2$$

and

$$\mathbf{V}_2 = \frac{-\mathbf{Z}_{AC}\mathbf{Z}_{BC}}{\mathbf{Z}_{AB} + \mathbf{Z}_{BC} + \mathbf{Z}_{AC}}\ \mathbf{I}_1 + \frac{\mathbf{Z}_{AB}\mathbf{Z}_{BC} + \mathbf{Z}_{AC}\mathbf{Z}_{BC}}{\mathbf{Z}_{AB} + \mathbf{Z}_{BC} + \mathbf{Z}_{AC}}\ \mathbf{I}_2$$

The mesh equations for the Y connection are

$$V_1 = (Z_A + Z_C)I_1 - Z_C I_2$$

and

$$V_2 = -Z_C I_1 + (Z_B + Z_C)I_2$$

Since the two connections are equivalent when the values of V_1 and V_2 and of I_1 and I_2 are the same for both cases, by equating coefficients we get

$$Z_A = \frac{Z_{AB} Z_{AC}}{Z_{AB} + Z_{BC} + Z_{AC}}$$

$$Z_B = \frac{Z_{AB} Z_{BC}}{Z_{AB} + Z_{BC} + Z_{AC}}$$

$$Z_C = \frac{Z_{AC} Z_{BC}}{Z_{AB} + Z_{BC} + Z_{AC}}$$

and these are the conditions for finding the equivalent Y connection given a Δ connection. By using dual arguments, we can derive the conditions for finding the equivalent Δ connection given a Y connection. (In this case we can use admittances instead of impedances.) They are

$$Y_{AB} = \frac{Y_A Y_B}{Y_A + Y_B + Y_C}$$

$$Y_{BC} = \frac{Y_B Y_C}{Y_A + Y_B + Y_C}$$

$$Y_{AC} = \frac{Y_A Y_C}{Y_A + Y_B + Y_C}$$

Alternatively, it is a simple matter to express these conditions in terms of impedances as follows:

$$Z_{AB} = \frac{Z_A Z_B + Z_B Z_C + Z_A Z_C}{Z_C}$$

$$Z_{BC} = \frac{Z_A Z_B + Z_B Z_C + Z_A Z_C}{Z_A}$$

$$Z_{AC} = \frac{Z_A Z_B + Z_B Z_C + Z_A Z_C}{Z_B}$$

Going from a Y connection to its equivalent Δ connection (or vice versa) is called a *Y-Δ (wye-delta) transformation* (or *Δ-Y (delta-wye) transformation*). Since the equivalence is with respect to the three terminals A, B, and C, the equivalence holds for any pair of them, also.

· · ·

Example

We can find the impedance (resistance) $\mathbf{Z}$ of the bridge circuit shown in Fig. 9.29 by using mesh, nodal, or loop analysis. Instead, let us use a Δ-Y transformation on the Δ formed by the $\frac{1}{5}$-Ω, $\frac{1}{2}$-Ω, and 1-Ω resistors. Doing so we get

$$\mathbf{Z}_A = \frac{(1)(\frac{1}{2})}{1 + \frac{1}{5} + \frac{1}{2}} = \tfrac{5}{17} \ \Omega$$

$$\mathbf{Z}_B = \frac{(1)(\frac{1}{5})}{1 + \frac{1}{5} + \frac{1}{2}} = \tfrac{2}{17} \ \Omega$$

$$\mathbf{Z}_C = \frac{(\frac{1}{2})(\frac{1}{5})}{1 + \frac{1}{5} + \frac{1}{2}} = \tfrac{1}{17} \ \Omega$$

fig. 9.29

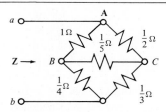

As far as terminals a and b are concerned, the bridge circuit in Fig. 9.29 is equivalent, therefore, to the configuration shown in Fig. 9.30. The $\frac{1}{17}$-Ω and $\frac{1}{3}$-Ω resistors in series have a resistance of $\frac{1}{17} + \frac{1}{3} = \frac{20}{51}$ Ω, and the $\frac{2}{17}$-Ω and $\frac{1}{4}$-Ω

fig. 9.30

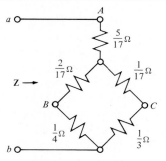

resistors in series have a resistance of $\frac{2}{17} + \frac{1}{4} = \frac{25}{68}$ Ω. The resulting parallel combination has a resistance of

$$\frac{(\frac{20}{51})(\frac{25}{68})}{\frac{20}{51} + \frac{25}{68}} = \frac{100}{527} \ \Omega$$

and this in series with $\frac{5}{17}$ Ω has a resistance of

$$\mathbf{Z} = \frac{100}{527} + \frac{5}{17} = \frac{255}{527} \ \Omega$$

· · ·

9.6   POWER MEASUREMENTS

The *wattmeter* is a device which measures the average power absorbed by a two-terminal load. Such a device is depicted in Fig. 9.31. The wattmeter, which is used typically at frequencies between a few hertz and a few hundred hertz, has two coils. One

fig. 9.31

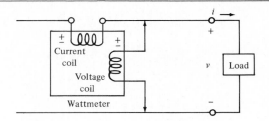

(the current coil) has a very low impedance (ideally zero), and the other (the voltage coil) has a very high impedance (ideally infinite). The current coil is connected in series with the load, and the voltage coil is connected in parallel with the load. When the current and the voltage shown for the load are both positive-valued, the wattmeter has an upscale deflection and reads the average power absorbed by the load. The same result occurs when both the current and voltage are negative-valued.

Now that we have determined how to measure the power absorbed by a single load, how can the wattmeter be used to measure the power absorbed by a three-phase load? For a balanced load, if we can connect a wattmeter to one phase of the load as shown in Fig. 9.31, then the total power absorbed is three times the reading. For an unbalanced load, if three separate wattmeters can be connected, then the total power absorbed is the sum of the three readings. However, a Y-connected load may not have its neutral terminal accessible. In such a case, although the current coil can be connected properly, the voltage coil cannot. On the other hand, for a Δ-connected load like a three-phase rotating machine, only the three terminals of the load are accessible. This means that the voltage coil can be connected appropriately, but the current coil cannot. Don't give up, however; there is a way to measure the power absorbed by a three-phase load, whether balanced or unbalanced, with the use of wattmeters. To see how, consider three wattmeters and a Δ-connected load, as shown in Fig. 9.32. In this situation, the current coils of the wattmeters are connected in series with the lines aA, bB, and cC. The voltage coils are connected between a line and some arbitrary point m. The average power read on wattmeter A is

$$P_A = \frac{1}{T} \int_0^T v_{am} i_{aA} \, dt$$

and similar expressions are obtained for wattmeters B and C. Summing these average powers, we get

$$P = P_A + P_B + P_C = \frac{1}{T} \int_0^T v_{am} i_{aA} \, dt + \frac{1}{T} \int_0^T v_{bm} i_{bB} \, dt + \frac{1}{T} \int_0^T v_{cm} i_{cC} \, dt$$

$$= \frac{1}{T} \int_0^T (v_{am} i_{aA} + v_{bm} i_{bB} + v_{cm} i_{cC}) \, dt$$

fig. 9.32

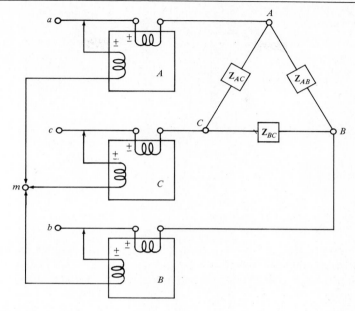

But, by KCL,

$$i_{aA} = i_{AB} + i_{AC} \qquad i_{bB} = -i_{AB} + i_{BC} \qquad i_{cC} = -i_{AC} - i_{BC}$$

Substituting these expressions for the line currents into the previous integral, we obtain

$$P = \frac{1}{T} \int_0^T \left[(v_{am} - v_{bm})i_{AB} + (v_{bm} - v_{cm})i_{BC} + (v_{am} - v_{cm})i_{AC} \right] dt$$

and, by KVL,

$$P = \frac{1}{T} \int_0^T \left(v_{ab}i_{AB} + v_{bc}i_{BC} + v_{ac}i_{AC} \right) dt$$

$$= \frac{1}{T} \int_0^T v_{ab}i_{AB} \, dt + \frac{1}{T} \int_0^T v_{bc}i_{BC} \, dt + \frac{1}{T} \int_0^T v_{ac}i_{AC} \, dt$$

which is the total power absorbed by the three-phase load.

Since the point m was chosen arbitrarily, we can place it anywhere without it affecting the end result: The sum of the three wattmeter readings is the power absorbed by the load. The judicious choice of point m to be on the line cC means that wattmeter C will have zero volts across its voltage coil, and hence will read 0 W. For such a case, wattmeter C is superfluous and is not needed. The resulting configuration, shown in Fig. 9.33, is an example of what is known as the *two-wattmeter method* for measuring power. It can also be applied when a three-phase load is Y-connected, and, as for the case of a Δ connection, the load can be unbalanced as well as balanced. Furthermore, it is immaterial whether or not the source is balanced—the parameters utilized are line voltages and line currents.

fig. 9.33

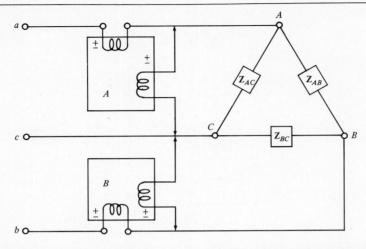

For the case that the load is balanced, it is also possible to determine the pf angle with the two-wattmeter method. Consider the Δ-connected load shown in Fig. 9.33, where $\mathbf{Z}_{AB} = \mathbf{Z}_{BC} = \mathbf{Z}_{AC} = \mathbf{Z} = |\mathbf{Z}|\,\underline{/\theta}$—that is, a balanced Δ-connected load. For the balanced source,

$$\mathbf{V}_{an} = V\,\underline{/0^\circ} \qquad \mathbf{V}_{bn} = V\,\underline{/-120^\circ} \qquad \mathbf{V}_{cn} = V\,\underline{/-240^\circ}$$

we have seen previously that

$$\mathbf{V}_{ab} = \sqrt{3}\,V\,\underline{/30^\circ} \qquad \mathbf{V}_{bc} = \sqrt{3}\,V\,\underline{/-90^\circ} \qquad \mathbf{V}_{ca} = \sqrt{3}\,V\,\underline{/150^\circ}$$

$$\mathbf{I}_{aA} = \frac{3V\,\underline{/0^\circ}}{\mathbf{Z}} \qquad \mathbf{I}_{bB} = \frac{3V\,\underline{/-120^\circ}}{\mathbf{Z}} \qquad \mathbf{I}_{cC} = \frac{3V\,\underline{/-240^\circ}}{\mathbf{Z}}$$

Since

$$P_A = |\mathbf{V}_{AC}|\,|\mathbf{I}_{aA}|\cos(\text{ang } \mathbf{V}_{AC} - \text{ang } \mathbf{I}_{aA})$$
$$= V_L I_L \cos(-30^\circ + \theta)$$
$$= V_L I_L \cos(30^\circ - \theta)$$

and

$$P_B = |\mathbf{V}_{BC}|\,|\mathbf{I}_{bB}|\cos(\text{ang } \mathbf{V}_{BC} - \text{ang } \mathbf{I}_{bB})$$
$$= V_L I_L \cos(-90^\circ - [-120^\circ - \theta])$$
$$= V_L I_L \cos(30^\circ + \theta)$$

Taking the ratio of P_A to P_B we get

$$\frac{P_A}{P_B} = \frac{\cos(30^\circ - \theta)}{\cos(30^\circ + \theta)}$$

and by the trigonometric identity

$$\cos(\alpha + \beta) = \cos \alpha \cos \beta - \sin \alpha \sin \beta$$

this ratio becomes

$$\frac{P_A}{P_B} = \frac{\cos 30° \cos \theta + \sin 30° \sin \theta}{\cos 30° \cos \theta - \sin 30° \sin \theta} = \frac{\dfrac{\sqrt{3}}{2} \cos \theta + \dfrac{1}{2} \sin \theta}{\dfrac{\sqrt{3}}{2} \cos \theta - \dfrac{1}{2} \sin \theta}$$

from which

$$\sqrt{3}(P_A - P_B)\cos \theta = (P_A + P_B)\sin \theta$$

or

$$\frac{\sin \theta}{\cos \theta} = \frac{\sqrt{3}(P_A - P_B)}{P_A + P_B} = \tan \theta$$

Thus the pf angle θ is

$$\theta = \tan^{-1} \frac{\sqrt{3}(P_A - P_B)}{P_A + P_B}$$

$\cdot$ $\cdot$ $\cdot$

Example

Suppose that a balanced Δ-connected load as shown in Fig. 9.33 has a line voltage of 240 V rms and wattmeter readings $P_A = 1000$ W and $P_B \doteq 300$ W. Thus the pf angle is

$$\theta = \tan^{-1} \frac{\sqrt{3}(1000 - 300)}{1000 + 300} = \tan^{-1}(0.93) = 43°$$

Since the total power absorbed by the load is $1000 + 300 = 1300$ W, the per-phase power absorbed is $(1300/3)$ W. Since

$$P_{AB} = |\mathbf{V}_{ab}| |\mathbf{I}_{AB}| \cos \theta$$

we get

$$|\mathbf{I}_{AB}| = \frac{P_{AB}}{|\mathbf{V}_{ab}| \cos \theta}$$

from which

$$|\mathbf{Z}| = \frac{|\mathbf{V}_{ab}|}{|\mathbf{I}_{AB}|} = \frac{|\mathbf{V}_{ab}|^2 \cos \theta}{P_{AB}} = \frac{(240)^2 \cos(43°)}{1300/3} = 97.2$$

Thus,

$$\mathbf{Z} = 97.2 \underline{/43°} = 71.1 + j66.3 \ \Omega$$

$\cdot$ $\cdot$ $\cdot$

PROBLEMS 9.28. For the single-phase three-wire system in Fig. P9.28, find the power supplied by each source if $Z_1 = 60$ Ω, $Z_2 = 80$ Ω, $Z_3 = 40$ Ω and (a) $R_g = R_n = 0$ Ω, (b) $R_g = 1$ Ω, $R_n = 2$ Ω.

fig. P9.28

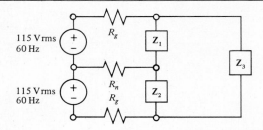

9.29. Repeat Problem 9.28 for the case that a 40-mH inductor is placed in series with Z_3.

*9.30. For the single-phase three-wire system given in Problem 9.28, $R_g = R_n = 0$ Ω. If Z_1 absorbs 500 W at a lagging pf of 0.8, Z_2 absorbs 1000 W at a lagging pf of 0.9, and Z_3 absorbs 1500 W at a leading pf of 0.95, find the currents through the sources.
Ans. $10.92 \underline{/-5.9°}$ A, $15.35 \underline{/-7.75°}$ A

*9.31. A balanced Y-Y three phase system has 130 V rms phase voltages and per-phase impedance $Z = 12 + j12$ Ω. Given that the line impedance is zero, find the line currents and the total power absorbed by the load.
Ans. 7.66 A, 2112.4 W

9.32. Repeat Problem 9.31 for the case that the line impedances are 1 Ω.

9.33. A balanced Y-Y three-phase system has line voltages of 210 V rms, 60 Hz. Assume that the total load absorbs 3 kW of power at a lagging pf of 0.85.
 (a) Find the per-phase impedance.
 (b) What value capacitors should be placed in parallel with the per-phase impedances to result in a 0.95 lagging pf?

*9.34. A balanced three-phase Y-connected source whose phase voltages are 115 V rms has an unbalanced Y-connected load. Given that $Z_{AN} = 3 + j4$ Ω, $Z_{BN} = 10$ Ω, and $Z_{CN} = 5 + j12$ Ω. For the case that the lines and the neutral wire have zero impedance, find the line currents and the total power absorbed by the load.
Ans. $23 \underline{/-53.13°}$ A, $11.5 \underline{/-120°}$ A, $8.85 \underline{/52.6°}$ A, 3301 W

9.35. Repeat Problem 9.34 for the case that there is no neutral wire.

*9.36. A balanced Y-Δ three-phase system has $V_{an} = 130 \underline{/0°}$ V rms and $Z = 4\sqrt{2} \underline{/45°}$ Ω. Given that the line impedance is zero, find the line currents and the total power absorbed by the load.
Ans. 68.9 A, 19, 013 W

9.37. Repeat Problem 9.36 for the case that the line impedances are 1 Ω.

9.38. A balanced, three-phase Y-connected source with 230-V-rms line voltages has an unbalanced Δ-connected load. The load impedances are $Z_{AB} = 8$ Ω, $Z_{BC} = 4 + j3$ Ω and $Z_{AC} = 12 - j5$ Ω. For the case that the lines have zero impedance, find the line currents and the total power absorbed by the load.

9.39.   Repeat Problem 9.33 for a balanced Y-Δ three-phase system.

9.40.   A balanced three-phase Y-connected source with 120-V-rms line voltages is loaded with a balanced Y connection having $3 + j4 \; \Omega$ per phase and a balanced Δ connection having $5 - j12 \; \Omega$ per phase. Find the total power absorbed by the load and the pf.

9.41.   A balanced three-phase Δ-connected load has a per-phase impedance $\mathbf{Z}$ $= 36 + j36 \; \Omega$. For the two-wattmeter connection given in the text (see Fig. 9.33), find the two readings for the case of line voltages of $130\sqrt{3}$ V rms produced by a balanced Y-connected source.

fig. P9.42

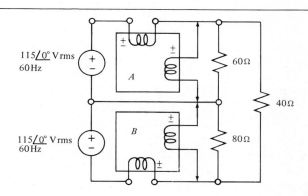

*9.42.   For the circuit shown in Fig. P9.42, find the wattmeter readings P_A and P_B.

   Ans.   881.7 W, 826.6 W

9.43.   Repeat Problem 9.42 for the case that a 40-mH inductor is placed in series with the 40-Ω resistor.

fig. P9.44

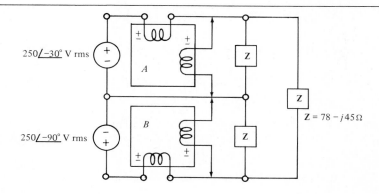

9.44.   Find the wattmeter readings P_A and P_B for the circuit in Fig. P9.44.

9.45.   Find the wattmeter readings P_A and P_B for the circuit in Fig. P9.45.

fig. P9.45

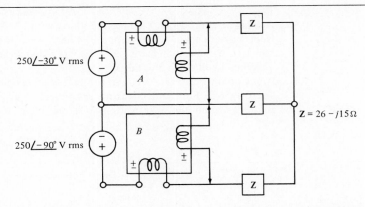

$250\underline{/-30°}$ V rms

$250\underline{/-90°}$ V rms

$\mathbf{Z} = 26 - j15\,\Omega$

9.46. The Y-Δ three-phase system given in Problem 9.38 has two wattmeters connected as described in the text (see Fig. 9.33). Find the readings of these wattmeters.

fig. P9.47

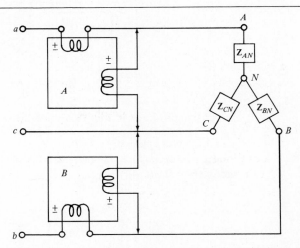

*9.47. For the Y-connected three-phase load shown in Fig. P9.47, find the wattmeter readings P_A and P_B when $\mathbf{Z}_{AN} = \mathbf{Z}_{BN} = \mathbf{Z}_{CN} = 12 + j12\,\Omega$ and the line voltage produced by a balanced three-phase Y-connected source is $130\sqrt{3}$ V rms.
Ans.   1666 W, 446.4 W

9.48. Repeat Problem 9.47 for the case that $\mathbf{Z}_{AN} = 3 + j4\,\Omega$, $\mathbf{Z}_{BN} = 10\,\Omega$, and $\mathbf{Z}_{CN} = 5 + j12\,\Omega$, and the line voltage produced by a balanced three-phase Y-connected source is $115\sqrt{3}$ V rms.

---

1. The instantaneous power absorbed by an element is equal to the product of the voltage across it and the current through it.

Summary

2.	The average power absorbed by a resistance R having a sinusoidal current of amplitude I and voltage of amplitude V is $P_R = VI/2 = RI^2/2 = V^2/2R$.

3.	The average power absorbed by a capacitance or an inductance is zero.

4.	A nonideal source whose internal impedance is $\mathbf{Z}_0$ or a circuit whose Thévenin equivalent (output) impedance is $\mathbf{Z}_0$ transfers maximum power to a load $\mathbf{Z}_L$ when $\mathbf{Z}_L$ is equal to the complex conjugate of $\mathbf{Z}_0$.

5.	For the case that $\mathbf{Z}_L$ is restricted to being purely resistive, maximum power is transferred when $\mathbf{Z}_L$ equals the magnitude of $\mathbf{Z}_0$.

6.	The effective or rms value of a sinusoid of amplitude A is $A/\sqrt{2}$.

7.	The average power absorbed by a resistance R having a current whose effective value is I_e and a voltage whose effective value is V_e is $P_R = V_e I_e = I_e^2 R = V_e^2/R$.

8.	The power factor (pf) is the ratio of average power to apparent power.

9.	If current lags voltage, the pf is lagging. If current leads voltage, the pf is leading.

10.	Average or real power can be generalized with the notion of complex power.

11.	The ordinary household uses a single-phase three-wire electric system.

12.	The most common polyphase electric system is the balanced three-phase system.

13.	Three-phase sources are generally Y-connected, and three-phase loads are generally Δ-connected.

14.	The device commonly used to measure power is the wattmeter.

15.	Three-phase load power measurements can be taken with the two-wattmeter method.

IMPORTANT ^{10} AC CONCEPTS

Since the impedance of capacitors and inductors is frequency-dependent, so is the impedance of an arbitrary *RLC* circuit. This dependence, which can be depicted graphically by plots of magnitude and phase angle versus frequency, constitute the "frequency response" of the impedance. Impedance, which is the ratio of voltage to current (phasors), is only one network function whose frequency response can be displayed. It is the frequency response of a network that describes the network's behavior to all sinusoidal forcing functions and nonsinusoidal forcing functions that are comprised of sinusoids.

An important frequency is that for which a network function or parameter reaches a maximum value. In certain simple networks, this occurs when an impedance or admittance is purely real—a condition known as "resonance." In such a situation we can quantitatively describe the "frequency selectivity" of the network with a figure of merit known as the "quality factor." We can also talk about the notion of "bandwidth."

In this chapter we shall also see how we can easily convert a circuit that consists of elements whose values are computationally convenient but impractical to one with practical values. This process is known as "scaling."

Just as the concept of complex sinusoids allowed us to treat sinusoidal circuits in a simple manner, the concept of complex frequency does a similar job for damped sinusoidal circuits. This generalization not only allows for the easy determination of forced responses to sinusoidal and damped sinusoidal forcing functions but indicates the form of the natural response as well.

460

10.1   FREQUENCY RESPONSE

For the parallel connection of a resistor R and capacitor C, shown in Fig. 10.1, the impedance $\mathbf{Z}$ of the combination is given by

$$\mathbf{Z} = \frac{\mathbf{Z}_R \mathbf{Z}_C}{\mathbf{Z}_R + \mathbf{Z}_C}$$

$$= \frac{R(1/j\omega C)}{R + (1/j\omega C)} = \frac{R}{1 + j\omega RC}$$

$$= \frac{R}{\sqrt{1 + (\omega RC)^2}} \underline{/-\tan^{-1}(\omega RC)}$$

$$= |\mathbf{Z}| \underline{/\theta}$$

fig. 10.1

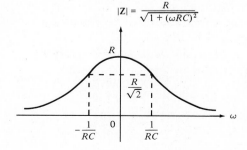

Since $\mathbf{Z}$ is a complex function—the notation $\mathbf{Z}(j\omega)$ is often used—and can therefore be expressed in terms of a magnitude and an angle as $\mathbf{Z} = |\mathbf{Z}| \underline{/\theta}$, both the magnitude $|\mathbf{Z}|$ and the angle θ are, in general, functions of ω. We therefore may plot $|\mathbf{Z}|$ versus ω and θ versus ω. In obtaining the former, note that when $\omega = 0$, the magnitude is $|\mathbf{Z}| = R$; when $\omega = 1/RC$, then $|\mathbf{Z}| = R/\sqrt{2} \approx 0.707R$. Also, when $\omega \to \infty$ then $|\mathbf{Z}| \to 0$. When we talk about real sinusoids having a frequency of ω radians per second, the implication is that $\omega \geq 0$. For the case of complex sinusoids, though, negative values of ω make just as much sense as do positive values. (Remember Euler's formula, which says that a real sinusoid of frequency ω is composed of complex sinusoids of frequency ω and $-\omega$.) Thus, in the plot of $|\mathbf{Z}|$ versus ω, we consider negative as well as positive values for ω. For this example, we have the plot shown in Fig. 10.2, which we call the *amplitude response* of $\mathbf{Z}$. The plot of θ versus ω is known as the *phase response* of $\mathbf{Z}$, and for this example it is shown in Fig. 10.3. In this plot we see that when $\omega = 0$, then $\theta = -\tan^{-1} \omega RC = 0$; when $\omega = 1/RC$, then $\theta = -\tan^{-1} \omega RC = -\pi/4$ rad.

fig. 10.2

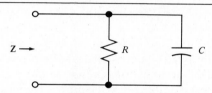

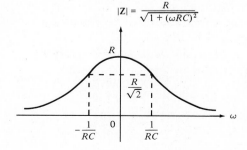

fig. 10.3

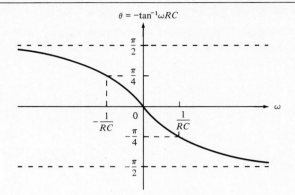

Furthermore, when $\omega \to \infty$, then $\theta \to -\pi/2$ rad. The amplitude and phase responses comprise the *frequency response* of the impedance **Z**.

The amplitude response of **Z** is just a pictorial description of how the magnitude of the impedance varies with frequency. It will be seen in the chapter on the Fourier transform that such a plot is always symmetrical about the vertical axis (called *even symmetry*). Thus, although the above magnitude is plotted for negative as well as positive values of frequency ω, quite often the amplitude response is plotted only for nonnegative values of ω. Furthermore, it is generally true that the negative portion of the phase response can be obtained by taking the mirror reflection around the vertical axis of the positive portion and then inverting it (called *odd symmetry*). Consequently, the phase response is also usually plotted only for nonnegative values of ω.

Although a frequency response consists of both an amplitude response and a phase response, quite often the amplitude response alone is referred to as the frequency response.

Consider the lag network shown in Fig. 10.4. By Ohm's law,

$$\mathbf{V} = \mathbf{ZI}$$

fig. 10.4

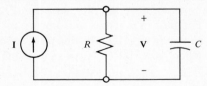

where **Z** is the impedance of the parallel RC combination in the circuit. From complex-number arithmetic,

$$|\mathbf{V}| = |\mathbf{ZI}| = |\mathbf{Z}|\,|\mathbf{I}|$$

Given the current **I**, then $|\mathbf{V}|$ is maximum when $|\mathbf{Z}|$ is maximum. This occurs when $\omega = 0$, in which case $|\mathbf{Z}| = R$ and the average power absorbed by the resistor is

$$P_0 = \frac{1}{2}\frac{|\mathbf{V}|^2}{R} = \frac{1}{2}\frac{R^2|\mathbf{I}|^2}{R} = \frac{1}{2}R|\mathbf{I}|^2$$

When $\omega = 1/RC = \omega_1$, however, then $|\mathbf{Z}| = R/\sqrt{2}$, so the power absorbed is

$$P_1 = \frac{1}{2}\frac{(R/\sqrt{2})^2|\mathbf{I}|^2}{R} = \frac{1}{2}\left(\frac{R|\mathbf{I}|^2}{2}\right)$$

$$= \frac{1}{2}P_0$$

That is, P_1 is half of the power P_0. For this reason, we say that $\omega_1 = 1/RC$ is a *half-power frequency* or *half-power point*; similarly, so is $\omega_{-1} = -1/RC$. In general, given an amplitude response whose maximum value is M, the half-power frequencies are those frequencies for which the magnitude is $M/\sqrt{2}$.

. . .

Example
Consider the lead network shown in Fig. 10.5.

---

fig. 10.5

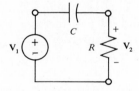

By voltage division,

$$\mathbf{V}_2 = \frac{R}{R + (1/j\omega C)}\,\mathbf{V}_1$$

from which

$$\frac{\mathbf{V}_2}{\mathbf{V}_1} = \frac{j\omega RC}{1 + j\omega RC}$$

Let us denote the ratio of output voltage to input voltage by $\mathbf{H}(j\omega)$; that is, $\mathbf{H}(j\omega) = \mathbf{V}_2/\mathbf{V}_1$. Then

$$|\mathbf{H}(j\omega)| = \left|\frac{\mathbf{V}_2}{\mathbf{V}_1}\right| = \frac{|\mathbf{V}_2|}{|\mathbf{V}_1|} = \frac{\omega RC}{\sqrt{1 + (\omega RC)^2}}$$

For $\omega = 0$, we have $|\mathbf{H}(j\omega)| = 0$. To determine the value of $|\mathbf{H}(j\omega)|$ as $\omega \to \infty$, note that

$$|\mathbf{H}(j\omega)| = \sqrt{\frac{(\omega RC)^2}{1 + (\omega RC)^2}} = \frac{1}{\sqrt{\dfrac{1}{(\omega RC)^2} + 1}}$$

Thus $|\mathbf{H}(j\omega)| \to 1$ as $\omega \to \infty$. Also, note that if $\omega = 1/RC$, then $|\mathbf{H}(j\omega)| = 1/\sqrt{2}$. The amplitude response of the function $\mathbf{H}(j\omega)$—called a *system function* or a

fig. 10.6

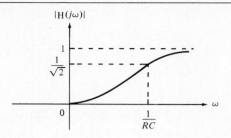

transfer function—is shown in Fig. 10.6. Clearly, $\omega = 1/RC$ is the half-power frequency.

It is possible to obtain a rough sketch of the amplitude response for the circuit above without writing an analytical expression for the transfer function $\mathbf{H}(j\omega)$. For $\omega = 0$—the dc case—the capacitor is an open circuit, so $\mathbf{V}_2$ and consequently $|\mathbf{H}(j\omega)| = 0$. For small values of ω, the impedance of the capacitor is still large relative to R, so by voltage division $\mathbf{V}_2$ is relatively small, as is $|\mathbf{H}(j\omega)|$. As frequency (ω) increases, however, the impedance of the capacitor decreases, and thus the values of $\mathbf{V}_2$ and $|\mathbf{H}(j\omega)|$ increase. In the limit, as frequency becomes infinite, the impedance of the capacitor becomes zero, so $\mathbf{V}_2 \rightarrow \mathbf{V}_1 \Rightarrow |\mathbf{H}(j\omega)| \rightarrow 1$.

· · ·

PROBLEMS   10.1.   Consider the circuit shown in Fig. P10.1.
  (a)   Sketch the amplitude response of $\mathbf{Y} = \mathbf{I}/\mathbf{V}_s$.
  (b)   Sketch the amplitude response of $\mathbf{V}_1/\mathbf{V}_s$.
  (c)   Sketch the amplitude response of $\mathbf{V}_2/\mathbf{V}_s$.

fig. P10.1

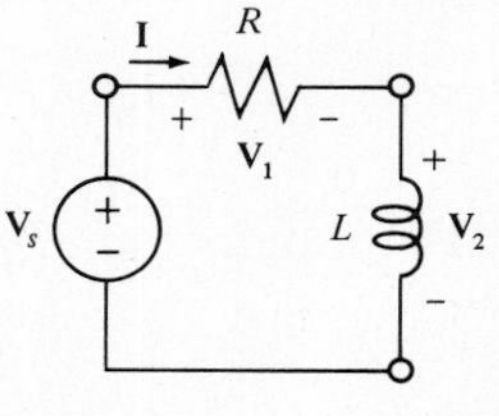

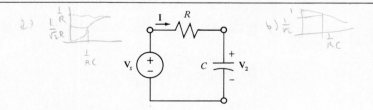

fig. P10.2

10.2.   Repeat Problem 10.1(a) and (c) for the circuit in Fig. P10.2.
10.3.   Sketch the amplitude response of $\mathbf{V}_2/\mathbf{V}_1$ for each of the op amp circuits shown in Fig. P10.3. Indicate the half-power frequency.

fig. P10.3

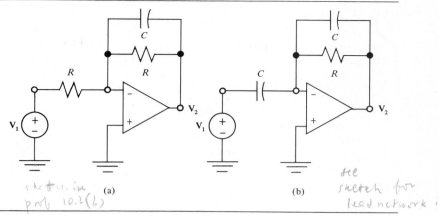

sketch in
prob 10.2(b)

(a)

see
sketch for
lead network in text

(b)

fig. P10.4

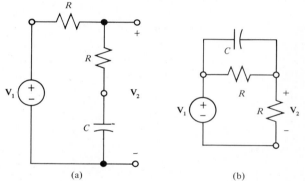

(a)

(b)

*10.4.   Repeat Problem 10.3 for each of the circuits shown in Fig. P10.4.

Ans.   (a) $\dfrac{1}{\sqrt{2}RC}$, (b) $\dfrac{\sqrt{2}}{RC}$

fig. P10.5

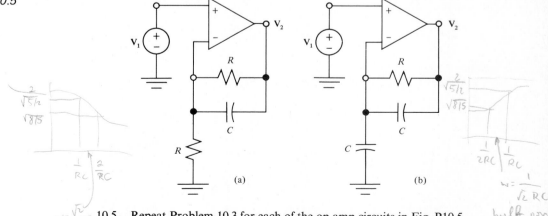

$\dfrac{2}{\sqrt{5}/2}$

$\sqrt{8}/5$

$\dfrac{1}{RC}$ $\dfrac{2}{RC}$

$w = \dfrac{\sqrt{2}}{RC}$

(a)

$\dfrac{2}{\sqrt{5}/2}$

$\sqrt{8}/5$

$\dfrac{1}{2RC}$ $\dfrac{1}{RC}$

$w = \dfrac{1}{\sqrt{2}RC}$

half power freq

(b)

10.5.   Repeat Problem 10.3 for each of the op amp circuits in Fig. P10.5.

fig. P10.6

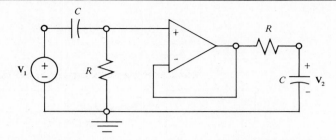

10.6. For the op amp circuit in Fig. P10.6 sketch the amplitude response of V_2/V_1, indicating the half-power frequencies.

fig. P10.7

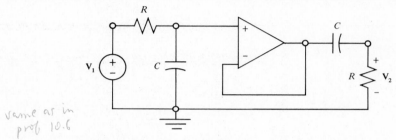

10.7. Repeat Problem 10.6 for the op amp circuit in Fig. P10.7.

10.2   RESONANCE

For the parallel *RLC* circuit shown in Fig. 10.7, the admittance "seen" by the current source is $\mathbf{Y} = \mathbf{I}/\mathbf{V}$, where

$$\mathbf{Y} = \mathbf{Y}_R + \mathbf{Y}_L + \mathbf{Y}_C$$

$$= \frac{1}{R} + \frac{1}{j\omega L} + j\omega C$$

$$= \frac{1}{R} + j\left(\omega C - \frac{1}{\omega L}\right)$$

fig. 10.7

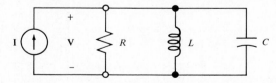

From this expression we can see that there is some frequency ω_r for which the imaginary part of $\mathbf{Y}$ will be zero; that is,

$$\omega_r C - \frac{1}{\omega_r L} = 0$$

from which

$$\omega_r C = \frac{1}{\omega_r L} \quad \Rightarrow \quad \omega_r{}^2 = \frac{1}{LC}$$

so that

$$\omega_r = \frac{1}{\sqrt{LC}}$$

We say that a network having at least one capacitor and one inductor is in *resonance* or is *resonant* when the imaginary part of its admittance (or impedance) is equal to zero. The circuit of Fig. 10.7 is in resonance when $\omega_r = 1/\sqrt{LC}$, and this is the *resonance frequency*.

Since

$$|\mathbf{V}| = \left|\frac{\mathbf{I}}{\mathbf{Y}}\right| = \frac{|\mathbf{I}|}{|\mathbf{Y}|}$$

from the fact that $|\mathbf{Y}|$ is minimum at resonance, we conclude that $|\mathbf{V}|$ is maximum at the resonance frequency $\omega_r = 1/\sqrt{LC}$ and is given by

$$|\mathbf{V}| = \frac{|\mathbf{I}|}{1/R} = R|\mathbf{I}|$$

For $\omega = 0$, the impedance of the inductor is zero; thus the voltage $\mathbf{V}$ is also zero. As $\omega \to \infty$, the impedance of the capacitor $1/j\omega C \to 0$, and again, $\mathbf{V} \to 0$. A sketch of $|\mathbf{V}|$ versus ω is shown in Fig. 10.8, where ω_1 and ω_2 are the "lower" and "upper" half-power frequencies, respectively.

fig. 10.8

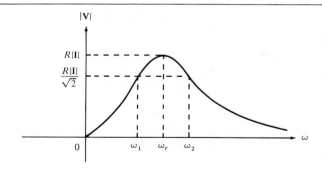

Since at resonance the parallel *RLC* combination acts simply as the resistance R, the parallel *LC* combination—known as a *tank circuit*—behaves as an open circuit.

We define the *bandwidth*, B, of the resonant circuit in terms of the half-power frequencies by

$$B = \omega_2 - \omega_1$$

Note that the smaller the bandwidth B is, the "sharper" or "narrower" is the amplitude response. Another conventional quantity that describes sharpness is the *quality factor*, designated Q, which is defined at the resonance frequency to be

$$Q = 2\pi \left(\frac{\text{maximum energy stored}}{\text{total energy lost in a period}} \right)$$

Since energy is stored in capacitors and inductors, the maximum energy stored is designated $[w_C(t) + w_L(t)]_{\max}$. Since energy is dissipated by resistors, the energy lost in a period T is $P_R T$, where P_R is the resistive power. Thus,

$$Q = \frac{2\pi [w_C(t) + w_L(t)]_{\max}}{P_R T}$$

Now let us determine the quality factor for the parallel *RLC* circuit shown in Fig. 10.7. Assume that the current source is given by $i(t) = I \cos \omega_r t$, where $\omega_r = 1/\sqrt{LC}$. At resonance, $\mathbf{Y} = 1/R$, so

$$v(t) = Ri(t) = RI \cos \omega_r t$$

Then the energy stored in the capacitor is

$$w_C(t) = \frac{1}{2} Cv^2(t)$$

$$= \frac{1}{2} C(RI \cos \omega_r t)^2$$

$$= \frac{1}{2} CR^2 I^2 \cos^2 \omega_r t$$

Furthermore, since the inductor current $\mathbf{I}_L$ (going down) is

$$\mathbf{I}_L = \frac{\mathbf{V}}{j\omega L}$$

then at resonance, since $v(t) = RI \cos \omega_r t$, we have

$$\mathbf{I}_L = \frac{RI \underline{/0^\circ}}{\omega_r L \underline{/90^\circ}} = \frac{RI}{\omega_r L} \underline{/-90^\circ}$$

so

$$i_L(t) = \frac{RI}{\omega_r L} \cos(\omega_r t - 90^\circ) = \frac{RI}{\omega_r L} \sin \omega_r t$$

at resonance. The fact that the *LC* tank portion of the circuit behaves as an open circuit at resonance does not mean that there is no current in the inductor and capacitor. As a matter of fact, the current (given above) circulates around the loop formed by the inductor and capacitor.

At resonance, the energy stored in the inductor is

$$w_L(t) = \frac{1}{2} L i_L{}^2(t) = \frac{1}{2} \frac{R^2 I^2}{\omega_r{}^2 L} \sin^2 \omega_r t$$

and, since $\omega_r{}^2 = 1/LC$,

$$w_L(t) = \frac{1}{2} CR^2 I^2 \sin^2 \omega_r t$$

Therefore, the total energy stored is

$$w_C(t) + w_L(t) = \frac{1}{2} CR^2 I^2 (\cos^2 \omega_r t + \sin^2 \omega_r t)$$

$$= \frac{1}{2} CR^2 I^2$$

which is a constant.

Since the power absorbed by a resistor is

$$P_R = \frac{1}{2} I^2 R$$

then the energy lost in a period is

$$P_R T = \frac{1}{2} I^2 RT$$

$$= \frac{1}{2} I^2 R\left(\frac{2\pi}{\omega_r}\right) = \frac{\pi I^2 R}{\omega_r}$$

Thus the quality factor of the given parallel RLC circuit is

$$Q = \frac{2\pi(\frac{1}{2} CR^2 I^2)}{\pi I^2 R/\omega_r} = \omega_r RC$$

Since $\omega_r C = 1/\omega_r L$, alternatively,

$$Q = \omega_r RC = \frac{R}{\omega_r L} = R\sqrt{\frac{C}{L}}$$

Again considering the expression for the admittance of the network in Fig. 10.7, we have

$$\mathbf{Y} = \frac{1}{R} + j\left(\omega C - \frac{1}{\omega L}\right)$$

$$= \frac{1}{R} + j\left(\frac{\omega C \omega_r R}{\omega_r R} - \frac{\omega_r R}{\omega L \omega_r R}\right)$$

$$= \frac{1}{R} + j\frac{1}{R}\left(\frac{\omega}{\omega_r} Q - \frac{\omega_r}{\omega} Q\right)$$

$$= \frac{1}{R}\left[1 + jQ\left(\frac{\omega}{\omega_r} - \frac{\omega_r}{\omega}\right)\right]$$

At the half-power frequencies ω_1 and ω_2,

$$|\mathbf{V}| = \frac{R|\mathbf{I}|}{\sqrt{2}} = \frac{|\mathbf{I}|}{|\mathbf{Y}|}$$

Thus

$$|\mathbf{Y}| = \frac{\sqrt{2}}{R}$$

which occurs when

$$Q\left(\frac{\omega}{\omega_r} - \frac{\omega_r}{\omega}\right) = \pm 1$$

First consider the case that

$$Q\left(\frac{\omega}{\omega_r} - \frac{\omega_r}{\omega}\right) = 1$$

Then

$$Q\left(\frac{\omega^2 - \omega_r{}^2}{\omega_r \omega}\right) = 1$$

from which

$$Q\omega^2 - \omega_r\omega - Q\omega_r{}^2 = 0$$

By the quadratic formula, we get a positive and a negative value that satisfies this equation. Since the half-power frequencies are positive quantities, we select the positive value. Thus, one of the half-power frequencies is

$$\frac{\omega_r}{2Q} + \omega_r \sqrt{\left(\frac{1}{2Q}\right)^2 + 1}$$

For the case that

$$Q\left(\frac{\omega}{\omega_r} - \frac{\omega_r}{\omega}\right) = -1$$

we get

$$Q\omega^2 + \omega_r\omega - Q\omega_r{}^2 = 0$$

and the resulting half-power frequency

$$-\frac{\omega_r}{2Q} + \omega_r \sqrt{\left(\frac{1}{2Q}\right)^2 + 1}$$

Thus we see that the latter half-power frequency is ω_1 and the former is ω_2; consequently, the bandwidth for the parallel RLC circuit is

$$B = \omega_2 - \omega_1 = \frac{\omega_r}{Q}$$

We also have

$$B = \frac{\omega_r}{\omega_r RC} = \frac{1}{RC}$$

Thus we see that when the Q of the circuit is relatively small, the bandwidth B is relatively large, whereas if Q is relatively large, then B is relatively small. The latter case indicates a sharper or more frequency-selective amplitude response.

From the next-to-last equation, we find that

$$Q = \frac{\omega_r}{B}$$

for the parallel RLC circuit. This ratio of resonance frequency to bandwidth is sometimes defined to be the *selectivity* of a circuit. For a parallel RLC circuit, we see that the circuit's quality factor and selectivity are equal. In general, though, the selectivity and quality factor of a circuit will not be the same.

Let us now consider the case of a series RLC circuit, as shown in Fig. 10.9. The impedance $\mathbf{Z} = \mathbf{V}/\mathbf{I}$ is given by

$$\mathbf{Z} = R + j\omega L + \frac{1}{j\omega C}$$

$$= R + j\left(\omega L - \frac{1}{\omega C}\right)$$

fig. 10.9

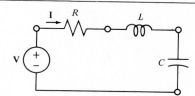

so the resonance frequency ω_r is that frequency for which

$$\omega_r L - \frac{1}{\omega_r C} = 0 \qquad \text{or} \qquad \omega_r L = \frac{1}{\omega_r C}$$

so that

$$\omega_r{}^2 = \frac{1}{LC} \qquad \text{or} \qquad \omega_r = \frac{1}{\sqrt{LC}}$$

as was the case for the parallel RLC circuit.

If the voltage source is a sinusoid whose frequency is equal to the resonance frequency, say $v(t) = V \cos \omega_r t$, then $\mathbf{Z} = R$ so that

$$i(t) = \frac{V}{R} \cos \omega_r t$$

Thus, the energy stored in the inductor is

$$w_L(t) = \frac{1}{2} Li^2(t)$$

$$= \frac{LV^2}{2R^2} \cos^2 \omega_r t$$

Although the series LC connection behaves as a short circuit at the resonance frequency ($j\omega_r L + 1/j\omega_r C = 0$), there is a nonzero voltage across each element. In particular, the capacitor voltage (with the $+$ at the top plate) is

$$\mathbf{V}_C = \frac{\mathbf{I}}{j\omega_r C} = \frac{\frac{V}{R}\underline{/0°}}{\omega_r C \underline{/90°}} = \frac{V}{\omega_r RC}\underline{/-90°}$$

so that

$$v_C(t) = \frac{V}{\omega_r RC} \cos(\omega_r t - 90°) = \frac{V}{\omega_r RC} \sin \omega_r t$$

Thus, the energy stored in the capacitor is

$$w_C(t) = \frac{1}{2} Cv_C^2(t) = \frac{1}{2} \frac{V^2}{\omega_r^2 R^2 C} \sin^2 \omega_r t$$

and since $\omega_r^2 = 1/LC$, then

$$w_C(t) = \frac{LV^2}{2R^2} \sin^2 \omega_r t$$

Thus the total energy stored is

$$w_C(t) + w_L(t) = \frac{LV^2}{2R^2} (\sin^2 \omega_r t + \cos^2 \omega_r t)$$

$$= \frac{LV^2}{2R^2}$$

which is a constant.

The power absorbed by the resistor is

$$P_R = \frac{1}{2} I^2 R = \frac{1}{2} \left(\frac{V}{R}\right)^2 R = \frac{V^2}{2R}$$

so that the energy lost in a period is

$$P_R T = \frac{V^2}{2R} \left(\frac{2\pi}{\omega_r}\right) = \frac{\pi V^2}{\omega_r R}$$

Therefore, the quality factor of the circuit is

$$Q = \frac{2\pi\left(\dfrac{LV^2}{2R^2}\right)}{\dfrac{\pi V^2}{\omega_r R}}$$

$$= \frac{\omega_r L}{R}$$

and since $\omega_r L = 1/\omega_r C$, then also

$$Q = \frac{1}{\omega_r RC} = \frac{1}{R}\sqrt{\frac{L}{C}}$$

In a process similar to that done for the parallel RLC circuit, it can be shown that the bandwidth for the series RLC circuit is again

$$B = \frac{\omega_r}{Q}$$

so that the selectivity and the quality factor are equal. Also

$$B = \frac{\omega_r}{\omega_r L/R} = \frac{R}{L}$$

$$\bullet \quad \bullet \quad \bullet$$

Example
Consider the series RLC circuit shown in Fig. 10.10. We have already analyzed this circuit for the case that $v(t) = 17 \cos 3t$. The result was that $v_C(t) = 5\sqrt{17} \cos (3t - 166°)$. For this circuit the resonance frequency is $\omega_r = 1/\sqrt{LC} = \sqrt{5}$ rad/s.

fig. 10.10

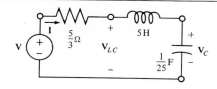

Thus, the quality factor is

$$Q = \frac{\omega_r L}{R} = \frac{5\sqrt{5}}{5/3} = 3\sqrt{5}$$

and the bandwidth is

$$B = \frac{R}{L} = \frac{5/3}{5} = \tfrac{1}{3} \text{ rad/s}$$

A derivation of the formulas for the half-power frequencies of the series *RLC* circuit can be accomplished as was done for the parallel *RLC* circuit. What result are the same expressions. Thus, the lower half-power frequency is

$$\omega_1 = -\frac{\omega_r}{2Q} + \omega_r \sqrt{\left(\frac{1}{2Q}\right)^2 + 1} = 2.076 \text{ rad/s}$$

and the upper half-power frequency is

$$\omega_2 = \frac{\omega_r}{2Q} + \omega_r \sqrt{\left(\frac{1}{2Q}\right)^2 + 1} = 2.409 \text{ rad/s}$$

and as a check on the above results, the bandwidth is

$$B = \omega_2 - \omega_1 = 0.333 \text{ rad/s}$$

Suppose now that the frequency of the voltage source equals the circuit's resonance frequency, say $v(t) = 17 \cos \sqrt{5}t$. Since

$$|\mathbf{I}| = \left|\frac{\mathbf{V}}{\mathbf{Z}}\right|$$

at resonance

$$|\mathbf{I}| = \frac{|\mathbf{V}|}{5/3} = \frac{3}{5}(17) = \tfrac{51}{5} \text{ A}$$

Then

$$|\mathbf{V}_C| = |\mathbf{Z}_C| |\mathbf{I}| = \frac{1}{\omega C} |\mathbf{I}|$$

$$= \frac{25}{\sqrt{5}}\left(\frac{51}{5}\right) = 51\sqrt{5} \text{ V}$$

Note that although the amplitude of the input sinusoid is $|\mathbf{V}| = 17$ V, the amplitude of the output voltage is $|\mathbf{V}_C| = 51\sqrt{5}$ V. That is, the output voltage magnitude is $Q = 3\sqrt{5}$ times larger than the input.

Furthermore, since

$$\mathbf{V}_{LC} = (\mathbf{Z}_L + \mathbf{Z}_C)\mathbf{I} = \left(j\omega L + \frac{1}{j\omega C}\right)\mathbf{I}$$

$$= \left(j5\sqrt{5} - \frac{j}{\sqrt{5/25}}\right)\mathbf{I} = 0$$

then the series combination of the inductor and capacitor acts as a short circuit, although the voltage across each element is nonzero.

$\bullet \quad \bullet \quad \bullet$

fig. 10.11

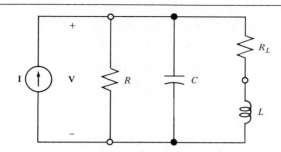

Now consider the series-parallel RLC circuit shown in Fig. 10.11. We have

$$\mathbf{Y} = \frac{\mathbf{I}}{\mathbf{V}} = \frac{1}{R} + j\omega C + \frac{1}{R_L + j\omega L}$$

$$= \frac{1}{R} + j\omega C + \frac{R_L - j\omega L}{R_L^2 + \omega^2 L^2}$$

$$= \frac{1}{R} + \frac{R_L}{R_L^2 + \omega^2 L^2} + j\left(\omega C - \frac{\omega L}{R_L^2 + \omega^2 L^2}\right)$$

and thus the imaginary part equals zero when $\omega = \omega_r$, where

$$\omega_r C - \frac{\omega_r L}{R_L^2 + \omega_r^2 L^2} = 0$$

from which

$$\omega_r^2 = \frac{1}{LC} - \frac{R_L^2}{L^2}$$

or

$$\omega_r = \sqrt{\frac{1}{LC} - \frac{R_L^2}{L^2}}$$

Suppose that $R = R_L = 2\ \Omega$, $L = 2$ H, and $C = \frac{1}{20}$ F. Then the resonance frequency is

$$\omega_r = \sqrt{10 - 1} = 3 \text{ rad/s}$$

and the admittance at this frequency is

$$\mathbf{Y}(j3) = \frac{1}{2} + \frac{2}{2^2 + 3^2 2^2} = \frac{11}{20} = 0.55\ \mho$$

Note, however, that the admittance at the frequency $\omega = 3.1$ rad/s is

$$\mathbf{Y}(j3.1) = \frac{1}{2} + \frac{2}{2^2 + (3.1)^2 2^2} + j\left(\frac{3.1}{20} - \frac{3.1(2)}{2^2 + (3.1)^2 2^2}\right)$$

$$= 0.547 + j0.0089\ \mho$$

Hence

$$|\mathbf{Y}(j3.1)| = 0.5471 \; \mho$$

For the parallel *RLC* circuit shown in Fig. 10.7, the magnitude of the admittance is minimum at resonance. (For the series *RLC* circuit in Fig. 10.9, the magnitude of the impedance is minimum at resonance.) However, for the circuit under consideration,

$$|\mathbf{Y}(j3.1)| < |\mathbf{Y}(j3)|$$

so that the magnitude of the admittance is not minimum at resonance. Finding the frequency at which the magnitude of the admittance is minimum requires a laborious process: Write the expression for $|\mathbf{Y}(j\omega)|$. Then take the derivative with respect to ω and equate the result to zero. From this, the value of ω which yields the minimum can be determined. For the circuit being discussed, the magnitude of the admittance is minimum when $\omega = 3.155$ rad/s, and its value is

$$|\mathbf{Y}(j3.155)| = 0.5458 \; \mho$$

For a series or parallel *RLC* circuit, we have seen that the resonance frequency is $\omega_r = 1/\sqrt{LC}$. Thus, for any positive values for L and C, we can easily determine ω_r. However, for the resonance frequency formula for the circuit in Fig. 10.11, when R, L, C, and R_L are positive numbers, we see that if

$$\frac{R_L{}^2}{L^2} > \frac{1}{LC} \Rightarrow R_L > \sqrt{\frac{L}{C}}$$

then there is no real-valued resonance frequency ω_r. Under this circumstance, the circuit never becomes resonant.

The term $1/\sqrt{LC}$ may seem familiar; it is the natural frequency ω_n for series and parallel *RLC* circuits, as well as the resonance frequency. However, in general, the resonance frequency and the natural frequency are not the same for an arbitrary *RLC* circuit.

. . .

Example
Consider the simple *RLC* circuit shown in Fig. 10.12.

fig. 10.12

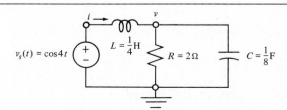

By KVL, for the mesh on the left,

$$L\frac{di}{dt} + v - v_s = 0$$

while, by KCL, at node v

$$i = C\frac{dv}{dt} + \frac{v}{R}$$

Substituting the second expression into the first, we get

$$LC\frac{d^2v}{dt^2} + \frac{L}{R}\frac{dv}{dt} + v - v_s = 0$$

from which

$$\frac{d^2v}{dt^2} + \frac{1}{RC}\frac{dv}{dt} + \frac{1}{LC}v = \frac{1}{LC}v_s$$

Thus we see that the natural frequency is

$$\omega_n = \frac{1}{\sqrt{LC}} = \frac{1}{\sqrt{\left(\frac{1}{4}\right)\left(\frac{1}{8}\right)}} = \sqrt{32} = 4\sqrt{2}\ \text{rad/s}$$

However, the impedance seen by the source is

$$\mathbf{Z} = \frac{j\omega}{4} + \frac{2(8/j\omega)}{2 + (8/j\omega)} = \frac{j\omega}{4} + \frac{16}{8 + j2\omega} = \frac{j\omega}{4} + \frac{8}{4 + j\omega}$$

$$= \frac{j\omega}{4} + \frac{8(4 - j\omega)}{16 + \omega^2}$$

$$= \frac{32}{16 + \omega^2} + j\omega\left(\frac{1}{4} - \frac{8}{16 + \omega^2}\right)$$

$$= \frac{32}{16 + \omega^2} + \frac{j\omega}{4}\left(\frac{\omega^2 - 16}{16 + \omega^2}\right)$$

Since the imaginary part of $\mathbf{Z}$ vanishes when $\omega^2 - 16 = 0$, we conclude that the resonance frequency is $\omega_r = 4$ rad/s $\neq \omega_n$. Since the frequency of the source is $\omega = 4$ rad/s, the circuit is in resonance. The circuit is shown in Fig. 10.13 in the frequency domain.

fig. 10.13

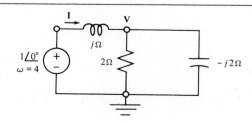

By nodal analysis,

$$\frac{1 - \mathbf{V}}{j} = \frac{\mathbf{V}}{2} + \frac{\mathbf{V}}{-j2}$$

from which

$$\mathbf{V} = \frac{2}{1 + j} = \frac{2}{\sqrt{2}\,\underline{/45°}} = \sqrt{2}\,\underline{/-45°}$$

and

$$\mathbf{I} = \frac{1 - \mathbf{V}}{j} = \frac{1 - 2/(1 + j)}{j} = \frac{(-1 + j)/(1 + j)}{j} = \frac{-1 + j}{-1 + j} = 1\,\underline{/0°}$$

so

$$v(t) = \sqrt{2}\,\cos(4t - 45°)$$

and

$$i(t) = \cos 4t$$

The energy stored in the inductor is

$$w_L(t) = \frac{1}{2}\,Li^2(t) = \frac{1}{8}\,\cos^2 4t$$

and the energy stored in the capacitor is

$$w_C(t) = \frac{1}{2}\,Cv^2(t) = \frac{1}{8}\,\cos^2(4t - 45°)$$

Using the trigonometric identity

$$\cos^2 x = \frac{1}{2}(1 + \cos 2x)$$

the total energy stored is

$$w_L(t) + w_C(t) = \frac{1}{16}\,[1 + \cos 8t] + \frac{1}{16}[1 + \cos(8t - 90°)]$$

$$= \frac{1}{16}\,[2 + \cos 8t + \sin 8t]$$

$$= \frac{1}{16}\,[2 + \sqrt{2}\,\cos(8t - 45°)]$$

and this has a maximum value of

$$[w_L(t) + w_C(t)]_{max} = \frac{1}{16}\,[2 + \sqrt{2}]\ \text{J}$$

The power dissipated by the resistor is

$$P_R = \frac{1}{2}\frac{|\mathbf{V}|^2}{R} = \frac{1}{2}\left(\frac{2}{2}\right) = \frac{1}{2}\ \text{W}$$

and the energy lost in a period is

$$P_R T = P_R\left(\frac{2\pi}{\omega_r}\right) = \frac{1}{2}\left(\frac{2\pi}{4}\right) = \frac{\pi}{4}\ \text{J}$$

Thus the Q of the circuit is

$$Q = 2\pi\ \frac{(1/16)(2 + \sqrt{2})}{\pi/4} = 1 + \frac{1}{\sqrt{2}} = 1.707$$

. . .

Just as we defined the quality factor Q for a circuit, we can define it for a reactance in series or parallel with a resistance. To demonstrate this fact, first consider the series case for an inductor L, as shown in Fig. 10.14.

---

fig. 10.14

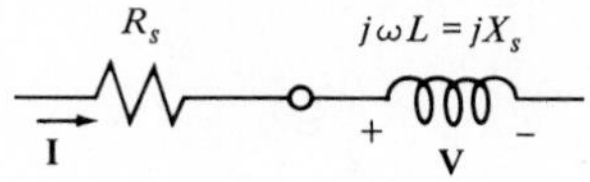

Let us assume that $\mathbf{I} = I\underline{/0°}$. Then the energy stored is

$$w(t) = \frac{1}{2}\ Li^2(t) = \frac{1}{2}\ LI^2\ \cos^2\ \omega t$$

which has the maximum value

$$w_m = \frac{1}{2}\ LI^2$$

and since $X_s = \omega L \Rightarrow L = X_s/\omega$, then

$$w_m = \frac{X_s}{2\omega}\ I^2$$

The energy lost per period is

$$P_s T = \frac{1}{2}\ I^2 R_s\left(\frac{2\pi}{\omega}\right) = \frac{\pi I^2 R_s}{\omega}$$

Therefore, the quality factor is

$$Q = \frac{2\pi(X_s I^2/2\omega)}{\pi I^2 R_s/\omega} = \frac{X_s}{R_s}$$

fig. 10.15

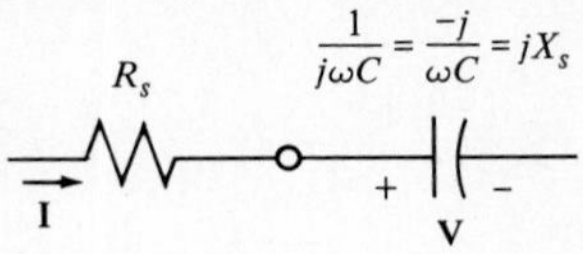

$$\frac{1}{j\omega C} = \frac{-j}{\omega C} = jX_s$$

Let us now consider the case of a series capacitor C, as shown in Fig. 10.15. Again, let $\mathbf{I} = I\underline{/0°}$. Then

$$\mathbf{V} = \frac{\mathbf{I}}{j\omega C} = \frac{I\underline{/0°}}{\omega C\underline{/90°}} = \frac{I}{\omega C}\underline{/-90°}$$

The energy stored in the capacitor is

$$w(t) = \frac{1}{2}Cv^2(t) = \frac{1}{2}C\left[\frac{I}{\omega C}\cos(\omega t - 90°)\right]^2$$

$$= \frac{1}{2}\frac{I^2}{\omega^2 C}\sin^2\omega t$$

which has the maximum value

$$w_m = \frac{I^2}{2\omega^2 C}$$

But since

$$-\frac{1}{\omega C} = X_s \Rightarrow |X_s| = \frac{1}{\omega C}$$

then

$$w_m = \frac{I^2|X_s|}{2\omega}$$

and thus

$$Q = \frac{2\pi(I^2|X_s|/2\omega)}{\pi I^2 R_s/\omega} = \frac{|X_s|}{R_s}$$

Hence, in general, for a series reactance X_s the quality factor is

$$\boxed{Q = \frac{|X_s|}{R_s}}$$

For the case of a parallel reactance X_p, as shown in Fig. 10.16, in a similar manner it can be shown that the quality factor is

$$\boxed{Q = \frac{R_p}{|X_p|}}$$

fig. 10.16

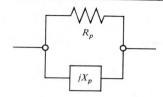

fig. 10.17

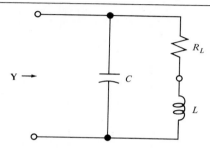

Consider the resonant circuit shown in Fig. 10.17, whose admittance is

$$\mathbf{Y} = j\omega C + \frac{1}{R_L + j\omega L}$$

Suppose that the series reactance, consisting of the inductor L and resistor R_L, has a large quality factor (such an inductor is sometimes called a "high-Q coil"). Then

$$\frac{X_L}{R_L} \gg 1 \qquad \text{or} \qquad \frac{\omega L}{R_L} \gg 1$$

so

$$\omega L \gg R_L$$

and the admittance is approximately

$$\mathbf{Y} \approx j\omega C + \frac{1}{j\omega L}$$

Thus the resonance frequency is approximately $\omega_r \approx 1/\sqrt{LC}$. The exact expression for the impedance is

$$\mathbf{Z} = \frac{(R_L + j\omega L)(1/j\omega C)}{R_L + j\omega L + (1/j\omega C)}$$

and therefore, for a high-Q series reactance,

$$\mathbf{Z} \approx \frac{(j\omega L)(1/j\omega C)}{R_L + j\omega L + (1/j\omega C)}$$

or

$$\mathbf{Y} \approx \frac{C}{L}\left(R_L + j\omega L + \frac{1}{j\omega C}\right)$$

$$\mathbf{Y} \approx \frac{R_L C}{L} + j\omega C + \frac{1}{j\omega L}$$

which is the expression for a parallel *RLC* circuit, where $R = L/R_L C$. Hence, for a high-*Q* coil, the two admittances shown in Fig. 10.18 are approximately equal.

fig. 10.18

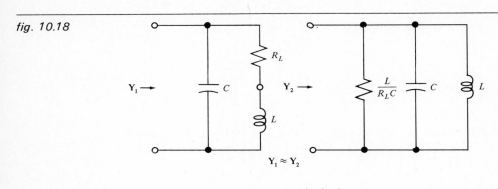

$$\mathbf{Y_1} \approx \mathbf{Y_2}$$

$\bullet$ $\bullet$ $\bullet$

Example
The practical tank circuit shown in Fig. 10.19 is to be resonant at 1 MHz; that is, $\omega_r = 2\pi(10^6)$ rad/s.

fig. 10.19

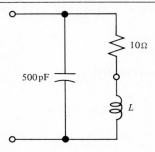

Assuming a high-*Q* coil, from

$$\omega_r \approx \frac{1}{\sqrt{LC}}$$

we obtain

$$L \approx \frac{1}{\omega_r^2 C} = \frac{1}{4\pi^2(10^{12})500(10^{-12})} = 50.7 \ \mu\text{H}$$

The quality factor of the series reactance at the resonance frequency is

$$\frac{X_s}{10} = \frac{\omega_r L}{10} = \frac{318}{10} = 31.8$$

which is much greater than one, as was assumed. Since

$$\frac{L}{R_L C} = 10{,}132 \approx 10{,}000$$

then the tank circuit in Fig. 10.19 can be approximated by the parallel *RLC* circuit shown in Fig. 10.20. The quality factor of this parallel *RLC* circuit is

$$Q = \omega_r RC = 2\pi(10^6)(10^4)(500)(10^{-12}) = 31.4$$

fig. 10.20

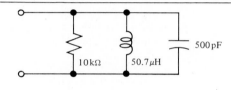

500 pF

10 kΩ   50.7 μH

• • •

PROBLEMS *10.8. Find the resonance frequency for each of the circuits shown in Fig. P10.8.

Ans. (a) $\dfrac{1}{\sqrt{LC - R^2 C^2}}$

fig. P10.8

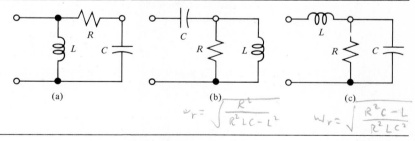

(a)

(b) $\omega_r = \sqrt{\dfrac{R^2}{R^2 LC - L^2}}$

(c) $W_r = \sqrt{\dfrac{R^2 C - L}{R^2 L C^2}}$

fig. P10.9

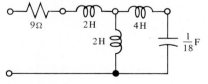

9 Ω   2 H   4 H

2 H $\frac{1}{18}$ F

$W_r = \dfrac{3\sqrt{10}}{5}$ rad/s

10.9. For the circuit in Fig. P10.9, find the resonance frequency.
*10.10. Find the quality factor for the circuit in Fig. P10.10.
Ans. 1.707

fig. P10.10

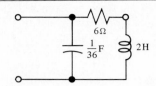

fig. P10.11

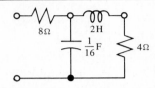

$Q = 0.85$

10.11. Find the quality factor for the circuit in Fig. P10.11.

*10.12. A 10-Ω resistor and a 2-H inductor are connected in series and $\omega = 50$ rad/s.
  (a) What is the Q of this series connection?
  (b) What parallel RL connection has the same admittance as the series connection at the given frequency?
  (c) What is the Q of this parallel connection?
  Ans.   (a) 10; (b) 1010 Ω, 2.02 H; (c) 10

10.13. A 10-Ω resistor and a 2-H inductor are connected in parallel and $\omega = 50$ rad/s.

a) $Q = 0.1$

b) $R = \dfrac{1000}{101}\ \Omega$ $L = \dfrac{2}{101}$ H

c) $Q = 0.1$

  (a) What is the Q of this parallel connection?
  (b) What series RL connection has the same impedance as the parallel connection at the given frequency?
  (c) What is the Q of this series connection?

10.14. Consider the series connection of a resistance R_s and a reactance X_s and the parallel connection of a resistance R_p and a reactance X_p.
  (a) Show that the admittance of the series connection is equal to the admittance of the parallel connection when

$$R_p = \frac{R_s^2 + X_s^2}{R_s} \qquad X_p = \frac{R_s^2 + X_s^2}{X_s}$$

  (b) Show that the quality factor Q_s of the series connection is equal to the quality factor Q_p of the parallel connection for the conditions given in (a).

fig. P10.15

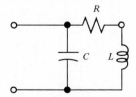

*10.15. Given the practical tank circuit in Fig. P10.15, if $R = 50\ \Omega$, $L = 50$ mH, and $C = 0.005\ \mu F$, approximate this circuit by a parallel RLC circuit. What is the quality factor of the parallel circuit?
  Ans.   63.2

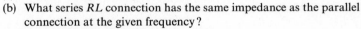

fig. P10.16

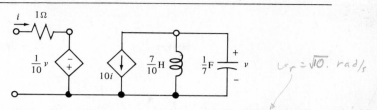

$\omega_r = \sqrt{10}.\ rad/s$

10.16.   Find the resonance frequency of the circuit shown in Fig. P10.16.

10.17.   Find the quality factor of the circuit given in Problem 10.9 given that $\omega_r = 6/\sqrt{10}$ rad/s.

$Q = 4.22$

10.3   SCALING

In the last example discussed, the element values given were practical in nature. This, however, has not been the trend previously, nor for the most part will it be in the remainder of this text. The reason for this is the simplification of computations. However, by a process known as "scaling," results obtained for circuits having simplified, nonrealistic element values can be extended easily to practical situations.

There are two ways to scale a circuit—*magnitude* or *impedance scaling* and *frequency scaling*. To magnitude-scale by the factor K_m, just multiply the impedance of each element by the real, positive number K_m. A resistor of R ohms is scaled to a resistor of $R' = K_m R$ ohms. An L-henry inductor of impedance $j\omega L$ ohms is scaled to an impedance of $j\omega K_m L$ ohms—that is, an $L' = K_m L$-henry inductor. A C-farad capacitor of impedance $1/j\omega C$ ohms is scaled to an impedance of $K_m/j\omega C = 1/j\omega(C/K_m)$ ohms—that is, a $C' = C/K_m$-farad capacitor. In summary, to scale an impedance by the factor K_m, scale R, L, and C as follows:

$$R \to K_m R$$
$$L \to K_m L$$
$$C \to \frac{C}{K_m}$$

• • •

Example

The parallel *RLC* circuit shown in Fig. 10.21 has the admittance

$$\mathbf{Y} = \frac{1}{\mathbf{Z}} = \frac{1}{R} + \frac{1}{j\omega L} + j\omega C = \frac{1}{2} + \frac{1}{j\omega} + \frac{j\omega}{25}$$

and the resonance frequency

$$\omega_r = \frac{1}{\sqrt{LC}} = 5 \text{ rad/s}$$

fig. 10.21

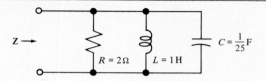

Scaling this circuit by $K_m = 5000$, we get $R' = 2K_m = 10 \text{ k}\Omega$, $L' = 1K_m = 5 \text{ kH}$, $C' = 1/25K_m = (1/125{,}000) \text{ F} = 8 \text{ }\mu\text{F}$ and an admittance $\mathbf{Y}'$ given by

$$\mathbf{Y}' = \frac{1}{K_m R} + \frac{1}{j\omega K_m L} + \frac{j\omega C}{K_m} = \frac{1}{10{,}000} + \frac{1}{j\omega 5000} + \frac{j\omega}{125{,}000}$$

and the resulting resonance frequency is

$$\omega_r' = \frac{1}{\sqrt{K_m L \dfrac{C}{K_m}}} = \frac{1}{\sqrt{LC}} = 5 \text{ rad/s}$$

That is, the resonance frequency is not affected by magnitude scaling. In actuality, the shape and the frequency axis of the amplitude response of an impedance are not affected by magnitude scaling, but the vertical axis is multiplied by K_m.

• • •

To frequency-scale by the factor K_f, since a resistor is not frequency-dependent, a resistor of R ohms is left as is. However, for an L-henry inductor of impedance $j\omega L$, at frequency $\omega = \omega_0$ the impedance is $j\omega_0 L$. What inductor L'' will have such an impedance at the scaled frequency $K_f \omega_0$? From

$$j\omega_0 L = j(K_f \omega_0) L''$$

we get

$$L'' = \frac{L}{K_f}$$

That is, to frequency-scale an inductor L by K_f, divide L by K_f.

For a C-farad capacitor of impedance $1/j\omega C$, at frequency $\omega = \omega_0$ the impedance is $1/j\omega_0 C$. The capacitor C'' that has such an impedance at the scaled frequency $K_f \omega_0$ is determined from

$$\frac{1}{j\omega_0 C} = \frac{1}{j(K_f \omega_0) C''}$$

Thus

$$C'' = \frac{C}{K_f}$$

That is, to frequency-scale a capacitor C by K_f, divide C by K_f. In summary, to frequency-scale by the factor K_f, scale R, L, and C as follows:

$$R \to R$$
$$L \to \frac{L}{K_f}$$
$$C \to \frac{C}{K_f}$$

$$\bullet \quad \bullet \quad \bullet$$

Example

To frequency-scale, by $K_f = 10^5$, the previously magnitude-scaled parallel RLC circuit given in the preceding example, we get $R'' = R' = 10$ kΩ, $L'' = L'/K_f$ = 5000/100,000 = 50 mH, $C'' = C'/K_f = 1/(125,000)(100,000) = 80$ pF and admittance $\mathbf{Y}''$ given by

$$\mathbf{Y}'' = \frac{1}{10^4} + \frac{1}{j\omega 50(10^{-3})} + j\omega 80(10^{-12})$$

and a resonance frequency

$$\omega_r'' = \frac{1}{\sqrt{(50)(10^{-3})(80)(10^{-12})}} = 5(10^5) \text{ rad/s} = 10^5 \, \omega_r = K_f \omega_r$$

That is, the resonance frequency is scaled by the factor $K_f = 10^5$. The vertical axis of the amplitude response of an impedance is not affected by frequency scaling, but the horizontal (frequency) axis is multiplied by K_f.

$$\bullet \quad \bullet \quad \bullet$$

For the original unscaled parallel RLC circuit, the bandwidth is

$$B = \frac{1}{RC} = \frac{1}{2(1/25)} = 12.5 \text{ rad/s}$$

For the magnitude-scaled parallel RLC circuit, the bandwidth is

$$B' = \frac{1}{R'C'} = \frac{1}{(10,000)(1/125,000)} = 12.5 \text{ rad/s} = B$$

However, after the circuit is frequency-scaled, the bandwidth is

$$B'' = \frac{1}{R''C''} = \frac{1}{(10^4)(80)(10^{-12})} = 1.25 \text{ Mrad/s} = 10^5 B = K_f B$$

Not only can elements such as resistors, inductors, and capacitors be magnitude-scaled, so can dependent sources. If a dependent source has a value whose units are ohms (i.e., a current-dependent voltage source), then to magnitude-scale we multiply the value by K_m; if the units are mhos (i.e., a voltage-dependent current source), we divide by K_m. Finally, a dependent source whose value is dimensionless (i.e., a voltage-dependent voltage source or a current-dependent current source) is left as is. Dependent sources of the type previously encountered are not affected by frequency scaling.

PROBLEMS   **10.18.**   Sketch the amplitude response (include the half-power frequencies) of the impedance of a parallel RLC circuit having $R = 2\ \Omega$, $L = 1$ H, and $C = \frac{1}{25}$ F, after it is magnitude-scaled by $K_m = 5(10^3)$ and frequency-scaled by $K_f = 10^5$. What are the scaled values of R, L, and C?

10.19.   Given a series RLC circuit with $R = \frac{5}{3}\ \Omega$, $L = 5$ H, and $C = \frac{1}{25}$ F. Suppose the input is V and the output is V_C, the voltage across the capacitor. Sketch the amplitude response of V_C/V after the circuit is magnitude-scaled by $K_m = 300$ and frequency-scaled by $K_f = 10^6/15$. What are the scaled values of R, L, and C?

fig. P10.20

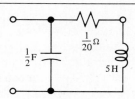

*10.20.   If the circuit shown in Fig. P10.20 is magnitude-scaled by $K_m = 10^3$ and frequency-scaled by $K_f = 10^5$, find the resulting values of R, L, and C.
Ans.   50 Ω, 50 mH, 0.005 μF

10.21.   What values of resistance, inductance, and capacitance, when magnitude-scaled by $K_m = 10^3$ and frequency-scaled by $K_f = 10^7$, will result in $R = 10$ kΩ, $L = 50\ \mu$H, and $C = 500$ pF?

fig. P10.22

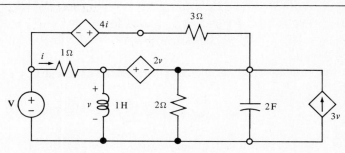

10.22.   The circuit shown in Fig. P10.22 is to be magnitude-scaled by $K_m = 2(10^3)$ and frequency-scaled by $K_f = 4(10^5)$. Determine all the resulting element values.

fig. P10.23

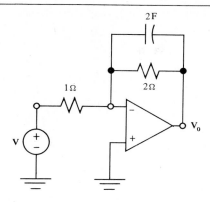

2F

1Ω

2Ω

V

V_o

*10.23.   Repeat Problem 10.22 for the op amp circuit in Fig. P10.23.
   Ans.   2 kΩ, 4 kΩ, 0.0025 μF

10.4   COMPLEX FREQUENCY

In the previous two and one-half chapters, our major preoccupation has been with the sinusoid. By generalizing real sinusoids to complex sinusoids we developed the notion of phasors, with which we were able to find forced responses of sinusoidal circuits without having to deal with differential equations. As a consequence, we were able to perform analyses in the frequency domain much more easily than in the time domain.

We are now in a position to go one step further. By extending the above-mentioned concepts, we can employ frequency domain analysis for damped sinusoidal circuits, that is, circuits having forcing functions of the form $Ae^{\sigma t} \cos(\omega t + \theta)$.

For the damped sinusoidal function $Ae^{\sigma t} \cos(\omega t + \theta)$, we know that the frequency ω has radians per second as its units. Since the power of the natural logarithm base e is to be dimensionless, the units of the (usually nonpositive) damping factor σ have been designated *nepers per second*, where the *neper** (Np) is a dimensionless unit. For this reason σ is referred to as the *neper frequency*.

Now suppose that the input to an *RLC* circuit has the form $Ae^{\sigma t} \cos(\omega t + \theta)$. Using an argument similar to that discussed for the real sinusoidal case, the forced response will have the form $Be^{\sigma t} \cos(\omega t + \phi)$. Thus, by the principle of superposition the response to

$$Ae^{\sigma t} \cos(\omega t + \theta) + jAe^{\sigma t} \sin(\omega t + \theta) = Ae^{\sigma t}e^{j(\omega t + \theta)}$$

$$= Ae^{j\theta}e^{(\sigma + j\omega)t}$$

$$= Ae^{j\theta}e^{st}$$

$$= \mathbf{V}_{in}e^{st}$$

* Named for the Scottish mathematician John Napier (1550–1617).

where $\mathbf{V}_{in} = Ae^{j\theta}$ and $s = \sigma + j\omega$, which is called *complex frequency*, has the form

$$Be^{\sigma t} \cos(\omega t + \phi) + jBe^{\sigma t} \sin(\omega t + \phi) = Be^{\sigma t} e^{j(\omega t + \phi)}$$
$$= Be^{j\phi} e^{(\sigma + j\omega)t}$$
$$= Be^{j\phi} e^{st}$$
$$= \mathbf{V}_o e^{st}$$

where $\mathbf{V}_o = Be^{j\phi}$. Hence, knowing the response to $Ae^{j\theta} e^{st}$ is tantamount to knowing the response to $Ae^{\sigma t} \cos(\omega t + \theta)$. That is, we can determine B and ϕ by finding the response either to $Ae^{j\theta} e^{st}$ or to $Ae^{\sigma t} \cos(\omega t + \theta)$.

$$\cdot \quad \cdot \quad \cdot$$

Example
To find the voltage $v_C(t)$ across the capacitor in the circuit shown in Fig. 10.22, note that

$$i = i_R + i_C$$
$$= \frac{v_C}{4} + \frac{1}{4}\frac{dv_C}{dt} \tag{10.1}$$

fig. 10.22

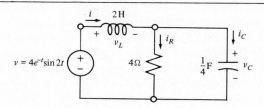

and

$$v_L = 2\frac{di}{dt}$$
$$v - v_C = 2\frac{di}{dt} \tag{10.2}$$

Substituting Equation (10.1) into (10.2) yields

$$v - v_C = 2\frac{d}{dt}\left(\frac{v_C}{4} + \frac{1}{4}\frac{dv_C}{dt}\right)$$

from which we get

$$\frac{d^2 v_C}{dt^2} + \frac{dv_C}{dt} + 2v_C = 2v \tag{10.3}$$

We are given that

$$v(t) = 4e^{-t} \sin 2t = 4e^{-t} \cos(2t - \pi/2)$$

Thus we recognize that we can find $v_C(t)$, which has the form

$$v_C(t) = Be^{\sigma t}\cos(\omega t + \phi)$$
$$= Be^{-t}\cos(2t + \phi)$$

by finding the response, instead, to

$$v_x(t) = Ae^{j\theta}e^{st}$$

which is

$$v_y(t) = Be^{j\phi}e^{st}$$

Substituting this into Equation (10.3) results in

$$\frac{d^2}{dt^2}[Be^{j\phi}e^{st}] + \frac{d}{dt}[Be^{j\phi}e^{st}] + 2[Be^{j\phi}e^{st}] = 2Ae^{j\theta}e^{st}$$

from which we get

$$Be^{j\phi}s^2e^{st} + Be^{j\phi}se^{st} + 2Be^{j\phi}e^{st} = 2Ae^{j\theta}e^{st}$$

Dividing both sides of this equation by e^{st} yields

$$Be^{j\phi}(s^2 + s + 2) = 2Ae^{j\theta}$$

so

$$Be^{j\phi} = \frac{2Ae^{j\theta}}{s^2 + s + 2}$$

Since $A = 4$, $s = \sigma + j\omega = -1 + j2$, and $\theta = -\pi/2$, then

$$Be^{j\phi} = \frac{2(4)e^{-j\pi/2}}{(-1 + j2)^2 + (-1 + j2) + 2}$$
$$= \frac{4e^{-j\pi/2}}{-1 - j} = \frac{4e^{-j\pi/2}}{\sqrt{2}e^{-3j\pi/4}}$$
$$= 2\sqrt{2}e^{j\pi/4}$$

In other words, $B = 2\sqrt{2}$ and $\phi = \pi/4$. Thus, the desired forced response is

$$v_C(t) = 2\sqrt{2}e^{-t}\cos\left(2t + \frac{\pi}{4}\right)$$

. . .

Just as we associated the real sinusoid $A\cos(\omega t + \theta)$ or the complex sinusoid $Ae^{j(\omega t + \theta)}$ with the phasor $\mathbf{A} = Ae^{j\theta}$, so we can associate the damped sinusoid $Ae^{\sigma t}\cos(\omega t + \theta)$ or the damped complex sinusoid $Ae^{\sigma t}e^{j(\omega t + \theta)} = Ae^{j\theta}e^{st}$ with the phasor $\mathbf{A} = Ae^{j\theta}$, also. However, for the latter case, $s = \sigma + j\omega$ must be implicit, whereas only ω need be for the former.

fig. 10.23

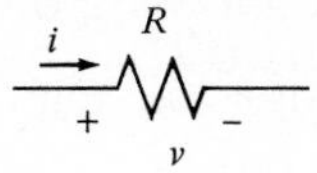

For a resistor (see Fig. 10.23), since $v = Ri$, if the current is a damped complex sinusoid, say $i = Ie^{\sigma t}e^{j(\omega t + \theta)} = Ie^{j\theta}e^{st}$, then the voltage has the form $v = Ve^{\sigma t}e^{j(\omega t + \phi)} = Ve^{j\phi}e^{st}$. Thus,

$$Ve^{j\phi}e^{st} = RIe^{j\theta}e^{st}$$

and dividing by e^{st} yields

$$Ve^{j\phi} = RIe^{j\theta}$$

from which we make the identities $V = RI$ and $\phi = \theta$. Using phasor notation,

$$V\underline{/\phi} = RI\underline{/\theta}$$

or

$$\mathbf{V} = R\mathbf{I}$$

where $\mathbf{V} = V\underline{/\phi}$ and $\mathbf{I} = I\underline{/\theta}$.

Defining the impedance $\mathbf{Z}_R$ of a resistor as the ratio of voltage phasor to current phasor, we therefore have

$$\boxed{\mathbf{Z}_R = \frac{\mathbf{V}}{\mathbf{I}} = R}$$

For an inductor (Fig. 10.24), if $i = Ie^{j\theta}e^{st}$ and $v = Ve^{j\phi}e^{st}$, since

$$v = L\frac{di}{dt}$$

fig. 10.24

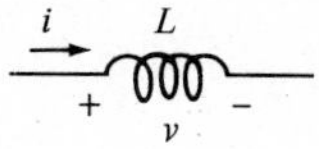

then

$$Ve^{j\phi}e^{st} = L\frac{d}{dt}(Ie^{j\theta}e^{st})$$

$$= LsIe^{j\theta}e^{st}$$

Dividing by e^{st} gives

$$Ve^{j\phi} = LsIe^{j\theta}$$

or

$$V\underline{/\phi} = LsI\underline{/\theta}$$

so

$$\mathbf{V} = Ls\mathbf{I}$$

Thus, the impedance $\mathbf{Z}_L$ of an inductor is

$$\boxed{\mathbf{Z}_L = \frac{\mathbf{V}}{\mathbf{I}} = Ls}$$

For a capacitor (Fig. 10.25), since

$$i = C\frac{dv}{dt}$$

fig. 10.25

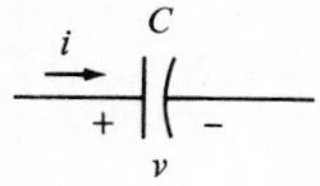

then

$$Ie^{j\theta}e^{st} = C\frac{d}{dt}(Ve^{j\phi}e^{st})$$

$$= CsVe^{j\phi}e^{st}$$

or

$$Ie^{j\theta} = CsVe^{j\phi}$$

so

$$I\underline{/\theta} = CsV\underline{/\phi}$$

and

$$\mathbf{I} = Cs\mathbf{V}$$

The impedance of a capacitor is, therefore,

$$\boxed{\mathbf{Z}_C = \frac{\mathbf{V}}{\mathbf{I}} = \frac{1}{Cs}}$$

From the definition of impedance, we have Ohm's law:

$$\mathbf{V} = \mathbf{ZI} \quad \text{and} \quad \mathbf{I} = \frac{\mathbf{V}}{\mathbf{Z}}$$

The second equation suggests the usefulness of the reciprocal of impedance. We, therefore, define the ratio of current phasor to voltage phasor as *admittance*, denoted **Y**. Thus, we have the element admittances

$$\mathbf{Y}_R = \frac{1}{R}$$

$$\mathbf{Y}_L = \frac{1}{Ls}$$

$$\mathbf{Y}_C = Cs$$

and the alternative forms of Ohm's law:

$$\mathbf{Y} = \frac{\mathbf{I}}{\mathbf{V}}$$

$$\mathbf{I} = \mathbf{YV}$$

$$\mathbf{V} = \frac{\mathbf{I}}{\mathbf{Y}}$$

Just as we analyzed sinusoidal circuits with the use of (sinusoidal) phasors and impedance, so can we analyze damped sinusoidal circuits. For the former case the impedance was a function of ω, denoted $\mathbf{Z}(j\omega)$, whereas for the latter it is a function of σ and ω, denoted $\mathbf{Z}(s)$.

$\cdot \quad \cdot \quad \cdot$

Example
The circuit in Fig. 10.22 is reproduced in Fig. 10.26 using phasor and impedance notation. The parallel *RC* combination has the impedance

$$\mathbf{Z}_{RC} = \frac{4(4/s)}{4 + 4/s} = \frac{4}{s+1}$$

fig. 10.26

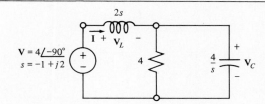

By voltage division, we have

$$\mathbf{V}_C = \frac{\mathbf{Z}_{RC}\mathbf{V}}{\mathbf{Z}_{RC} + \mathbf{Z}_L}$$

$$= \frac{\dfrac{4}{s+1}\mathbf{V}}{\dfrac{4}{s+1} + 2s} = \frac{4\mathbf{V}}{2s^2 + 2s + 4} = \frac{2\mathbf{V}}{s^2 + s + 2}$$

From the fact that $\mathbf{V} = 4 \underline{/-90^\circ}$ and $s = -1 + j2$, we get

$$\mathbf{V}_C = \frac{2(4\underline{/-90^\circ})}{(-1+j2)^2 + (-1+j2) + 2} = 2\sqrt{2}\,\underline{/45^\circ}$$

as before.

By removing the capacitor and applying Thévenin's theorem, we have the circuit shown in Fig. 10.27. By voltage division,

$$\mathbf{V}_{oc} = \frac{4\mathbf{V}}{4 + 2s} = \frac{2\mathbf{V}}{s+2}$$

fig. 10.27 (left)
fig. 10.28 (right)

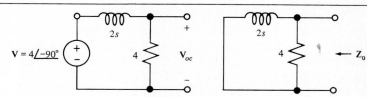
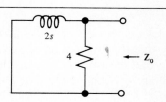

We find $\mathbf{Z}_0$ from Fig. 10.28 as follows:

$$\mathbf{Z}_0 = \frac{2s(4)}{2s + 4} = \frac{4s}{s+2}$$

Hence, we have the circuit shown in Fig. 10.29. Again, by voltage division,

$$\mathbf{V}_C = \frac{\dfrac{4}{s}\mathbf{V}_{oc}}{\dfrac{4}{s} + \dfrac{4s}{s+2}} = \frac{(s+2)\mathbf{V}_{oc}}{(s+2) + s^2}$$

$$= \frac{2\mathbf{V}}{s^2 + s + 2}$$

as was obtained previously.

fig. 10.29

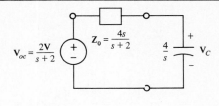

$$V_{oc} = \frac{2V}{s+2} \qquad Z_o = \frac{4s}{s+2} \qquad \frac{4}{s} \qquad V_C$$

. . .

For the circuit given in the preceding example (Fig. 10.26), if we define the voltage transfer function $H_C(s)$ to be the ratio of V_C to V, we then have that

$$H_C(s) = \frac{V_C}{V} = \frac{2}{s^2 + s + 2}$$

Factoring the denominator, we can rewrite $H_C(s)$ as

$$H_C(s) = \frac{2}{\left(s + \frac{1}{2} - j\frac{\sqrt{7}}{2}\right)\left(s + \frac{1}{2} + j\frac{\sqrt{7}}{2}\right)}$$

Thus, we see for the special cases that

$$s = -\frac{1}{2} + j\frac{\sqrt{7}}{2} \qquad \text{or} \qquad s = -\frac{1}{2} - j\frac{\sqrt{7}}{2}$$

the transfer function becomes infinite, which implies that the forced response is infinite. We say that the function has a *pole* at each of these two complex frequencies.

To find the voltage transfer function $H_L(s) = V_L/V$ for this circuit, we can again use voltage division as follows: From

$$V_L = \frac{2sV}{2s + 4/(s + 1)} = \frac{2s(s + 1)V}{2s(s + 1) + 4}$$

we get

$$H_L(s) = \frac{V_L}{V} = \frac{s(s + 1)}{s^2 + s + 2}$$

Therefore, $H_L(s)$ has the same poles as $H_C(s)$. However, there are two frequencies, $s = 0$ and $s = -1$, for which $H_L(s) = 0$. For this reason we say that the network function $H_L(s)$ has *zeros* at $s = 0$ and at $s = -1$.

Since $s = \sigma + j\omega$ is a complex quantity, we can indicate any value of s as a point in a complex-number plane. If we label the horizontal axis σ and the vertical axis $j\omega$, the result is known as the *s-plane*. If we denote a pole by the symbol "$\times$" and a zero by the symbol "$\bigcirc$," then depicting the poles and zeros of a function in the s-plane is referred to as a *pole-zero plot*. The pole-zero plot for $H_L(s)$ is shown in Fig. 10.30. The pole-zero plot for $H_C(s)$ can be obtained from this by removing the two zeros on the σ (real) axis; that is, the pole-zero plot for $H_C(s)$ does not have any (finite) zeros.

fig. 10.30

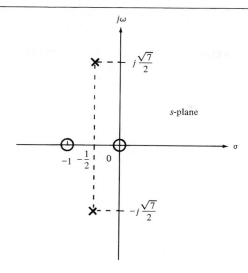

Return now to Equation (10.3), the differential equation describing the voltage across the capacitor. The solution to this equation when $v(t) = 0$ is the natural response. As was done in the chapter on second-order circuits, we get the natural response by assuming a solution of the form Ae^{st}. Doing so, we obtain two values for s that satisfy the equation

$$s^2 + s + 2 = 0$$

Using the quadratic formula, we get

$$s = \frac{-1 \pm \sqrt{1 - 8}}{2} = -\frac{1}{2} \pm j\frac{\sqrt{7}}{2}$$

In other words, the natural response $v_n(t)$ has the form

$$v_n(t) = A_1 e^{s_1 t} + A_2 e^{s_2 t}$$

where

$$s_1 = -\frac{1}{2} + j\frac{\sqrt{7}}{2} \qquad \text{and} \qquad s_2 = -\frac{1}{2} - j\frac{\sqrt{7}}{2}$$

But note, these are precisely the poles of the voltage transfer function $\mathbf{H}_C(s) = \mathbf{V}_C/\mathbf{V}$. Thus, and this is true in general, the poles of the transfer function specify the natural response—provided that there was no cancellation of a common pole and zero. In this example the complex values for s_1 and s_2—that is, the complex poles—indicate that this second-order circuit is underdamped. Thus, the natural response has the more common alternative forms

$$v_n(t) = B_1 e^{-\alpha t} \cos \omega_d t + B_2 e^{-\alpha t} \sin \omega_d t$$

$$= B e^{-\alpha t} \cos(\omega_d t - \phi)$$

and, from Equation (10.3), $\alpha = 1/2$ and $\omega_n = \sqrt{2}$, so

$$\omega_d = \sqrt{\omega_n{}^2 - \alpha^2} = \sqrt{2 - \frac{1}{4}} = \frac{\sqrt{7}}{2}$$

Hence, for an input of $v(t) = 4e^{-t} \sin 2t\, u(t)$, the complete response has the form

$$v_C(t) = Be^{-t/2} \cos\left(\frac{\sqrt{7}}{2} t - \phi\right) + 2\sqrt{2} e^{-t} \cos\left(2t + \frac{\pi}{4}\right)$$

The constants B and ϕ (or B_1 and B_2) are determined from the initial conditions $i(0)$ and $v_C(0)$. However, this requires the cumbersome approach discussed in Chapter 6. Therefore, we shall postpone the topic of obtaining the complete response until the chapter on Laplace transforms, where we will take a greatly simplified approach.

Again let us return to the circuit given in the preceding example (Fig. 10.26) and consider yet another network function for the same input. This time, suppose that the output variable is the inductor current **I**. Since the impedance seen by the voltage source is

$$\mathbf{Z} = \mathbf{Z}_L + \mathbf{Z}_{RC} = 2s + \frac{4}{s + 1} = \frac{2(s^2 + s + 2)}{s + 1}$$

and since $\mathbf{Z} = \mathbf{V}/\mathbf{I}$, then the transfer function of interest $\mathbf{I}/\mathbf{V}$ is just the admittance $\mathbf{Y}(s)$ seen by the source; that is,

$$\mathbf{Y}(s) = \frac{\mathbf{I}}{\mathbf{V}} = \frac{\frac{1}{2}(s + 1)}{s^2 + s + 2}$$

and again the poles of $\mathbf{Y}(s)$ are the poles of $\mathbf{H}_C(s)$ and $\mathbf{H}_L(s)$. The pole-zero plot for $\mathbf{Y}(s)$ can be obtained from the pole-zero plot for $\mathbf{H}_L(s)$ by removing the zero at the origin of the s-plane.

It's not just coincidence that the poles for the network functions $\mathbf{V}_C/\mathbf{V}$, $\mathbf{V}_L/\mathbf{V}$, and $\mathbf{I}/\mathbf{V}$ determined above are the same. Provided that one portion of a circuit is not physically separated from the rest, each transfer function (defined as the ratio of an output to a given input) will have the same poles regardless of which voltage or current is chosen as the output variable. This should not be surprising, however, for the poles are those values of s, called *natural frequencies*, that determine the natural response. And since the form of the natural response is the same throughout a (nonseparated) circuit, the poles of any transfer function (with respect to a given input variable) will be the same.

$$\cdot \quad \cdot \quad \cdot$$

Example
A circuit which contains two dependent sources is shown in Fig. 10.31 and again in the frequency domain in Fig. 10.32.

fig. 10.31

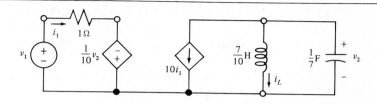

fig. 10.32

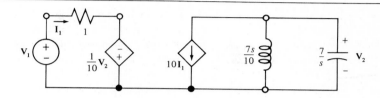

By nodal analysis,

$$10\mathbf{I}_1 + \frac{\mathbf{V}_2}{7s/10} + \frac{\mathbf{V}_2}{7/s} = 0$$

and

$$\mathbf{I}_1 = \frac{\mathbf{V}_1 + \frac{1}{10}\mathbf{V}_2}{1}$$

Combining these two equations, we get

$$10\mathbf{V}_1 + \mathbf{V}_2 + \frac{10}{7s}\mathbf{V}_2 + \frac{s}{7}\mathbf{V}_2 = 0$$

from which

$$(s^2 + 7s + 10)\mathbf{V}_2 = -70s\mathbf{V}_1$$

and thus the voltage transfer function $\mathbf{H}(s) = \mathbf{V}_2/\mathbf{V}_1$ is given by

$$\mathbf{H}(s) = \frac{\mathbf{V}_2}{\mathbf{V}_1} = \frac{-70s}{s^2 + 7s + 10} = \frac{-70s}{(s+2)(s+5)}$$

In this case all the poles and zeros are on the nonpositive real axis of the s-plane. There is a zero at the origin and poles at $s = -2$ and $s = -5$.

Solving for $\mathbf{I}_1$ in terms of $\mathbf{V}_1$, we have

$$\mathbf{I}_1 = \mathbf{V}_1 + \frac{1}{10}\mathbf{V}_2 = \mathbf{V}_1 + \frac{1}{10}\left(\frac{-70s\mathbf{V}_1}{s^2 + 7s + 10}\right)$$

$$= \left(1 - \frac{7s}{s^2 + 7s + 10}\right)\mathbf{V}_1 = \frac{s^2 + 10}{s^2 + 7s + 10}\mathbf{V}_1$$

from which the transfer function $I_1/V_1 = Y(s)$ is

$$Y(s) = \frac{I_1}{V_1} = \frac{s^2 + 10}{s^2 + 7s + 10} = \frac{(s - j\sqrt{10})(s + j\sqrt{10})}{(s + 2)(s + 5)}$$

and although the poles are the same as for $H(s)$, there is a pair of purely imaginary complex conjugate zeros (i.e., zeros on the $j\omega$-axis) as shown in Fig. 10.33—the pole-zero plot for the admittance $Y(s)$.

fig. 10.33

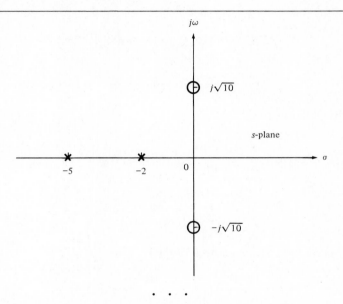

So far, we have seen examples only of network functions with distinct poles. There are cases in which a network function contains more than one pole at a particular point in the s-plane (a critically damped circuit is an example). In such a case we say that the function contains a *multiple pole*. Again, we shall defer the discussion of the natural response for such a situation to the chapter on Laplace transforms.

PROBLEMS   10.24.   For the series RLC circuit shown in Fig. P10.24, suppose that $R = \frac{5}{3}\Omega$, $L = 5\,\text{H}$, and $C = \frac{1}{25}\,\text{F}$. Draw a pole-zero plot for (a) I/V, (b) V_R/V, (c) V_L/V, and (d) V_C/V.

fig. P10.24

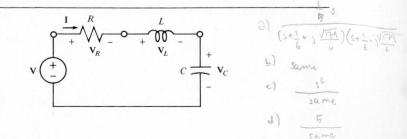

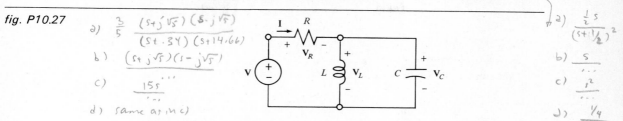

10.25. Repeat Problem 10.24 for the case that $R = 2\,\Omega$, $L = \frac{1}{2}\,H$, and $C = 2\,F$.
10.26. Repeat Problem 10.24 for the case that $R = 2\,\Omega$, $L = 2\,H$, and $C = 2\,F$.

Handwritten (top right):

a) $\dfrac{2s}{(s+3.73)(s+,26...)}$

b) $\dfrac{4s}{...}$

c) $\dfrac{4s^2}{...}$

d) $\dfrac{1}{...}$

fig. P10.27

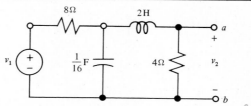

Handwritten (left):

a) $\dfrac{3}{5}\dfrac{(s+j\sqrt5)(s-j\sqrt5)}{(s+.34)(s+14.66)}$

b) $(s+j\sqrt5)(s-j\sqrt5)$

c) $\dfrac{15s}{...}$

d) same as in c)

Handwritten (right):

a) $\dfrac{\frac{1}{2}s}{(s+1\frac{1}{2})^2}$

b) $\dfrac{s}{...}$

c) $\dfrac{s^2}{...}$

d) $\dfrac{1}{4}$

10.27. Repeat Problem 10.24 for the RLC circuit in Fig. P10.27.

fib. P10.28

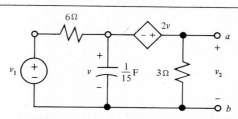

10.28. Consider the circuit in Fig. P10.28.
  (a) Find the impedance seen by the voltage source.
  (b) Find the voltage transfer function $\mathbf{V_2}/\mathbf{V_1}$.
  (c) Find the Thévenin equivalent circuit with respect to terminals a and b.
  (d) Draw the pole-zero plots for parts (a) and (b).

Handwritten:

$Z = \dfrac{8(s^2+4s+12)}{s^2+2s+8}$

$V_{oc}=\dfrac{4V_1}{s^2+4s+12}$

$Z_0 = 4(s^2+2s+8)$

fig. P10.29

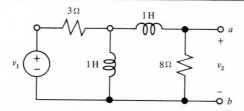

*10.29. Repeat Problem 10.28 for the circuit shown in Fig. P10.29.

Ans. $\dfrac{6s+105}{s+15}$, $\dfrac{15}{2s+35}$

Handwritten: $\dfrac{15V}{2s+35}$ $\dfrac{90}{2s+35}$

fig. P10.30

3Ω   1H

v_1 1H   8Ω v_2

a   b

Handwritten:

a) $\dfrac{s^2+14s+24}{2(s+4)}$

b) $\dfrac{8s}{s^2+14s+24}$

c) $V_{oc}=\dfrac{8sV_1}{s^2+14s+24}$

$Z_0=\dfrac{8s(s+6)}{s^2+14s+24}$

10.30. Repeat Problem 10.28 for the circuit in Fig. P10.30.

fig. P10.31

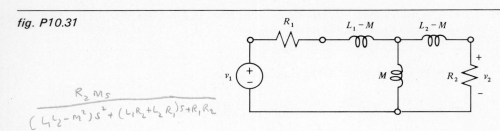

$$\frac{R_2 M s}{(L_1 L_2 - M^2)s^2 + (L_1 R_2 + L_2 R_1)s + R_1 R_2}$$

10.31.   Find the voltage transfer function $\mathbf{V}_2/\mathbf{V}_1$ for the circuit in Fig. P10.31.

fig. P10.32

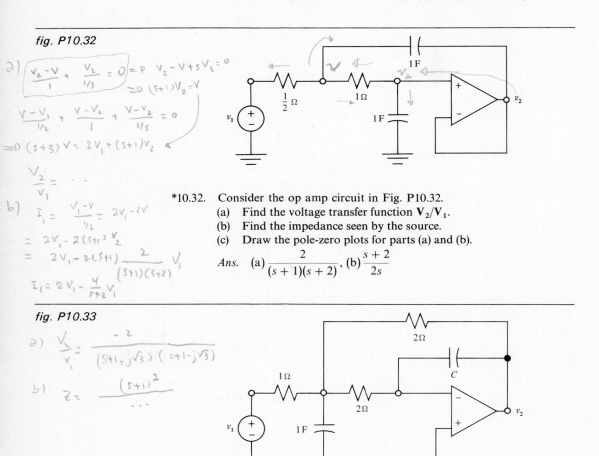

a) $\boxed{\dfrac{V_2 - V}{1} + \dfrac{V_2}{1/s} = 0} = P \quad V_2 - V + sV_2 = 0$

$\Rightarrow (s+1)V_2 = V$

$\dfrac{V - V_1}{1/2} + \dfrac{V - V_2}{1} + \dfrac{V - V_2}{1/s} = 0$

$\Rightarrow (s+3)V = 2V_1 + (s+1)V_2$

$\dfrac{V_2}{V_1} = \cdots$

b) $I_1 = \dfrac{V_1 - V}{1/2} = 2V_1 - 2V$

$= 2V_1 - 2(s+1)V_2$

$= 2V_1 - 2(s+1)\dfrac{2}{(s+1)(s+2)}V_1$

$I_1 = 2V_1 - \dfrac{4}{s+2}V_1$

*10.32.   Consider the op amp circuit in Fig. P10.32.
(a)   Find the voltage transfer function $\mathbf{V}_2/\mathbf{V}_1$.
(b)   Find the impedance seen by the source.
(c)   Draw the pole-zero plots for parts (a) and (b).

Ans.   (a) $\dfrac{2}{(s+1)(s+2)}$, (b) $\dfrac{s+2}{2s}$

fig. P10.33

a) $\dfrac{V_2}{V_1} = \dfrac{-2}{(s+1+j\sqrt{3})(s+1-j\sqrt{3})}$

b) $Z = \dfrac{(s+1)^2}{\cdots}$

10.33.   Repeat Problem 10.32 for the op amp circuit in Fig. P10.33 given that
(a) $C = \frac{1}{4}$ F and (b) $C = \frac{1}{16}$ F.

10.5 GRAPHICAL DETERMINATION OF FREQUENCY RESPONSES

As was demonstrated previously, a circuit having an input, say $\mathbf{V}_{in}e^{st}$, will have a forced response of the form $\mathbf{V}_o e^{st}$. Substituting these terms into the general form of the differential equation describing the circuit we get

$$a_n \frac{d^n v_o}{dt^n} + a_{n-1} \frac{d^{n-1} v_o}{dt^{n-1}} + \cdots + a_1 \frac{dv_o}{dt} + a_0 v_o$$

$$= b_m \frac{d^m v_{in}}{dt^m} + b_{m-1} \frac{d^{m-1} v_{in}}{dt^{m-1}} + \cdots + b_1 \frac{dv_{in}}{dt} + b_0 v_{in}$$

which results in

$$(a_n s^n + a_{n-1} s^{n-1} + \cdots + a_1 s + a_0)\mathbf{V}_o e^{st} = (b_m s^m + b_{m-1} s^{m-1} + \cdots + b_1 s + b_0)\mathbf{V}_{in} e^{st}$$

From this expression, we obtain the following general form of the transfer function $\mathbf{H}(s)$.

$$\mathbf{H}(s) = \frac{\mathbf{V}_o}{\mathbf{V}_{in}} = \frac{b_m s^m + b_{m-1} s^{m-1} + \cdots + b_1 s + b_0}{a_n s^n + a_{n-1} s^{n-1} + \cdots + a_1 s + a_0}$$

that is, $\mathbf{V}_o / \mathbf{V}_{in}$ is a ratio of polynomials in s with real coefficients.

A result from mathematics theory is that a polynomial with real coefficients can be factored into a product of quadratics of the form $as^2 + bs + c$, where $a, b,$ and c are real numbers. From the quadratic formula, such a term can further be factored into $(s - s_1)(s - s_2)$, where

$$s_1 = \frac{-b + \sqrt{b^2 - 4ac}}{2a} \quad \text{and} \quad s_2 = \frac{-b - \sqrt{b^2 - 4ac}}{2a}$$

For the case that $b^2 \geq 4ac$, the numbers s_1 and s_2 are real. However, if $b^2 < 4ac$, then s_1 and s_2 are complex numbers and are conjugates. As a consequence of this, we can express the transfer function, or any other network function for that matter, as

$$\mathbf{H}(s) = \frac{K(s - z_1)(s - z_2)\cdots(s - z_m)}{(s - p_1)(s - p_2)\cdots(s - p_n)}$$

where $z_1, z_2, \ldots, z_m$ are the zeros and $p_1, p_2, \ldots, p_n$ are the poles. The zeros and poles can be either real or complex, but complex zeros or poles occur in conjugate pairs.

For the case of distinct poles (the case of multiple poles will be discussed in Chapter 14), the natural response has the form

$$v_n(t) = A_1 e^{p_1 t} + A_2 e^{p_2 t} + \cdots + A_n e^{p_n t}$$

If pole p_i is at the origin of the s-plane, that is, if $p_i = 0$, then the term in the natural response corresponding to it is $A_i e^0 = A_i$, a constant. If p_i is on the negative real axis or the positive real axis of the s-plane, then $A_i e^{p_i t}$ is a decaying exponential or increasing exponential, respectively, as shown in Fig. 10.34. If p_i is a complex number, then there is a pole $p_j = p_i^*$. As we have seen in Chapter 6, when $p_i = \sigma + j\omega$, the corresponding

fig. 10.34

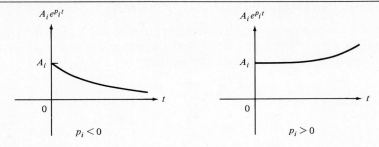

terms in the natural response can be combined in the form $Ae^{\sigma t}\cos(\omega t + \theta)$. If the pair of conjugate poles is in the left half of the s-plane (i.e., $\sigma < 0$), this term is a damped sinusoid. If the pair is on the $j\omega$-axis ($\sigma = 0$), the term is a sinusoid. If the pair is in the right half of the s-plane, the term is an increasing sinusoid. These three cases are depicted in Fig. 10.35.

fig. 10.35

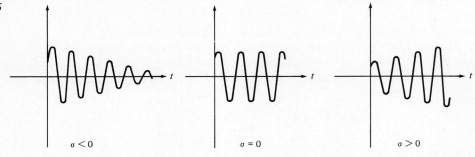

As can be surmised from the above discussion, a circuit that has a stable operation does not have a network function with poles in the right half of the s-plane (RHP). In some applications, poles on the imaginary axis are undesirable, and only poles in the left half of the s-plane (LHP) are allowable.

Suppose that s_0 is a complex number. Then s_0 is not only a point $s_0 = \sigma_0 + j\omega_0$ in the s-plane, it also represents the vector from the origin to the point $\sigma_0 + j\omega_0$, as shown in Fig. 10.36. Also shown is the vector s_1 from the origin to the point $\sigma_1 + j\omega_1$. If we place a vector s' directed from point s_0 to point s_1, then by vector addition, we have $s_0 + s' = s_1$. Thus $s' = s_1 - s_0$. We shall now use this result to graphically determine frequency responses.

We have seen that the sinusoid $A\cos(\omega t + \theta)$ is a special case of the damped sinusoid $Ae^{\sigma t}\cos(\omega t + \theta)$. We can obtain the former from the latter by setting $\sigma = 0$. In analyzing a circuit, using $s = \sigma + j\omega$ allows us to determine forced responses to damped sinusoids. If we set $\sigma = 0$ (that is, $s = j\omega$), we get the sinusoidal case.

Given the network function

$$\mathbf{H}(s) = \frac{K(s - z_1)(s - z_2)\cdots(s - z_m)}{(s - p_1)(s - p_2)\cdots(s - p_n)}$$

fig. 10.36

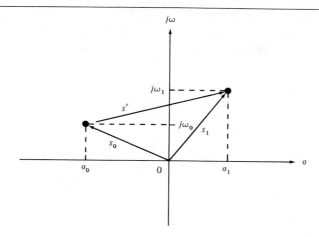

for the sinusoidal case we set $s = j\omega$. Thus

$$\mathbf{H}(j\omega) = \frac{K(j\omega - z_1)(j\omega - z_2)\cdots(j\omega - z_m)}{(j\omega - p_1)(j\omega - p_2)\cdots(j\omega - p_n)}$$

From the discussion above, $j\omega - p_i$ is the vector from pole p_i to the point $j\omega$ on the imaginary axis. Also, $j\omega - z_j$ is the vector from zero z_j to $j\omega$. Let us denote the former complex number by $D_i \,\underline{/\theta_i}$ and the latter by $N_j \,\underline{/\phi_j}$. Thus, we have

$$\mathbf{H}(j\omega) = \frac{K(N_1\underline{/\phi_1})(N_2\underline{/\phi_2})\cdots(N_m\underline{/\phi_m})}{(D_1\underline{/\theta_1})(D_2\underline{/\theta_2})\cdots(D_n\underline{/\theta_n})}$$

$$= \frac{KN_1N_2\cdots N_m}{D_1D_2\cdots D_n}\,\underline{/(\phi_1 + \phi_2 + \cdots + \phi_m - \theta_1 - \theta_2 - \cdots - \theta_n)}$$

In other words

$$|\mathbf{H}(j\omega)| = \frac{KN_1N_2\cdots N_m}{D_1D_2\cdots D_n}$$

and

$$\text{ang } \mathbf{H}(j\omega) = \phi_1 + \phi_2 + \cdots + \phi_m - \theta_1 - \theta_2 - \cdots - \theta_n$$

With these results we can graphically determine points for the amplitude and phase responses of the network function $\mathbf{H}(j\omega)$.

$$\bullet \quad \bullet \quad \bullet$$

Example
The transfer function

$$\mathbf{H}(s) = \frac{4(s + 2)}{s^2 + 2s + 5} = \frac{4(s + 2)}{(s + 1 - j2)(s + 1 + j2)}$$

has a zero at $s = -2$ and poles at $s = -1 + j2$ and $s = -1 - j2$. We can form $\mathbf{H}(j\omega)$ from $\mathbf{H}(s)$ simply by setting $s = j\omega$. Thus

$$\mathbf{H}(j\omega) = \frac{4(j\omega + 2)}{(j\omega + 1 - j2)(j\omega + 1 + j2)} = \frac{4(2 + j\omega)}{(5 - \omega^2) + j2\omega}$$

To obtain the amplitude and phase responses of $\mathbf{H}(j\omega)$, we need to plot $|\mathbf{H}(j\omega)|$ and ang $\mathbf{H}(j\omega)$ versus ω, respectively. We can determine values for these plots graphically from the s-plane as follows:

Let $\omega = 1$. Draw vectors from all the poles and zeros to $j\omega = j1$ on the imaginary axis as shown in Fig. 10.37. The vector from pole p_1 to $j1$ is $(j1 - p_1)$ $= \sqrt{2}\,\underline{/-45°}$, while the vector from pole p_2 to $j1$ is $(j1 - p_2) = \sqrt{10}\,\underline{/71.6°}$. The vector from zero z_1 to $j1$ is $(j1 - z_1) = \sqrt{5}\,\underline{/26.6°}$. Thus

$$\mathbf{H}(j1) = \frac{4(\sqrt{5}\,\underline{/26.6°})}{(\sqrt{2}\,\underline{/-45°})(\sqrt{10}\,\underline{/71.6°})} = 2\,\underline{/0°}$$

fig. 10.37

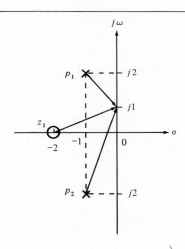

from which, when $\omega = 1$,

$$|\mathbf{H}(j\omega)| = 2 \qquad \text{and} \qquad \text{ang } \mathbf{H}(j\omega) = 0°$$

For $\omega = 2$ we get Fig. 10.38, and in this case we have

$$(j2 - p_1) = 1\,\underline{/0°}$$

$$(j2 - p_2) = \sqrt{17}\,\underline{/76°}$$

$$(j2 - z_1) = 2\sqrt{2}\,\underline{/45°}$$

fig. 10.38

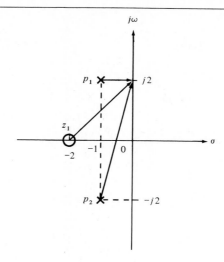

and therefore

$$\mathbf{H}(j2) = \frac{4(2\sqrt{2}\ \underline{/45°})}{(1\ \underline{/0°})(\sqrt{17}\ \underline{/76°})} = \frac{8\sqrt{34}}{17}\ \underline{/-31°} = 2.7\ \underline{/-31°}$$

Thus, when $\omega = 2$,

$$|\mathbf{H}(j\omega)| = 2.7 \qquad \text{and} \qquad \text{ang } \mathbf{H}(j\omega) = -31°$$

For $\omega = 3$ we have the situation shown in Fig. 10.39, and

$$(j3 - p_1) = \sqrt{2}\ \underline{/45°}$$

$$(j3 - p_2) = \sqrt{26}\ \underline{/78.7°}$$

$$(j3 - z_1) = \sqrt{13}\ \underline{/56.3°}$$

and hence

$$\mathbf{H}(j3) = \frac{4(\sqrt{13}\ \underline{/56.3°})}{(\sqrt{2}\ \underline{/45°})(\sqrt{26}\ \underline{/78.7°})} = 2\ \underline{/-67.4°}$$

Thus, when $\omega = 3$,

$$|\mathbf{H}(j\omega)| = 2 \qquad \text{and} \qquad \text{ang } \mathbf{H}(j\omega) = -67.4°$$

fig. 10.39

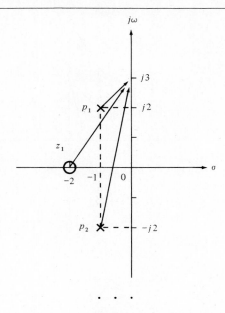

PROBLEMS *10.34. A transfer function $\mathbf{H}(s)$ has the pole-zero plot shown in Fig. P10.34. Find $\mathbf{H}(s)$ given that (a) $\mathbf{H}(0) = 2$, (b) $\mathbf{H}(1) = 2$, (c) $\mathbf{H}(-2) = 3$.

Ans. (a) $\dfrac{10(s + 1)}{s^2 + 4s + 5}$ b) $\dfrac{10(s+1)}{\cdots}$ c) $\dfrac{-3(s+1)}{\cdots}$

fig. P10.34

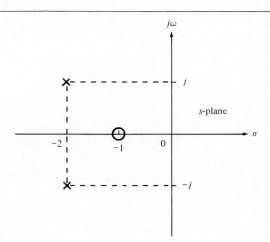

s-plane

a) $\dfrac{\frac{10}{3}(s^2 + 4s + 3)}{s^2 + 4s + 5}$

b) $\dfrac{\frac{5}{2}(s^2 + 4s + 3)}{\cdots}$

c) $\dfrac{-3(s^2 + 4s + 3)}{\cdots}$

10.35. Repeat Problem 10.34 given that there is an additional zero at $s = -3$.
10.36. Repeat Problem 10.34 for the case in which there is a double zero at $s = -1$.

$\dfrac{10(s^2 + 2s + 1)}{s^2 + 4s + 5}$, $5(\cdots)$, $-3(\cdots)$

*10.37. Given the transfer function

$$H(s) = \frac{10(s + 1)}{s^2 + 4s + 5}$$

graphically determine the magnitude and angle of $H(s)$ for the case that
(a) $s = j1$, (b) $s = j2$, and (c) $s = -2$.
Ans. (a) 2.5 $\underline{/0°}$, (b) 2.77 $\underline{/-19.45°}$, (c) 10 $\underline{/180°}$

10.38. Given the transfer function

$$H(s) = \frac{2(s + 1)}{s^2 + 5s + 6}$$

graphically determine the magnitude and angle of $H(j\omega)$ for the case that
(a) $s = j1$, (b) $s = j2$, and (c) $s = j3$.

(handwritten) $H(j1) = 0.4$
$H(j2) = 0.439$
$\underline{/-15.26}$

10.39. Repeat Problem 10.38 for

$$H(s) = \frac{s^2 + 4s + 5}{s^2 + 2s + 1}$$

(handwritten) $H(j1) = 2.83 \underline{/-45°}$
$H(j2) = 1.61 \underline{/-43.95°}$
$H(j3) = 1.26 \underline{/-34.7°}$
$H(j3) = 0.413 \underline{/29}$

Summary

1. The frequency response of a network function consists of the amplitude response and the phase response.
2. The frequencies at which the amplitude response drops down to $1/\sqrt{2}$ of its maximum value are the half-power points.
3. The frequencies at which an impedance (or admittance) is purely real are the resonance frequencies of the impedance (or admittance).
4. The bandwidth and the quality factor are measures of the sharpness of an amplitude response.
5. The amplitude response is not necessarily maximum at a resonance frequency.
6. An impedance can be expressed as a function of the complex frequency s, as can other parameters such as the voltage transfer function.
7. The poles and zeros of a ratio of polynomials in s can be depicted in the s-plane with a pole-zero plot.
8. If there are no cancellations of common poles and zeros, the poles of a network function indicate the form of the natural response.
9. The pole-zero plot of a network function can be used to graphically determine the frequency response of the network function.

TWO-PORT 11 NETWORKS

Introduction

With the exception of the operational amplifier, we have been dealing exclusively with two-terminal circuit elements such as resistors, inductors, capacitors, and voltage and current sources (both independent and dependent). In this chapter we consider a new circuit element, the transformer, which is a four-terminal device that consists, in essence, of two inductors placed in physical proximity.

Transformers come in various sizes and have a variety of uses. Relatively small transformers are used in radios and televisions to couple amplifier stages, while larger transformers are used in the power supply sections of these and numerous types of electronic equipment. Large, massive transformers are employed by electric utilities in the distribution of electric power.

Since a pair of terminals can be thought of as a "port," a four-terminal device or circuit is often referred to as a "two-port" network. Just as we could model a one-port network by its Thévenin or Norton equivalent circuit, so can we characterize and model two-port networks—including the special case of a transformer.

11.1 TRANSFORMERS

We know that a current i flowing through a coil produces a magnetic field around the coil. If the current is time-varying, then so will be the resulting magnetic field. This magnetic field passes through the coil and in turn induces a voltage v across it. The relationship between the voltage and the current is the familiar $v = L\, di/dt$, where L is the inductance of the coil. If there is a second coil near the first coil, the magnetic field also passes through the second coil and consequently induces a voltage across it as well. Under this circumstance, we say that the two coils are *magnetically coupled*.

Specifically, let us now consider the two inductors L_1 and L_2 shown in Fig. 11.1. For the case that $i_1 \neq 0$ and $i_2 = 0$, the voltage across inductor L_2 is

$$v_2(t) = M_{21} \frac{di_1(t)}{dt}$$

510

fig. 11.1

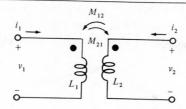

while if $i_1 = 0$ and $i_2 \neq 0$, then the voltage across inductor L_1 is

$$v_1(t) = M_{12} \frac{di_2(t)}{dt}$$

We call M_{12} and M_{21} the *coefficients of mutual inductance* or *mutual inductance*, for short. As with the case of the *self-inductances* L_1 and L_2, the units of mutual inductance are henries. The dots shown in Fig. 11.1 indicate the phase relationship between the current in one inductor and the resulting voltage induced in the other inductor, and this can be stated as follows:

Given two magnetically coupled inductors, a current going into the dotted end of one results in an induced voltage in the second, where the dotted end of the second is positive ($+$). Conversely, if the current in one inductor is going out of the dotted end, then the dotted end of the other is negative ($-$).

. . .

Example

Given a pair of magnetically coupled inductors, suppose the current in one inductor is produced by a current source and the other inductor has no current through it. Then the voltage induced across one inductor due to the current in the other inductor is depicted by the four cases shown in Fig. 11.2.

fig. 11.2

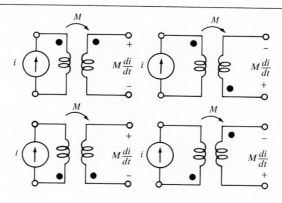

. . .

Returning now to the original pair of magnetically coupled inductors L_1 and L_2, having mutual inductances M_{12} and M_{21} (see Fig. 11.2), the general case is to have nonzero currents i_1 and i_2. By the principle of superposition, we then have

$$v_1(t) = L_1 \frac{di_1(t)}{dt} + M_{12} \frac{di_2(t)}{dt}$$

and

$$v_2(t) = M_{21} \frac{di_1(t)}{dt} + L_2 \frac{di_2(t)}{dt}$$

Although L_1 and L_2 need not be the same, we shall now see that $M_{12} = M_{21}$.

Suppose that the currents $i_1(t)$ and $i_2(t)$ are zero for $t < 0$. What is the energy required to bring these two currents, respectively, up to the positive constant values I_1 and I_2? There is more than one way to calculate this energy. One way is to have $i_1(t)$ go from 0 to I_1 between times $t = 0$ and $t = t_1$ while $i_2(t)$ remains at zero, and then have $i_2(t)$ go from 0 to I_2 between times $t = t_1$ and $t = t_2$ while $i_1(t)$ remains at I_1. These conditions are shown pictorially in Fig. 11.3.

fig. 11.3

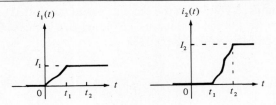

The amount of energy required to reach this state is

$$w = \int_0^{t_2} v_1 i_1 \, dt + \int_0^{t_2} v_2 i_2 \, dt$$

$$= \int_0^{t_1} v_1 i_1 \, dt + \int_{t_1}^{t_2} v_1 i_1 \, dt + \int_0^{t_1} v_2 i_2 \, dt + \int_{t_1}^{t_2} v_2 i_2 \, dt$$

Looking at each of these four integrals individually, we find that the first integral is

$$\int_0^{t_1} v_1 i_1 \, dt = \int_0^{t_1} \left[L_1 \frac{di_1}{dt} + M_{12} \frac{di_2}{dt} \right] i_1 \, dt$$

$$= \int_0^{t_1} L_1 i_1 \frac{di_1}{dt} \, dt + \int_0^{t_1} M_{12} i_1 \frac{di_2}{dt} \, dt$$

Since $di_2/dt = 0$ for $0 \le t < t_1$, then

$$\int_0^{t_1} v_1 i_1 \, dt = \int_{i_1(0)}^{i_1(t_1)} L_i i_1 \, di_1 + \int_0^{t_1} 0 \, dt$$

But $i_1(0) = 0$ and $i_1(t_1) = I_1$. Thus

$$\int_0^{t_1} v_1 i_1 \; dt = \int_0^{I_1} L_1 i_1 \; di_1 = \frac{1}{2} L_1 i_1{}^2 \bigg|_0^{I_1} = \frac{1}{2} L_1 I_1{}^2$$

Since $i_2(t) = 0$ for $0 \le t \le t_1$, then the third integral is

$$\int_0^{t_1} v_2 i_2 \; dt = \int_0^{t_1} 0 \; dt = 0$$

The second integral is

$$\int_{t_1}^{t_2} v_1 i_1 \; dt = \int_{t_1}^{t_2} \left[L_1 \frac{di_1}{dt} + M_{12} \frac{di_2}{dt} \right] i_1 \; dt$$

$$= \int_{t_1}^{t_2} L_1 i_1 \frac{di_1}{dt} \; dt + \int_{t_1}^{t_2} M_{12} i_1 \frac{di_2}{dt} \; dt$$

Since $di_1/dt = 0$ and $i_1(t) = I_1$ for $t_1 < t \le t_2$, then

$$\int_{t_1}^{t_2} v_1 i_1 \; dt = \int_{t_1}^{t_2} 0 \; dt + \int_{i_2(t_1)}^{i_2(t_2)} M_{12} I_1 \; di_2$$

But $i_2(t_1) = 0$ and $i_2(t_2) = I_2$, so

$$\int_{t_1}^{t_2} v_1 i_1 \; dt = \int_0^{I_2} M_{12} I_1 \; di_2 = M_{12} I_1 i_2 \bigg|_0^{I_2} = M_{12} I_1 I_2$$

Finally, the fourth integral is

$$\int_{t_1}^{t_2} v_2 i_2 \; dt = \int_{t_1}^{t_2} \left[M_{21} \frac{di_1}{dt} + L_2 \frac{di_2}{dt} \right] i_2 \; dt$$

$$= \int_{t_1}^{t_2} M_{21} i_2 \frac{di_1}{dt} \; dt + \int_{t_1}^{t_2} L_2 i_2 \frac{di_2}{dt} \; dt$$

Since $di_1/dt = 0$ for $t_1 < t \le t_2$, then

$$\int_{t_1}^{t_2} v_2 i_2 \; dt = \int_{t_1}^{t_2} 0 \; dt + \int_{i_2(t_1)}^{i_2(t_2)} L_2 i_2 \; di_2$$

$$= \int_0^{I_2} L_2 i_2 \; di_2 = \frac{1}{2} L_2 i_2{}^2 \bigg|_0^{I_2} = \frac{1}{2} L_2 I_2{}^2$$

Summing these four integrals, we get

$$w = \frac{1}{2} L_1 I_1{}^2 + \frac{1}{2} L_2 I_2{}^2 + M_{12} I_1 I_2$$

An alternative way of obtaining the final currents, I_1 and I_2, is to first increase $i_2(t)$ and then $i_1(t)$. Doing so, as above, we get that the energy required is

$$w = \frac{1}{2} L_1 I_1{}^2 + \frac{1}{2} L_2 I_2{}^2 + M_{21} I_1 I_2$$

fig. 11.4

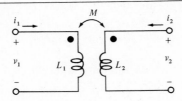

Since the preceding two expressions must be equal, we conclude that $M_{12} = M_{21}$, so we use the single symbol M for mutual inductance, as shown in Fig. 11.4. We call such a four-terminal device a *transformer*. One side of the transformer is often called the *primary* winding and the other side is the *secondary* winding. The pair of equations describing this circuit element in the time domain is

$$v_1 = L_1 \frac{di_1}{dt} + M \frac{di_2}{dt}$$

$$v_2 = M \frac{di_1}{dt} + L_2 \frac{di_2}{dt}$$

In a manner similar to the energy discussion above, it can be shown that the energy stored in the transformer depicted in Fig. 11.4 is

$$w(t) = \frac{1}{2} L_1 i_1{}^2(t) + \frac{1}{2} L_2 i_2{}^2(t) + M i_1(t) i_2(t)$$

If the dot on either the primary or secondary (but not both) is moved to the other side, the expression becomes

$$w(t) = \frac{1}{2} L_1 i_1{}^2(t) + \frac{1}{2} L_2 i_2{}^2(t) - M i_1(t) i_2(t)$$

For an inductor L, in the time domain,

$$v = L \frac{di}{dt}$$

For the case of sinusoids,

$$\mathbf{V} = j\omega L \mathbf{I}$$

whereas, for damped sinusoids,

$$\mathbf{V} = Ls \mathbf{I}$$

Just as for inductors, for the transformer we find, in the case of sinusoids, that

$$\mathbf{V}_1 = j\omega L_1 \mathbf{I}_1 + j\omega M \mathbf{I}_2$$
$$\mathbf{V}_2 = j\omega M \mathbf{I}_1 + j\omega L_2 \mathbf{I}_2$$

whereas, for damped sinusoids,

$$\mathbf{V}_1 = L_1 s \mathbf{I}_1 + M s \mathbf{I}_2$$
$$\mathbf{V}_2 = M s \mathbf{I}_1 + L_2 s \mathbf{I}_2$$

· · ·

Example
The sinusoidal circuit shown in Fig. 11.5 contains a transformer. The circuit is represented in the frequency domain in Fig. 11.6.

fig. 11.5

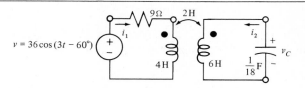

fig. 11.6

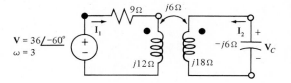

Writing mesh equations, we get

$$\mathbf{V} = 9\mathbf{I}_1 + j12\mathbf{I}_1 + j6\mathbf{I}_2$$
$$= (9 + j12)\mathbf{I}_1 + j6\mathbf{I}_2$$

and

$$0 = j6\mathbf{I}_1 + j18\mathbf{I}_2 - j6\mathbf{I}_2$$
$$= \mathbf{I}_1 + 2\mathbf{I}_2$$

From the latter equation,

$$\mathbf{I}_1 = -2\mathbf{I}_2$$

Substituting this into the former equation yields

$$\mathbf{V} = -2(9 + j12)\mathbf{I}_2 + j6\mathbf{I}_2$$
$$= -18\mathbf{I}_2 - j18\mathbf{I}_2$$
$$= -18(1 + j)\mathbf{I}_2$$

and thus

$$\mathbf{I}_2 = \frac{\mathbf{V}}{-18(1 + j)}$$

Since

$$\mathbf{V}_C = (-j6)(-\mathbf{I}_2)$$

$$= \frac{j6\mathbf{V}}{-18(1 + j)}$$

$$= \frac{(6\underline{/90°})(36\underline{/-60°})}{(18\underline{/180°})(\sqrt{2}\underline{/45°})}$$

$$= 6\sqrt{2}\underline{/-195°} = 6\sqrt{2}\underline{/165°}$$

For the more general case of damped sinusoids, the circuit representation is shown in Fig. 11.7. By mesh analysis,

$$\mathbf{V} = 9\mathbf{I}_1 + 4s\mathbf{I}_1 + 2s\mathbf{I}_2$$
$$= (9 + 4s)\mathbf{I}_1 + 2s\mathbf{I}_2$$

fig. 11.7

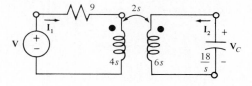

and

$$0 = 2s\mathbf{I}_1 + 6s\mathbf{I}_2 + \frac{18}{s}\mathbf{I}_2$$

$$= s\mathbf{I}_1 + \left(3s + \frac{9}{s}\right)\mathbf{I}_2$$

From the latter equation,

$$-s\mathbf{I}_1 = 3\left(\frac{s^2 + 3}{s}\right)\mathbf{I}_2$$

or

$$\mathbf{I}_1 = -3\left(\frac{s^2 + 3}{s^2}\right)\mathbf{I}_2 \qquad (11.1)$$

Substituting this expression into the former equation gives

$$\mathbf{V} = -3\left(\frac{s^2 + 3}{s^2}\right)(4s + 9)\mathbf{I}_2 + 2s\mathbf{I}_2$$

from which

$$\mathbf{V} = \frac{-(10s^3 + 27s^2 + 36s + 81)}{s^2}\mathbf{I}_2 \qquad (11.2)$$

Thus

$$\mathbf{I}_2 = \frac{-s^2\mathbf{V}}{10s^3 + 27s^2 + 36s + 81}$$

From this equation and the fact that

$$\mathbf{V}_C = \frac{18}{s}(-\mathbf{I}_2)$$

we obtain the voltage transfer function

$$\frac{\mathbf{V}_C}{\mathbf{V}} = \frac{18s}{10s^3 + 27s^2 + 36s + 81}$$

Furthermore, from Equation (11.1),

$$\mathbf{I}_2 = \frac{-s^2}{3(s^2 + 3)}\mathbf{I}_1$$

which, when substituted into Equation (11.2), yields the impedance seen by the source

$$\frac{\mathbf{V}}{\mathbf{I}_1} = \frac{10s^3 + 27s^2 + 36s + 81}{3(s^2 + 3)}$$

• • •

Although a transformer is in essence a pair of inductors that are magnetically coupled, when the bottom terminals are connected we can model such a device with circuit elements that are not magnetically coupled. Shown in Fig. 11.8 is a transformer

fig. 11.8

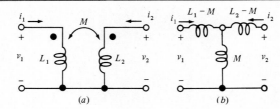

(a)   (b)

having mutual inductance M and a "T" connection of three inductors. Applying KVL to both circuits results in the same pair of equations:

$$v_1 = L_1 \frac{di_1}{dt} + M \frac{di_2}{dt}$$

$$v_2 = M \frac{di_1}{dt} + L_2 \frac{di_2}{dt}$$

and these equations are the same as when the two bottom terminals in Fig. 11.8(a) are not connected. Thus, the circuit given in (b) is a model of the transformer shown in (a). The "T-equivalent" circuit in (b) is only one possible model of a transformer.

. . .

Example

In the transformer circuit shown in Fig. 11.5, $L_1 = 4$ H, $L_2 = 6$ H, and $M = 2$ H. Suppose that we connect the bottom two terminals. Since $L_1 - M = 2$ H and $L_2 - M = 4$ H, replacing the transformer by its T-equivalent, we get the time-domain circuit shown in Fig. 11.9.

fig. 11.9

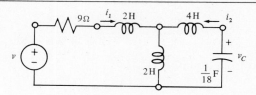

fig. 11.10

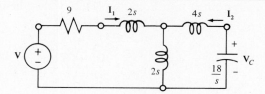

In the frequency domain the circuit is as shown in Fig. 11.10. By mesh analysis,

$$\mathbf{V} = 9\mathbf{I}_1 + 2s\mathbf{I}_1 + 2s(\mathbf{I}_1 + \mathbf{I}_2)$$
$$= (9 + 4s)\mathbf{I}_1 + 2s\mathbf{I}_2$$

and

$$0 = 4s\mathbf{I}_2 + 2s(\mathbf{I}_1 + \mathbf{I}_2) + \frac{18}{s}\mathbf{I}_2$$

$$0 = 2s\mathbf{I}_1 + \left(6s + \frac{18}{s}\right)\mathbf{I}_2$$

$$= s\mathbf{I}_1 + \left(3s + \frac{9}{s}\right)\mathbf{I}_2$$

which are the two equations obtained previously, but this time we didn't have to worry about dots!

· · ·

PROBLEMS *11.1. Consider the sinusoidal transformer circuit given in Fig. 11.5.
(a) Find the impedance seen by the source.
(b) Find the energy stored in the transformer at time $t = 0$.
Ans. (a) $9\sqrt{2}\ \underline{/45°}\ \Omega$, (b) 0.94 J

11.2. Consider the sinusoidal transformer circuit in Fig. 11.5. Find the resonance frequency.

$\dfrac{V}{I_1} = \dfrac{9(18 - 6\omega^2) + j(72\omega - 20\omega^3)}{18 - 6\omega^2}$

$\omega_r = \dfrac{6}{\sqrt{10}}$ rad/s

fig. P11.3

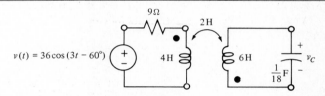

$v(t) = 36\cos(3t - 60°)$ 9Ω $2\,\mathrm{H}$ $4\,\mathrm{H}$ $6\,\mathrm{H}$ $\frac{1}{18}\,\mathrm{F}$ v_C

$9\sqrt{2}\ \underline{/45°} = z$
$w(0) = 0.94\,\mathrm{J}$

11.3. Repeat Problem 11.1 for the transformer circuit shown in Fig. P11.3.
*11.4. For the circuit given in Problem 11.3, find $v_C(t)$.
Ans. $6\sqrt{2}\cos(3t - 15°)$

fig. P11.5

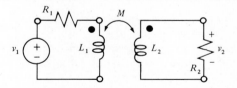

R_1 M v_1 L_1 L_2 v_2 R_2

11.5. Find the voltage transfer function $\mathbf{H}(s) = \mathbf{V}_2/\mathbf{V}_1$ for the transformer circuit shown in Fig. P11.5.

$\dfrac{R_2 Ms}{(L_1 L_2 - M^2)s^2 + (R_1 L_2 + R_2 L_1)s + R_1 R_2}$

fig. P11.6

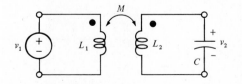

M v_1 L_1 L_2 C v_2

$\dfrac{M}{C(L_1 L_2 - M^2)s^2 + L_1}$

11.6. Find the voltage transfer function $\mathbf{H}(s) = \mathbf{V}_2/\mathbf{V}_1$ for the magnetically coupled circuit in Fig. P11.6.

11.7. Consider the transformer circuit given in Problem 11.6.
(a) Find the impedance seen by the source.
(b) Find a formula for the resonance frequency.

$j\omega L_1 + \dfrac{j\omega^3 M^2 C}{1 - \omega^2 L_2 C}$

$\omega_r = \sqrt{\dfrac{L_1}{(L_1 L_2 - M^2)C}}$

fig. P11.8

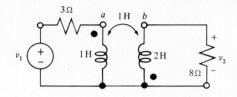

*11.8.   Consider the transformer circuit shown in Fig. P11.8.
    (a)   Find the voltage transfer function $H(s) = V_2/V_1$.
    (b)   Find the impedance seen by the source.

    Ans.   (a) $\dfrac{-8s}{(s+2)(s+12)}$,   (b) $\dfrac{(s+2)(s+12)}{2(s+4)}$

11.9.   In the circuit given in Problem 11.8, the 8-Ω resistor is the load. Replace the remainder of the circuit by its Thévenin equivalent, and then find $H(s) = V_2/V_1$.

11.10.   For the transformer circuit given in Problem 11.8, place a 1-F capacitor between nodes *a* and *b*, and find V_2/V_1.

$\dfrac{-8s}{(s+2)(s+12)}$

$\begin{bmatrix} s+3 & s & 3 \\ s & 2s+8 & 8 \\ 2s^2 & 3s^2 & -1 \end{bmatrix} \begin{bmatrix} I_1 \\ I_2 \\ \tfrac{I_2}{s} \end{bmatrix} = \begin{bmatrix} V_1 \\ 0 \\ 0 \end{bmatrix}$

$\dfrac{V_2}{V_1} = \dfrac{8s\,(4s^2 - 3s - 1)}{11s^3 + 121s^2 + 146s + 24}$

fig. P11.11

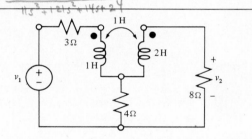

$\dfrac{8(s+4)}{s^2 + 18s + 68}$

11.11.   Find the voltage transfer function $H(s) = V_2/V_1$ for the circuit in Fig. P11.11.

11.12.   For the circuit given in Problem 11.11 replace the 4-Ω resistor with a 1-F capacitor, and repeat the problem.

$\dfrac{8(s^2+1)}{s^3 + 14s^2 + 25s + 11}$

fig. P11.13

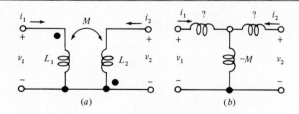

(a)   (b)

*11.13.   Given the transformer circuit in Fig. P11.13 (a), what inductance values should be chosen in order to make the circuit in (b) the T-equivalent of (a)?

    Ans. $L_1 + M,\ L_2 + M$

11.14.   Use the result obtained in Problem 11.13 to solve Problem 11.8.

fig. P11.15

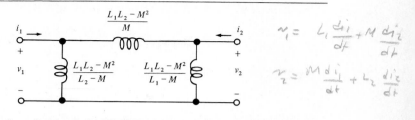

$$v_1 = L_1 \frac{di_1}{dt} + M \frac{di_2}{dt}$$

$$v_2 = M \frac{di_1}{dt} + L_2 \frac{di_2}{dt}$$

11.15.   Verify that the circuit in Fig. P11.15 is a "π-equivalent" circuit of a transformer by expressing v_1 and v_2 in terms of i_1, i_2, L_1, L_2, and M.

11.16.   For the sinusoidal transformer circuit given in Fig. 11.5, use the π-equivalent circuit shown in Problem 11.15 to determine $v_C(t)$.

$$v_c(t) = 6\sqrt{2} \cos(3t + 165°)$$

11.2   THE IDEAL TRANSFORMER

Since the energy stored in a transformer is

$$w = \frac{1}{2}L_1 i_1^2 + \frac{1}{2}L_2 i_2^2 + M i_1 i_2$$

$$= \frac{1}{2}[L_1 i_1^2 + L_2 i_2^2] + M i_1 i_2$$

then, by completing the square,

$$w = \frac{1}{2}[(\sqrt{L_1}\,i_1 + \sqrt{L_2}\,i_2)^2 - 2\sqrt{L_1 L_2}\,i_1 i_2] + M i_1 i_2$$

$$= \frac{1}{2}(\sqrt{L_1}\,i_1 + \sqrt{L_2}\,i_2)^2 + (M - \sqrt{L_1 L_2})i_1 i_2$$

Consider the case in which

$$\sqrt{L_1}\,i_1 + \sqrt{L_2}\,i_2 = 0$$

This occurs when

$$\sqrt{L_1}\,i_1 = -\sqrt{L_2}\,i_2$$

or

$$i_2 = -\sqrt{\frac{L_1}{L_2}}\,i_1$$

and the resulting energy stored is

$$w = (M - \sqrt{L_1 L_2})i_1\left(-\sqrt{\frac{L_1}{L_2}}\,i_1\right)$$

$$= (\sqrt{L_1 L_2} - M)\left(\sqrt{\frac{L_1}{L_2}}\,i_1^2\right)$$

Since the energy stored is a nonnegative quantity, we must have

$$0 \le \sqrt{L_1 L_2} - M$$

or

$$M \le \sqrt{L_1 L_2}$$

Thus we have an upper bound for the mutual inductance M.

If we define the *coefficient of coupling* of a transformer, denoted by k, as

$$k = \frac{M}{\sqrt{L_1 L_2}}$$

we find, by the previous inequality, that

$$0 \le k \le 1$$

The coupling coefficient k of a physical transformer is determined by a number of factors: the magnetic properties of the core on which the primary and secondary coils are wound, the number of turns of each coil, and the relative positions and physical dimensions of the coils. If $0 < k \le 0.5$, we say that the coils are *loosely coupled*, and if $0.5 < k \le 1$, we say that they are *tightly coupled*. Air-core transformers are typically loosely coupled, whereas iron-core transformers are usually tightly coupled. As a matter of fact, iron-core transformers can have coupling coefficients approaching unity.

Consider now the case of a transformer with *perfect coupling*, that is, $k = 1$. We represent such a device in Fig. 11.11.

fig. 11.11

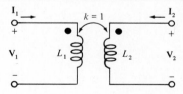

Since

$$\mathbf{V}_1 = j\omega L_1 \mathbf{I}_1 + j\omega M \mathbf{I}_2$$

then

$$j\omega L_1 \mathbf{I}_1 = \mathbf{V}_1 - j\omega M \mathbf{I}_2$$

or

$$\mathbf{I}_1 = \frac{\mathbf{V}_1 - j\omega M \mathbf{I}_2}{j\omega L_1} \tag{11.3}$$

Also,

$$\mathbf{V}_2 = j\omega M \mathbf{I}_1 + j\omega L_2 \mathbf{I}_2$$

$$= j\omega M \left(\frac{\mathbf{V}_1 - j\omega M \mathbf{I}_2}{j\omega L_1} \right) + j\omega L_2 \mathbf{I}_2$$

$$= \frac{M\mathbf{V}_1}{L_1} - \frac{j\omega M^2 \mathbf{I}_2}{L_1} + j\omega L_2 \mathbf{I}_2$$

But, for perfect coupling,

$$\frac{M}{\sqrt{L_1 L_2}} = 1$$

or

$$M^2 = L_1 L_2$$

and hence

$$\mathbf{V}_2 = \frac{\sqrt{L_1 L_2}\, \mathbf{V}_1}{L_1} - \frac{j\omega L_1 L_2 \mathbf{I}_2}{L_1} + j\omega L_2 \mathbf{I}_2$$

$$= \sqrt{\frac{L_2}{L_1}}\, \mathbf{V}_1 = n\mathbf{V}_1$$

where $n = \sqrt{L_2/L_1}$.

Recall from freshman physics that inductance L is proportional to the square of the number of turns N; that is, $L = KN^2$. The primary and secondary self-inductances are

$$L_1 = KN_1^2 \quad \text{and} \quad L_2 = KN_2^2$$

Thus

$$n = \sqrt{\frac{L_2}{L_1}} = \sqrt{\frac{KN_2^2}{KN_1^2}} = \frac{N_2}{N_1}$$

so we call n the *turns ratio* of the transformer.

From Equation (11.3),

$$\mathbf{I}_1 = \frac{\mathbf{V}_1}{j\omega L_1} - \frac{M\mathbf{I}_2}{L_1}$$

so, for perfect coupling,

$$\mathbf{I}_1 = \frac{\mathbf{V}_1}{j\omega L_1} - \frac{\sqrt{L_1 L_2}\, \mathbf{I}_2}{L_1}$$

$$= \frac{\mathbf{V}_1}{j\omega L_1} - \sqrt{\frac{L_2}{L_1}}\, \mathbf{I}_2 \tag{11.4}$$

For the case that L_1 and L_2 approach infinity such that the turns ratio remains constant, along with perfect coupling, we say that the transformer is *ideal*. For an ideal transformer, Equation (11.4) becomes

$$\mathbf{I}_1 = -n\mathbf{I}_2 \quad \text{or} \quad \mathbf{I}_2 = -\frac{\mathbf{I}_1}{n}$$

Since good coupling is achieved by transformers with iron cores, ideal transformers are often represented as in Fig. 11.12, and the relationships between the voltages and the currents are given by

$$\boxed{\mathbf{V}_2 = n\mathbf{V}_1 \quad \text{and} \quad \mathbf{I}_2 = -\frac{\mathbf{I}_1}{n}}$$

fig. 11.12

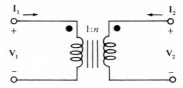

The instantaneous power absorbed by the primary winding of the ideal transformer is $p_1 = v_1 i_1$, and that absorbed by the secondary winding is $p_2 = v_2 i_2$. Thus, the total instantaneous power absorbed by the transformer is

$$p = p_1 + p_2 = v_1 i_1 + v_2 i_2 = v_1 i_1 + n v_1 \left(\frac{-i_1}{n} \right) = 0$$

Since the instantaneous power absorbed is zero, so is the average power; moreover, the energy stored must also be zero. This fact can be confirmed from the formula for the energy stored in a transformer by using the relationships for perfect coupling $\left(M = \sqrt{L_1 L_2} \right)$ and turns ratio $\left(n = \sqrt{L_2/L_1} \right)$ along with the ideal transformer formulas.

Since an ideal transformer is described by the pair of expressions

$$\boxed{\begin{aligned} \mathbf{V}_2 = n\mathbf{V}_1 \quad &\Leftrightarrow \quad \mathbf{V}_1 = \frac{\mathbf{V}_2}{n} \\ \mathbf{I}_2 = \frac{-\mathbf{I}_1}{n} \quad &\Leftrightarrow \quad \mathbf{I}_1 = -n\mathbf{I}_2 \end{aligned}}$$

we can model this device by either of the two equivalent circuits shown in Fig. 11.13. For the circuit given in (a), $\mathbf{V}_2 = n\mathbf{V}_1$ and $\mathbf{I}_1 = -n\mathbf{I}_2$; for the circuit in (b), $\mathbf{V}_1 = \mathbf{V}_2/n$ and $\mathbf{I}_2 = -\mathbf{I}_1/n$.

fig. 11.13

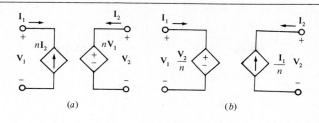

(a)   (b)

• • •

Example

The equivalent circuit of a simple "class A" transistor power amplifier is shown in Fig. 11.14.

Using the ideal transformer equivalent circuit given in Fig. 11.13(a), we obtain the circuit shown in Fig. 11.15. For the mesh on the left,

$$\mathbf{V}_g = 600\mathbf{I} + 50(21\mathbf{I}) = 1650\mathbf{I}$$

fig. 11.14

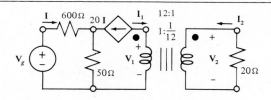

fig. 11.15

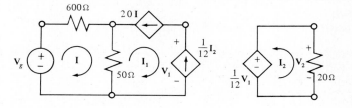

from which

$$\mathbf{I} = \frac{\mathbf{V}_g}{1650}$$

For the mesh in the middle,

$$\mathbf{I}_1 = -20\mathbf{I} = -\frac{1}{12}\mathbf{I}_2$$

Since

$$\frac{1}{12}\mathbf{I}_2 = 20\mathbf{I} \Rightarrow \mathbf{I}_2 = 240\mathbf{I}$$

then, for the mesh on the right,

$$\frac{1}{12}\mathbf{V}_1 = \mathbf{V}_2 = -20\mathbf{I}_2 = -20(240\mathbf{I}) = -4800\left(\frac{\mathbf{V}_g}{1650}\right) = -\frac{32}{11}\mathbf{V}_g$$

Therefore, the voltage transfer function $\mathbf{V}_2/\mathbf{V}_g$ is

$$\frac{\mathbf{V}_2}{\mathbf{V}_g} = -\frac{32}{11} = -2.91$$

When the voltage transfer function is a real number, it is often called the *voltage gain*.

The instantaneous power supplied by the independent voltage source is

$$p_g = v_g i = v_g\left(\frac{v_g}{1650}\right) = \frac{v_g^2}{1650}$$

and the power absorbed by the 20-Ω load resistor is

$$p_2 = \frac{v_2^2}{20} = \frac{\left(-\frac{32}{11}v_g\right)^2}{20} = \frac{256}{605}v_g^2$$

Thus, the power gain p_2/p_g is

$$\frac{p_2}{p_g} = \frac{(256/605)v_g^2}{v_g^2/1650} = \frac{256(1650)}{605} = \frac{7680}{11} = 698.2$$

. . .

Example

Consider the ideal transformer circuit shown in Fig. 11.16. Since

$$\mathbf{V}_g = \mathbf{Z}_g\mathbf{I}_1 + \mathbf{V}_1 \qquad \text{and} \qquad \mathbf{V}_2 = -\mathbf{Z}_L\mathbf{I}_2$$

fig. 11.16

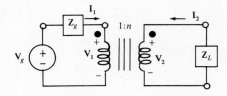

from the latter equation,

$$n\mathbf{V}_1 = -\mathbf{Z}_L\left(-\frac{\mathbf{I}_1}{n}\right) \qquad \text{or} \qquad \mathbf{V}_1 = \frac{\mathbf{Z}_L}{n^2}\mathbf{I}_1$$

Substitution of the last expression into the equation for $\mathbf{V}_g$ yields

$$\mathbf{V}_g = \mathbf{Z}_g\mathbf{I}_1 + \frac{\mathbf{Z}_L}{n^2}\mathbf{I}_1$$

Thus the impedance seen by the source $\mathbf{V}_g$ is

$$\frac{\mathbf{V}_g}{\mathbf{I}_1} = \mathbf{Z}_g + \frac{\mathbf{Z}_L}{n^2}$$

which is $\mathbf{Z}_g$ in series with the load impedance $\mathbf{Z}_L$ scaled by the factor $1/n^2$. The term $\mathbf{Z}_L/n^2$ is often called the *reflected impedance*.

From the preceding equation,

$$\mathbf{I}_1 = \frac{\mathbf{V}_g}{\mathbf{Z}_g + (\mathbf{Z}_L/n^2)}$$

Thus,

$$\mathbf{V}_2 = -\mathbf{Z}_L\mathbf{I}_2 = -\mathbf{Z}_L\left(-\frac{\mathbf{I}_1}{n}\right)$$

$$= \frac{\mathbf{Z}_L}{n}\left[\frac{\mathbf{V}_g}{\mathbf{Z}_g + (\mathbf{Z}_L/n^2)}\right]$$

$$= \frac{n\mathbf{Z}_L}{n^2\mathbf{Z}_g + \mathbf{Z}_L}\mathbf{V}_g$$

Alternatively, we can apply Thévenin's theorem to the circuit under consideration. First, we obtain the circuit of Fig. 11.17.

fig. 11.17

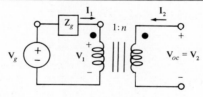

Since $\mathbf{I}_2 = 0$, then

$$\mathbf{I}_1 = -n\mathbf{I}_2 = 0$$

and

$$\mathbf{V}_g = \mathbf{Z}_g\mathbf{I}_1 + \mathbf{V}_1 = \mathbf{V}_1$$

Thus,

$$\mathbf{V}_{oc} = \mathbf{V}_2 = n\mathbf{V}_1 = n\mathbf{V}_g$$

To find the Thévenin equivalent impedance, set $\mathbf{V}_g = 0$ and take the ratio of $\mathbf{V}_0$ to $\mathbf{I}_0$ in the circuit shown in Fig. 11.18.

fig. 11.18

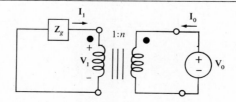

Since

$$V_1 = -Z_g I_1 = -Z_g(-nI_0) = nZ_g I_0$$

and

$$V_0 = nV_1$$

we get

$$\frac{V_0}{n} = nZ_g I_0$$

and hence

$$Z_0 = \frac{V_0}{I_0} = n^2 Z_g$$

Thus, using the Thévenin equivalent, we have the circuit shown in Fig. 11.19 and, by voltage division,

$$V_2 = \frac{Z_L}{Z_L + n^2 Z_g}(nV_g)$$

$$= \frac{nZ_L}{Z_L + n^2 Z_g} V_g$$

as was obtained above.

---

fig. 11.19

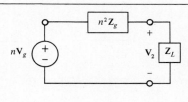

· · ·

PROBLEMS   11.17.   Consider the ideal transformer circuit shown in Fig. P11.17.

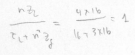

(a)   What is the impedance seen by the voltage source? $4\,\Omega$

(b)   What is the voltage gain V_2/V_g? $= 1$

(c)   To what value should the 16-Ω load resistance be changed such that it will absorb the maximum power? $48\,\Omega$

---

fig. P11.17

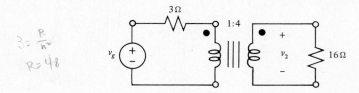

*11.18.   For the circuit given in Problem 11.17, assume a turns ratio of $1:n$. What value of n will result in the 16-Ω load resistor absorbing the maximum amount of power?

   Ans.   2.31

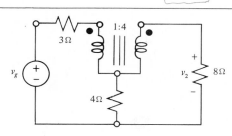

$\leftarrow Z_0 = 3 \times n^2 = 16$

$n^2 = \frac{16}{3} \Rightarrow n = 2.31$

fig. P11.19

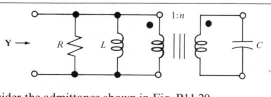

$a) \; \frac{V_g}{I_1} = \frac{23}{4}\,\Omega$

11.19.   Repeat Problem 11.17 for the ideal transformer circuit in Fig. P11.19 where the 8-Ω resistor is the load.

$b) \; \frac{V_2}{V_g} = \frac{8}{23}$

$c) \; 8.4\,\Omega$

fig. P11.20

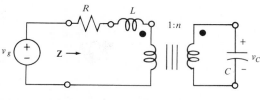

$admittance \; \frac{n^2}{\frac{1}{2L}}$

$Y = \frac{1}{R} + \frac{1}{j\omega L} + n^2 j\omega C$

$a) \; w_r = \frac{1}{n\sqrt{LC}}$

11.20.   Consider the admittance shown in Fig. P11.20.
   (a)   Find the resonance frequency.
   (b)   Find the quality factor.
   (c)   Find the bandwidth.

$b) \; \equiv \parallel RLC \; circuit$

$Q = R\sqrt{\frac{n^2 C}{L}}$

$2) \; Y = \frac{1}{R} + \frac{n^2}{j\omega L} + j\omega C$

$w_r = \frac{1}{\sqrt{\frac{L}{n^2}C}}$

$b) \; Q = R\sqrt{\frac{C}{L/n^2}}$

11.21.   For the admittance given in Problem 11.20, interchange the inductor and the capacitor, and then repeat the problem.

$c) \; B = \frac{1}{R(n^2 C)}$

fig. P11.22

$c) \; B = \frac{1}{RC}$

$Z = R + j\omega L + \frac{1}{j\omega (n^2)}$

$\equiv series \; RLC \; circuit$

$a) \; w_r = \frac{1}{\sqrt{Ln^2 C}}$

$b) \; Q = \frac{1}{R}\sqrt{\frac{L}{n^2 C}}$

$c) = B = R/L$

*11.22.   Consider the impedance indicated in Fig. P11.22.
   (a)   Find the resonance frequency.
   (b)   Find the quality factor.
   (c)   Find the bandwidth.

equivalent ind.

$Z = R + \frac{1}{j\omega C} + \frac{j\omega L}{n^2}$

$w_r = \frac{1}{\sqrt{\frac{L}{n^2}C}}$

$b) \; Q = \frac{1}{R}\sqrt{\frac{L/n^2}{C}}$

$c) \; B = \frac{n^2 R}{L}$

   Ans.   (a) $1/n\sqrt{LC}$, (b) $(1/nR)\sqrt{L/C}$, (c) R/L

11.23.   For the impedance given in Problem 11.22, interchange the inductor and the capacitor; then repeat the problem.

11.24.   Find the voltage transfer function $\mathbf{H}(s) = \mathbf{V}_C/\mathbf{V}_g$ for the circuit shown in Problem 11.22 when $R = 2\,\Omega$, $L = 3$ H, $C = \frac{1}{4}$ F, and $n = 5$.

11.25.   For the circuit given in Problem 11.22, find $v_C(t)$ when $R = 2\,\Omega$, $L = 2$ H, $C = 2$ F, $n = \frac{1}{10}$, and (a) $v_g(t) = 10 \cos 5t$, (b) $v_g(t) = 10 \cos 10t$.

$5 \sin 5t$

$.33 \cos(10t - 172.4°)$

$\frac{V_c}{V_g} = \frac{20}{7s^2 + 50s + 4}$

fig. P11.26

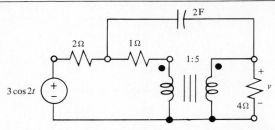

11.26. Find $v(t)$ in the circuit in Fig. P11.26. $1.08 \cos (2t + 1.2°)$

*11.27. For the power amplifier given in Fig. 11.14 use the alternative equivalent circuit of the ideal transformer to determine the power gain p_2/p_g.

Ans.   698.2

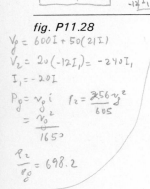

fig. P11.28

$V_g = 600I + 50(21I)$

$V_2 = 20(-12I_1) = -240I_1$

$I_1 = -20I$

$P_g = v_g i$ $P_2 = \dfrac{256 v_g^2}{605}$

$= \dfrac{v_0}{1650}$

$\dfrac{R_2}{P_g} = 698.2$

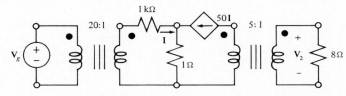

11.28. Figure P11.28 represents the equivalent circuit of a simple "class B" transistor power amplifier.

(a)   Find the voltage gain $\mathbf{V}_2/\mathbf{V}_g$. $- .095$

(b)   If p_g is the power supplied by the voltage source and p_2 is the power absorbed by the 8-Ω load resistor, find the power gain p_2/p_g. $= 475.7$

fig. P11.29

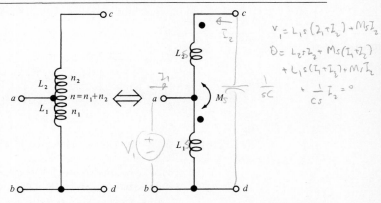

$v_1 = L_1 s(I_1 + I_2) + M_s I_2$

$0 = L_2 s I_2 + M_s(I_1 + I_2)$
$\quad + L_1 s(I_1 + I_2) + M_s I_2$
$\quad + \dfrac{1}{sC} I_2 = 0$

$\dfrac{1}{sC}$

$\dfrac{1}{C_s} I_2 = 0$

11.29. If a coil with n turns is "tapped," the result is called an *autotransformer*. This connection is shown, along with its equivalent circuit, in Fig. P11.29. Place a capacitor C between terminals c and d, and assume that the autotransformer has perfect coupling ($k = 1$).

(a)   Show that the admittance between terminals a and b is

$$\dfrac{I_1}{V_1} = sC(1 + n)^2 + 1/L_1 s$$

(b)   Show that $L_1 = L/(1 + n)^2$, where $L = L_1 + L_2 + 2M$.

fig. P11.30

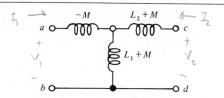

11.30. Verify that the autotransformer given in Problem 11.29 can be modeled by the equivalent circuit shown in Fig. P11.30.

11.31. For the autotransformer given in Problem 11.29, place a capacitor C between terminals a and b. Assuming perfect coupling, show that the admittance between terminals c and d is $sC/(1 + n)^2 + 1/Ls$, where $L = L_1 + L_2 + 2M$.

11.32. Find the resonance frequency for the admittance given in (a) Problem 11.29 and (b) Problem 11.31.

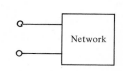

11.3 TWO-PORT ADMITTANCE PARAMETERS

Such circuit elements as resistors, inductors, capacitors, and dependent and independent sources are two-terminal devices, which can be represented in either of the two ways shown in Fig. 11.20. More generally, an interconnection of circuit elements may have a single pair of terminals accessible as shown in Fig. 11.21. We can say that the two terminals constitute a *port*, and that the network is a *one-port network*. For the

fig. 11.20 (left)
fig. 11.21 (right)

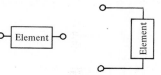

case that a one-port network contains no independent sources, it can be characterized by its impedance. For the case that independent sources are contained in a one-port network, it can be characterized by its Thévenin or Norton equivalent.

Now consider the situation where a network has two accessible pairs of terminals instead of one. Such a network, which is depicted in Fig. 11.22, is referred to as a *two-port network*. Just as a one-port network (or one-port, for short) can consist of a single

fig. 11.22

element or many elements, so can a two-port network, or two-port. For example, a transformer is a simple two-port. So is the simple op amp circuit shown in Fig. 11.23. In this case the device, the op amp, has three terminals, but one terminal is common to both ports.

fig. 11.23

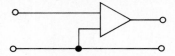

The characterization of two-port networks is both important and useful. Even a simple two-port consisting of a single bipolar transistor is typically described in terms of its two-port characteristics.

In the following discussion, we shall consider two-ports that possibly contain dependent sources but do not contain independent sources. We begin by considering a two-port network (in the frequency domain) that has an independent voltage source applied to each port. This situation is depicted in Fig. 11.24.

fig. 11.24

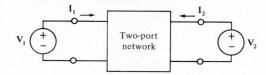

Given the voltages V_1 and V_2, the currents I_1 and I_2 can be measured or, if the network is explicitly known, calculated. In either case I_1 and I_2 can be determined by means of the principle of superposition. For instance, that portion of I_1 due solely to V_1 is KV_1. Since KV_1 must have amperes as units, then K must have amperes/volts = mhos as units. In other words, K must be an admittance, say y_{11}. Similarly, the portion of I_1 due solely to V_2 is $y_{12}V_2$. Thus, by the principle of superposition, $I_1 = y_{11}V_1 + y_{12}V_2$. Using an analogous argument, we deduce that $I_2 = y_{21}V_1 + y_{22}V_2$.

Although the voltages V_1 and V_2 were produced by independent sources, the same results hold no matter how these voltages arise when the two-port is part of a larger circuit. In summary, we can describe the two-port network shown in Fig. 11.25 by the two equations

$$I_1 = y_{11}V_1 + y_{12}V_2 \tag{11.5}$$
$$I_2 = y_{21}V_1 + y_{22}V_2 \tag{11.6}$$

fig. 11.25

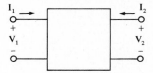

or by the single matrix equation

$$\begin{bmatrix} I_1 \\ I_2 \end{bmatrix} = \begin{bmatrix} y_{11} & y_{12} \\ y_{21} & y_{22} \end{bmatrix} \begin{bmatrix} V_1 \\ V_2 \end{bmatrix} \tag{11.7}$$

From Equation (11.5), if we set $V_2 = 0$, then we have

$$y_{11} = \frac{I_1}{V_1}\bigg|_{V_2=0}$$

(read $y_{11} = I_1/V_1$ with $V_2 = 0$). We can set $V_2 = 0$ by placing a short circuit across port 2 (the port with V_2 and I_2) as shown in Fig. 11.26. Similarly, we have

$$y_{12} = \frac{I_1}{V_2}\bigg|_{V_1=0}$$

$$y_{21} = \frac{I_2}{V_1}\bigg|_{V_2=0}$$

$$y_{22} = \frac{I_2}{V_2}\bigg|_{V_1=0}$$

fig. 11.26

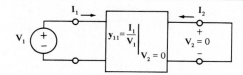

The admittance y_{21} also can be determined from the configuration shown in Fig. 11.26. The remaining two admittances, y_{12} and y_{22}, can be obtained from Fig. 11.27.

Because the admittances y_{11}, y_{12}, y_{21}, and y_{22} can be determined by short-circuiting one of the ports, they are called the *short-circuit admittance parameters* of the two-port. If port 1 is considered the input and port 2 the output, then y_{11} is called the *short-circuit input admittance*, y_{22} is the *short-circuit output admittance*, and y_{12} and y_{21} are *short-circuit transfer admittances*.

fig. 11.27

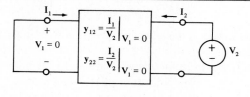

$\cdots$

Example
Let us determine the short-circuit admittance parameters for the resistive two-port shown in Fig. 11.28.

fig. 11.28

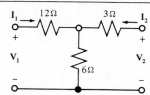

We can determine the short-circuit input admittance y_{11} and transfer admittance y_{21} from Fig. 11.29. Since the 3-Ω and 6-Ω resistors are in parallel, they form a 2-Ω resistance. Thus, by voltage division,

$$V = \frac{2V_1}{2 + 12} = \frac{V_1}{7}$$

fig. 11.29

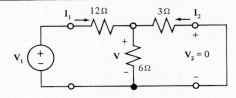

and therefore

$$I_1 = \frac{V_1 - V}{12} = \frac{V_1 - (V_1/7)}{12} = \frac{V_1}{14}$$

so

$$y_{11} = \left. \frac{I_1}{V_1} \right|_{V_2 = 0} = \tfrac{1}{14} \, \mho$$

Also,

$$I_2 = \frac{V_2 - V}{3} = \frac{0 - (V_1/7)}{3} = -\frac{V_1}{21}$$

from which

$$y_{21} = \left. \frac{I_2}{V_1} \right|_{V_2 = 0} = -\tfrac{1}{21} \, \mho$$

Don't get upset because y_{21} is a negative quantity. Although the input admittance y_{11} is in this case the admittance of the connection of three positive-valued resistors, the transfer admittance is the ratio of a current at one port and the voltage at another. The term "admittance" is used because the ratio is current to voltage, but y_{21} is not the admittance of a particular port.

To find y_{12} and y_{22}, we short-circuit port 1, as shown in Fig. 11.30. In this case the parallel combination of the 6-Ω and 12-Ω resistors is effectively a 4-Ω resistance. Thus, by voltage division,

$$V = \frac{4V_2}{4 + 3} = \tfrac{4}{7}V_2$$

From

$$I_1 = \frac{V_1 - V}{12} = \frac{0 - \tfrac{4}{7}V_2}{12} = -\frac{V_2}{21}$$

fig. 11.30

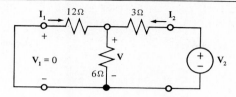

we get

$$y_{12} = \left.\frac{I_1}{V_2}\right|_{V_1=0} = -\tfrac{1}{21}\ \mho$$

and from

$$I_2 = \frac{V_2 - V}{3} = \frac{V_2 - \tfrac{4}{7}V_2}{3} = \frac{V_2}{7}$$

we obtain

$$y_{22} = \frac{I_2}{V_2} = \tfrac{1}{7}\ \mho$$

. . .

Having seen an example of a resistive two-port in a "T" configuration, let us now consider a "π" network consisting of arbitrary admittances $\mathbf{Y}_a$, $\mathbf{Y}_b$, and $\mathbf{Y}_c$ as shown in Fig. 11.31.

fig. 11.31

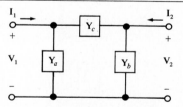

In order to obtain $\mathbf{y}_{11}$ and $\mathbf{y}_{21}$, we place a short circuit across port 2. From the viewpoint of port 1, $\mathbf{Y}_a$ and $\mathbf{Y}_c$ are then in parallel. Thus,

$$\mathbf{y}_{11} = \mathbf{Y}_a + \mathbf{Y}_c$$

Since the short circuit across $\mathbf{Y}_b$ means that the voltage across $\mathbf{Y}_b$ is zero and that therefore there is no current through $\mathbf{Y}_b$, by current division we have that

$$\mathbf{I}_2 = -\frac{\mathbf{Y}_c}{\mathbf{Y}_c + \mathbf{Y}_a}\mathbf{I}_1$$

But since (when there is a short circuit across port 2)

$$\mathbf{I}_1 = \mathbf{y}_{11}\mathbf{V}_1 = (\mathbf{Y}_a + \mathbf{Y}_c)\mathbf{V}_1 \Rightarrow \frac{\mathbf{I}_1}{\mathbf{Y}_a + \mathbf{Y}_c} = \mathbf{V}_1$$

then

$$\mathbf{I}_2 = -\mathbf{Y}_c\mathbf{V}_1$$

Hence

$$\mathbf{y}_{21} = -\mathbf{Y}_c$$

Similarly, we can obtain

$$\mathbf{y}_{12} = -\mathbf{Y}_c \quad \text{and} \quad \mathbf{y}_{22} = \mathbf{Y}_b + \mathbf{Y}_c$$

Note that in this example, as in the previous one, $\mathbf{y}_{12} = \mathbf{y}_{21}$. This is not sheer luck, but is due to a special property. Resistors, inductors, and capacitors are elements that are "bilateral"; that is, they can be placed in a circuit in either direction and the result will be the same. A circuit that contains only bilateral elements is known as a *bilateral network*. For such a two-port it can be shown that $\mathbf{y}_{12} = \mathbf{y}_{21}$. Yet there are two-ports containing nonbilateral elements, such as dependent sources, that also have this property. We say that any two-port having the property that $\mathbf{y}_{12} = \mathbf{y}_{21}$ is a *reciprocal network*.

In the foregoing discussion we had a reciprocal network (Fig. 11.31) consisting of three arbitrary admittances $\mathbf{Y}_a$, $\mathbf{Y}_b$, and $\mathbf{Y}_c$. Having determined the short-circuit admittance parameters in terms of these admittances, we find that

$$\mathbf{y}_{11} + \mathbf{y}_{12} = \mathbf{Y}_a$$
$$\mathbf{y}_{22} + \mathbf{y}_{12} = \mathbf{Y}_b$$
$$-\mathbf{y}_{12} = \mathbf{Y}_c$$

Hence, given an arbitrary reciprocal two-port, we can model it with the π-equivalent circuit shown in Fig. 11.32. For the T-connected, resistive two-port given in Fig. 11.28

$$\mathbf{y}_{11} + \mathbf{y}_{12} = \frac{1}{14} + \left(-\frac{1}{21}\right) = \tfrac{1}{42} \, \mho$$

$$\mathbf{y}_{22} + \mathbf{y}_{12} = \frac{1}{7} + \left(-\frac{1}{21}\right) = \tfrac{2}{21} \, \mho$$

$$-\mathbf{y}_{12} = \tfrac{1}{21} \, \mho$$

so its π-equivalent circuit is as shown in Fig. 11.33.

fig. 11.32

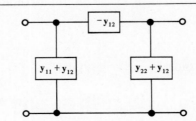

fig. 11.33

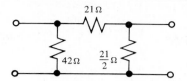

The π-equivalent circuit shown in Fig. 11.32 can be used for reciprocal two-ports. For nonreciprocal two-ports, as well as reciprocal two-ports, we can use the equivalent circuit shown in Fig. 11.34, which employs two voltage-dependent current sources. Applying KCL at ports 1 and 2, we get Equations (11.5) and (11.6), respectively.

fig. 11.34

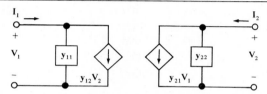

Knowing the short-circuit admittance parameters (**y**-parameters, for short) of a two-port, we can express other parameters of the two-port in terms of the **y**-parameters. For example, let us find the voltage transfer function $\mathbf{V}_2/\mathbf{V}_1$ with no load connected to port 2, as shown in Fig. 11.35. Although we can replace the two-port by its equivalent

fig. 11.35

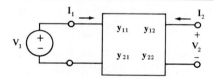

circuit, since Equations (11.5) and (11.6) describe the two-port regardless of what is connected to it, setting $\mathbf{I}_2 = 0$ we get, from Equation (11.6),

$$0 = \mathbf{y}_{21}\mathbf{V}_1 + \mathbf{y}_{22}\mathbf{V}_2$$

Thus

$$\frac{\mathbf{V}_2}{\mathbf{V}_1} = -\frac{\mathbf{y}_{21}}{\mathbf{y}_{22}}$$

For the case that port 2 is terminated with a 1-Ω resistor, as shown in Fig. 11.36, we have

$$\mathbf{V}_2 = -1\mathbf{I}_2 \Rightarrow \mathbf{I}_2 = -\mathbf{V}_2$$

fig. 11.36

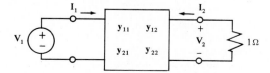

Substituting this into Equation (11.6) we obtain

$$-\mathbf{V}_2 = \mathbf{y}_{21}\mathbf{V}_1 + \mathbf{y}_{22}\mathbf{V}_2$$

and therefore

$$\frac{\mathbf{V}_2}{\mathbf{V}_1} = \frac{-\mathbf{y}_{21}}{1 + \mathbf{y}_{22}}$$

It may be that some or all of the y-parameters of a two-port do not exist. Consider the two-port shown in Fig. 11.37. In order to calculate $\mathbf{y}_{11}$ and $\mathbf{y}_{21}$, a short circuit is placed across port 2. This means that $3\mathbf{I} = 0$, and hence $\mathbf{I} = 0$. However, a nonzero

fig. 11.37

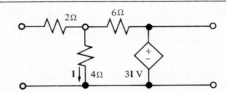

voltage applied to port 1 implies that $\mathbf{I} \neq 0$. The apparent paradox indicates that no solution exists, and consequently, neither does $\mathbf{y}_{11}$ and $\mathbf{y}_{21}$. The same conclusion can be drawn about $\mathbf{y}_{12}$ and $\mathbf{y}_{22}$ if you attempt to determine these parameters.

PROBLEMS   11.33.   Find the y-parameters for the resistive two-port T-network in Fig. P11.33.

fig. P11.33

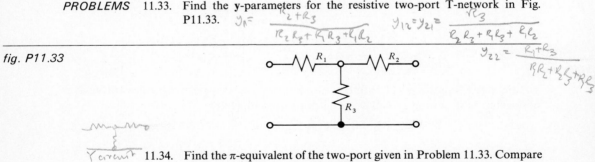

11.34.   Find the π-equivalent of the two-port given in Problem 11.33. Compare your results with the Y-Δ transformation studied previously.

fig. P11.35

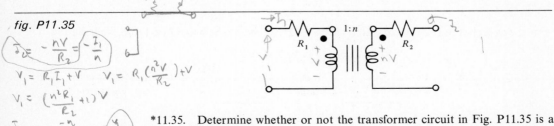

*11.35.   Determine whether or not the transformer circuit in Fig. P11.35 is a reciprocal network.
   Ans.   reciprocal

fig. P11.36

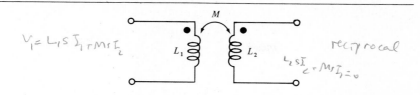

$V_1 = L_1 s I_1 + M s I_2$

reciprocal

$L_2 s I_2 + M s I_1 = 0$

11.36.   Repeat Problem 11.35 for the transformer shown in Fig. P11.36.

fig. P11.37

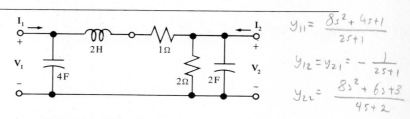

$y_{11} = \dfrac{8s^2 + 4s + 1}{2s + 1}$

$y_{12} = y_{21} = -\dfrac{1}{2s + 1}$

$y_{22} = \dfrac{8s^2 + 6s + 3}{4s + 2}$

11.37.   Find the y-parameters for the π-network in Fig. P11.37.
11.38.   For the circuit shown in Problem 11.37, given that $I_2 = 0$, (a) find V_2/V_1,
and (b) change the 2-Ω resistor to 3 Ω and find V_2/V_1.

$\dfrac{2}{8s^2 + 6s + 3}$

$\dfrac{3}{12s^2 + 8s + 4}$

fig. P11.39

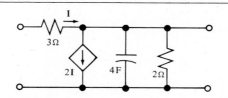

*11.39.   Determine whether or not the two-port shown in Fig. P11.39 is a
reciprocal network.
   Ans.   not reciprocal

$y_{21} = \dfrac{1}{3}$ reciprocal

$y_{12} = -\dfrac{1}{3}$

fig. P11.40

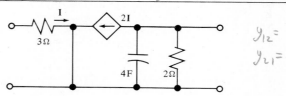

$y_{12} = 0$ not rec.

$y_{21} = \dfrac{2}{3}$

11.40.   Repeat Problem 11.39 for the circuit in Fig. P11.40.
*11.41.   Find the values of the dependent voltage sources so that the circuit
shown in Fig. P11.41 is the equivalent circuit of a two-port having the
y-parameters y_{11}, y_{12}, y_{21}, y_{22}.

$V_1 = \dfrac{I_1}{y_{11}} - V_a \Rightarrow \dfrac{I_1}{y_{11}} \neq V_1 + V_a \Rightarrow I_1 = y_{11} V_1 + y_{11} V_a$

$= y_{11} V_1 + y_{12} V_2 \qquad V_a = \dfrac{y_{12}}{y_{11}} V_2$

   Ans. $\dfrac{y_{12}}{y_{11}} V_2, \dfrac{y_{21}}{y_{22}} V_1$

$V_2 = \dfrac{I_2}{y_{22}} - V_b \Rightarrow \dfrac{I_2}{y_{22}} = V_2 + V_b$

$\Rightarrow I_2 = y_{22} V_2 + y_{22} V_b = y_{21} V_1 + y_{22} V_2 \quad V_b = \dfrac{y_{21}}{y_{22}} V_1$

fig. P11.41

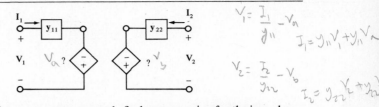

For an arbitrary two-port network, find an expression for the impedance

$$\frac{V_1}{I_1}\bigg|_{I_2=0} = \frac{y_{22}}{y_{11}y_{22} - y_{12}y_{21}}$$

in terms of the **y**-parameters of the two-port.

fig. P11.43

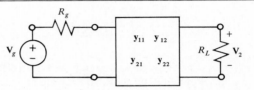

*11.43.   For the network shown in Fig. P11.43, find V_2/V_g given that $y_{11} = 3\ \mho$, $y_{12} = -1\ \mho$, $y_{21} = 20\ \mho$, $y_{22} = 2\ \mho$, $R_g = 1\ \Omega$, and $R_L = 0.1\ \Omega$. *Ans.* $-\frac{5}{17}$

$$\frac{V_2}{V_g} = \frac{25}{8}$$

11.44.   Repeat Problem 11.43 for the case that $y_{11} = y_{22} = 0.5\ \mho$, $y_{12} = y_{21} = -1\ \mho$, $R_g = 0.4\ \Omega$, and $R_L = 10\Omega$.

fig. P11.45

$$y_{11} = \frac{z_{22}}{z_{11}z_{22} - z_{12}z_{21}}$$

$$y_{12} = \frac{-z_{12}}{z_{11}z_{22} - z_{12}z_{21}}$$

$$y_{21} = \frac{-z_{21}}{z_{11}z_{22} - z_{12}z_{21}}$$

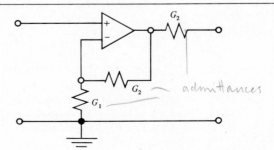

11.45.   In the circuit in Fig. P11.45, z_{11}, z_{12}, z_{21}, and z_{22} are impedances. Find the **y**-parameters for this two-port.

fig. P11.46

$$y_{22} = \frac{z_{11}}{z_{11}z_{22} - z_{12}z_{21}}$$

$y_{11} = 0$

$y_{21} = -(G_1 + G_2)$

$y_{12} = 0$

$y_{22} = G_2$

11.46.   Find the **y**-parameters for the two-port shown in Fig. P11.46, which contains an operational amplifier.

fig. P11.47

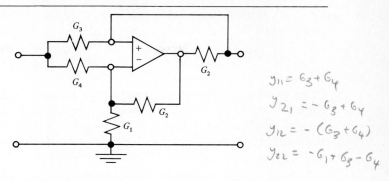

$$y_{11} = G_3 + G_4$$
$$y_{21} = -G_3 + G_4$$
$$y_{12} = -(G_3 + G_4)$$
$$y_{22} = -G_1 + G_3 - G_4$$

11.47.   Repeat Problem 11.46 for the op amp circuit in Fig. P11.47.

11.4   OTHER TWO-PORT PARAMETERS

Our development of the **y**-parameters of a two-port was obtained by letting the port voltages be the independent variables and the port currents be the dependent variables. If we look at the converse situation, we get a pair of equations of the form

$$V_1 = z_{11}I_1 + z_{12}I_2 \qquad (11.8)$$
$$V_2 = z_{21}I_1 + z_{22}I_2 \qquad (11.9)$$

which can be written as the single matrix equation

$$\begin{bmatrix} V_1 \\ V_2 \end{bmatrix} = \begin{bmatrix} z_{11} & z_{12} \\ z_{21} & z_{22} \end{bmatrix} \begin{bmatrix} I_1 \\ I_2 \end{bmatrix} \qquad (11.10)$$

Given a two-port network, we can determine the impedances z_{11} and z_{21} by open-circuiting port 2 as shown in Fig. 11.38. From Equation (11.8),

$$z_{11} = \left.\frac{V_1}{I_1}\right|_{I_2=0}$$

fig. 11.38

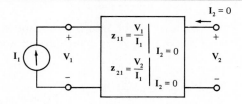

and, from Equation (11.9),

$$z_{21} = \left.\frac{V_2}{I_1}\right|_{I_2=0}$$

Similarly, applying a current source I_2 at port 2 and open-circuiting port 1 ($I_1 = 0$), we easily obtain

$$z_{12} = \frac{V_1}{I_2}\bigg|_{I_1 = 0} \quad \text{and} \quad z_{22} = \frac{V_2}{I_2}\bigg|_{I_1 = 0}$$

From the fact that the impedances z_{11}, z_{12}, z_{21}, and z_{22} can be determined by open-circuiting one of the ports, they are called the *open-circuit impedance parameters* or *z-parameters* of the two-port. If port 1 is considered the input and port 2 the output, then z_{11} is called the *open-circuit input impedance*, z_{22} is the *open-circuit output impedance*, and z_{12} and z_{21} are *open-circuit transfer impedances*.

· · ·

Example

Let us find the z-parameters of the resistive π-network shown in Fig. 11.39.

fig. 11.39

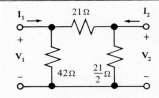

fig. 11.40

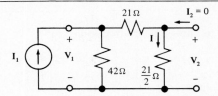

We determine z_{11} and z_{21} from the configuration shown in Fig. 11.40. Since the 21-Ω and $\frac{21}{2}$-Ω resistors are in series, they form a $\frac{63}{2}$-Ω resistance. This in turn is in parallel with 42 Ω. Thus,

$$z_{11} = \frac{42(63/2)}{42 + (63/2)} = 18\ \Omega$$

Of course, $z_{11} = V_1/I_1$ can also be obtained by using mesh or nodal analysis.

One way to obtain $z_{21} = V_2/I_1$ is by using current division to determine I. Since

$$I = \frac{42}{42 + (63/2)} I_1 = \frac{4}{7} I_1$$

then

$$V_2 = \frac{21}{2} I = 6I_1$$

from which

$$z_{21} = 6\,\Omega$$

We can find z_{12} and z_{22} from the circuit shown in Fig. 11.41. In this case,

$$I = \frac{21/2}{(21/2) + 63}\,I_2 = \frac{1}{7}\,I_2$$

fig. 11.41

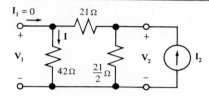

and

$$V_1 = 42I = 6I_2$$

and therefore

$$z_{12} = 6\,\Omega$$

Also,

$$z_{22} = \frac{(21/2)63}{(21/2) + 63} = 9\,\Omega$$

Note that in this example that $z_{12} = z_{21}$. Since the network is obviously bilateral, and hence reciprocal, then $y_{12} = y_{21}$. Does this imply that $z_{12} = z_{21}$?

Return now to Equation (11.7). By Cramer's rule, we have

$$V_1 = \frac{\begin{vmatrix} I_1 & y_{12} \\ I_2 & y_{22} \end{vmatrix}}{\begin{vmatrix} y_{11} & y_{12} \\ y_{21} & y_{22} \end{vmatrix}} = \frac{y_{22}}{\Delta_y}\,I_1 - \frac{y_{12}}{\Delta_y}\,I_2$$

where $\Delta_y = y_{11}y_{22} - y_{12}y_{21}$. Also,

$$V_2 = \frac{\begin{vmatrix} y_{11} & I_1 \\ y_{21} & I_2 \end{vmatrix}}{\begin{vmatrix} y_{11} & y_{12} \\ y_{21} & y_{22} \end{vmatrix}} = \frac{y_{11}}{\Delta_y}\,I_2 - \frac{y_{21}}{\Delta_y}\,I_1$$

Comparing these two expressions with Equations (11.8) and (11.9), we conclude that

$$z_{11} = \frac{y_{22}}{\Delta_y} \qquad z_{12} = \frac{-y_{12}}{\Delta_y} \qquad z_{21} = \frac{-y_{21}}{\Delta_y} \qquad z_{22} = \frac{y_{11}}{\Delta_y}$$

Thus if $\mathbf{y}_{12} = \mathbf{y}_{21}$, then $\mathbf{z}_{12} = \mathbf{z}_{21}$, and vice versa. Furthermore, we see that if the **y**-parameters of a two-port have been determined, the above relationships can be used to obtain the **z**-parameters.

For the two-port given in Fig. 11.39 we previously determined (see Fig. 11.33) that

$$\mathbf{y}_{11} = \tfrac{1}{14}\,\text{Ʊ} \qquad \mathbf{y}_{12} = -\tfrac{1}{21}\,\text{Ʊ} = \mathbf{y}_{21} \qquad \mathbf{y}_{22} = \tfrac{1}{7}\,\text{Ʊ}$$

Hence

$$\Delta_{\mathbf{y}} = \mathbf{y}_{11}\mathbf{y}_{22} - \mathbf{y}_{12}\mathbf{y}_{21} = \left(\frac{1}{14}\right)\left(\frac{1}{7}\right) - \left(-\frac{1}{21}\right)^{2} = \frac{1}{126}$$

so

$$\mathbf{z}_{11} = \frac{\mathbf{y}_{22}}{\Delta_{\mathbf{y}}} = \frac{1/7}{1/126} = 18\,\Omega$$

$$\mathbf{z}_{12} = \frac{-\mathbf{y}_{12}}{\Delta_{\mathbf{y}}} = \frac{1/21}{1/126} = 6\,\Omega = \mathbf{z}_{21}$$

$$\mathbf{z}_{22} = \frac{\mathbf{y}_{11}}{\Delta_{\mathbf{y}}} = \frac{1/14}{1/126} = 9\,\Omega$$

$\cdot \quad \cdot \quad \cdot$

Having seen an example of a resistive π-network, let us now consider the T-network consisting of arbitrary impedances $\mathbf{Z}_a$, $\mathbf{Z}_b$, and $\mathbf{Z}_c$ as shown in Fig. 11.42. In order to obtain $\mathbf{z}_{11}$ and $\mathbf{z}_{21}$, we leave port 2 open-circuited. Then, clearly,

$$\mathbf{z}_{11} = \mathbf{Z}_a + \mathbf{Z}_c$$

fig. 11.42

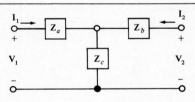

Since $\mathbf{I}_2 = 0$, then

$$\mathbf{V}_2 = \mathbf{Z}_c\mathbf{I}_1$$

so

$$\mathbf{z}_{21} = \mathbf{Z}_c$$

Similarly,

$$\mathbf{z}_{12} = \mathbf{Z}_c$$

and

$$\mathbf{z}_{22} = \mathbf{Z}_b + \mathbf{Z}_c$$

From the **z**-parameters of this reciprocal two-port, we have

$$z_{11} - z_{12} = Z_a$$

$$z_{22} - z_{12} = Z_b$$

$$z_{12} = Z_c$$

Thus we can deduce that a reciprocal two-port can be modeled by the T-equivalent network shown in Fig. 11.43. For the resistive π-network given in Fig. 11.39, we have

$$z_{11} - z_{12} = 18 - 6 = 12\,\Omega$$

$$z_{22} - z_{12} = 9 - 6 = 3\,\Omega$$

$$z_{12} = 6\,\Omega$$

and therefore its T-equivalent circuit is as shown in Fig. 11.44.

fig. 11.43

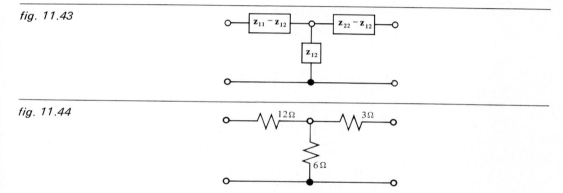

fig. 11.44

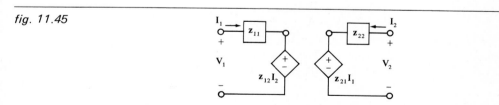

A general equivalent circuit of a two-port, whether reciprocal or not, whose **z**-parameters are z_{11}, z_{12}, z_{21}, and z_{22} is shown in Fig. 11.45. To verify that this configuration indeed yields Equations (11.8) and (11.9), we need only apply KVL at ports 1 and 2.

fig. 11.45

As with the case of **y**-parameters, there are special cases of two-ports for which the **z**-parameters do not exist (are undefined).

Given a two-port, by selecting the port voltages to be the independent variables, we obtained the **y**-parameters; by selecting the port currents as the independent variables, we got the **z**-parameters. However, there is no reason why we can't pick the voltage at one port and the current at the other to be the independent variables. For example, if we select $\mathbf{I}_1$ and $\mathbf{V}_2$ to be the independent variables, we get a pair of equations of the form

$$\mathbf{V}_1 = \mathbf{h}_{11}\mathbf{I}_1 + \mathbf{h}_{12}\mathbf{V}_2 \qquad (11.11)$$

$$\mathbf{I}_2 = \mathbf{h}_{21}\mathbf{I}_1 + \mathbf{h}_{22}\mathbf{V}_2 \qquad (11.12)$$

which can be written as the single matrix equation

$$\begin{bmatrix} \mathbf{V}_1 \\ \mathbf{I}_2 \end{bmatrix} = \begin{bmatrix} \mathbf{h}_{11} & \mathbf{h}_{12} \\ \mathbf{h}_{21} & \mathbf{h}_{22} \end{bmatrix} \begin{bmatrix} \mathbf{I}_1 \\ \mathbf{V}_2 \end{bmatrix} \qquad (11.13)$$

Since the independent variables are mixed (one current and one voltage), we refer to these $\mathbf{h}$'s as the *hybrid parameters* or *$\mathbf{h}$-parameters* of the two-port. It is in terms of these parameters that ordinary (bipolar) transistors are typically described.

If we set $\mathbf{V}_2 = 0$ (short-circuit port 2), then, from Equation (11.11),

$$\mathbf{h}_{11} = \left. \frac{\mathbf{V}_1}{\mathbf{I}_1} \right|_{\mathbf{V}_2 = 0}$$

which is called the *short-circuit input impedance* and consequently has ohms as units. Furthermore, from Equation (11.12),

$$\mathbf{h}_{21} = \left. \frac{\mathbf{I}_2}{\mathbf{I}_1} \right|_{\mathbf{V}_2 = 0}$$

which is a dimensionless quantity called the *short-circuit forward current gain.*

Conversely, if we set $\mathbf{I}_1 = 0$ (open circuit port 1), then from Equation (11.11),

$$\mathbf{h}_{12} = \left. \frac{\mathbf{V}_1}{\mathbf{V}_2} \right|_{\mathbf{I}_1 = 0}$$

which is a dimensionless quantity called the *open-circuit reverse voltage gain.* Finally, from Equation (11.12),

$$\mathbf{h}_{22} = \left. \frac{\mathbf{I}_2}{\mathbf{V}_2} \right|_{\mathbf{I}_1 = 0}$$

which is called the *open-circuit output admittance*, and its units are mhos.

In transistor applications, the symbols $\mathbf{h}_{11}, \mathbf{h}_{21}, \mathbf{h}_{12}, \mathbf{h}_{22}$ are replaced by $\mathbf{h}_i, \mathbf{h}_f, \mathbf{h}_r, \mathbf{h}_o$, respectively, to denote *input, forward, reverse, output.*

· · ·

Example

Let us calculate the $\mathbf{h}$-parameters for the T-network shown in Fig. 11.44. We calculate $\mathbf{h}_{11}$ from Fig. 11.46. Instead of writing mesh or nodal equations, note that the 3-Ω and 6-Ω resistors in parallel are effectively 2 Ω, and this in turn is in series with 12 Ω. Thus

$$\mathbf{h}_{11} = 12 + 2 = 14 \, \Omega$$

fig. 11.46

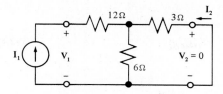

We can also determine $\mathbf{h}_{21}$ from this circuit. By current division,

$$-I_2 = \frac{6}{6+3} I_1$$

so

$$\mathbf{h}_{21} = -\tfrac{2}{3}$$

We can determine $\mathbf{h}_{12}$ and $\mathbf{h}_{22}$ from the circuit shown in Fig. 11.47.

fig. 11.47

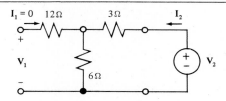

Since

$$I_2 = \frac{V_2}{3+6}$$

then

$$\mathbf{h}_{22} = \tfrac{1}{9} \, \mho$$

Also

$$V_1 = 6I_2 = 6\left(\frac{V_2}{9}\right)$$

so

$$\mathbf{h}_{12} = \tfrac{2}{3}$$

$$\bullet \quad \bullet \quad \bullet$$

Note that, for the reciprocal network given in Fig. 11.44, $\mathbf{h}_{12} \neq \mathbf{h}_{21}$. But is there a relationship between these two parameters for reciprocal networks?
From Equation (11.9), we have

$$I_2 = \frac{V_2 - \mathbf{z}_{21}I_1}{\mathbf{z}_{22}}$$

Substituting this into Equation (11.8), we get

$$\mathbf{V}_1 = \mathbf{z}_{11}\mathbf{I}_1 + \mathbf{z}_{12}\left(\frac{\mathbf{V}_2 - \mathbf{z}_{21}\mathbf{I}_1}{\mathbf{z}_{22}}\right)$$

$$= \left(\mathbf{z}_{11} - \frac{\mathbf{z}_{12}\mathbf{z}_{21}}{\mathbf{z}_{22}}\right)\mathbf{I}_1 + \frac{\mathbf{z}_{12}}{\mathbf{z}_{22}}\mathbf{V}_2$$

$$= \frac{\Delta_\mathbf{z}}{\mathbf{z}_{22}}\mathbf{I}_1 + \frac{\mathbf{z}_{12}}{\mathbf{z}_{22}}\mathbf{V}_2$$

where $\Delta_\mathbf{z} = \mathbf{z}_{11}\mathbf{z}_{22} - \mathbf{z}_{12}\mathbf{z}_{21}$. Comparing this result with Equation (11.11) we conclude that

$$\mathbf{h}_{11} = \frac{\Delta_\mathbf{z}}{\mathbf{z}_{22}} \qquad \mathbf{h}_{12} = \frac{\mathbf{z}_{12}}{\mathbf{z}_{22}}$$

In addition, from Equation (11.9), we have

$$\mathbf{I}_2 = \frac{\mathbf{V}_2 - \mathbf{z}_{21}\mathbf{I}_1}{\mathbf{z}_{22}}$$

$$= -\frac{\mathbf{z}_{21}}{\mathbf{z}_{22}}\mathbf{I}_1 + \frac{1}{\mathbf{z}_{22}}\mathbf{V}_2$$

Comparing this result with Equation (11.12) we conclude that

$$\mathbf{h}_{21} = -\frac{\mathbf{z}_{21}}{\mathbf{z}_{22}} \qquad \mathbf{h}_{22} = \frac{1}{\mathbf{z}_{22}}$$

Thus, for a reciprocal network ($\mathbf{z}_{12} = \mathbf{z}_{21}$), we see that $\mathbf{h}_{12} = -\mathbf{h}_{21}$, and vice versa.

A general equivalent circuit of a two-port whose **h**-parameters are known is shown in Fig. 11.48. To verify that this equivalent circuit yields Equations (11.11) and (11.12) we need only apply KVL at port 1 and KCL at port 2.

fig. 11.48

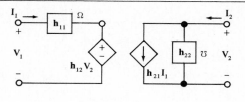

· · ·

Example

A two-port consisting of a transistor is terminated with a 10-kΩ resistor, as shown in Fig. 11.49. Typical **h**-parameters are $\mathbf{h}_{11} = 1\,\text{k}\Omega$, $\mathbf{h}_{12} = 2.5(10^{-4})$, $\mathbf{h}_{21} = 50$, and $\mathbf{h}_{22} = 25\,\mu\mho \Rightarrow 1/\mathbf{h}_{22} = 40\,\text{k}\Omega$. By KCL,

$$50\mathbf{I}_1 + \frac{\mathbf{V}_2}{40,000} + \frac{\mathbf{V}_2}{10,000} = 0$$

fig. 11.49

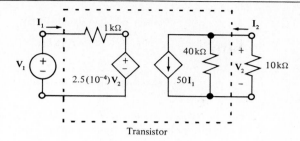

Transistor

from which we obtain

$$2(10^6)\mathbf{I}_1 + \mathbf{V}_2 + 4\mathbf{V}_2 = 0$$
$$2(10^6)\mathbf{I}_1 + 5\mathbf{V}_2 = 0$$

But

$$\mathbf{I}_1 = \frac{\mathbf{V}_1 - 2.5(10^{-4})\mathbf{V}_2}{1000} \tag{11.14}$$

and substituting this expression into the previous one, we get

$$2000[\mathbf{V}_1 - 2.5(10^{-4})\mathbf{V}_2] + 5\mathbf{V}_2 = 0$$
$$2000\mathbf{V}_1 - 0.5\mathbf{V}_2 + 5\mathbf{V}_2 = 0$$
$$2000\mathbf{V}_1 = -4.5\mathbf{V}_2$$

Thus

$$\frac{\mathbf{V}_2}{\mathbf{V}_1} = -444.44$$

The input impedance is obtained from Equation (11.14):

$$\mathbf{I}_1 = \frac{\mathbf{V}_1 + 2.5(10^{-4})(\tfrac{4}{9})(10^3)\mathbf{V}_1}{1000}$$

$$= \frac{1}{900}\mathbf{V}_1$$

Hence,

$$\frac{\mathbf{V}_1}{\mathbf{I}_1} = 900 \ \Omega$$

• • •

PROBLEMS   11.48.   Find the **z**-parameters for the resistive two-port π-network in Fig. P11.48.

fig. P11.48

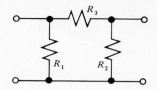

11.49. Find the T-equivalent of the two-port given in Problem 11.48. Compare your results with the Δ-Y transformation studied previously.

fig. P11.50

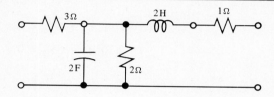

*11.50. Find the **z**-parameters of the T-network shown in Fig. P11.50.

Ans. $\dfrac{12s + 5}{4s + 1}, \dfrac{2}{4s + 1}, \dfrac{8s^2 + 6s + 3}{4s + 1}$

11.51. Find the **z**-parameters of the two-port given in Problem 11.39 in the preceding problem set.

11.52. Find the **z**-parameters of the two-port given in Problem 11.40.

11.53. A two-port network has **z**-parameters z_{11}, z_{12}, z_{21}, and z_{22}. Find the **y**-parameters of this two-port in terms of the **z**-parameters.

*11.54. Given a two-port network, find

$$\left. \frac{\mathbf{V}_2}{\mathbf{V}_1} \right|_{\mathbf{I}_2 = 0}$$

in terms of the **z**-parameters of the two-port.

Ans. $\dfrac{\mathbf{z}_{21}}{\mathbf{z}_{11}}$

11.55. A two-port network is terminated with a 1-Ω load resistor. Find an expression for $\mathbf{V}_2/\mathbf{V}_1$ in terms of the two-port's **z**-parameters.

fig. P11.56

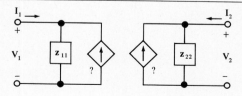

11.56. Find the values of the dependent current sources so that the circuit shown in Fig. P11.56 is the equivalent circuit of a two-port having the **z**-parameters z_{11}, z_{12}, z_{21}, z_{22}.

fig. P11.57

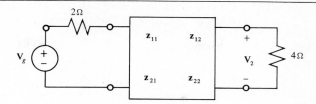

*11.57. For the network shown in Fig. P11.57, find V_2/V_g given that $z_{11} = 2\ \Omega$, $z_{12} = 3\ \Omega = z_{21}$, and $z_{22} = 3\ \Omega$.
Ans. $\frac{12}{19}$

11.58. Repeat Problem 11.57 for the case that z_{21} is changed to $z_{21} = 5\ \Omega$.

fig. P11.59

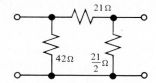

11.59. Find the **h**-parameters of the two-port shown in Fig. P11.59.

11.60. Consider a two-port network whose y-parameters are y_{11}, y_{12}, y_{21}, and y_{22}. Express the **h**-parameters of this network in terms of its **y**-parameters.

11.61. Given a two-port network, find an expression for

$$\left.\frac{V_2}{V_1}\right|_{I_2 = 0}$$

in terms of the network's **h**-parameters.

11.62. A two-port network is terminated with a 1-Ω load resistor. Find an expression for V_2/V_1 in terms of the **h**-parameters of the two-port.

*11.63. The two-port given in Problem 11.57 has **h**-parameters $h_{11} = 2\ \Omega$, $h_{12} = 0.6$, $h_{21} = 4$, and $h_{22} = 0.5\ \mho$. Find V_2/V_g.
Ans. $-\frac{20}{3}$

11.64. The **g**-*parameters* of a two-port are defined by the pair of equations

$$I_1 = g_{11}V_1 + g_{12}I_2$$

$$V_2 = g_{21}V_1 + g_{22}I_2$$

Find the **g**-parameters for the π-network given in Problem 11.59.

11.65. Draw an equivalent circuit for a two-port described by the **g**-parameters defined in Problem 11.64.

11.66. Express the **g**-parameters defined in Problem 11.64 in terms of the **z**-parameters of a two-port.

11.67. The *transmission parameters* or **A**, **B**, **C**, **D** *parameters* of a two-port are defined by the pair of equations

$$V_1 = AV_2 - BI_2$$

$$I_1 = CV_2 - DI_2$$

Express **A**, **B**, **C**, and **D** in terms of the **z**-parameters of a two-port.

*11.68. A two-port network is terminated with a 1-Ω load resistor. Find an expression for V_2/V_1 in terms of the transmission parameters defined in Problem 11.67.

Ans. $1/(A + B)$

---

1. A transformer is constructed by placing two inductors in physical proximity.
2. An ideal transformer has perfect coupling and infinite self-inductances.
3. The energy stored in an ideal transformer is zero.
4. A one-port network containing no independent sources can be described by its impedance or admittance.
5. A two-port network containing no independent sources may be described by certain two-port parameters.
6. A two-port for which $y_{12} = y_{21}$ (equivalently, $z_{12} = z_{21}$ or $h_{12} = -h_{21}$) is a reciprocal network.

Summary

FOURIER SERIES 12

Previously we mentioned that the sinusoid is a very important function. In this chapter we shall discuss a major reason why: A repetitive or periodic function is equal to a (probably infinite) sum of sinusoids! As a consequence of the principle of superposition, this result, known as Fourier's theorem, means that circuits with nonsinusoidal inputs that are periodic can be analyzed with sinusoidal phasor techniques.

The expression of a periodic function as a sum of sinusoids—known as its Fourier series representation—can be either trigonometric or exponential in form. It is the latter form that suggests a generalization, known as the Fourier transform, that enables the analysis of circuits having inputs that are nonperiodic.

We start this chapter by stating (without proof) Fourier's theorem. Next we discuss the direct approach for obtaining the trigonometric form of the Fourier series representation of a periodic function. When speaking of the trigonometric Fourier series, we can use one of two forms; one consists of sines and cosines, whereas the other consists of sinusoids with phase angles. We then see how to determine the response of a circuit to a periodic function by using the Fourier series representation of the function, the principle of superposition, and sinusoidal circuit analysis. By using symmetry properties of certain functions we then discuss the subject of reducing the work required for determining the Fourier series representations of these functions. Furthermore, we see that it may be quite a bit simpler to find the Fourier series representation of a function by utilizing one of its derivatives rather than dealing directly with the function. The key to this shortcut is the delta function and its sampling property. Finally, we use Euler's formula to develop the complex or exponential form of the Fourier series. Although less intuitive (since complex sinusoids are less intuitive than real sinusoids), this form is more compact than the trigonometric form. More important, though, we shall see in the next chapter how the exponential form can be extended to the concept of the Fourier transform, which allows the analysis of circuits whose inputs are nonperiodic.

12.1   THE TRIGONOMETRIC FOURIER SERIES

We begin by considering functions that have a very special property—periodicity.

A function $f(t)$ is said to be *periodic* if there exists a positive, real number T such that

$$f(t + T) = f(t)$$

for all t. If $f(t)$ is periodic, then we say that T is the *period* of $f(t)$.

According to this definition, if $f(t)$ is periodic with period T, it is also periodic with period $2T$ (or $3T$, or $4T$, etc.). However, when it is stated that the period function $f(t)$ has period T, the usual implication is that there is no positive $T' < T$ for which $f(t)$ is periodic.

· · ·

Example
For the function $f(t) = \sin \omega t$, since

$$\sin \omega\left(t + \frac{2\pi}{\omega}\right) = \sin(\omega t + 2\pi)$$

$$= \sin \omega t \qquad \text{for all } t$$

then $\sin \omega t$ is periodic, with period $T = 2\pi/\omega$.

Similarly, $\cos \omega t$ and $\cos(\omega t + \phi)$ are periodic and have period $T = 2\pi/\omega$ (see pages 373–374).

Furthermore, since

$$e^{j\omega(t + 2\pi/\omega)} = e^{j\omega t + j2\pi}$$

$$= e^{j\omega t}e^{j2\pi}$$

$$= e^{j\omega t} \qquad \text{for all } t$$

then $e^{j\omega t}$ is periodic, with period $T = 2\pi/\omega$.

· · ·

From this example, it should be clear that sinusoids are periodic functions. What may not be clear is the remarkable result that "any" periodic function (exceptions are certain types of exotic functions that are rarely, if ever, encountered) can be expressed as a sum of sinusoids of appropriate amplitudes and frequencies. This result, known as Fourier's theorem,* can be summarized as follows.

---

* Named for the French mathematician Joseph Fourier (1768–1830).

Fourier's Theorem

Given the real function $f(t)$, which is periodic with period T seconds, there exist real numbers $a_0, a_1, b_1, a_2, b_2, a_3, b_3, \ldots$ such that

$$f(t) = a_0 + a_1 \cos \omega_0 t + b_1 \sin \omega_0 t + a_2 \cos 2\omega_0 t$$

$$+ b_2 \sin 2\omega_0 t + a_3 \cos 3\omega_0 t + b_3 \sin 3\omega_0 t + \cdots$$

or

$$f(t) = a_0 + \sum_{n=1}^{\infty} (a_n \cos n\omega_0 t + b_n \sin n\omega_0 t)$$

where $\omega_0 = 2\pi/T$ radians per second is called the *fundamental frequency* or *first harmonic* of $f(t)$. The frequency $n\omega_0$ is called the nth *harmonic*. The above formula (either form) is known as the *trigonometric Fourier series* of $f(t)$. The constants $a_0, a_1, b_1, a_2, b_2, a_3, b_3, \ldots$ are called the (*trigonometric*) *Fourier coefficients* of $f(t)$.

Although the proof of Fourier's theorem is beyond the scope of this book, the application is not. The problem therefore is, given a periodic function $f(t)$, what is its Fourier series representation? In other words, how are the Fourier coefficients determined and what are they?

In order to answer these questions, we shall need to employ some trigonometric properties. It is a matter of elementary integral calculus to show for $\omega_0 = 2\pi/T$ that

$$\int_0^T \sin \omega_0 t \, dt = 0$$

and

$$\int_0^T \cos \omega_0 t \, dt = 0$$

Although the above integrations are performed over the period from 0 to T, the same results are obtained when the integrations are performed over an arbitrary period, say from t_1 to $t_1 + T$, for any t_1.

Furthermore, it is not much more involved to verify the more general results

$$\int_0^T \sin n\omega_0 t \, dt = 0$$

and

$$\int_0^T \cos n\omega_0 t \, dt = 0$$

for all $n = 1, 2, 3, \ldots$. Again, when the integrations are performed over any interval of T seconds, the same results are obtained.

By using the trigonometric identity

$$\sin A \cos B = \frac{1}{2}[\sin(A + B) + \sin(A - B)]$$

it can be shown that for $n = 1, 2, 3, \ldots$ and $m = 1, 2, 3, \ldots$ that

$$\int_0^T \sin m\omega_0 t \cos n\omega_0 t \, dt = 0$$

Furthermore with the use of

$$\cos A \cos B = \frac{1}{2}[\cos(A + B) + \cos(A - B)]$$

it follows that

$$\int_0^T \cos m\omega_0 t \cos n\omega_0 t \, dt = \begin{cases} 0 & \text{for } m \neq n \\ \dfrac{T}{2} & \text{for } m = n \end{cases}$$

Finally, with

$$\sin A \sin B = \frac{1}{2}[\cos(A - B) - \cos(A + B)]$$

it follows that

$$\int_0^T \sin m\omega_0 t \sin n\omega_0 t \, dt = \begin{cases} 0 & \text{for } m \neq n \\ \dfrac{T}{2} & \text{for } m = n \end{cases}$$

Returning to the Fourier series

$$f(t) = a_0 + \sum_{n=1}^{\infty} (a_n \cos n\omega_0 t + b_n \sin n\omega_0 t)$$

integrating both sides of this equation over a period yields

$$\int_0^T f(t) \, dt = \int_0^T \left[a_0 + \sum_{n=1}^{\infty} (a_n \cos n\omega_0 t + b_n \sin n\omega_0 t) \right] dt$$

But since the integral of a sum is the sum of the integrals,

$$\int_0^T f(t) \, dt = \int_0^T a_0 \, dt + \sum_{n=1}^{\infty} \left[\int_0^T a_n \cos n\omega_0 t \, dt + \int_0^T b_n \sin n\omega_0 t \, dt \right]$$

But, as mentioned above, the two integrals inside the summation vanish. The result is

$$\int_0^T f(t) \, dt = a_0 \int_0^T dt = a_0 t \Big|_0^T = a_0 T$$

from which

$$a_0 = \frac{1}{T} \int_0^T f(t)\, dt$$

Thus, we have a formula for the Fourier coefficient a_0—called the *average value* of $f(t)$ or the *dc component* of $f(t)$. The limits on the integral can be replaced with any interval of T seconds.

Suppose now that in the formula for the Fourier series, both sides are first multiplied by the factor $\cos m\omega_0 t$ and then integrated with respect to t. Then

$$\int_0^T f(t) \cos m\omega_0 t\, dt$$

$$= \int_0^T \left[a_0 + \sum_{n=1}^{\infty} (a_n \cos n\omega_0 t + b_n \sin n\omega_0 t) \right] \cos m\omega_0 t\, dt$$

$$= \int_0^T a_0 \cos m\omega_0 t\, dt + \sum_{n=1}^{\infty} \left[\int_0^T a_n \cos n\omega_0 t \cos m\omega_0 t\, dt \right.$$

$$\left. + \int_0^T b_n \sin n\omega_0 t \cos m\omega_0 t\, dt \right]$$

From the previous trigonometric properties, on the right-hand side of this equation the first integral is zero for all $m = 1, 2, 3, \ldots$, the second integral is zero for $m \neq n$ and equal to $a_n(T/2)$ when $m = n$, and the third integral is zero for all $m = 1, 2, 3, \ldots$ and $n = 1, 2, 3, \ldots$. Thus,

$$\int_0^T f(t) \cos m\omega_0 t\, dt = a_n \left(\frac{T}{2} \right) \qquad \text{for } n = m$$

Hence

$$a_n = \frac{2}{T} \int_0^T f(t) \cos n\omega_0 t\, dt \qquad \text{for } n = 1, 2, 3, \ldots$$

is a formula for the coefficient a_n; the limits of integration can be over an arbitrary period as well.

By multiplying both sides of the Fourier series formula by $\sin m\omega_0 t$ and integrating with respect to t, in a similar manner we obtain the formula

$$b_n = \frac{2}{T} \int_0^T f(t) \sin n\omega_0 t\, dt \qquad \text{for } n = 1, 2, 3, \ldots$$

where the integral can be over an arbitrary period as well.

. . .

Example

The periodic function shown in Fig. 12.1 is known as a half-wave rectified sine wave. For this function $T = 2$ s. Thus, $\omega_0 = 2\pi/T = \pi$ rad/s. The dc component is

$$a_0 = \frac{1}{T} \int_0^T f(t) \, dt = \frac{1}{2} \left[\int_0^1 f(t) \, dt + \int_1^2 f(t) \, dt \right]$$

$$= \frac{1}{2} \left[\int_0^1 \sin \pi t \, dt + 0 \right]$$

---

fig. 12.1

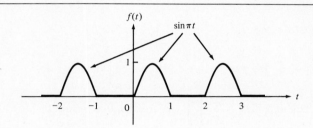

Thus

$$a_0 = \frac{1}{\pi}$$

In addition,

$$a_n = \frac{2}{T} \int_0^T f(t) \cos n\omega_0 t \, dt$$

$$= \frac{2}{T} \left[\int_0^1 f(t) \cos n\omega_0 t \, dt + \int_1^2 f(t) \cos n\omega_0 t \, dt \right]$$

$$= \frac{2}{2} \int_0^1 \sin \pi t \cos n\pi t \, dt$$

Thus

$$a_n = \int_0^1 \frac{1}{2} [\sin \pi(1 + n)t + \sin \pi(1 - n)t] \, dt$$

For $n = 1$, we get

$$a_1 = \int_0^1 \frac{1}{2} \sin 2\pi t \, dt = \frac{1}{2} \left(\frac{1}{2\pi} \cos 2\pi t \right) \Big|_0^1 = 0$$

For $n = 2, 3, 4, \ldots,$

$$a_n = \frac{1 - \cos(1 + n)\pi}{2\pi(1 + n)} + \frac{1 - \cos(1 - n)\pi}{2\pi(1 - n)}$$

Since for n even ($n = 2, 4, 6, \ldots$), $n + 1$ and $n - 1$ are odd, then

$$\cos(1 + n)\pi = -1$$

$$\cos(1 - n)\pi = \cos(n - 1)\pi = -1$$

so

$$\cos(1 + n)\pi = \cos(1 - n)\pi$$

and since for n odd ($n = 3, 5, 7, \ldots$), $n + 1$ and $n - 1$ are even, then

$$\cos(1 + n)\pi = \cos(n - 1)\pi = \cos(1 - n)\pi = 1$$

Therefore

$$a_n = \frac{1 - \cos(1 + n)\pi}{2\pi(1 + n)} + \frac{1 - \cos(1 + n)\pi}{2\pi(1 - n)}$$

or

$$a_n = \frac{1 - \cos(1 + n)\pi}{\pi(1 + n)(1 - n)} = \frac{1 - \cos(1 + n)\pi}{\pi(1 - n^2)} \qquad \text{for } n = 2, 3, 4, \ldots$$

Next,

$$b_n = \frac{2}{T} \int_0^T f(t) \sin n\omega_0 t \, dt$$

$$= \int_0^1 \sin \pi t \sin n\pi t \, dt$$

$$= \int_0^1 \frac{1}{2}[\cos \pi(1 - n)t - \cos \pi(1 + n)t] \, dt$$

For the case $n = 1$, we get

$$b_1 = \int_0^1 \frac{1}{2}[1 - \cos 2\pi t] \, dt$$

$$= \frac{1}{2} \int_0^1 dt - \frac{1}{2} \int_0^1 \cos 2\pi t \, dt$$

$$= \frac{1}{2} t \Big|_0^1 - \frac{1}{2} \left(\frac{1}{2\pi} \sin 2\pi t \right) \Big|_0^1 = \frac{1}{2}$$

For $n = 2, 3, 4, \ldots$,

$$b_n = \frac{\sin(1 - n)\pi}{2\pi(1 - n)} - \frac{\sin(1 + n)\pi}{2\pi(1 + n)}$$

But

$$\sin(1 - n)\pi = \sin(1 + n)\pi = 0$$

so

$$b_n = 0 \qquad \text{for } n = 2, 3, 4, \ldots$$

Thus, the trigonometric Fourier series for $f(t)$ is

$$f(t) = a_0 + \sum_{n=1}^{\infty} (a_n \cos n\omega_0 t + b_n \sin n\omega_0 t)$$

$$= \frac{1}{\pi} + \sum_{n=2}^{\infty} \frac{1 - \cos(1 + n)\pi}{\pi(1 - n^2)} \cos n\pi t + \frac{1}{2} \sin \pi t$$

Writing the first few terms in this sum we have

$$f(t) = \frac{1}{\pi} + \frac{1}{2} \sin \pi t - \frac{2}{3\pi} \cos 2\pi t - \frac{2}{15\pi} \cos 4\pi t$$

$$- \frac{2}{35\pi} \cos 6\pi t - \frac{2}{63\pi} \cos 8\pi t - \cdots$$

$$= 0.32 + 0.5 \sin \pi t - 0.21 \cos 2\pi t - 0.04 \cos 4\pi t$$

$$- 0.02 \cos 6\pi t - 0.01 \cos 8\pi t - \cdots$$

Therefore, the half-wave rectified sine wave is comprised of a dc component, a sinusoid having the fundamental frequency, and sinusoids whose frequencies are the even harmonics. In order to get an idea of the relative amount of the components constituting $f(t)$, we can plot the magnitude of each sinusoid (the dc component is a sinusoid of zero frequency) versus frequency. Such a plot is shown in Fig. 12.2.

· · ·

fig. 12.2

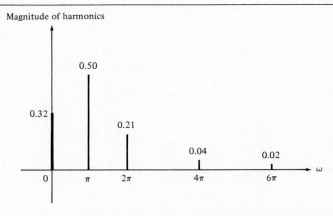

In the foregoing example, the nth harmonic of $f(t)$ consists either entirely of a pure sine or a cosine, but not both. Thus, the magnitude of the nth harmonic is equal to the amplitude of the corresponding sinusoid.

In general, however, the nth term in the summation of the Fourier series is

$$a_n \cos n\omega_0 t + b_n \sin n\omega_0 t$$

Since this can be rewritten as

$$A_n \cos(n\omega_0 t - \phi_n)$$

where

$$A_n = \sqrt{a_n{}^2 + b_n{}^2} \quad \text{and} \quad \phi_n = \tan^{-1}\frac{b_n}{a_n}$$

then the amplitude of the nth harmonic is $\sqrt{a_n{}^2 + b_n{}^2}$.

Consequently, an alternative form of the trigonometric Fourier series is

$$f(t) = a_0 + \sum_{n=1}^{\infty} A_n \cos(n\omega_0 t - \phi_n)$$

In the example just given, the plot of the magnitude of the harmonics versus frequency is the same as a plot of the amplitude A_n of the harmonics versus frequency, which we shall call the *amplitude spectrum* of $f(t)$. In a similar vein, we can plot the phase angle ϕ_n of the harmonics versus frequency. Such a plot is called the *phase spectrum* of $f(t)$. The amplitude and phase spectra constitute the *frequency spectrum* of $f(t)$. Despite this definition, people often refer to the amplitude spectrum alone as the frequency spectrum, for in many situations phase is not an important factor.

For the previous example, we can write $f(t)$ as

$$f(t) = 0.32 + 0.5 \cos(\pi t - 90°) + 0.21 \cos(2\pi t - 180°) + 0.04 \cos(4\pi t - 180°)$$

$$+ 0.02 \cos(6\pi t - 180°) + 0.01 \cos(8\pi t - 180°) + \cdots$$

Thus a plot of the phase spectrum of $f(t)$ is as shown in Fig. 12.3.

fig. 12.3

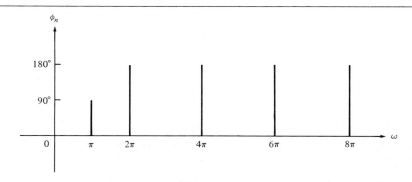

Suppose now that the function $f(t)$ given in the previous example is a voltage that is the input to the simple RC network shown in Fig. 12.4.

fig. 12.4

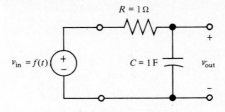

Since

$$v_{\text{in}} = \frac{1}{\pi} + \frac{1}{2} \sin \pi t + \sum_{n=2}^{\infty} a_n \cos n\pi t$$

$$= v_0 + v_1 + \sum_{n=2}^{\infty} v_n$$

then by the principle of superposition

$$v_{\text{out}} = v_0^1 + v_1^1 + \sum_{n=2}^{\infty} v_n^1$$

where v_i^1 is the output when the input is v_i.

By phasor analysis and voltage division,

$$\mathbf{V}_{\text{out}} = \frac{1/j\omega C}{(1/j\omega C) + R} \mathbf{V}_{\text{in}} = \frac{1}{1 + j\omega RC} \mathbf{V}_{\text{in}}$$

$$= \frac{1}{1 + j\omega} \mathbf{V}_{\text{in}} = \frac{\mathbf{V}_{\text{in}}}{\sqrt{1 + \omega^2}\ \underline{/\tan^{-1} \omega}}$$

The response to the input

$$\mathbf{V}_0 = \frac{1}{\pi} \qquad (\omega = 0)$$

is

$$\mathbf{V}_0^1 = \frac{1}{1 + j0}\left(\frac{1}{\pi}\right) = \frac{1}{\pi}$$

(Remember that a capacitor is an open circuit to dc, and therefore, for the dc part, the output voltage equals the input voltage.) The phasor representation of the first harmonic is

$$\mathbf{V}_1 = \frac{1}{2}\ \underline{/-90^\circ} \qquad (\omega = \pi)$$

The response (in phasor form) to this input is

$$\mathbf{V}_1^1 = \frac{\frac{1}{2} \big/ -90°}{\sqrt{1 + \pi^2} \big/ \tan^{-1} \pi}$$

$$= \frac{1}{2\sqrt{1 + \pi^2}} \big/ (-90° - \tan^{-1} \pi)$$

$$= 0.15 \big/ -162.3°$$

The nth harmonic ($n \geq 2$) of the input is

$$\mathbf{V}_n = a_n \big/ 0° = a_n \qquad (\omega = n\pi)$$

and its response is

$$\mathbf{V}_n^1 = \frac{a_n}{\sqrt{1 + (n\pi)^2} \big/ \tan^{-1}(n\pi)}$$

$$= \frac{a_n}{\sqrt{1 + (n\pi)^2}} \big/ -\tan^{-1}(n\pi)$$

Hence, the output voltage is

$$v_{\text{out}} = \frac{1}{\pi} + 0.15 \cos(\pi t - 162.3°) + \sum_{n=2}^{\infty} \frac{a_n}{\sqrt{1 + (n\pi)^2}} \cos[n\pi t - \tan^{-1}(n\pi)]$$

Writing the first two terms in the summation, we get

$$v_{\text{out}} = 0.32 + 0.15 \cos(\pi t - 162.3°) - 0.033 \cos(2\pi t - 81°)$$
$$- 0.0034 \cos(4\pi t - 85.4°) - \cdots$$

The amplitude spectrum of v_{out} is shown in Fig. 12.5. Note that the entire dc component of the input appears at the output. However, only 30 percent of the first

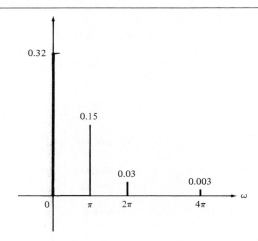

fig. 12.5

harmonic reaches the output, only about 14 percent of the second harmonic, and so on. In the expression for v_{out} we see that the *RC* circuit diminishes the amount of higher harmonics (second and above) by the factor $1/\sqrt{1 + (n\pi)^2}$. Such a circuit is therefore called a "low-pass filter."

Having seen a use for the Fourier series, let us return to the problem of determining the Fourier series of a periodic function $f(t)$. Needless to say, this can be accomplished by substituting the expression for $f(t)$ into the formulas for a_0, a_n, and b_n. However, for even some rather innocuous-looking functions, an inordinate amount of calculation may result. Fortunately, it is often possible to take shortcuts by using different mathematical operations and properties. To begin with, we define the following:

A function $f(t)$ is said to be an *even function*, or simply *even*, if

$$f(-t) = f(t) \qquad \text{for all } t$$

From the definition, it is apparent that an even function is one whose plot is symmetrical around the vertical axis.

. . .

Example
The function $f(t) = t^2$ is an even function.
Since $\cos(-\theta) = \cos\theta$, the function

$$f(t) = \cos\omega t$$

is an even, periodic function of t.

. . .

We shall now see how to reduce the amount of work required to determine the Fourier series of an even, periodic function.

In the formula for the coefficient a_n, let us select the period of integration to be $-T/2$ to $T/2$. Then

$$a_n = \frac{2}{T} \int_{-T/2}^{T/2} f(t) \cos n\omega_0 t \, dt$$

$$= \frac{2}{T} \int_{-T/2}^{0} f(t) \cos n\omega_0 t \, dt + \frac{2}{T} \int_{0}^{T/2} f(t) \cos n\omega_0 t \, dt$$

Replacing t by $-t$ in the first integral yields

$$a_n = \frac{2}{T} \int_{-T/2}^{0} f(-t) \cos n\omega_0(-t) \, d(-t) + \frac{2}{T} \int_{0}^{T/2} f(t) \cos n\omega_0 t \, dt$$

Since

$$\cos n\omega_0(-t) = \cos(-n\omega_0 t) = \cos n\omega_0 t$$

$$d(-t) = -dt$$

$$\int_{-T/2}^{0} d(-t) = -\int_{T/2}^{0} dt = \int_{0}^{T/2} dt$$

then

$$a_n = \frac{2}{T} \int_{0}^{T/2} f(-t) \cos n\omega_0 t \, dt + \frac{2}{T} \int_{0}^{T/2} f(t) \cos n\omega_0 t \, dt$$

If $f(t)$ is even, then $f(-t) = f(t)$, and thus

$$a_n = \frac{4}{T} \int_{0}^{T/2} f(t) \cos n\omega_0 t \, dt$$

Therefore, the coefficient a_n can be determined by integrating over half a period, but this half-period must be the interval from $t = 0$ to $t = T/2$.

Although this is possibly time-saving, the significant result about even periodic functions deals with the coefficient b_n. From the formula given previously,

$$b_n = \frac{2}{T} \int_{-T/2}^{T/2} f(t) \sin n\omega_0 t \, dt$$

$$= \frac{2}{T} \int_{-T/2}^{0} f(t) \sin n\omega_0 t \, dt + \frac{2}{T} \int_{0}^{T/2} f(t) \sin n\omega_0 t \, dt$$

Again, replacing t by $-t$ in the first integral, since $\sin n\omega_0(-t) = \sin(-n\omega_0 t) = -\sin n\omega_0 t$, then

$$b_n = -\frac{2}{T} \int_{0}^{T/2} f(-t) \sin n\omega_0 t \, dt + \frac{2}{T} \int_{0}^{T/2} f(t) \sin n\omega_0 t \, dt$$

If $f(t)$ is an even function, then $f(-t) = f(t)$ and

$$b_n = 0 \qquad \text{for } n = 1, 2, 3, \ldots$$

These results can also be justified intuitively. Since the cosine function is even, a sum of cosines will also be even. However, the sine function is not even, so the addition of sine components will yield a function that is not even.

· · ·

Example
Consider the "square-wave" function shown in Fig. 12.6. The period of this function is $T = 2$ s. Thus $\omega_0 = 2\pi/T = \pi$ rad/s.

fig. 12.6

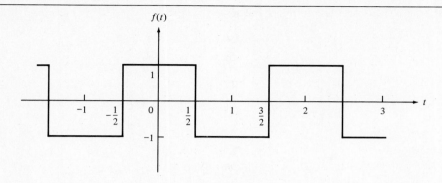

In a manner similar to that described for a_n, we can also show that for an even function

$$a_0 = \frac{2}{T} \int_0^{T/2} f(t)\, dt$$

Since $f(t)$ is even,

$$a_0 = \frac{2}{2} \int_0^1 f(t)\, dt = \int_0^{1/2} 1\, dt + \int_{1/2}^1 -1\, dt = \frac{1}{2} - \frac{1}{2} = 0$$

Also,

$$a_n = \frac{4}{T} \int_0^{T/2} f(t) \cos n\omega_0 t\, dt$$

$$= 2 \int_0^1 f(t) \cos n\pi t\, dt$$

$$= 2\left[\int_0^{1/2} \cos n\pi t\, dt + \int_{1/2}^1 - \cos n\pi t\, dt \right]$$

$$= \frac{4}{n\pi} \sin \frac{n\pi}{2}$$

Furthermore, since $f(t)$ is an even function,

$$b_n = 0 \qquad \text{for } n = 1, 2, 3, \ldots$$

Thus

$$f(t) = a_0 + \sum_{n=1}^{\infty} (a_n \cos n\omega_0 t + b_n \sin n\omega_0 t)$$

$$= \sum_{n=1}^{\infty} \frac{4}{n\pi} \sin \frac{n\pi}{2} \cos n\pi t$$

Writing out the first few terms of the series, we get

$$f(t) = \frac{4}{\pi} \cos \pi t - \frac{4}{3\pi} \cos 3\pi t + \frac{4}{5\pi} \cos 5\pi t - \frac{4}{7\pi} \cos 7\pi t + \cdots$$

$$= 1.27 \cos \pi t - 0.42 \cos 3\pi t + 0.25 \cos 5\pi t - 0.18 \cos 7\pi t + \cdots$$

$\bullet \quad \bullet \quad \bullet$

In addition to even function symmetry, we have the following definition: A function $f(t)$ is said to be an *odd function*, or simply *odd*, if

$$f(-t) = -f(t) \qquad \text{for all } t$$

From this definition, we see that the plot of an odd function is antisymmetrical around the vertical axis. That is, if the reflection of the positive portion of the plot is inverted, then the negative portion is obtained.

$\bullet \quad \bullet \quad \bullet$

Example
The function

$$f(t) = t^3$$

is odd, as is the function

$$f(t) = t$$

Since $\sin(-\theta) = -\sin \theta$, the function

$$f(t) = \sin \omega t$$

is also an odd function.

$\bullet \quad \bullet \quad \bullet$

In a manner almost identical to that done for even functions, it can be shown that when $f(t)$ is an odd function,

$$a_n = \frac{2}{T} \int_{-T/2}^{T/2} f(t) \cos n\omega_0 t \, dt = 0 \qquad \text{for } n = 1, 2, 3, \ldots$$

and

$$b_n = \frac{4}{T} \int_0^{T/2} f(t) \sin n\omega_0 t \, dt$$

Furthermore,

$$a_0 = \frac{1}{T} \int_{-T/2}^{T/2} f(t)\, dt = \frac{1}{T} \int_{-T/2}^{0} f(t)\, dt + \frac{1}{T} \int_{0}^{T/2} f(t)\, dt$$

$$= \frac{1}{T} \int_{-T/2}^{0} f(-t)\, d(-t) + \frac{1}{T} \int_{0}^{T/2} f(t)\, dt$$

$$= \frac{1}{T} \int_{0}^{T/2} f(-t)\, dt + \frac{1}{T} \int_{0}^{T/2} f(t)\, dt$$

and, since $f(-t) = -f(t)$,

$$a_0 = -\frac{1}{T} \int_{0}^{T/2} f(t)\, dt + \frac{1}{T} \int_{0}^{T/2} f(t)\, dt = 0$$

for an odd function.

$\bullet \quad \bullet \quad \bullet$

Example

Consider another square-wave function, as shown in Fig. 12.7. Inspection reveals that is an odd function, and thus we can determine its Fourier series representation

fig. 12.7

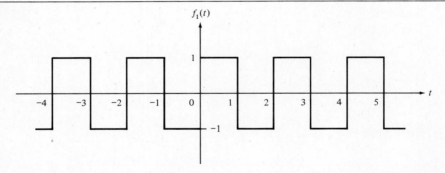

as was done for the even square-wave function $f(t)$ given in Fig. 12.6. However, using the fact that $f_1(t) = f(t - 1/2)$, we have

$$f_1(t) = \frac{4}{\pi} \cos \pi\left(t - \frac{1}{2}\right) - \frac{4}{3\pi} \cos 3\pi\left(t - \frac{1}{2}\right)$$

$$+ \frac{4}{5\pi} \cos 5\pi\left(t - \frac{1}{2}\right) - \frac{4}{7\pi} \cos 7\pi\left(t - \frac{1}{2}\right) + \cdots$$

$$= \frac{4}{\pi} \cos\left(\pi t - \frac{\pi}{2}\right) - \frac{4}{3\pi} \cos\left(3\pi t - \frac{3\pi}{2}\right)$$

$$+ \frac{4}{5\pi} \cos\left(5\pi t - \frac{5\pi}{2}\right) - \frac{4}{7\pi} \cos\left(7\pi t - \frac{7\pi}{2}\right) + \cdots$$

$$= 1.27 \sin \pi t + 0.42 \sin 3\pi t + 0.25 \sin 5\pi t + 0.18 \sin 7\pi t + \cdots$$

$\bullet \quad \bullet \quad \bullet$

Note that for both square waves just mentioned, the Fourier series representations contain only odd harmonics. This fact is due to another type of symmetry that both functions have.

A periodic function $f(t)$ having period T is said to be *half-wave symmetric* if

$$f\left(t - \frac{T}{2}\right) = -f(t) \qquad \text{for all } t$$

If a plot of $f(t)$ is shifted half a period and then inverted, the original plot is recovered.

For a periodic function $f(t)$,

$$a_n = \frac{2}{T} \int_{-T/2}^{T/2} f(t) \cos n\omega_0 t \, dt$$

$$= \frac{2}{T} \int_{-T/2}^{0} f(t) \cos n\omega_0 t \, dt + \frac{2}{T} \int_{0}^{T/2} f(t) \cos n\omega_0 t \, dt$$

For the first integral, define the new variable x by

$$x = t + \frac{T}{2}$$

Thus

$$t = x - \frac{T}{2}$$

and

$$dt = dx$$

Therefore

$$\frac{2}{T} \int_{-T/2}^{0} f(t) \cos n\omega_0 t \, dt = \frac{2}{T} \int_{0}^{T/2} f\left(x - \frac{T}{2}\right) \cos n\omega_0\left(x - \frac{T}{2}\right) dx$$

$$= \frac{2}{T} \int_{0}^{T/2} f\left(x - \frac{T}{2}\right) \left[\cos n\omega_0 x \cos \frac{n\omega_0 T}{2} + \sin n\omega_0 x \sin \frac{n\omega_0 T}{2} \right] dx$$

Since

$$\cos \frac{n\omega_0 T}{2} = \cos n \frac{2\pi}{T} \frac{T}{2} = \cos n\pi$$

$$\sin \frac{n\omega_0 T}{2} = \sin n \frac{2\pi}{T} \frac{T}{2} = \sin n\pi = 0 \qquad \text{for } n = 1, 2, 3, \ldots$$

we have

$$\frac{2}{T} \int_{-T/2}^{0} f(t) \cos n\omega_0 t \, dt = \frac{2}{T} \int_{0}^{T/2} f\left(x - \frac{T}{2}\right) \cos n\omega_0 x \cos n\pi \, dx$$

$$= \frac{2}{T} \cos n\pi \int_{0}^{T/2} f\left(x - \frac{T}{2}\right) \cos n\omega_0 x \, dx$$

$$= \frac{2}{T} \cos n\pi \int_{0}^{T/2} f\left(t - \frac{T}{2}\right) \cos n\omega_0 t \, dt$$

Substituting this into the expression for a_n given above, we get

$$a_n = \frac{2}{T} \cos n\pi \int_{0}^{T/2} f\left(t - \frac{T}{2}\right) \cos n\omega_0 t \, dt + \frac{2}{T} \int_{0}^{T/2} f(t) \cos n\omega_0 t \, dt$$

If $f(t)$ is a half-wave symmetric function, then

$$f\left(t - \frac{T}{2}\right) = -f(t)$$

and

$$a_n = \frac{2}{T} \cos n\pi \int_{0}^{T/2} -f(t) \cos n\omega_0 t \, dt + \frac{2}{T} \int_{0}^{T/2} f(t) \cos n\omega_0 t \, dt$$

$$= (1 - \cos n\pi) \frac{2}{T} \int_{0}^{T/2} f(t) \cos n\omega_0 t \, dt$$

Since

$$(1 - \cos n\pi) = \begin{cases} 2 & \text{for } n = 1, 3, 5, \dots \\ 0 & \text{for } n = 2, 4, 6, \dots \end{cases}$$

then

$$a_n = \begin{cases} \dfrac{4}{T} \int_{0}^{T/2} f(t) \cos n\omega_0 t \, dt & \text{for } n \text{ odd} \\ \\ 0 & \text{for } n \text{ even} \end{cases}$$

Similarly, we can show that

$$b_n = \begin{cases} \dfrac{4}{T} \int_{0}^{T/2} f(t) \sin n\omega_0 t \, dt & \text{for } n \text{ odd} \\ \\ 0 & \text{for } n \text{ even} \end{cases}$$

In other words, a half-wave symmetric function contains only odd harmonics.

For functions possessing even, odd, and half-wave symmetries we can use the results given previously to eliminate some of the work required to determine the Fourier series. In addition, we shall now discuss a time-saving technique that can be applied to many types of functions—including functions not having any of the symmetries mentioned above. This technique usually employs the sampling property of delta functions in the evaluations of integrals.

Now suppose that the periodic function $f(t)$ has the trigonometric Fourier series representation

$$f(t) = a_0 + \sum_{n=1}^{\infty} (a_n \cos n\omega_0 t + b_n \sin n\omega_0 t)$$

where T is the period of $f(t)$ and $\omega_0 = 2\pi/T$. Taking the derivative of this equation, since the derivative of a sum equals the sum of the derivatives, we get

$$\frac{df(t)}{dt} = 0 + \sum_{n=1}^{\infty} (-n\omega_0 a_n \sin n\omega_0 t + n\omega_0 b_n \cos n\omega_0 t)$$

or

$$f'(t) = \sum_{n=1}^{\infty} (n\omega_0 b_n \cos n\omega_0 t - n\omega_0 a_n \sin n\omega_0 t)$$

But since $f(t)$ is periodic with period T, its derivative $f'(t)$ must be periodic with period T. Therefore, $f'(t)$ has a Fourier series representation

$$f'(t) = a_0' + \sum_{n=1}^{\infty} (a_n' \cos n\omega_0 t + b_n' \sin n\omega_0 t)$$

where $\omega_0 = 2\pi/T$. Comparing this equation with the preceding one, we have

$$a_0' = 0$$

$$a_n' = n\omega_0 b_n$$

$$b_n' = -n\omega_0 a_n$$

Thus, if we know the Fourier coefficients of $f'(t)$, it is a simple matter to determine the Fourier coefficients of $f(t)$—with the exception of a_0. Of course, the coefficients a_n' and b_n' can be obtained from the formulas

$$a_n' = \frac{2}{T} \int_{-T/2}^{T/2} f'(t) \cos n\omega_0 t \, dt$$

$$b_n' = \frac{2}{T} \int_{-T/2}^{T/2} f'(t) \sin n\omega_0 t \, dt$$

Again, the limits of integration can be changed to any interval constituting one period.

$\cdot \quad \cdot \quad \cdot$

Example

Consider the "sawtooth" waveform shown in Fig. 12.8. Since $T = 1$, then $\omega_0 = 2\pi/T = 2\pi$. Also,

$$a_0 = \frac{1}{T} \int_0^T f(t) \, dt$$

$$= \frac{1}{1} \int_0^1 t \, dt = \frac{t^2}{2} \Big|_0^1 = \frac{1}{2}$$

fig. 12.8

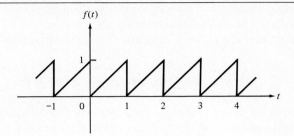

fig. 12.9

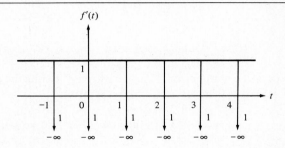

Instead of determining a_n and b_n directly, let us take the derivative of $f(t)$. The result is shown in Fig. 12.9. Thus

$$a'_n = \frac{2}{T} \int_{-T/2}^{T/2} f'(t) \cos n\omega_0 t \, dt$$

$$n\omega_0 b_n = \frac{2}{1} \int_{-1/2}^{1/2} [1 - \delta(t)] \cos n\omega_0 t \, dt$$

$$n(2\pi)b_n = 2 \int_{-1/2}^{1/2} \cos n(2\pi)t \, dt - 2 \int_{-1/2}^{1/2} \delta(t) \cos n(2\pi)t \, dt$$

$$2\pi n b_n = \frac{2}{2\pi n} \sin 2\pi n t \Big|_{-1/2}^{1/2} - 2 \cos n(2\pi)0$$

$$2\pi n b_n = 0 - 2$$

$$b_n = -\frac{1}{n\pi}$$

Also,

$$b'_n = \frac{2}{T} \int_{-T/2}^{T/2} f'(t) \sin n\omega_0 t \, dt$$

and, as above,

$$-2\pi n a_n = \frac{2}{2\pi n} \cos 2\pi n t \bigg|_{-1/2}^{1/2} - 2 \sin n(2\pi)0$$

$$-2\pi n a_n = \frac{2}{2\pi n} [\cos n\pi - \cos(-n\pi)] - 0$$

$$a_n = 0$$

Thus

$$f(t) = \frac{1}{2} + \sum_{n=1}^{\infty} \left(-\frac{1}{n\pi}\right) \sin 2n\pi t$$

or

$$f(t) = \frac{1}{2} - \frac{1}{\pi} \sin 2\pi t - \frac{1}{2\pi} \sin 4\pi t - \frac{1}{3\pi} \sin 6\pi t - \frac{1}{4\pi} \sin 8\pi t - \cdots$$

$$= 0.5 - 0.32 \sin 2\pi t - 0.16 \sin 4\pi t - 0.11 \sin 6\pi t - 0.08 \sin 8\pi t - \cdots$$

The fact that $a_n = 0$ is not surprising for the following reason. Although $f(t)$ is not an odd function, $f_1(t) = f(t) - \frac{1}{2}$ is. Thus, there are no cosine terms in the Fourier series of $f_1(t)$. Consequently, the series representation of $f(t) = f_1(t) + \frac{1}{2}$ also contains no cosine terms; that is, $a_n = 0$.

· · ·

In the example above, the coefficients a_n' and b_n' were determined by using the limits $t = -\frac{1}{2}$ to $t = \frac{1}{2}$. Suppose instead that the limits used went from $t = 0$ to $t = 1$. Under this circumstance, one may be tempted to include the impulse at $t = 0$ and the impulse at $t = 1$ in the integral. However, to verify that this is not the case, simply subtract an arbitrarily small positive quantity, say ε, from both limits. The resulting limits, $t = -\varepsilon$ and $t = 1 - \varepsilon$, indicate that the impulse at $t = 0$ is included in the integral, but the one at $t = 1$ is not.

One additional point. It was indicated that when the formulas (not based on one of the symmetries) were being used, the integral limits were irrelevant provided that the interval was one period. When different periods are used for the same coefficient, however, the results will be apparently different formulas. But although these formulas appear different, they are equivalent; that is, plugging the same value of n in both formulas will yield the same number.

Suppose now that we take the second derivative of the Fourier series representation of $f(t)$. Then

$$f''(t) = \frac{d^2 f(t)}{dt^2} = \sum_{n=1}^{\infty} \left(-(n\omega_0)^2 a_n \cos n\omega_0 t - (n\omega_0)^2 b_n \sin n\omega_0 t\right)$$

But since $f(t)$ is periodic, so is $f''(t)$; and the series representation of $f''(t)$ is

$$f''(t) = a_0'' + \sum_{n=1}^{\infty} (a_n'' \cos n\omega_0 t + b_n'' \cos n\omega_0 t)$$

Comparing the last two equations, and from the formulas for the coefficients we have

$$a_0'' = 0$$

$$-(n\omega_0)^2 a_n = a_n'' = \frac{2}{T} \int_{-T/2}^{T/2} f''(t) \cos n\omega_0 t \, dt$$

and

$$-(n\omega_0)^2 b_n = b_n'' = \frac{2}{T} \int_{-T/2}^{T/2} f''(t) \sin n\omega_0 t \, dt$$

where $\omega_0 = 2\pi/T$ and the integrations can equally be performed over an arbitrary period.

Therefore, knowing the coefficients a_n'' and b_n'' of $f''(t)$, it is a simple matter to determine a_n and b_n, the coefficients of $f(t)$.

This approach can be extended to higher-order derivatives.

· · ·

Example
Consider the triangular waveform shown in Fig. 12.10. Since $T = 2$, then $\omega_0 = 2\pi/T = \pi$. Also,

$$a_0 = \frac{1}{T} \int_{-1/2}^{3/2} f(t) \, dt = \frac{1}{2} \int_{-1/2}^{1/2} 2t \, dt + \frac{1}{2} \int_{1/2}^{3/2} (-2t + 2) \, dt$$

and, after a few lines of computation,

$$a_0 = 0$$

This result can be obtained by inspection when one sees that $f(t)$ is an odd function.

---

fig. 12.10

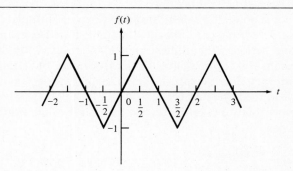

It would take a few pages of messy calculations to determine expressions for a_n and b_n by directly using their formulas—try it and see. However, by taking derivatives, the complexity of the solution can be reduced significantly. First we form $f'(t)$ as shown in Fig. 12.11.

fig. 12.11

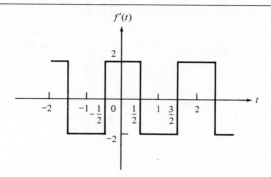

fig. 12.12

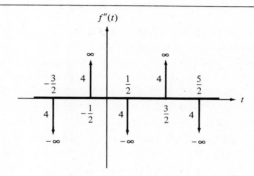

To get $f''(t)$ we take the derivative of $f'(t)$. The result is shown in Fig. 12.12. Since $f''(t)$ is an odd function, then

$$-(n\omega_0)^2 a_n = a_n'' = 0 \Rightarrow a_n = 0$$

which is to be expected since $f(t)$ is also an odd function. Furthermore,

$$b_n'' = \frac{4}{T} \int_0^{T/2} f''(t) \sin n\omega_0 t \, dt$$

$$-(n\omega_0)^2 b_n = \frac{4}{2} \int_0^1 -4\delta\left(t - \frac{1}{2}\right) \sin n\omega_0 t \, dt$$

$$-(n\pi)^2 b_n = 2\left[-4 \sin n\pi\left(\frac{1}{2}\right)\right]$$

$$b_n = \frac{8}{(n\pi)^2} \sin \frac{n\pi}{2}$$

Thus, for the original function,

$$f(t) = \sum_{n=1}^{\infty} \frac{8}{(n\pi)^2} \sin \frac{n\pi}{2} \sin n\pi t$$

$$= \frac{8}{\pi^2} \sum_{n=1}^{\infty} \frac{\sin(n\pi/2)}{n^2} \sin n\pi t$$

Writing the first few terms in the series, we have

$$f(t) = \frac{8}{\pi^2} \sin \pi t - \frac{8}{9\pi^2} \sin 3\pi t + \frac{8}{25\pi^2} \sin 5\pi t - \frac{8}{49\pi^2} \sin 7\pi t + \cdots$$

$$= 0.81 \sin \pi t - 0.09 \sin 3\pi t + 0.032 \sin 5\pi t - 0.017 \sin 7\pi t + \cdots$$

Note that because

$$\sin \frac{n\pi}{2} = 0 \qquad \text{for } n = 2, 4, 6, \ldots$$

$f(t)$ contains no even harmonics. This is a consequence of the fact that $f(t)$, as well as its derivatives, is a half-wave symmetric function.

. . .

Upon inspection of the foregoing example, we see that $f(t)$ is an odd function, its derivative $f'(t)$ is an even function, and its derivative $f''(t)$ is again an odd function. This results from the fact that the derivative of an even periodic function is odd, and vice versa. Since an even periodic function consists only of pure cosines (and possibly a constant), the derivative of such a function will be comprised only of pure sines and thus must be an odd function. Conversely, since an odd periodic function consists only of pure sines, its derivative will be comprised only of pure cosines; thus it must be an even function (with no dc component). Implicit in this argument is the fact that a sum of even functions is again even, and the sum of odd functions is again odd. (See Problem 12.10.)

In the next example we shall see a more subtle use of derivatives for the purpose of obtaining the Fourier series representation of a periodic function.

. . .

Example
Reconsider the half-wave rectified sine wave given by Fig. 12.13. Again, $T = 2 \Rightarrow \omega_0 = 2\pi/T = \pi$. Before, we saw that $a_0 = 1/\pi$. Taking the derivative of

fig. 12.13

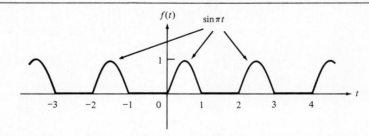

$f(t)$ results in the waveform shown in Fig. 12.14, whereas the second derivative of $f(t)$ is shown in Fig. 12.15. But note—and this is the crux of the matter—that the second derivative $f''(t)$ contains the function $f(t)$. Specifically, for the interval $0 \leq t < 2$, we have

$$f''(t) = \pi\delta(t) + \pi\delta(t-1) - \pi^2 f(t)$$

fig. 12.14

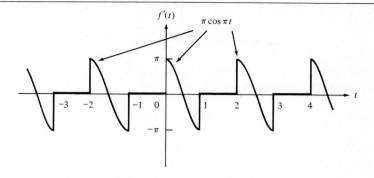

fig. 12.15

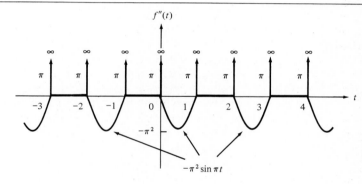

Thus,

$$a_n'' = \frac{2}{T} \int_0^T f''(t) \cos n\omega_0 t \, dt$$

$$-(n\omega_0)^2 a_n = \frac{2}{T} \int_0^T [\pi\delta(t) + \pi\delta(t-1) - \pi^2 f(t)] \cos n\omega_0 t \, dt$$

$$-(n\pi)^2 a_n = \pi \cos n\pi(0) + \pi \cos n\pi(1) + \frac{2}{T} \int_0^T - \pi^2 f(t) \cos n\omega_0 t \, dt$$

$$= \pi(1 + \cos n\pi) - \pi^2 \left(\frac{2}{T} \int_0^T f(t) \cos n\omega_0 t \, dt \right)$$

But

$$\frac{2}{T} \int_0^T f(t) \cos n\omega_0 t \, dt = a_n$$

Hence

$$-(n\pi)^2 a_n = \pi(1 + \cos n\pi) - \pi^2 a_n$$

so

$$\pi^2 a_n - n^2 \pi^2 a_n = \pi(1 + \cos n\pi)$$

and

$$a_n = \frac{1 + \cos n\pi}{\pi(1 - n^2)}$$

as was obtained before for $n = 2, 3, 4, \ldots$. But for $n = 1$, $a_n = 0/0$—which is undefined. By L'Hopital's rule, however,

$$a_1 = \lim_{n \to 1} a_n = \lim_{n \to 1} \frac{\dfrac{d}{dn}(1 + \cos n\pi)}{\dfrac{d}{dn}\pi(1 - n^2)} = \lim_{n \to 1} \frac{-\pi \sin n\pi}{-2n\pi}$$

and

$$a_1 = 0$$

Continuing,

$$b_n'' = \frac{2}{T} \int_0^T f''(t) \sin n\omega_0 t \, dt$$

$$-(n\omega_0)^2 b_n = \frac{2}{T} \int_0^1 [\pi\delta(t) + \pi\delta(t - 1) - \pi^2 f(t)] \sin n\pi t \, dt$$

and, as above,

$$-(n\pi)^2 b_n = \pi \sin n\pi(0) + \pi \sin n\pi(1) - \pi^2 b_n$$

from which

$$\pi^2 b_n - n^2\pi^2 b_n = \pi(0) + \pi \sin n\pi$$

or

$$b_n = \frac{\sin n\pi}{\pi(1 - n^2)}$$

so

$$b_n = 0 \qquad \text{for } n = 2, 3, 4, \ldots$$

When $n = 1$, $b_n = 0/0$, and therefore, by L'Hopital's rule,

$$b_1 = \lim_{n \to 1} b_n = \lim_{n \to 1} \frac{\dfrac{d}{dn} \sin n\pi}{\dfrac{d}{dn}\pi(1 - n^2)} = \lim_{n \to 1} \frac{\pi \cos n\pi}{-2n\pi}$$

and

$$b_1 = \frac{\cos \pi}{-2} = \frac{1}{2}$$

and was obtained previously.

$$\cdot \quad \cdot \quad \cdot$$

PROBLEMS   12.1.   (a) Find the Fourier series representation of the function shown in Fig. P12.1.
(b) Plot the amplitude spectrum for this function.

fig. P12.1

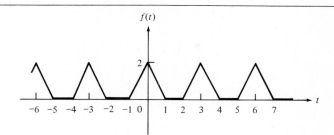

fig. P12.2

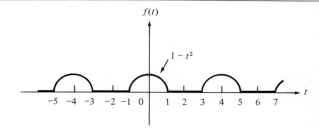

12.2.   Repeat Problem 12.1 for the function shown in Fig. P12.2

fig. P12.3

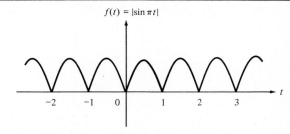

12.3.   Repeat Problem 12.1 for the full-wave rectified sine wave in Fig. P12.3.

fig. P12.4

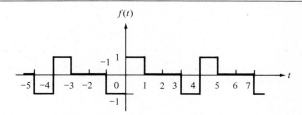

*12.4.   Repeat Problem 12.1 for the function shown in Fig. P12.4.

$$Ans. \quad \frac{2}{\pi} \sin \frac{\pi}{2} t + \frac{2}{\pi} \sin \pi t + \frac{2}{3\pi} \sin \frac{3\pi}{2} t + \frac{2}{5\pi} \sin \frac{5\pi}{2} t + \frac{2}{3\pi} \sin 3\pi t + \cdots$$

fig. P12.5

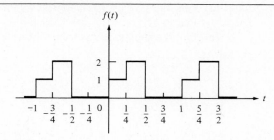

*12.5. Repeat Problem 12.1 for the function shown in Fig. P12.5.

Ans. $\dfrac{3}{4} - \dfrac{1}{\pi} \cos 2\pi t + \dfrac{3}{\pi} \sin 2\pi t - \dfrac{1}{\pi} \sin 4\pi t$

$$+ \dfrac{1}{3\pi} \cos 6\pi t + \dfrac{1}{\pi} \sin 6\pi t + \cdots$$

fig. P12.6

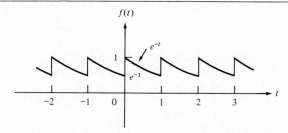

12.6. Repeat Problem 12.1 for the function shown in Fig. P12.6.

fig. P12.7

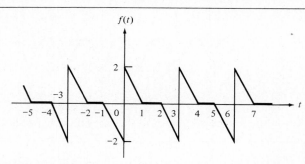

12.7. Repeat Problem 12.1 for the function shown in Fig. P12.7.

fig. P12.8

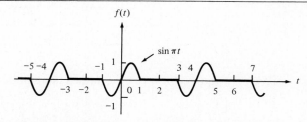

12.8. Repeat Problem 12.1 for the function shown in Fig. P12.8.

fig. P12.9

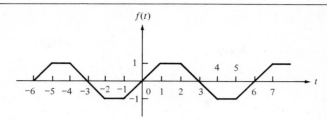

*12.9.   Repeat Problem 12.1 for the function shown in Fig. P12.9.

Ans. $\dfrac{6\sqrt{3}}{\pi^2}\sin\dfrac{\pi}{3}t - \dfrac{6\sqrt{3}}{25\pi^2}\sin\dfrac{5\pi}{3}t$

$$+\dfrac{6\sqrt{3}}{49\pi^2}\sin\dfrac{7\pi}{3}t - \dfrac{6\sqrt{3}}{121\pi^2}\sin\dfrac{11\pi}{3}t + \cdots$$

12.10.   (a)   Show that the sum of two even functions is an even function.
(b)   Show that the sum of two odd functions is an odd function.
(c)   Show that the product of two even functions is an even function.
(d)   Show that the product of two odd functions is an even function.
(e)   Show that the product of an odd function and an even function is an
odd function.

12.11.   Show that the amplitude spectra of the periodic functions $f(t)$ and
$f(t - a)$ are the same.

fig. P12.12

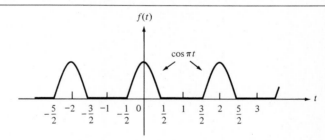

*12.12.   Use the Fourier series of the half-wave rectified sine wave given in
Fig. 12.1 to find the Fourier series representation of the function shown
in Fig. P12.12.
Ans. $0.32 + 0.5\cos\pi t + 0.21\cos 2\pi t - 0.04\cos 4\pi t$
$$+ 0.02\cos 6\pi t - 0.01\cos 8\pi t + \cdots$$

fig. P12.13

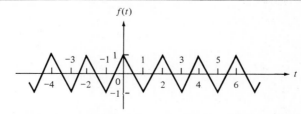

12.13.   Use the Fourier series of the triangular waveform given in Fig. 12.10
to find the Fourier series representation of the function shown in
Fig. P12.13.

fig. P12.14

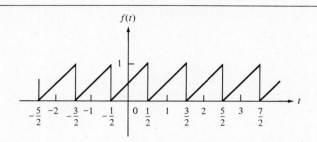

12.14.   Use the Fourier series of the sawtooth waveform given in Fig. 12.8 to find the Fourier series representation of the function shown in Fig. P12.14.

fig. P12.15

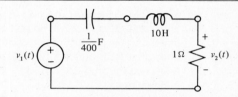

12.15.   Suppose that the half-wave rectified sine wave given in Fig. 12.1 is the input voltage $v_1(t)$ in the circuit shown in Fig. P12.15.
(a)   Find the amplitude spectrum of the output voltage $v_2(t)$.
(b)   The output $v_2(t)$ is approximately what function?
12.16.   Repeat Problem 12.15 for the case that $v_1(t)$ is the function given in Fig. 12.8.
12.17.   Repeat Problem 12.15 for the case that $v_1(t)$ is the function given in Problem 12.5.

12.2   THE COMPLEX FOURIER SERIES

Although the trigonometric Fourier series is a relatively intuitive concept (we can picture sines and cosines in our minds), by using an alternative series representation known as the *exponential* or *complex* Fourier series we can (1) mathematically shorten the series representation and (2) generalize the concept to the Fourier transform, which will enable us to deal with nonperiodic functions.

The key equation in the development of the complex Fourier series is Euler's formula. From

$$e^{j\theta} = \cos\theta + j\sin\theta \tag{12.1}$$

replacing θ by $-\theta$, we get

$$e^{-j\theta} = \cos\theta - j\sin\theta \tag{12.2}$$

Adding Equations (12.1) and (12.2), we obtain

$$\cos\theta = \frac{e^{j\theta} + e^{-j\theta}}{2} \tag{12.3}$$

On the other hand, subtracting Equation (12.2) from Equation (12.1) results in

$$\sin \theta = \frac{e^{j\theta} - e^{-j\theta}}{2j}$$

(12.4)

Suppose now that $f(t)$ is a periodic function having period T and the trigonometric Fourier series representation given by

$$f(t) = a_0 + \sum_{n=1}^{\infty} (a_n \cos n\omega_0 t + b_n \sin n\omega_0 t)$$

where $\omega_0 = 2\pi/T$. Then, by Equations (12.3) and (12.4),

$$f(t) = a_0 + \sum_{n=1}^{\infty} \left[a_n \left(\frac{e^{jn\omega_0 t} + e^{-jn\omega_0 t}}{2} \right) + b_n \left(\frac{e^{jn\omega_0 t} - e^{-jn\omega_0 t}}{2j} \right) \right]$$

$$= a_0 + \sum_{n=1}^{\infty} \left[\left(\frac{a_n}{2} + \frac{b_n}{2j} \right) e^{jn\omega_0 t} + \left(\frac{a_n}{2} - \frac{b_n}{2j} \right) e^{-jn\omega_0 t} \right]$$

$$= a_0 + \sum_{n=1}^{\infty} \frac{a_n - jb_n}{2} e^{jn\omega_0 t} + \sum_{n=1}^{\infty} \frac{a_n + jb_n}{2} e^{-jn\omega_0 t}$$

If we define $m = -n$, then $n = -m$ and

$$f(t) = a_0 + \sum_{n=1}^{\infty} \frac{a_n - jb_n}{2} e^{jn\omega_0 t} + \sum_{-m=1}^{\infty} \frac{a_{-m} + jb_{-m}}{2} e^{jm\omega_0 t}$$

But, from the formulas for the Fourier coefficients,

$$a_{-m} = \frac{2}{T} \int_0^T f(t) \cos(-m)\omega_0 t \, dt$$

$$= \frac{2}{T} \int_0^T f(t) \cos m\omega_0 t \, dt$$

$$= a_m$$

Similarly,

$$b_{-m} = -b_m$$

But

$$\sum_{-m=1}^{\infty} = \sum_{m=-1}^{-\infty} = \sum_{m=-\infty}^{-1}$$

Thus

$$f(t) = a_0 + \sum_{n=1}^{\infty} \frac{a_n - jb_n}{2} e^{jn\omega_0 t} + \sum_{m=-\infty}^{-1} \frac{a_m - jb_m}{2} e^{jm\omega_0 t}$$

$$= a_0 + \sum_{n=1}^{\infty} \frac{a_n - jb_n}{2} e^{jn\omega_0 t} + \sum_{n=-\infty}^{-1} \frac{a_n - jb_n}{2} e^{jn\omega_0 t}$$

Now define the coefficient c_n by

$$c_n = \frac{a_n - jb_n}{2} \qquad \text{for all } n \neq 0$$

and

$$c_0 = a_0$$

Then

$$f(t) = \sum_{n=-\infty}^{\infty} c_n e^{jn\omega_0 t}$$

where $\omega_0 = 2\pi/T$. This is the complex Fourier series representation of $f(t)$. In other words, we can express $f(t)$ as a sum of complex exponentials. Note the simple form of this series representation. But, if we first have to calculate a_n and b_n in order to find c_n, this representation would not be any simpler to determine. Fortunately, we can derive a formula for obtaining c_n directly. This is done as follows.

From the definition of c_n,

$$c_n = \frac{1}{2}(a_n - jb_n)$$

$$= \frac{1}{2}\left(\frac{2}{T}\int_0^T f(t)\cos n\omega_0 t \, dt - j\frac{2}{T}\int_0^T f(t)\sin n\omega_0 t \, dt\right)$$

$$= \frac{1}{2}\left(\frac{2}{T}\int_0^T f(t)(\cos n\omega_0 t - j\sin n\omega_0 t)\,dt\right)$$

By Euler's formula,

$$c_n = \frac{1}{T}\int_0^T f(t)e^{-jn\omega_0 t}\,dt$$

which is a formula for c_n. Note that if $n = 0$, we obtain

$$c_0 = \frac{1}{T}\int_0^T f(t)e^0\,dt = \frac{1}{T}\int_0^T f(t)\,dt = a_0$$

and thus the formula for c_n is valid for all n. Again, in the formula for c_n the integration can be performed over any period, not only from $t = 0$ to $t = T$; different periods will result in equivalent expressions for c_n.

For the trigonometric Fourier series, a_n and b_n are coefficients of real sinusoids—respectively, of cosines and sines having nonnegative frequencies. The nth harmonic component ($n \neq 0$) of $f(t)$ is $\sqrt{a_n^2 + b_n^2}$, and a plot of $\sqrt{a_n^2 + b_n^2}$ along with $|a_0|$ is the (real) amplitude spectrum of $f(t)$. On the other hand, for the complex Fourier series, c_n is the coefficient of a complex exponential—called a *complex sinusoid* (recall Euler's formula). Furthermore, since n takes on values from $-\infty$ to ∞, these complex sinusoids have negative frequencies as well as nonnegative frequencies.

Since for $n \neq 0$, $c_n = \frac{1}{2}(a_n - jb_n)$, then

$$|c_n| = \frac{1}{2}\sqrt{a_n^2 + b_n^2}$$

Thus, we see that the magnitude of c_n is equal to half of the nth harmonic component of $f(t)$. In addition,

$$c_{-n} = \frac{1}{2}(a_{-n} - jb_{-n})$$

But, we have seen that $a_{-n} = a_n$ and $b_{-n} = -b_n$, so

$$c_{-n} = \frac{1}{2}(a_n + jb_n) = c_n{}^*$$

Therefore

$$|c_{-n}| = \frac{1}{2}\sqrt{a_n^2 + b_n^2} = |c_n|$$

which accounts for the other half of the nth harmonic component of $f(t)$. In other words, $|c_n| + |c_{-n}| = \sqrt{a_n^2 + b_n^2}$. Also, since $c_0 = a_0$, then $|c_0| = |a_0|$.

For these reasons, a plot of $|c_n|$ versus ω is called the *(complex) amplitude spectrum* of $f(t)$. A plot of the angle of c_n versus ω is called the *phase spectrum* of $f(t)$. These two spectra constitute the *frequency spectrum* of $f(t)$.

• • •

Example
Let us find the complex Fourier series of the pulse train given in Fig. 12.16, where the period is T and $\omega_0 = 2\pi/T$.

fig. 12.16

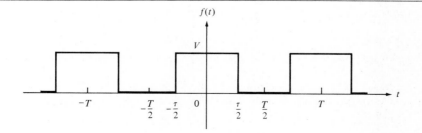

From the formula for c_n,

$$c_n = \frac{1}{T} \int_{-T/2}^{T/2} f(t)e^{-jn\omega_0 t}\,dt$$

$$= \frac{1}{T} \int_{-\tau/2}^{\tau/2} Ve^{-jn\omega_0 t}\,dt = \frac{V}{T} \frac{-e^{-jn\omega_0 t}}{jn\omega_0}\bigg|_{-\tau/2}^{\tau/2}$$

$$= \frac{V}{jn\omega_0 T}(e^{jn\omega_0\tau/2} - e^{-jn\omega_0\tau/2})$$

$$= \frac{V\tau}{T} \frac{\sin(n\omega_0\tau/2)}{n\omega_0\tau/2} = \frac{V\tau}{T} \frac{\sin(n\pi\tau/T)}{n\pi\tau/T}$$

Thus,

$$f(t) = \sum_{n=-\infty}^{\infty} \frac{V\tau}{T} \frac{\sin(n\pi\tau/T)}{n\pi\tau/T} e^{j2\pi nt/T}$$

In determining the amplitude spectrum of $f(t)$, we use the fact that c_n is the product of a constant and a function of the form $(\sin x)/x$. A plot of this well-known function is shown in Fig. 12.17. As a consequence, the amplitude spectrum of $f(t)$ is given by Fig. 12.18. Note that the pulse width τ determines the sharpness of the $(\sin x)/x$ envelope of the amplitude spectrum—the narrower

fig. 12.17

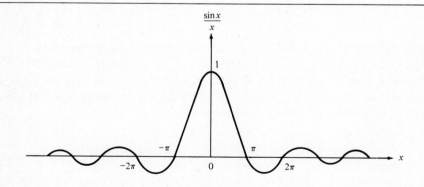

fig. 12.18

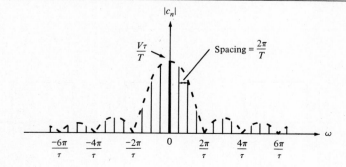

the pulse width, the broader the envelope and hence the more harmonics of large magnitudes comprise $f(t)$. Also, note that if the period increases (pulse repetition rate decreases), then the spacing between the harmonics decreases. In the limit as the period approaches infinity (the repetition rate goes to zero), the discrete spectrum becomes a continuous spectrum.

. . .

If $f(t)$ is a periodic function having period T with a complex Fourier series representation

$$f(t) = \sum_{n=-\infty}^{\infty} c_n e^{jn\omega_0 t}$$

where $\dot{\omega}_0 = 2\pi/T$ and

$$c_n = \frac{1}{T} \int_{-T/2}^{T/2} f(t) e^{-jn\omega_0 t} \, dt$$

then, taking the derivative of $f(t)$, we get

$$f'(t) = \frac{df(t)}{dt} = \sum_{n=-\infty}^{\infty} jn\omega_0 c_n e^{jn\omega_0 t}$$

But, since $f'(t)$ is also periodic, with period T, it has a complex Fourier series representation

$$f'(t) = \sum_{n=-\infty}^{\infty} c'_n e^{jn\omega_0 t}$$

Comparing this equation with the one preceding it, and by the formula for the complex Fourier series coefficients, we have

$$jn\omega_0 c_n = c'_n = \frac{1}{T} \int_{-T/2}^{T/2} f'(t) e^{-jn\omega_0 t} \, dt$$

Thus, knowing c'_n, it is a simple matter to determine c_n.

Continuing to the second derivative,

$$f''(t) = \frac{d^2 f(t)}{dt^2} = \sum_{n=-\infty}^{\infty} - (n\omega_0)^2 c_n e^{jn\omega_0 t}$$

and

$$f''(t) = \sum_{n=-\infty}^{\infty} c''_n e^{jn\omega_0 t}$$

implies that

$$-(n\omega_0)^2 c_n = c''_n = \frac{1}{T} \int_{-T/2}^{T/2} f''(t) e^{-jn\omega_0 t} \, dt$$

Thus, from c_n'' we can easily obtain c_n.

Of course, this idea can be extended to higher-order derivatives, and an arbitrary period can be used for the limits of the integrals.

$$\cdot \quad \cdot \quad \cdot$$

Example
For the triangular waveform given in a previous example (Fig. 12.10),

$$c_n'' = \frac{1}{T} \int_{-T/2}^{T/2} f''(t) e^{-jn\omega_0 t} \, dt$$

$$-(n\omega_0)^2 c_n = \frac{1}{2} \int_{-1}^{1} \left[4\delta\left(t + \frac{1}{2}\right) - 4\delta\left(t - \frac{1}{2}\right) \right] e^{-jn\omega_0 t} \, dt$$

$$-(n\pi)^2 c_n = \frac{4e^{jn\pi/2} - 4e^{-jn\pi/2}}{2}$$

or

$$c_n = -\frac{4j}{(n\pi)^2} \frac{e^{jn\pi/2} - e^{-jn\pi/2}}{2j}$$

and, by Equation (12.4),

$$c_n = -\frac{4j}{(n\pi)^2} \sin \frac{n\pi}{2}$$

Since

$$c_n = \frac{a_n - jb_n}{2}$$

we have

$$a_n = 0$$

and

$$\frac{b_n}{2} = \frac{4}{(n\pi)^2} \sin \frac{n\pi}{2} \qquad \text{or} \qquad b_n = \frac{8}{(n\pi)^2} \sin \frac{n\pi}{2}$$

which agrees with the answer obtained previously.

Note that by L'Hopital's rule,

$$a_0 = c_0 = \lim_{n \to 0} \frac{-4j \sin \dfrac{n\pi}{2}}{(n\pi)^2} = \lim_{n \to 0} \frac{-j2\pi \cos \dfrac{n\pi}{2}}{2n\pi^2}$$

$$= \lim_{n \to 0} \frac{j\pi^2 \sin \dfrac{n\pi}{2}}{2\pi^2} = 0$$

$$\cdot \quad \cdot \quad \cdot$$

PROBLEMS *12.18. Use Equation (12.3) to determine the complex Fourier series representation of $f(t) = 5 \cos 7t$. Plot the amplitude spectrum of $f(t)$.

Ans. $\frac{5}{2}e^{-j7t} + \frac{5}{2}e^{j7t}$

12.19. Repeat Problem 12.18 for $f(t) = 5 \sin 7t$, using Equation (12.4).

12.20. Repeat Problem 12.18 for $f(t) = 5 \cos(7t - \pi/2)$.

12.21. Suppose that $f(t)$ is an even periodic function. Show that

$$c_n = \frac{2}{T} \int_0^{T/2} f(t) \cos n\omega_0 t \, dt$$

12.22. Suppose that $f(t)$ is an odd periodic function. Show that

$$c_n = \frac{-j2}{T} \int_0^{T/2} f(t) \sin n\omega_0 t \, dt$$

12.23. Find the complex Fourier series for the function given in Problem 12.1 in the preceding problem set. (Do not use a_n and b_n determined previously to calculate c_n.)

12.24. Repeat Problem 12.23 for the function given in Problem 12.2.

*12.25. Repeat Problem 12.23 for the function given in Problem 12.3.

Ans. $\dfrac{2}{\pi} - \dfrac{2}{3\pi}e^{-j2\pi t} - \dfrac{2}{3\pi}e^{j2\pi t} - \dfrac{2}{15\pi}e^{-j4\pi t} - \dfrac{2}{15\pi}e^{j4\pi t} - \cdots$

*12.26. Repeat Problem 12.23 for the function given in Problem 12.4.

Ans. $\dfrac{j}{\pi}e^{-j\pi t/2} - \dfrac{j}{\pi}e^{j\pi t/2} + \dfrac{j}{\pi}e^{-j\pi t} - \dfrac{j}{\pi}e^{j\pi t} + \cdots$

12.27. Repeat Problem 12.23 for the function given in Problem 12.5.

*12.28. Repeat Problem 12.23 for the function given in Problem 12.6.

Ans. $(1 - e^{-1}) + \dfrac{1 - e^{-1}}{1 - j2\pi}e^{-j2\pi t} + \dfrac{1 - e^{-1}}{1 + j2\pi}e^{j2\pi t}$

$\qquad\qquad + \dfrac{1 - e^{-1}}{1 - j4\pi}e^{-j4\pi t} + \dfrac{1 - e^{-1}}{1 + j4\pi}e^{j4\pi t} + \cdots$

12.29. Repeat Problem 12.23 for the function given in Problem 12.7.

12.30. Repeat Problem 12.23 for the function given in Problem 12.8.

Summary

1. A periodic function $f(t)$ with period T has the trigonometric Fourier series representation

$$f(t) = a_0 + \sum_{n=1}^{\infty} (a_n \cos n\omega_0 t + b_n \sin n\omega_0 t)$$

where $\omega_0 = 2\pi/T$.

2. An alternative trigonometric Fourier series representation is

$$f(t) = a_0 + \sum_{n=1}^{\infty} A_n \cos(n\omega_0 t - \phi_n)$$

where $A_n = \sqrt{a_n^2 + b_n^2}$ and $\phi_n = \tan^{-1}(b_n/a_n)$.

3. The frequency spectrum of $f(t)$ consists of two discrete spectra. The amplitude spectrum is a plot of A_n (along with a_0) versus frequency. The phase spectrum is a plot of ϕ_n versus frequency.

4. A circuit's (forced) response to a periodic function $f(t)$ is found by expressing $f(t)$ as a sum of sinusoids (its Fourier series representation), finding the response to each individual sinusoid, and then summing all the individual responses.

5. The trigonometric Fourier series of an even periodic function contains no sine terms. The Fourier series of an odd function contains no cosine terms and has an average value of zero. A half-wave symmetric function contains only odd harmonics. (The dc term is the zero harmonic and thus is even.)

6. The derivative $f'(t)$ of a function $f(t)$ having period T also has period T. The Fourier coefficients of $f'(t)$ are often easier to calculate than the coefficients of $f(t)$. If we know the coefficients of $f'(t)$, it is extremely easy to determine the coefficients of $f(t)$—except for a_0.

7. A periodic function can also be expressed as a sum of complex sinusoids: the exponential Fourier series representation is

$$f(t) = \sum_{n=-\infty}^{\infty} c_n e^{jn\omega_0 t}$$

where $\omega_0 = 2\pi/T$.

THE FOURIER TRANSFORM

Introduction

We have seen that for a periodic function $f(t)$ having period T, we can represent $f(t)$ as a sum of sinusoids—for real sinusoids we have the trigonometric Fourier series, and for complex sinusoids we have the exponential Fourier series.

Although periodic functions are very important, so are nonperiodic functions. We can extend the concept of a series representation of $f(t)$ to the case of a nonperiodic function by considering such a function to be periodic with period $T = \infty$.

We begin this chapter with the heuristic development of the Fourier transform pair of integrals. One of these, the Fourier transform, takes a function of time $f(t)$ and converts or transforms it into a function of frequency $\mathbf{F}(j\omega)$. The other, the inverse Fourier transform, takes the function of frequency $\mathbf{F}(j\omega)$ and transforms it back into the function of time $f(t)$. We see that both of these transforms are linear transformations.

We then determine the Fourier transforms of a number of important functions of time and, as in the case of Fourier series, may utilize the operation of time differentiation to simplify the computations required to find transforms. We also derive Parseval's theorem, which gives us an integral equation relating $f(t)$ and $\mathbf{F}(j\omega)$. It is this relation that gives us an important physical interpretation in terms of energy of the Fourier transform. More important, we see how to use the Fourier transform to describe and analyze systems and circuits. By transforming circuits to the frequency domain (i.e., taking Fourier transforms) we get a circuit analysis problem that can be handled algebraically, as was done for the case of resistive circuits and sinusoidal (phasor approach) circuits. The result obtained is an expression in the frequency domain. By determining the inverse Fourier transform we get the corresponding function in the time domain.

13.1   THE FOURIER INTEGRAL

To begin with, let $f(t)$ be periodic, with period T. Then the complex Fourier series representation is

$$f(t) = \sum_{n=-\infty}^{\infty} c_n e^{jn\omega_0 t}$$

where $\omega_0 = 2\pi/T$ and

$$c_n = \frac{1}{T} \int_{-T/2}^{T/2} f(t)e^{-jn\omega_0 t}\, dt = \frac{1}{T} \int_{-T/2}^{T/2} f(x)e^{-jn\omega_0 x}\, dx$$

Substituting this into the series representation of $f(t)$ yields

$$f(t) = \sum_{n=-\infty}^{\infty} \left[\frac{1}{T} \int_{-T/2}^{T/2} f(x)e^{-jn\omega_0 x}\, dx \right] e^{jn\omega_0 t}$$

Since $\omega_0 = 2\pi/T$, then $1/T = \omega_0/2\pi$ and

$$f(t) = \sum_{n=-\infty}^{\infty} \left[\frac{1}{2\pi} \int_{-T/2}^{T/2} f(x)e^{-jn\omega_0 x}\, dx \right] e^{jn\omega_0 t} \omega_0$$

If we let T get large, then the quantity $\Delta\omega$ defined by $\omega_0 = 2\pi/T = \Delta\omega$ gets small, and

$$f(t) = \sum_{n=-\infty}^{\infty} \left[\frac{1}{2\pi} \int_{-T/2}^{T/2} f(x)e^{-jn\,\Delta\omega x}\, dx \right] e^{jn\,\Delta\omega t}\, \Delta\omega$$

As $T \to \infty$, then $\Delta\omega \to d\omega$ and the discrete harmonics $n\omega_0$ become a continuous frequency variable, say ω; that is, $n\omega_0 = n\,\Delta\omega \to \omega$. Furthermore, the discrete sum $\sum_{n=-\infty}^{\infty}$ becomes the continuous sum $\int_{-\infty}^{\infty}$. Thus, in the limit, the previous equation becomes

$$f(t) = \int_{-\infty}^{\infty} \frac{1}{2\pi} \left[\int_{-\infty}^{\infty} f(x)e^{-j\omega x}\, dx \right] e^{j\omega t}\, d\omega$$

and, by changing variables,

$$f(t) = \frac{1}{2\pi} \int_{-\infty}^{\infty} \left[\int_{-\infty}^{\infty} f(t)e^{-j\omega t}\, dt \right] e^{j\omega t}\, d\omega$$

which has the form

$$f(t) = \frac{1}{2\pi} \int_{-\infty}^{\infty} \mathbf{F}(j\omega)e^{j\omega t}\, d\omega \qquad (13.1)$$

where

$$\mathbf{F}(j\omega) = \int_{-\infty}^{\infty} f(t)e^{-j\omega t}\, dt \tag{13.2}$$

The function $\mathbf{F}(j\omega)$ is called the *Fourier transform* of $f(t)$. The Fourier transform of the function $f(t)$ is often denoted $\mathscr{F}[f(t)]$; that is,

$$\mathscr{F}[f(t)] = \mathbf{F}(j\omega)$$

Equation (13.1) is a formula for obtaining $f(t)$ from the transform $\mathbf{F}(j\omega)$. Thus, this integral is called the *inverse Fourier transform* of $\mathbf{F}(j\omega)$, denoted $\mathscr{F}^{-1}[\mathbf{F}(j\omega)]$; that is,

$$\mathscr{F}^{-1}[\mathbf{F}(j\omega)] = f(t)$$

Equations (13.1) and (13.2) are known as the *Fourier transform pair*.

· · ·

Example
The function $f(t)$ shown in Fig. 13.1 consists of a single pulse having width τ and height V. From the Fourier transform formula,

$$
\begin{aligned}
\mathbf{F}(j\omega) &= \int_{-\infty}^{\infty} f(t)e^{-j\omega t}\, dt \\[4pt]
&= \int_{-\tau/2}^{\tau/2} Ve^{-j\omega t}\, dt \\[4pt]
&= -\frac{V}{j\omega}\, e^{-j\omega t}\Big|_{-\tau/2}^{\tau/2} \\[4pt]
&= \frac{2V}{\omega}\left[\frac{e^{j\omega\tau/2} - e^{-j\omega\tau/2}}{2j}\right] = \frac{2V}{\omega}\left(\frac{\omega\tau}{2}\right)\left(\frac{\sin(\omega\tau/2)}{\omega\tau/2}\right) \\[4pt]
&= V\tau\,\frac{\sin(\omega\tau/2)}{\omega\tau/2}
\end{aligned}
$$

fig. 13.1

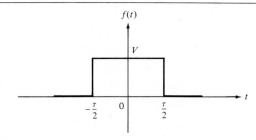

· · ·

We have determined the frequency spectrum for a train of pulses of width τ and height V and found that, as the period of this pulse train approaches infinity, the frequency spectrum becomes continuous.

For the case of Fourier transforms, the *amplitude spectrum* of $f(t)$ is the plot of $|\mathbf{F}(j\omega)|$ versus ω, and the *phase spectrum* is the plot of the angle of $\mathbf{F}(j\omega)$ versus frequency. These two spectra constitute the *frequency spectrum* of $f(t)$. Quite often the amplitude spectrum alone is referred to as the frequency spectrum.

In the previous example, the Fourier transform $\mathbf{F}(j\omega)$ is the product of a constant and a function of the form $(\sin x)/x$, and the (amplitude portion of the) frequency spectrum is shown in Fig. 13.2. For the case of a periodic function, the discrete amplitude spectrum signifies the amplitudes (for the trigonometric case) or the magnitudes (for the complex case) of the sinusoids comprising the function. However, for a nonperiodic function, the frequency spectrum is continuous and no longer signifies the amplitudes or magnitudes of sinusoids.

fig. 13.2

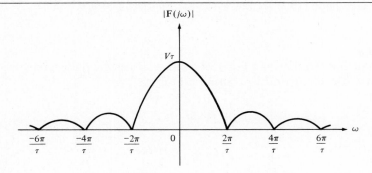

In order to obtain an interpretation of the frequency spectrum, we proceed as follows: From

$$\mathbf{F}(j\omega) = \int_{-\infty}^{\infty} f(t)e^{-j\omega t}\, dt$$

by Euler's formula,

$$\mathbf{F}(j\omega) = \int_{-\infty}^{\infty} f(t)(\cos \omega t - j \sin \omega t)dt$$

$$= \int_{-\infty}^{\infty} f(t) \cos \omega t\, dt - j \int_{-\infty}^{\infty} f(t) \sin \omega t\, dt$$

Replacing ω by $-\omega$, we get

$$\mathbf{F}(-j\omega) = \int_{-\infty}^{\infty} f(t) \cos(-\omega)t\, dt - j \int_{-\infty}^{\infty} f(t) \sin(-\omega)t\, dt$$

$$= \int_{-\infty}^{\infty} f(t) \cos \omega t\, dt + j \int_{-\infty}^{\infty} f(t) \sin \omega t\, dt$$

and for real functions $f(t)$,

$$\mathbf{F}(-j\omega) = \mathbf{F}^*(j\omega)$$

where $\mathbf{F}^*(j\omega)$ is the complex conjugate of $\mathbf{F}(j\omega)$.

We also obtain

$$\int_{-\infty}^{\infty} f^2(t)\, dt = \int_{-\infty}^{\infty} f(t)[f(t)]\, dt$$

$$= \int_{-\infty}^{\infty} f(t)\left[\frac{1}{2\pi}\int_{-\infty}^{\infty} \mathbf{F}(j\omega)e^{j\omega t}\, d\omega\right] dt$$

$$= \frac{1}{2\pi}\int_{-\infty}^{\infty}\int_{-\infty}^{\infty} f(t)\mathbf{F}(j\omega)e^{j\omega t}\, d\omega\, dt$$

$$= \frac{1}{2\pi}\int_{-\infty}^{\infty} \mathbf{F}(j\omega)\left[\int_{-\infty}^{\infty} f(t)e^{j\omega t}\, dt\right] d\omega$$

$$= \frac{1}{2\pi}\int_{-\infty}^{\infty} \mathbf{F}(j\omega)\left[\int_{-\infty}^{\infty} f(t)e^{-j(-\omega)t}\, dt\right] d\omega$$

$$= \frac{1}{2\pi}\int_{-\infty}^{\infty} \mathbf{F}(j\omega)\mathbf{F}(-j\omega)\, d\omega$$

$$= \frac{1}{2\pi}\int_{-\infty}^{\infty} \mathbf{F}(j\omega)\mathbf{F}^*(j\omega)\, d\omega$$

But if $c = a + jb$, then

$$cc^* = (a + jb)(a - jb) = a^2 + b^2 = |c|^2$$

Thus

$$\boxed{\int_{-\infty}^{\infty} f^2(t)\, dt = \frac{1}{2\pi}\int_{-\infty}^{\infty} |\mathbf{F}(j\omega)|^2\, d\omega}$$

This result is known as *Parseval's theorem.*

If $f(t)$ is the voltage across or current through a 1-Ω resistor, then

$$\int_{-\infty}^{\infty} f^2(t)\, dt$$

is the energy absorbed by the resistor. Thus, we can think of $|\mathbf{F}(j\omega)|^2$—the square of the amplitude spectrum of $f(t)$—as the energy density of $f(t)$.

13.2   TRANSFORMS OF IMPORTANT FUNCTIONS

Before we apply Fourier transform techniques to electric circuits, let us determine the transforms of various important functions.

Given a unit impulse function $\delta(t - a)$, then

$$\mathscr{F}[\delta(t - a)] = \int_{-\infty}^{\infty} \delta(t - a)e^{-j\omega t}\, dt$$

and, by the sampling property of the delta function,

$$\mathscr{F}[\delta(t - a)] = e^{-j\omega a}$$

For the special case $a = 0$,

$$\mathscr{F}[\delta(t)] = 1$$

The amplitude spectrum of the delta function is obtained from

$$|\mathscr{F}[\delta(t - a)]| = |e^{-j\omega a}|$$
$$= |\cos \omega a - j \sin \omega a|$$
$$= \sqrt{\cos^2 \omega a + \sin^2 \omega a} = 1$$

Since the amplitude spectrum (and hence, energy density) of a delta function is constant, the energy content of a delta function is infinite! This explains why impulse functions cannot be physically realized.

Since

$$\mathscr{F}[\delta(t - a)] = e^{-j\omega a}$$

then

$$\mathscr{F}^{-1}[e^{-j\omega a}] = \delta(t - a)$$

and, from the inverse Fourier transform formula [Equation (13.1)],

$$\frac{1}{2\pi}\int_{-\infty}^{\infty} e^{-j\omega a}e^{j\omega t}\, d\omega = \delta(t - a)$$

$$\frac{1}{2\pi}\int_{-\infty}^{\infty} e^{j\omega(t - a)}\, d\omega = \delta(t - a)$$

Thus

$$\int_{-\infty}^{\infty} e^{j\omega(t - a)}\, d\omega = 2\pi\delta(t - a)$$

For the case $a = 0$,

$$\int_{-\infty}^{\infty} e^{j\omega t}\, d\omega = 2\pi\delta(t)$$

Interchanging variables, we also have

$$\int_{-\infty}^{\infty} e^{j\omega t}\, dt = 2\pi\delta(\omega)$$

We use this result to determine the Fourier transform of a complex sinusoid of frequency ω_1 as follows:

$$\mathscr{F}[e^{j\omega_1 t}] = \int_{-\infty}^{\infty} e^{j\omega_1 t} e^{-j\omega t}\, dt$$

$$= \int_{-\infty}^{\infty} e^{j(\omega_1 - \omega)t}\, dt$$

$$= 2\pi\delta(\omega_1 - \omega)$$

But since a delta function is zero everywhere except where its argument is zero, $\delta(\omega_1 - \omega) = \delta(\omega - \omega_1)$. Thus

$$\mathscr{F}[e^{j\omega_1 t}] = 2\pi\delta(\omega - \omega_1)$$

The amplitude spectrum of $e^{j\omega_1 t}$ is therefore as shown in Fig. 13.3. By setting $\omega_1 = 0$, we have the special case

$$\mathscr{F}[e^{j0}] = 2\pi\delta(\omega - 0)$$

fig. 13.3

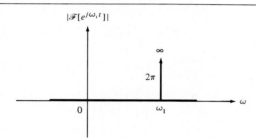

or

$$\mathscr{F}[1] = 2\pi\delta(\omega)$$

Suppose that $f(t) = f_1(t) + f_2(t)$. Then

$$\mathscr{F}[f(t)] = \int_{-\infty}^{\infty} f(t)e^{-j\omega t}\, dt$$

$$= \int_{-\infty}^{\infty} [f_1(t) + f_2(t)]e^{-j\omega t}\, dt$$

$$= \int_{-\infty}^{\infty} f_1(t)e^{-j\omega t}\, dt + \int_{-\infty}^{\infty} f_2(t)e^{-j\omega t}\, dt$$

$$= \mathscr{F}[f_1(t)] + \mathscr{F}[f_2(t)]$$

Also,

$$\mathcal{F}[Kf(t)] = \int_{-\infty}^{\infty} Kf(t)e^{-j\omega t}\, dt$$

$$= K \int_{-\infty}^{\infty} f(t)e^{-j\omega t}\, dt$$

$$= K\mathcal{F}[f(t)]$$

Because of these two properties, we say that the Fourier transform is a *linear transform.* This result is based on the fact that integration is a *linear operation*; that is,

$$\int [f_1(x) + f_2(x)]\, dx = \int f_1(x)\, dx + \int f_2(x)\, dx$$

and

$$\int Kf(x)\, dx = K \int f(x)\, dx$$

By the linearity of the Fourier transform,

$$\mathcal{F}[K] = 2\pi K\delta(\omega)$$

Now we can use linearity to determine the Fourier transform of a real sinusoid of frequency ω_1. By Equation (12.3)

$$\mathcal{F}[\cos \omega_1 t] = \mathcal{F}\left[\frac{e^{j\omega_1 t} + e^{-j\omega_1 t}}{2}\right]$$

$$= \frac{1}{2}\,\mathcal{F}[e^{j\omega_1 t}] + \frac{1}{2}\,\mathcal{F}[e^{-j\omega_1 t}]$$

$$= \pi\delta(\omega - \omega_1) + \pi\delta(\omega + \omega_1)$$

The amplitude spectrum of $\cos \omega_1 t$ is shown in Fig. 13.4.

 fig. 13.4

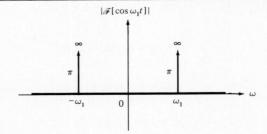

To find the Fourier transform of $e^{-at}u(t)$, where $a > 0$, we can proceed as follows:

$$\mathscr{F}[e^{-at}u(t)] = \int_{-\infty}^{\infty} e^{-at}u(t)e^{-j\omega t}\, dt$$

$$= \int_{0}^{\infty} e^{-(j\omega + a)t}\, dt$$

$$= \frac{-1}{j\omega + a}\, e^{-j\omega t}e^{-at}\,\Big|_{0}^{\infty}$$

Now

$$\lim_{t \to \infty} e^{-j\omega t} = \lim_{t \to \infty}(\cos \omega t - j \sin \omega t)$$

But $\cos \omega t$ and $\sin \omega t$ have no limit as $t \to \infty$. However, since

$$-1 \le \cos \omega t \le 1$$

$$-1 \le \sin \omega t \le 1$$

then $e^{-j\omega t}$ is a complex number whose magnitude is bounded (equal to unity). Thus, since

$$\lim_{t \to \infty} e^{-at} = 0$$

then

$$\mathscr{F}[e^{-at}u(t)] = \frac{-1}{j\omega + a}\,[0 - 1] = \frac{1}{j\omega + a}$$

$$\cdot \quad \cdot \quad \cdot$$

Example
Suppose that $v(t) = e^{-at}u(t)$ is the voltage across a 1-Ω resistor. Then the total energy absorbed by the resistor—that is, the energy content of the function $v(t)$—is

$$\int_{-\infty}^{\infty} v^2(t)\, dt = \int_{-\infty}^{\infty} [e^{-at}u(t)]^2\, dt$$

$$= \int_{-\infty}^{\infty} e^{-2at}u(t)\, dt$$

$$= \int_{0}^{\infty} e^{-2at}\, dt$$

$$= \frac{-1}{2a}\, e^{-2at}\,\Big|_{0}^{\infty}$$

$$= -\frac{1}{2a}\,(0 - 1) = \frac{1}{2a} \qquad \text{joules}$$

A sketch of the amplitude spectrum of $v(t)$ is shown in Fig. 13.5.

fig. 13.5

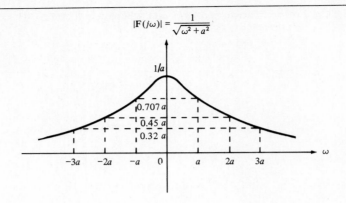

To confirm Parseval's theorem for this specific example, we have

$$\frac{1}{2\pi} \int_{-\infty}^{\infty} |\mathbf{F}(j\omega)|^2 \, d\omega = \frac{1}{2\pi} \int_{-\infty}^{\infty} \frac{1}{\omega^2 + a^2} \, d\omega$$

and, from a table of integrals,

$$\frac{1}{2\pi} \int_{-\infty}^{\infty} |\mathbf{F}(j\omega)|^2 \, d\omega = \frac{1}{2a\pi} \tan^{-1} \frac{\omega}{a} \Big|_{-\infty}^{\infty}$$

$$= \frac{1}{2a\pi} \left[\frac{\pi}{2} - \left(-\frac{\pi}{2} \right) \right]$$

$$= \frac{1}{2a} \text{ joules}$$

as was obtained above.

In this example, the half-power frequencies are $\omega = -a$ and $\omega = a$. To determine the amount of energy contained in the frequency band $-a \leq \omega \leq a$, we evaluate the integral

$$\frac{1}{2\pi} \int_{-a}^{a} |\mathbf{F}(j\omega)|^2 \, d\omega = \frac{1}{2a\pi} \tan^{-1} \frac{\omega}{a} \Big|_{-a}^{a}$$

$$= \frac{1}{2a\pi} \left[\frac{\pi}{4} - \left(-\frac{\pi}{4} \right) \right]$$

$$= \frac{1}{4a} \quad \text{joules}$$

Thus, 50 percent of the energy content of the function is in the band between the half-power frequencies. In a similar manner, we can calculate energy content of other frequency bands. The results are summarized in Table 13.1.

TABLE 13.1   Energy Content of Various Frequency Bands

Frequency Band	Percent of Total Energy
$-a \leq \omega \leq a$	50.0
$-2a \leq \omega \leq 2a$	70.5
$-3a \leq \omega \leq 3a$	79.5
$-4a \leq \omega \leq 4a$	84.4
$-5a \leq \omega \leq 5a$	87.4
$-7a \leq \omega \leq 7a$	91.0
$-10a \leq \omega \leq 10a$	93.7
$-20a \leq \omega \leq 20a$	96.8
$-30a \leq \omega \leq 30a$	97.9
$-50a \leq \omega \leq 50a$	98.7
$-100a \leq \omega \leq 100a$	99.4
$-\infty \leq \omega \leq \infty$	100.0

· · ·

From the function $e^{-at}u(t)$, by setting $a = 0$ we get a unit step function $u(t)$. This may lead us to believe that the Fourier transform of $u(t)$ is $1/j\omega$. However, because

$$\int_{-\infty}^{\infty} f(t)\, dt$$

is not finite when $f(t) = u(t)$, this conclusion cannot be justified. This problem does not arise when $f(t)$ is the *signum function*, designated sgn(t), which is defined by

$$\text{sgn}(t) = \begin{cases} -1 & \text{for } t < 0 \\ 1 & \text{for } t \geq 0 \end{cases}$$

A sketch of this function is shown in Fig. 13.6. However,

$$\mathscr{F}[\text{sgn}(t)] = \int_{-\infty}^{\infty} \text{sgn}(t)e^{-j\omega t}\, dt$$

$$= \int_{-\infty}^{0} -1e^{-j\omega t}\, dt + \int_{0}^{\infty} 1e^{-j\omega t}\, dt$$

fig. 13.6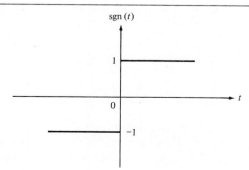

sgn (t)

which cannot be evaluated. Therefore consider the function $f(t)$ shown in Fig. 13.7, in which $a > 0$. In this case,

$$\mathscr{F}[f(t)] = \int_{-\infty}^{\infty} f(t)e^{-j\omega t}\,dt$$

$$= \int_{-\infty}^{0} -e^{at}e^{-j\omega t}\,dt + \int_{0}^{\infty} e^{-at}e^{-j\omega t}\,dt$$

$$= \frac{-1}{a - j\omega} e^{(a - j\omega)t}\Big|_{-\infty}^{0} + \frac{-1}{a + j\omega} e^{-(a + j\omega)t}\Big|_{0}^{\infty}$$

$$= -\frac{1}{a - j\omega} + \frac{1}{a + j\omega} = \frac{-2j\omega}{a^2 + \omega^2}$$

fig. 13.7

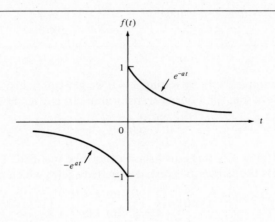

Since

$$\lim_{a \to 0} f(t) = \operatorname{sgn}(t)$$

then

$$\mathscr{F}[\operatorname{sgn}(t)] = \lim_{a \to 0} \frac{-2j\omega}{a^2 + \omega^2}$$

$$= \frac{-2j}{\omega} = \frac{2}{j\omega}$$

Consequently, since

$$u(t) = \frac{1}{2} + \frac{1}{2}\operatorname{sgn}(t)$$

then

$$\mathscr{F}[u(t)] = \mathscr{F}\left[\frac{1}{2} + \frac{1}{2}\,\mathrm{sgn}(t)\right]$$

$$= \mathscr{F}\left[\frac{1}{2}\right] + \frac{1}{2}\,\mathscr{F}[\mathrm{sgn}(t)]$$

$$= \left(\frac{1}{2}\right)2\pi\delta(\omega) + \frac{1}{2}\left(\frac{2}{j\omega}\right)$$

$$= \pi\delta(\omega) + \frac{1}{j\omega}$$

Given the periodic function $f(t)$ having period T, its complex Fourier series representation is

$$f(t) = \sum_{n=-\infty}^{\infty} c_n e^{jn\omega_0 t}$$

where $\omega_0 = 2\pi/T$ and c_n are the complex Fourier coefficients. Then

$$\mathscr{F}[f(t)] = \int_{-\infty}^{\infty}\left(\sum_{n=-\infty}^{\infty} c_n e^{jn\omega_0 t}\right)e^{-j\omega t}\,dt$$

$$= \sum_{n=-\infty}^{\infty}\left(\int_{-\infty}^{\infty} c_n e^{jt(n\omega_0-\omega)}\,dt\right)$$

$$= \sum_{n=-\infty}^{\infty} c_n 2\pi\delta(n\omega_0 - \omega)$$

$$= \sum_{n=-\infty}^{\infty} 2\pi c_n\delta(\omega - n\omega_0)$$

Thus the Fourier transform of a periodic function is a sum of impulses whose values are determined by the complex Fourier series coefficients.

A technique for determining the Fourier transform of a function $f(t)$ by employing its derivative(s)—as was done for Fourier series—is based on the following derivation:

Given a function $f(t)$, then

$$\mathscr{F}[f'(t)] = \mathscr{F}\left[\frac{df(t)}{dt}\right] = \int_{-\infty}^{\infty} \frac{df(t)}{dt}\,e^{-j\omega t}\,dt$$

For integration by parts,

$$\int_a^b u\,dv = uv\Big|_a^b - \int_a^b v\,du$$

If we let

$$u = e^{-j\omega t} \qquad \text{and} \qquad dv = \frac{df(t)}{dt}\,dt = df(t)$$

then

$$du = -j\omega e^{-j\omega t}\, dt \qquad \text{and} \qquad v = f(t)$$

and

$$\mathscr{F}[f'(t)] = e^{-j\omega t} f(t)\Big|_{-\infty}^{\infty} - \int_{-\infty}^{\infty} f(t)[-j\omega e^{-j\omega t}]\, dt$$

For the case in which

$$\lim_{t \to -\infty} f(t) = \lim_{t \to \infty} f(t) = 0$$

we have

$$\mathscr{F}[f'(t)] = j\omega \int_{-\infty}^{\infty} f(t)e^{-j\omega t}\, dt$$

$$= j\omega \mathscr{F}[f(t)] = j\omega \mathbf{F}(j\omega)$$

Reapplying this result, we obtain

$$\mathscr{F}[f''(t)] = \mathscr{F}\left[\frac{d^2 f(t)}{dt^2}\right] = (j\omega)^2 \mathbf{F}(j\omega)$$

and, in general,

$$\mathscr{F}\left[\frac{d^n f(t)}{dt^n}\right] = (j\omega)^n \mathbf{F}(j\omega)$$

. . .

Example
To find the Fourier transform of the function $f(t) = te^{-at}u(t)$, take the derivative of $f(t)$. Thus

$$\frac{df(t)}{dt} = \frac{d}{dt}[te^{-at}u(t)]$$

$$= t\frac{d}{dt}[e^{-at}u(t)] + e^{-at}u(t)$$

$$= t[e^{-at}\delta(t) - ae^{-at}u(t)] + e^{-at}u(t)$$

$$= te^{-at}\delta(t) - ate^{-at}u(t) + e^{-at}u(t)$$

$$= 0\delta(t) - af(t) + e^{-at}u(t)$$

Taking the Fourier transform, since $\lim_{t \to -\infty} f(t) = \lim_{t \to \infty} f(t) = 0$, we obtain

$$j\omega \mathbf{F}(j\omega) = -a\mathbf{F}(j\omega) + \frac{1}{j\omega + a}$$

or

$$(j\omega + a)\mathbf{F}(j\omega) = \frac{1}{j\omega + a}$$

so that

$$\mathbf{F}(j\omega) = \frac{1}{(j\omega + a)^2}$$

In other words,

$$\mathscr{F}\left[te^{-at}u(t)\right] = \frac{1}{(j\omega + a)^2}$$

• • •

Example
Given the function $f(t)$ shown in Fig. 13.8. Since

$$\lim_{t \to -\infty} f(t) = \lim_{t \to \infty} f(t) = 0$$

we can apply the differentiation property developed above.

fig. 13.8

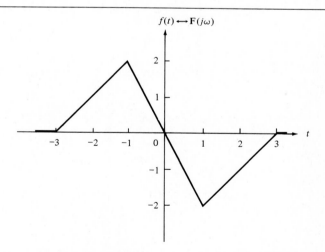

The derivative of $f(t)$ is shown in Fig. 13.9. Taking the second derivative, we get the function shown in Fig. 13.10. Since

$$f''(t) = \delta(t + 3) - 3\delta(t + 1) + 3\delta(t - 1) - \delta(t - 3)$$

fig. 13.9

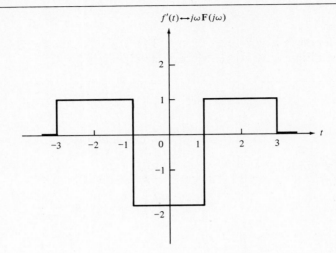

$$f'(t) \longleftrightarrow j\omega\, \mathbf{F}(j\omega)$$

fig. 13.10

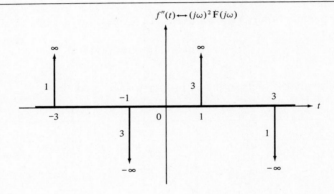

$$f''(t) \longleftrightarrow (j\omega)^2\, \mathbf{F}(j\omega)$$

then

$$(j\omega)^2\mathbf{F}(j\omega) = e^{j\omega 3} - 3e^{j\omega 1} + 3e^{-j\omega 1} - e^{-j\omega 3}$$

$$-\omega^2\mathbf{F}(j\omega) = 2j\left(\frac{e^{j3\omega} - e^{-j3\omega}}{2j}\right) - 2j\left(\frac{3e^{j\omega} - 3e^{-j\omega}}{2j}\right)$$

$$\mathbf{F}(j\omega) = \frac{2j}{\omega^2}(3\sin\omega - \sin 3\omega)$$

· · ·

If you don't appreciate the differentiation property of the Fourier transform by now, try to find $\mathbf{F}(j\omega)$ for the function in the previous example by directly substituting $f(t)$ into the Fourier transform formula.

A more subtle use of the differentiation property is shown in the following example.

• • •

Example

Consider the function shown in Fig. 13.11, whose first and second derivatives are shown in Figs. 13.12 and 13.13, respectively.

fig. 13.11

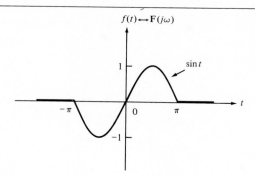

$$f(t) \longleftrightarrow \mathbf{F}(j\omega)$$

fig. 13.12

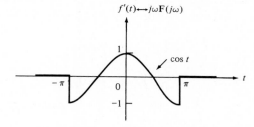

$$f'(t) \longleftrightarrow j\omega \mathbf{F}(j\omega)$$

fig. 13.13

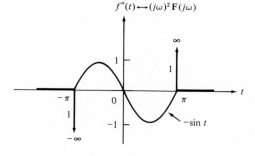

$$f''(t) \longleftrightarrow (j\omega)^2 \mathbf{F}(j\omega)$$

Since

$$f''(t) = -\delta(t + \pi) + \delta(t - \pi) - f(t)$$

then

$$(j\omega)^2 \mathbf{F}(j\omega) = -e^{j\omega\pi} + e^{-j\omega\pi} - \mathbf{F}(j\omega)$$

or

$$(1 - \omega^2)\mathbf{F}(j\omega) = -e^{j\omega\pi} + e^{-j\omega\pi}$$

and

$$\mathbf{F}(j\omega) = \frac{2j(e^{j\omega\pi} - e^{-j\omega\pi})}{(\omega^2 - 1)(2j)} = \frac{2j \sin \pi\omega}{\omega^2 - 1}$$

$$\cdot \quad \cdot \quad \cdot$$

We have just seen that taking the derivative in the time domain corresponds to multiplication by $j\omega$ in the frequency domain. The proof presented was for functions $f(t)$ having the property

$$\lim_{t \to \pm\infty} f(t) = 0$$

However, for some other functions not having this property (e.g., a sinusoid or a step function) it is also true that

$$\mathscr{F}\left[\frac{d}{dt} f(t)\right] = j\omega \mathscr{F}[f(t)]$$

To demonstrate this fact, consider the complex sinusoid $e^{j\omega_1 t}$. We have

$$\mathscr{F}\left[\frac{d}{dt} e^{j\omega_1 t}\right] = \mathscr{F}[j\omega_1 e^{j\omega_1 t}]$$

$$= j\omega_1 \mathscr{F}[e^{j\omega_1 t}]$$

$$= j\omega_1[2\pi\delta(\omega - \omega_1)] = 2\pi j\omega_1 \delta(\omega - \omega_1)$$

However, by the sampling property of the delta function,

$$2\pi j\omega_1 \delta(\omega - \omega_1) = 2\pi j\omega\delta(\omega - \omega_1)$$

Thus

$$\mathscr{F}\left[\frac{d}{dt} e^{j\omega_1 t}\right] = 2\pi j\omega\delta(\omega - \omega_1)$$

$$= j\omega[2\pi\delta(\omega - \omega_1)] = j\omega \mathscr{F}[e^{j\omega_1 t}]$$

Now consider the real sinusoid $\cos \omega_1 t$. We have

$$\mathscr{F}\left[\frac{d}{dt} \cos \omega_1 t\right] = \mathscr{F}[-\omega_1 \sin \omega_1 t]$$

$$= \mathscr{F}\left[-\frac{\omega_1}{2j}(e^{j\omega_1 t} - e^{-j\omega_1 t})\right]$$

$$= -\frac{\omega_1}{2j}[2\pi\delta(\omega - \omega_1) - 2\pi\delta(\omega + \omega_1)]$$

$$= j\omega_1\pi\delta(\omega - \omega_1) - j\omega_1\pi\delta(\omega + \omega_1)$$

Using the sampling property of the delta function,

$$\mathscr{F}\left[\frac{d}{dt}\cos\omega_1 t\right] = j\omega\pi\delta(\omega-\omega_1) + j\omega\pi\delta(\omega+\omega_1)$$

$$= j\omega[\pi\delta(\omega-\omega_1) + \pi\delta(\omega+\omega_1)]$$

$$= j\omega\mathscr{F}[\cos\omega_1 t]$$

Now consider the case of a step function. Since

$$\mathscr{F}\left[\frac{du(t)}{dt}\right] = \mathscr{F}[\delta(t)] = 1$$

and

$$j\omega\mathscr{F}[u(t)] = j\omega\left[\frac{1}{j\omega} + \pi\delta(\omega)\right]$$

$$= 1 + j\omega\pi\delta(\omega)$$

$$= 1 + 0\delta(\omega) = 1$$

then

$$\mathscr{F}\left[\frac{du(t)}{dt}\right] = j\omega\mathscr{F}[u(t)]$$

In essence, we shall only be considering functions that have the property

$$\mathscr{F}\left[\frac{df(t)}{dt}\right] = j\omega\mathscr{F}[f(t)]$$

We have previously seen that

$$\mathscr{F}[e^{-at}u(t)] = \frac{1}{j\omega + a}$$

where a is a positive real number. The same result holds when a is a complex number with a positive real part. Using this fact, we have

$$\mathscr{F}[e^{-\alpha t}\sin\omega_1 t\, u(t)] = \mathscr{F}\left[e^{-\alpha t}\left(\frac{e^{j\omega_1 t} - e^{-j\omega_1 t}}{2j}\right)u(t)\right]$$

$$= \frac{1}{2j}\mathscr{F}[e^{-(\alpha-j\omega_1)t}u(t) - e^{-(\alpha+j\omega_1)t}u(t)]$$

$$= \frac{1}{2j}\left[\frac{1}{j\omega + (\alpha-j\omega_1)} - \frac{1}{j\omega + (\alpha+j\omega_1)}\right]$$

$$= \frac{1}{2j}\left[\frac{(j\omega+\alpha+j\omega_1) - (j\omega+\alpha-j\omega_1)}{(j\omega)^2 + j\omega(\alpha+j\omega_1) + j\omega(\alpha-j\omega_1) + (\alpha-j\omega_1)(\alpha+j\omega_1)}\right]$$

$$= \frac{1}{2j}\left(\frac{2j\omega_1}{(j\omega)^2 + 2\alpha j\omega + \alpha^2 + \omega_1{}^2}\right)$$

$$= \frac{\omega_1}{(j\omega + \alpha)^2 + \omega_1{}^2}$$

TABLE 13.2 Table of Fourier Transforms

$f(t)$	$\mathbf{F}(j\omega)$
$\dfrac{1}{2\pi}\displaystyle\int_{-\infty}^{\infty}\mathbf{F}(j\omega)e^{j\omega t}\,d\omega$	$\displaystyle\int_{-\infty}^{\infty}f(t)e^{-j\omega t}\,dt$
$\delta(t)$	1
$\delta(t-a)$	$e^{-j\omega a}$
K	$2\pi K\delta(\omega)$
$e^{j\omega_1 t}$	$2\pi\delta(\omega-\omega_1)$
$\cos \omega_1 t$	$\pi\delta(\omega-\omega_1)+\pi\delta(\omega+\omega_1)$
$\sin \omega_1 t$	$j\pi\delta(\omega+\omega_1)-j\pi\delta(\omega-\omega_1)$
$e^{-\alpha t}u(t)$	$\dfrac{1}{j\omega+\alpha}$
$\mathrm{sgn}(t)$	$\dfrac{2}{j\omega}$
$u(t)$	$\dfrac{1}{j\omega}+\pi\delta(\omega)$
$\cos \omega_1 t\ u(t)$	$\dfrac{j\omega}{(j\omega)^2+\omega_1{}^2}$
$\sin \omega_1 t\ u(t)$	$\dfrac{\omega_1}{(j\omega)^2+\omega_1{}^2}$
$e^{-\alpha t}\cos \omega_1 t\ u(t)$	$\dfrac{j\omega+\alpha}{(j\omega+\alpha)^2+\omega_1{}^2}$
$e^{-\alpha t}\sin \omega_1 t\ u(t)$	$\dfrac{\omega_1}{(j\omega+\alpha)^2+\omega_1{}^2}$
$f_1(t)+f_2(t)$	$\mathbf{F}_1(j\omega)+\mathbf{F}_2(j\omega)$
$Kf(t)$	$K\mathbf{F}(j\omega)$
$\dfrac{df(t)}{dt}$	$j\omega\mathbf{F}(j\omega)$
$f(t-a)$	$e^{-j\omega a}\mathbf{F}(j\omega)$
$e^{-\alpha t}f(t)$	$\mathbf{F}(j\omega+\alpha)$
$e^{j\omega_1 t}f(t)$	$\mathbf{F}(j[\omega-\omega_1])$
$te^{-\alpha t}u(t)$	$\dfrac{1}{(j\omega+\alpha)^2}$
$u(t+a)-u(t-a)$	$2a\left(\dfrac{\sin a\omega}{a\omega}\right)$
$\displaystyle\int_{-\infty}^{\infty}f_1(\tau)f_2(t-\tau)\,d\tau$	$\mathbf{F}_1(j\omega)\mathbf{F}_2(j\omega)$
$f_1(t)f_2(t)$	$\dfrac{1}{2\pi}\displaystyle\int_{-\infty}^{\infty}\mathbf{F}_1(j\lambda)\mathbf{F}_2(j[\omega-\lambda])\,d\lambda$

In a similar manner we can show that

$$\mathscr{F}\left[e^{-\alpha t}\cos \omega_1 t\, u(t)\right] = \frac{j\omega + \alpha}{(j\omega + \alpha)^2 + \omega_1{}^2}$$

For the special case that $\alpha = 0$, the preceding two formulas reduce to

$$\mathscr{F}\left[\sin \omega_1 t\, u(t)\right] = \frac{\omega_1}{(j\omega)^2 + \omega_1{}^2}$$

and

$$\mathscr{F}\left[\cos \omega_1 t\, u(t)\right] = \frac{j\omega}{(j\omega)^2 + \omega_1{}^2}$$

respectively.

A number of common Fourier transforms are summarized in Table 13.2.

PROBLEMS   13.1.   Find the Fourier transform of $f(t) = \sin \omega_1 t$.

---

fig. P13.2

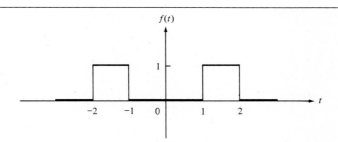

*13.2.   Find the Fourier transform of the function shown in Fig. P13.2.

Ans. $\dfrac{2}{\omega}(\sin 2\omega - \sin \omega)$

---

fig. P13.3

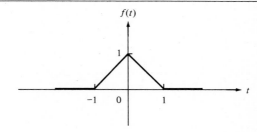

13.3.   Find the Fourier transform of the function shown in Fig. P13.3.

fig. P13.4

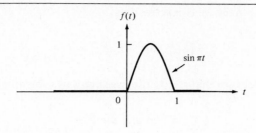

13.4.   Find the Fourier transform of the function shown in Fig. P13.4.

fig. P13.5

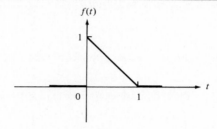

*13.5.   Find the Fourier transform of the function shown in Fig. P13.5.

Ans. $\dfrac{1}{j\omega} - \dfrac{1}{\omega^2}(e^{-j\omega} - 1)$

13.6.   Given that $\mathscr{F}[f(t)] = \mathbf{F}(j\omega)$, show that

$$\mathscr{F}[f(t - a)] = e^{-j\omega a}\,\mathbf{F}(j\omega)$$

This result is known as the *real translation property* of the Fourier transform.

13.7.   Use the property given in Problem 13.6 to find the Fourier transform of

$$f(t) = Vu\left(t + \frac{\tau}{2}\right) - Vu\left(t - \frac{\tau}{2}\right)$$

*13.8.   Find the Fourier transform of

$$f(t) = \cos(\omega_1 t - \phi) = \cos\omega_1\left(t - \frac{\phi}{\omega_1}\right)$$

Ans. $\pi e^{-j\phi}\,\delta(\omega - \omega_1) + \pi e^{j\phi}\,\delta(\omega + \omega_1)$

13.9.   Suppose that $f(t)$ is an even function. Show that

$$\mathbf{F}(j\omega) = 2\int_0^\infty f(t)\cos\omega t\,dt$$

13.10.   Use the result given in Problem 13.9 to find the Fourier transform of the function shown in Fig. P13.10.

fig. P13.10

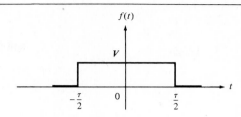

13.11. Suppose that $f(t)$ is an odd function. Show that

$$\mathbf{F}(j\omega) = -2j \int_0^\infty f(t) \sin \omega t \, dt$$

fig. P13.12

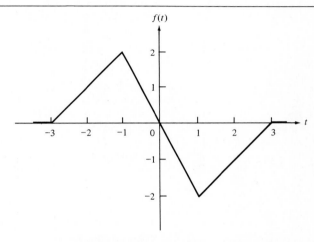

*13.12. Use the result given in Problem 13.11 to find the Fourier transform of the function shown in Fig. P13.12.

Ans. $\dfrac{j2}{\omega^2}(3 \sin \omega - \sin 3\omega)$

13.13. Given that $\mathscr{F}[f(t)] = \mathbf{F}(j\omega)$, show that

$$\mathscr{F}[e^{-\alpha t}f(t)] = \mathbf{F}(j\omega + \alpha)$$

That is, given $\mathbf{F}(j\omega)$—the Fourier transform of $f(t)$—then the Fourier transform of $e^{-\alpha t}f(t)$ can be obtained from $\mathbf{F}(j\omega)$ by replacing $j\omega$ with $j\omega + \alpha$. The constant α can be real or complex. This result is known as the *complex translation property* of the Fourier transform.

13.14. (a) Show that

$$\mathscr{F}[e^{j\omega_1 t}f(t)] = \mathbf{F}(j[\omega - \omega_1])$$

This result is known as the *frequency translation property*.

(b) Show that

$$\mathscr{F}[\cos \omega_1 t \, f(t)] = \frac{1}{2}\mathbf{F}(j[\omega - \omega_1]) + \frac{1}{2}\mathbf{F}(j[\omega + \omega_1])$$

This result is known as *modulation*.

fig. P13.15

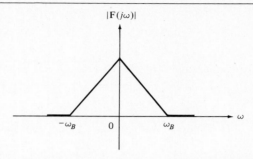

$|F(j\omega)|$

$-\omega_B$ 0 ω_B ω

13.15. Suppose that the frequency spectrum (amplitude portion) of $f(t)$ is as shown in Fig. P13.15.
  (a) Sketch the amplitude spectrum of $e^{j\omega_c t} f(t)$ for the case that $\omega_c > \omega_B$.
  (b) Sketch the amplitude spectrum of $\cos \omega_c t \, f(t)$ for the case that $\omega_c > \omega_B$.

13.16. Suppose that $f(t)$ is a real function.
  (a) Show that $|F(j\omega)|$ is an even function of ω. (It's for this reason that amplitude spectra of real functions are often plotted for non-negative values of ω; the negative side of the plot is just the mirror reflection of the positive side.)
  (b) Show that the angle of $F(j\omega)$ is an odd function of ω.

13.17. Suppose that the voltage across a 1-Ω resistor is $v(t) = te^{-at}u(t)$.
  (a) What is the total energy absorbed by the resistor?
  (b) What percent of this energy is in the frequency band $-a \le \omega \le a$?

  $Hint:$ $\displaystyle \int \frac{d\omega}{(\omega^2 + a^2)^2} = \frac{\omega}{2(\omega^2 + a^2)} + \frac{1}{2a^2} \int \frac{d\omega}{\omega^2 + a^2}$

13.18. Given that $F(j\omega)$ is the Fourier transform of $f(t)$. In the expression $F(j\omega)$, replacing ω by t results in a time function $F(jt)$, which may be complex. In the expression $f(t)$, replacing t by ω results in a function $f(\omega)$.
  (a) Show that

  $$\mathscr{F}[F(jt)] = 2\pi \mathbf{f}(-\omega)$$

  (b) Show that

  $$\mathscr{F}^{-1}[\mathbf{f}(\omega)] = \frac{1}{2\pi} F(-jt)$$

These results are known as the *symmetry property* of the Fourier transform.

13.3 APPLICATION OF THE FOURIER TRANSFORM

Suppose that we have a linear time-invariant system with input $x(t)$ and output $y(t)$, as shown in Fig. 13.14, which is described by a linear differential equation with constant coefficients. If the initial conditions are zero, in general we have that

$$A_n \frac{d^n y}{dt^n} + A_{n-1} \frac{d^{n-1} y}{dt^{n-1}} + \cdots + A_1 \frac{dy}{dt} + A_0 y$$

$$= B_m \frac{d^m x}{dt^m} + B_{m-1} \frac{d^{m-1} x}{dt^{m-1}} + \cdots + B_1 \frac{dx}{dt} + B_0 x$$

fig. 13.14

$$x(t) \longrightarrow \boxed{} \longrightarrow y(t)$$

If we take the Fourier transform of both sides of this equation we obtain

$$(j\omega)^n A_n Y(j\omega) + (j\omega)^{n-1} A_{n-1} Y(j\omega) + \cdots + j\omega A_1 Y(j\omega) + A_0 Y(j\omega)$$
$$= (j\omega)^m B_m X(j\omega) + \cdots + j\omega B_1 X(j\omega) + B_0 X(j\omega)$$

or

$$[(j\omega)^n A_n + (j\omega)^{n-1} A_{n-1} + \cdots + j\omega A_1 + A_0] Y(j\omega)$$
$$= [(j\omega)^m B_m + (j\omega)^{m-1} B_{m-1} + \cdots + j\omega B_1 + B_0] X(j\omega)$$

Thus

$$Y(j\omega) = \frac{(j\omega)^m B_m + \cdots + j\omega B_1 + B_0}{(j\omega)^n A_n + \cdots + j\omega A_1 + A_0} X(j\omega)$$

or

$$\boxed{Y(j\omega) = H(j\omega)X(j\omega)}$$

where $H(j\omega)$ is called the *system function* or *transfer function* of the system. We therefore have that

$$\boxed{H(j\omega) = \frac{Y(j\omega)}{X(j\omega)}}$$

Note that if we know the system function and the transform of the input, it is a simple matter to determine the transform of the output; by taking the inverse transform, we get the output.

Suppose that the input is $x(t) = \delta(t)$. Then

$$X(j\omega) = 1$$

Thus

$$Y(j\omega) = H(j\omega)X(j\omega) = H(j\omega)$$

and the output is

$$y(t) = \mathscr{F}^{-1}[Y(j\omega)]$$
$$= \mathscr{F}^{-1}[H(j\omega)]$$
$$= h(t)$$

Therefore $h(t)$, the inverse transform of the system function, is called the *impulse response* of the system.

. . .

Example

Given a system whose impulse response is $h(t) = 2e^{-t}u(t)$, let us determine the step response.

Since

$$\mathbf{X}(j\omega) = \mathscr{F}[u(t)] = \pi\delta(\omega) + \frac{1}{j\omega}$$

and

$$\mathbf{H}(j\omega) = \mathscr{F}[h(t)] = \frac{2}{j\omega + 1}$$

then

$$\mathbf{Y}(j\omega) = \mathbf{H}(j\omega)\mathbf{X}(j\omega)$$

$$= \frac{2}{j\omega + 1}\left[\pi\delta(\omega) + \frac{1}{j\omega}\right]$$

$$= \frac{2\pi}{j\omega + 1}\delta(\omega) + \frac{2}{(j\omega)(j\omega + 1)} = 2\pi\delta(\omega) + \frac{2}{(j\omega)(j\omega + 1)}$$

Just as we showed that the Fourier transform is linear, we can show that the inverse transform is also. Thus,

$$\mathscr{F}^{-1}[\mathbf{F}_1(j\omega) + \mathbf{F}_2(j\omega)] = \mathscr{F}^{-1}[\mathbf{F}_1(j\omega)] + \mathscr{F}^{-1}[\mathbf{F}_2(j\omega)]$$

and

$$\mathscr{F}^{-1}[K\mathbf{F}(j\omega)] = K\mathscr{F}^{-1}[\mathbf{F}(j\omega)]$$

Thus

$$y(t) = \mathscr{F}^{-1}[2\pi\delta(\omega)] + \mathscr{F}^{-1}\left[\frac{2}{(j\omega)(j\omega + 1)}\right]$$

The inverse transform of the first term is 1. To obtain the inverse transform of the second term we can decompose the fraction into a sum of smaller fractions; that is,

$$\frac{2}{(j\omega)(j\omega + 1)} = \frac{K_1}{j\omega} + \frac{K_2}{j\omega + 1} \tag{13.3}$$

Although there is more than one way, the simplest approach to determine the constants K_1 and K_2 is as follows:

To find K_1, multiply both sides of Equation (13.3) by $j\omega$. The result is

$$\frac{2}{j\omega + 1} = K_1 + \frac{j\omega K_2}{j\omega + 1}$$

This expression holds for all $j\omega$. In particular, it holds for $j\omega = 0$. By selecting this specific value we can eliminate the term containing K_2. The result is

$$\frac{2}{0 + 1} = K_1 + \frac{0K_2}{0 + 1} \quad \text{or} \quad K_1 = 2$$

To determine K_2, multiply both sides of Equation (13.3) by $j\omega + 1$. Thus

$$\frac{2}{j\omega} = \frac{(j\omega + 1)K_1}{j\omega} + K_2$$

By choosing $j\omega = -1$, we eliminate the term containing K_1 and obtain

$$\frac{2}{-1} = \frac{(-1 + 1)K_1}{-1} + K_2 \quad \text{or} \quad K_2 = -2$$

Substituting the values just obtained for K_1 and K_2 into Equation (13.3) yields

$$\frac{2}{(j\omega)(j\omega + 1)} = \frac{2}{j\omega} - \frac{2}{j\omega + 1}$$

Thus

$$\mathscr{F}^{-1}\left[\frac{2}{(j\omega)(j\omega + 1)}\right] = \mathscr{F}^{-1}\left[\frac{2}{j\omega}\right] - \mathscr{F}^{-1}\left[\frac{2}{j\omega + 1}\right] = \text{sgn}(t) - 2e^{-t}u(t)$$

and the step response is

$$y(t) = 1 + \text{sgn}(t) - 2e^{-t}u(t)$$
$$= 2u(t) - 2e^{-t}u(t)$$
$$= 2(1 - e^{-t})u(t)$$

· · ·

For the case that $x(t)$ is not a delta function, knowing the impulse response $h(t)$, we can determine the output without resorting to transforms. To obtain this result, we proceed as follows:

From the fact that

$$\mathbf{Y}(j\omega) = \mathbf{H}(j\omega)\mathbf{X}(j\omega)$$

and

$$\mathbf{X}(j\omega) = \int_{-\infty}^{\infty} x(t)e^{-j\omega t}\, dt = \int_{-\infty}^{\infty} x(\tau)e^{-j\omega \tau}\, d\tau$$

we have

$$\mathbf{Y}(j\omega) = \mathbf{H}(j\omega) \int_{-\infty}^{\infty} x(\tau)e^{-j\omega\tau} \, d\tau$$

$$= \int_{-\infty}^{\infty} x(\tau)e^{-j\omega\tau}\mathbf{H}(j\omega) \, d\tau$$

$$= \int_{-\infty}^{\infty} x(\tau)e^{-j\omega\tau}\left[\int_{-\infty}^{\infty} h(\lambda)e^{-j\omega\lambda} \, d\lambda\right] d\tau$$

$$= \int_{-\infty}^{\infty} x(\tau)\left[\int_{-\infty}^{\infty} h(\lambda)e^{-j\omega(\lambda+\tau)} \, d\lambda\right] d\tau$$

In evaluating the integral in the brackets, let

$$t = \lambda + \tau$$

then

$$dt = d\lambda$$

and

$$\lambda = t - \tau$$

Thus

$$\lambda = -\infty \Rightarrow t = -\infty$$

$$\lambda = \infty \Rightarrow t = \infty$$

Therefore, changing variables,

$$\mathbf{Y}(j\omega) = \int_{-\infty}^{\infty} x(\tau)\left[\int_{-\infty}^{\infty} h(t-\tau)e^{-j\omega t} \, dt\right] d\tau$$

$$= \int_{-\infty}^{\infty}\int_{-\infty}^{\infty} x(\tau)\,h(t-\tau)e^{-j\omega t} \, d\tau \, dt$$

$$= \int_{-\infty}^{\infty}\left[\int_{-\infty}^{\infty} x(\tau)h(t-\tau) \, d\tau\right] e^{-j\omega t} \, dt$$

Comparing this equation with the fact that

$$\mathbf{Y}(j\omega) = \int_{-\infty}^{\infty} y(t)e^{-j\omega t} \, dt$$

we conclude that

$$\boxed{y(t) = \int_{-\infty}^{\infty} x(\tau)h(t-\tau) \, d\tau}$$

If we interchange the roles of $x(t)$ and $h(t)$, in a similar manner we can obtain

$$y(t) = \int_{-\infty}^{\infty} h(\tau)x(t - \tau) \, d\tau$$

Each of these last two integrals describes a process called *convolution*, and therefore each is called the *convolution integral*. This process (the evaluation of the integral) enables us to determine the output from the input and the impulse response without taking transforms. Thus, we see that multiplication in the frequency (or transform) domain corresponds to convolution in the time domain.

Although the evaluation of the convolution integral is not difficult, it can be quite subtle. We shall postpone our general discussion on how to evaluate this integral until we come across it again in the chapter on Laplace transforms. However, for the very special case that either the impulse response $h(t)$ or the input $x(t)$ is a delta function, the convolution integral is easily evaluated by using the sampling property of the delta function.

$$\bullet \quad \bullet \quad \bullet$$

Example
Suppose that $x(t) = 3\delta(t - 4)$ is the input to a system whose impulse response is $h(t) = 2e^{-5t}u(t)$. Then the output is

$$y(t) = \int_{-\infty}^{\infty} x(\tau)h(t - \tau) \, d\tau$$

$$= \int_{-\infty}^{\infty} [3\delta(\tau - 4)][2e^{-5(t-\tau)}u(t - \tau)] \, d\tau$$

$$= \int_{-\infty}^{\infty} [\delta(\tau - 4)][6e^{-5(t-\tau)}u(t - \tau)] \, d\tau$$

and by the sampling property

$$y(t) = 6e^{-5(t-4)}u(t - 4)$$

This result is by no means surprising, for if $2e^{-5t}u(t)$ is the impulse response (i.e., the response to an input of $\delta(t)$), then by linearity the response to $3\delta(t)$ is $6e^{-5t}u(t)$. Furthermore, because of the time-invariance property, the response to $3\delta(t - 4)$ is $6e^{-5(t-4)}u(t - 4)$.

Also note that

$$y(t) = \int_{-\infty}^{\infty} h(\tau)x(t - \tau) \, d\tau$$

$$= \int_{-\infty}^{\infty} [2e^{-5\tau}u(\tau)][3\delta(t - \tau - 4)] \, d\tau$$

$$= \int_{-\infty}^{\infty} [6e^{-5\tau}u(\tau)][\delta(t - \tau - 4)] \, d\tau$$

and by the sampling property (since the variable of integration is τ, the delta function is nonzero when $\tau = t - 4$), we get that

$$y(t) = 6e^{-5(t-4)}u(t-4)$$

as was obtained above.

· · ·

From the fact that

$$\mathbf{Y}(j\omega) = \mathbf{H}(j\omega)\mathbf{X}(j\omega)$$

we have

$$|\mathbf{Y}(j\omega)| = |\mathbf{H}(j\omega)\mathbf{X}(j\omega)|$$

and, by complex arithmetic,

$$|\mathbf{Y}(j\omega)| = |\mathbf{H}(j\omega)|\,|\mathbf{X}(j\omega)|$$

that is, the amplitude spectrum of the output function is the product of the amplitude spectrum of the input function and $|\mathbf{H}(j\omega)|$, which we call the *amplitude response* of the system.

Consider the system whose amplitude response is shown in Fig. 13.15. Mathematically, we can write

$$|\mathbf{H}(j\omega)| = \begin{cases} 1 & \text{for } 0 \le \omega \le \omega_c \\ 0 & \text{for } \omega_c < \omega < \infty \end{cases}$$

fig. 13.15

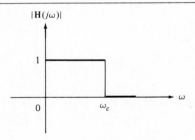

If the input of this system is $x(t)$, then the amplitude spectrum of the output is given by

$$|\mathbf{Y}(j\omega)| = |\mathbf{H}(j\omega)|\,|\mathbf{X}(j\omega)| = \begin{cases} |\mathbf{X}(j\omega)| & \text{for } 0 \le \omega \le \omega_c \\ 0 & \text{for } \omega_c < \omega < \infty \end{cases}$$

Hence this system passes that portion of the input below the frequency ω_c, but stops that portion above ω_c. We call the frequency band $0 \le \omega \le \omega_c$ the *pass band* and $\omega_c < \omega < \infty$ the *stop band*. Since it is the low frequencies that are passed and the high frequencies that are not passed, we call such a system a *low-pass filter*. Because all of the frequencies in the pass band are passed equally and all the frequencies in the stop band are completely stopped, this low-pass filter is said to be *ideal*. A general low-pass filter is sometimes represented as in Fig. 13.16.

fig. 13.16

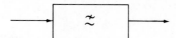

The frequency response of an *ideal high-pass filter* and a symbolic representation of a general high-pass filter are shown in Fig. 13.17. In this case the stop band is the frequency band $0 \leq \omega < \omega_b$ and the pass band is $\omega_b \leq \omega < \infty$.

fig. 13.17

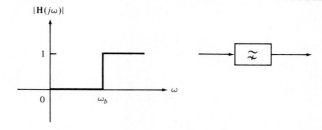

The frequency response of an *ideal band-pass filter* and a representation of a general band-pass filter are shown in Fig. 13.18. In this case the pass band consists of the interval $\omega_1 \leq \omega \leq \omega_2$, whereas $0 \leq \omega < \omega_1$ and $\omega_2 < \omega < \infty$ are both stop bands.

fig. 13.18

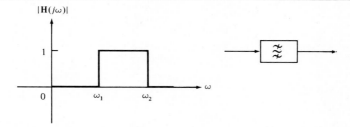

Another type of filter is an *ideal band-elimination* (or *rejection*) *filter.** The amplitude response of such a filter along with a representation of a general band-elimination filter are seen in Fig. 13.19. The stop band is $\omega_\alpha < \omega < \omega_\beta$, whereas $0 \leq \omega \leq \omega_\alpha$ and $\omega_\beta \leq \omega < \infty$ are both pass bands.

fig. 13.19

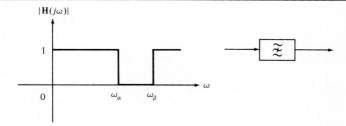

* Also called a *trap*.

13.4   APPLICATION TO CIRCUITS

In order to apply Fourier transforms to circuits, first consider a resistor (Fig. 13.20). Since

$$v(t) = Ri(t)$$

then taking the Fourier transform yields

$$\mathbf{V}(j\omega) = R\mathbf{I}(j\omega)$$

fig. 13.20

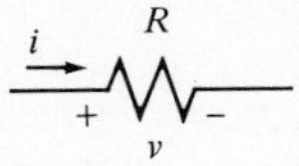

Defining impedance to be the ratio of the voltage transform to the current transform, the impedance of the resistor $\mathbf{Z}_R(j\omega)$ is therefore

$$\mathbf{Z}_R(j\omega) = \frac{\mathbf{V}(j\omega)}{\mathbf{I}(j\omega)} = R$$

For an inductor (Fig. 13.21), since

$$v(t) = L\frac{di(t)}{dt}$$

fig. 13.21

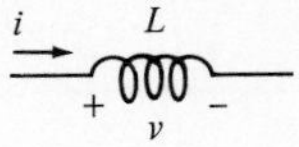

then

$$\mathbf{V}(j\omega) = Lj\omega\mathbf{I}(j\omega)$$

and thus

$$\mathbf{Z}_L(j\omega) = \frac{\mathbf{V}(j\omega)}{\mathbf{I}(j\omega)} = j\omega L$$

In addition, for a capacitor (Fig. 13.22), since

$$i(t) = C\frac{dv(t)}{dt}$$

fig. 13.22

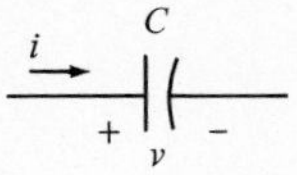

then

$$\mathbf{I}(j\omega) = Cj\omega\mathbf{V}(j\omega)$$

and thus

$$\mathbf{Z}_C(j\omega) = \frac{\mathbf{V}(j\omega)}{\mathbf{I}(j\omega)} = \frac{1}{j\omega C}$$

For each of the elements described above, we have the three forms of Ohm's law:

$$\mathbf{Z}(j\omega) = \frac{\mathbf{V}(j\omega)}{\mathbf{I}(j\omega)}$$

$$\mathbf{V}(j\omega) = \mathbf{Z}(j\omega)\mathbf{I}(j\omega)$$

$$\mathbf{I}(j\omega) = \frac{\mathbf{V}(j\omega)}{\mathbf{Z}(j\omega)}$$

Thus we see that we get the same expressions for the impedances of resistors, capacitors, and inductors as was obtained for phasors (the sinusoidal case). In addition, the expressions describing transformers are the same as for phasors. Hence we can employ the circuit analysis techniques we used previously for sinusoidal circuits—the difference being that instead of phasors, we use Fourier transforms.

· · ·

Example
We have seen that the circuit shown in Fig. 13.23, whose input is $v_s(t)$ and whose output is $v_2(t)$, approximates an integrator for certain types of inputs.

fig. 13.23 (left)
fig. 13.24 (right)

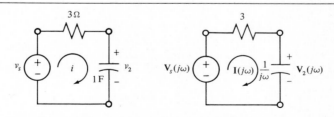

The circuit is represented in the frequency domain by Fig. 13.24. By voltage division,

$$\mathbf{V}_2(j\omega) = \frac{1/j\omega}{(1/j\omega) + 3} \, \mathbf{V}_s(j\omega)$$

$$= \frac{1}{1 + 3j\omega} \, \mathbf{V}_s(j\omega)$$

Thus, the system function is given by

$$\mathbf{H}(j\omega) = \frac{\mathbf{V}_2(j\omega)}{\mathbf{V}_s(j\omega)} = \frac{\frac{1}{3}}{j\omega + \frac{1}{3}}$$

For very small values of ω the value of $|\mathbf{H}(j\omega)|$ is approximately unity (when $\omega = 0$ then $|\mathbf{H}(j\omega)| = 1$), whereas for very large values of ω the value of $|\mathbf{H}(j\omega)|$ is very small. A plot of the amplitude response is shown in Fig. 13.25. (Note that the half-power frequency is $\frac{1}{3}$ rad/s.) Thus we see that the circuit is a (nonideal) low-pass filter.

fig. 13.25

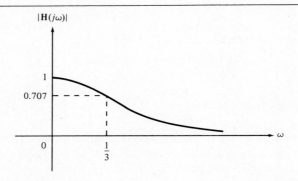

The impulse response is

$$h(t) = \mathscr{F}^{-1}[\mathbf{H}(j\omega)] = \tfrac{1}{3}e^{-t/3}u(t)$$

that is, this is the output $v_2(t)$ when the input is $v_s(t) = \delta(t)$.

To find the step response, that is, the output $v_2(t)$ when the input is $v_s(t) = u(t)$, we use

$$\mathbf{V}_2(j\omega) = \frac{\frac{1}{3}}{j\omega + \frac{1}{3}}\mathbf{V}_s(j\omega)$$

$$= \frac{\frac{1}{3}}{j\omega + \frac{1}{3}}\left[\frac{1}{j\omega} + \pi\delta(\omega)\right] = \frac{\frac{1}{3}}{j\omega(j\omega + \frac{1}{3})} + \frac{\pi\delta(\omega)/3}{j\omega + \frac{1}{3}}$$

so

$$\mathbf{V}_2(j\omega) = \frac{\frac{1}{3}}{j\omega(j\omega + \frac{1}{3})} + \pi\delta(\omega)$$

$$= \mathbf{V}_a(j\omega) + \mathbf{V}_b(j\omega)$$

The function $\mathbf{V}_a(j\omega)$ can be rewritten as follows:

$$\frac{\frac{1}{3}}{j\omega(j\omega + \frac{1}{3})} = \frac{K_1}{j\omega} + \frac{K_2}{j\omega + \frac{1}{3}} \tag{13.4}$$

To find K_1, multiply both sides of Equation (13.4) by $j\omega$. The result is

$$\frac{\frac{1}{3}}{j\omega + \frac{1}{3}} = K_1 + \frac{j\omega K_2}{j\omega + \frac{1}{3}}$$

Now set $j\omega = 0$ (this eliminates the term containing K_2). The result is

$$K_1 = 1$$

To find K_2, multiply both sides of Equation (13.4) by $j\omega + \frac{1}{3}$. The result is

$$\frac{\frac{1}{3}}{j\omega} = \frac{(j\omega + \frac{1}{3})K_1}{j\omega} + K_2$$

To eliminate the term containing K_1 from this expression, set $j\omega = -\frac{1}{3}$. The result is

$$K_2 = -1$$

Substituting the values of K_1 and K_2 just determined into Equation (13.4) yields

$$\mathbf{V}_a(j\omega) = \frac{1}{j\omega} - \frac{1}{j\omega + \frac{1}{3}}$$

It is now apparent that the inverse Fourier transform of $\mathbf{V}_a(j\omega)$ is

$$\mathscr{F}^{-1}[\mathbf{V}_a(j\omega)] = \frac{1}{2}\,\text{sgn}(t) - e^{-t/3}u(t)$$

Clearly,

$$\mathscr{F}^{-1}[\mathbf{V}_b(j\omega)] = \frac{1}{2}$$

Hence the step response is

$$\mathscr{F}^{-1}[\mathbf{V}_2(j\omega)] = \mathscr{F}^{-1}[\mathbf{V}_a(j\omega)] + \mathscr{F}^{-1}[\mathbf{V}_b(j\omega)]$$

$$= \frac{1}{2}\,\text{sgn}(t) - e^{-t/3}u(t) + \frac{1}{2}$$

$$= u(t) - e^{-t/3}u(t)$$

$$= (1 - e^{-t/3})u(t)$$

Suppose now that the input is changed to be $v_s(t) = 3e^{-t}u(t)$ and we wish to determine the resulting current $i(t)$. Designate the Fourier transform of $i(t)$ by $\mathbf{I}(j\omega)$. The transform of $v_s(t)$ is

$$\mathscr{F}[v_s(t)] = \mathscr{F}[3e^{-t}u(t)] = \frac{3}{j\omega + 1}$$

By mesh analysis,

$$\mathbf{V}_s(j\omega) = 3\mathbf{I}(j\omega) + \frac{1}{j\omega}\mathbf{I}(j\omega)$$

$$\frac{3}{j\omega + 1} = \left(3 + \frac{1}{j\omega}\right)\mathbf{I}(j\omega)$$

$$\mathbf{I}(j\omega) = \frac{j\omega}{(j\omega + \frac{1}{3})(j\omega + 1)} = \frac{-\frac{1}{2}}{j\omega + \frac{1}{3}} + \frac{\frac{3}{2}}{j\omega + 1}$$

Hence,

$$i(t) = \left(-\frac{1}{2}e^{-t/3} + \frac{3}{2}e^{-t}\right)u(t)$$

Also,

$$\mathbf{V}_2(j\omega) = \frac{1}{j\omega}\mathbf{I}(j\omega)$$

$$= \frac{1}{(j\omega + \frac{1}{3})(j\omega + 1)} = \frac{\frac{3}{2}}{j\omega + \frac{1}{3}} - \frac{\frac{3}{2}}{j\omega + 1}$$

so

$$v_2(t) = \frac{3}{2}(e^{-t/3} - e^{-t})u(t).$$

$$\bullet \quad \bullet \quad \bullet$$

Example

For the circuit shown in Fig. 13.26 the input is $v_1(t)$ and the output is $v_2(t)$. Representing this circuit in the frequency domain and employing nodal analysis, we obtain Fig. 13.27.

fig. 13.26

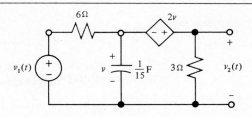

fig. 13.27

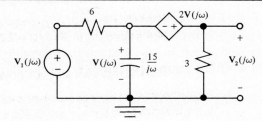

Summing currents at the supernode (let's get lazy and abbreviate $\mathbf{V}_1(j\omega)$ by $\mathbf{V}_1$, $\mathbf{V}_2(j\omega)$ by $\mathbf{V}_2$, and $\mathbf{V}(j\omega)$ by $\mathbf{V}$), we get

$$\frac{\mathbf{V}_1 - \mathbf{V}}{6} = \frac{\mathbf{V}}{15/j\omega} + \frac{\mathbf{V}_2}{3}$$

from which

$$5\mathbf{V}_1 - 10\mathbf{V}_2 = (5 + j2\omega)\mathbf{V}$$

From the dependent voltage source we get

$$2\mathbf{V} = \mathbf{V}_2 - \mathbf{V} \quad \text{or} \quad 3\mathbf{V} = \mathbf{V}_2 \Rightarrow \mathbf{V} = \frac{\mathbf{V}_2}{3}$$

Substituting this result into the second equation yields

$$5\mathbf{V}_1 - 10\mathbf{V}_2 = (5 + j2\omega)\frac{\mathbf{V}_2}{3}$$

from which

$$15\mathbf{V}_1 = (35 + j2\omega)\mathbf{V}_2$$

so that the system function is

$$\mathbf{H}(j\omega) = \frac{\mathbf{V}_2}{\mathbf{V}_1} = \frac{15/2}{j\omega + (35/2)}$$

Thus the impulse response is

$$h(t) = \mathscr{F}^{-1}[\mathbf{H}(j\omega)] = \frac{15}{2}e^{-35t/2}u(t)$$

Suppose now that the input of the circuit is the real sinusoid $v_1(t) = 9 \cos 5t$. To find the resulting output, although it is less work to use the phasor technique we studied previously, let us use Fourier transforms to demonstrate its validity.

We have

$$\mathbf{V}_1(j\omega) = 9\pi\delta(\omega - 5) + 9\pi\delta(\omega + 5)$$

Thus

$$\mathbf{V}_2(j\omega) = \frac{15/2}{j\omega + (35/2)}\mathbf{V}_1(j\omega)$$

$$= \frac{(135/2)\pi\delta(\omega - 5)}{j\omega + (35/2)} + \frac{(135/2)\pi\delta(\omega + 5)}{j\omega + (35/2)}$$

$$= \frac{(135/4)}{j5 + (35/2)}2\pi\delta(\omega - 5) + \frac{(135/4)}{-j5 + (35/2)}2\pi\delta(\omega + 5)$$

From Table 13.2,

$$v_2(t) = \frac{135/4}{j5 + (35/2)}e^{j5t} + \frac{135/4}{-j5 + (35/2)}e^{-j5t}$$

$$= \frac{\left(-j5 + \frac{35}{2}\right)\left(\frac{135}{4}e^{j5t}\right) + \left(j5 + \frac{35}{2}\right)\left(\frac{135}{4}e^{-j5t}\right)}{25 + (1225/4)}$$

$$= \frac{135}{1325}\left(\frac{35}{2}e^{j5t} + \frac{35}{2}e^{-j5t} - j5e^{j5t} + j5e^{-j5t}\right)$$

$$= \frac{27}{265}(35\cos 5t + 10\sin 5t) = \frac{27}{53}(7\cos 5t + 2\sin 5t)$$

$$= 3.71\cos(5t - 15.95°)$$

. . .

Example

In the *RLC* circuit shown in Fig. 13.28 the input is $v_1(t)$ and the output is $v_2(t)$. This circuit is represented in the frequency domain by Fig. 13.29.

fig. 13.28

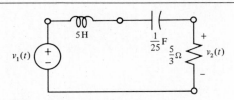

fig. 13.29

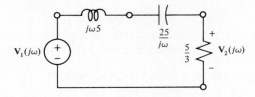

By voltage division,

$$\mathbf{V}_2(j\omega) = \frac{\frac{5}{3}}{\frac{5}{3} + (25/j\omega) + j\omega5}\mathbf{V}_1(j\omega)$$

from which the system function is

$$\mathbf{H}(j\omega) = \frac{\mathbf{V}_2(j\omega)}{\mathbf{V}_1(j\omega)} = \frac{\frac{1}{3}j\omega}{(j\omega)^2 + \frac{1}{3}j\omega + 5}$$

In obtaining a plot of the amplitude response for this system, note that when $\omega = 0$, $\mathbf{H}(j\omega) = 0$. (Remember that a capacitor is an open circuit for the dc case, and thus no current flows. The resulting dc output voltage is therefore zero.) As $\omega \to \infty$, $\mathbf{H}(j\omega) \to 0$. (The impedance of the series inductor increases with frequency and thus the current decreases. In the limit, the impedance of the inductor becomes infinite, and again the current becomes zero.) At the resonance frequency, $\omega_r = 1/\sqrt{LC} = \sqrt{5}$,

$$\mathbf{H}(j\sqrt{5}) = \frac{j\sqrt{5}/3}{-5 + (j\sqrt{5}/3) + 5} = 1$$

This result is no surprise. At the resonance frequency, the series LC combination behaves as a short circuit, so the output voltage equals the input voltage. The resulting current is the input voltage divided by the resistance. At any frequency other than resonance, the impedance seen by the source has a magnitude that is larger than the impedance at resonance. Thus the current (and hence the output voltage) will have a magnitude that is less than that at resonance. A sketch of the frequency response of the circuit is shown in Fig. 13.30.

 fig. 13.30

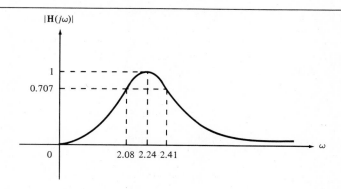

In order to find the impulse response of this circuit, we must take the inverse Fourier transform of $\mathbf{H}(j\omega)$. However, the previous expression for $\mathbf{H}(j\omega)$ is not in a readily recognizable form. Perhaps, if we factor the denominator, we can write $\mathbf{H}(j\omega)$ as a sum of simpler fractions. However, applying the quadratic formula to the denominator, we conclude that we will obtain complex roots. Thus, in this approach, the inverse Fourier transform would take the form of complex exponentials, which could then be combined to yield damped sinusoids. Alternatively, we can complete the square as follows:

$$\mathbf{H}(j\omega) = \frac{\frac{1}{3}j\omega}{(j\omega)^2 + \frac{1}{3}j\omega + 5} = \frac{\frac{1}{3}j\omega}{(j\omega + \frac{1}{6})^2 - \frac{1}{36} + 5}$$

$$= \frac{\frac{1}{3}j\omega}{(j\omega + \frac{1}{6})^2 + (179/36)} = \frac{\frac{1}{3}j\omega}{(j\omega + \frac{1}{6})^2 + (\sqrt{179}/6)^2}$$

With $\mathbf{H}(j\omega)$ in this form, damped sinusoids are suggested since

$$\mathscr{F}[K_1 e^{-\alpha t} \cos \omega_1 t \, u(t)] = \frac{K_1(j\omega + \alpha)}{(j\omega + a)^2 + \omega_1{}^2} \tag{13.5}$$

and

$$\mathscr{F}[K_2 e^{-\alpha t} \sin \omega_1 t \, u(t)] = \frac{K_2 \omega_1}{(j\omega + \alpha)^2 + \omega_1{}^2} \tag{13.6}$$

Although $\mathbf{H}(j\omega)$ is not quite in either of these two forms, we recognize that $\alpha = 1/6$ and $\omega_1 = \sqrt{179}/6$. Furthermore, we can write the numerator of $\mathbf{H}(j\omega)$ as

$$\frac{1}{3}j\omega = \frac{1}{3}\left(j\omega + \frac{1}{6}\right) - \frac{1}{18}$$

$$= \frac{1}{3}\left(j\omega + \frac{1}{6}\right) - \frac{1}{18}\left(\frac{6}{\sqrt{179}}\right)\left(\frac{\sqrt{179}}{6}\right)$$

$$= \frac{1}{3}\left(j\omega + \frac{1}{6}\right) - \frac{1}{3\sqrt{179}}\left(\frac{\sqrt{179}}{6}\right)$$

$$= \frac{1}{3}(j\omega + \alpha) - \frac{1}{3\sqrt{179}}\omega_1$$

Thus

$$\mathbf{H}(j\omega) = \frac{\frac{1}{3}(j\omega + \frac{1}{6})}{(j\omega + \frac{1}{6})^2 + (\sqrt{179}/6)^2} - \frac{(1/3\sqrt{179})(\sqrt{179}/6)}{(j\omega + \frac{1}{6})^2 + (\sqrt{179}/6)^2}$$

from which we obtain the impulse response

$$h(t) = \frac{1}{3}e^{-t/6}\cos\frac{\sqrt{179}}{6}t\,u(t) - \frac{1}{3\sqrt{179}}e^{-t/6}\sin\frac{\sqrt{179}}{6}t\,u(t)$$

$$= \frac{1}{3}e^{-t/6}\left(\cos\frac{\sqrt{179}}{6}t - \frac{1}{\sqrt{179}}\sin\frac{\sqrt{179}}{6}t\right)u(t)$$

$$= 0.334e^{-t/6}\cos(2.23t + 4.3°)u(t)$$

We can find the step response of the circuit from

$$\mathbf{V}_2(j\omega) = \frac{\frac{1}{3}j\omega}{(j\omega)^2 + \frac{1}{3}j\omega + 5}\mathbf{V}_1(j\omega)$$

$$= \frac{\frac{1}{3}j\omega}{(j\omega)^2 + \frac{1}{3}j\omega + 5}\left[\frac{1}{j\omega} + \pi\delta(\omega)\right]$$

$$= \frac{\frac{1}{3}}{(j\omega)^2 + \frac{1}{3}j\omega + 5} + \frac{(\pi/3)j\omega\delta(\omega)}{(j\omega)^2 + \frac{1}{3}j\omega + 5}$$

The inverse transform of the term with the delta function in it is zero—due to the sampling property. The other term, however, can be written as

$$\frac{\dfrac{1}{3}}{(j\omega)^2 + \dfrac{1}{3}j\omega + 5} = \frac{\dfrac{1}{3}\left(\dfrac{6}{\sqrt{179}}\right)\left(\dfrac{\sqrt{179}}{6}\right)}{\left(j\omega + \dfrac{1}{6}\right)^2 + \left(\dfrac{\sqrt{179}}{6}\right)^2}$$

which has the form in Equation (13.6). Hence the step response is

$$v_2(t) = \frac{2}{\sqrt{179}} e^{-t/6} \sin \frac{\sqrt{179}}{6} t \, u(t)$$

$$= 0.15 \, e^{-t/6} \sin 2.23t \, u(t)$$

· · ·

Example

Consider the "parallel" *RLC* circuit shown in Fig. 13.31. (Remember, as far as the *LC* tank portion is concerned, the voltage source in series with the resistance is equivalent to a current source in parallel with that same resistance.) In the frequency domain the circuit is as shown in Fig. 13.32.

fig. 13.31

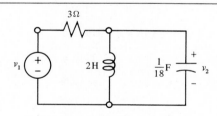

fig. 13.32

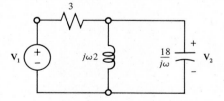

By nodal analysis,

$$\frac{V_2 - V_1}{3} + \frac{V_2}{2j\omega} + \frac{j\omega V_2}{18} = 0$$

Multiplying both sides of this equation by $18j\omega$ yields

$$6j\omega V_2 - 6j\omega V_1 + 9V_2 + (j\omega)^2 V_2 = 0$$

from which

$$[(j\omega)^2 + 6j\omega + 9]\mathbf{V}_2 = 6j\omega\mathbf{V}_1$$

Thus the system function (which describes a band-pass filter) is

$$\mathbf{H}(j\omega) = \frac{\mathbf{V}_2(j\omega)}{\mathbf{V}_1(j\omega)} = \frac{6j\omega}{(j\omega)^2 + 6j\omega + 9}$$

$$= \frac{6j\omega}{(j\omega + 3)^2}$$

We obtain the step response of this circuit from

$$\mathbf{V}_2(j\omega) = \frac{6j\omega}{(j\omega + 3)^2}\left[\frac{1}{j\omega} + \pi\delta(\omega)\right]$$

$$= \frac{6}{(j\omega + 3)^2} + \frac{6\pi j\omega\delta(\omega)}{(j\omega + 3)^2}$$

$$= \frac{6}{(j\omega + 3)^2}$$

Hence, the step response is

$$v_2(t) = 6te^{-3t}u(t)$$

Writing the system function in the form

$$\mathbf{H}(j\omega) = j\omega\left[\frac{6}{(j\omega + 3)^2}\right]$$

$$= j\omega\mathbf{F}(j\omega)$$

we then find that the inverse transform is the impulse response

$$h(t) = \frac{d}{dt}f(t)$$

$$= \frac{d}{dt}[6te^{-3t}u(t)]$$

Of course, we have long known that the impulse response is the derivative of the step response. Thus,

$$h(t) = 6t\frac{d}{dt}[e^{-3t}u(t)] + 6e^{-3t}u(t)$$

$$= 6te^{-3t}\delta(t) - 18te^{-3t}u(t) + 6e^{-3t}u(t)$$

$$= 6(1 - 3t)e^{-3t}u(t)$$

For those who feel that first finding $f(t)$ and then taking its derivative, in order to obtain the inverse transform of $\mathbf{H}(j\omega)$, is cheating (because we employed a time-domain operation), note that

$$\mathbf{H}(j\omega) = \frac{6j\omega}{(j\omega + 3)^2} = \frac{6(j\omega + 3) - 18}{(j\omega + 3)^2}$$

$$= \frac{6}{j\omega + 3} - \frac{18}{(j\omega + 3)^2}$$

and thus

$$h(t) = 6e^{-3t}u(t) - 18te^{-3t}u(t)$$

$$= 6(1 - 3t)e^{-3t}u(t)$$

as was obtained above.

It should be evident from the form of the impulse and step responses that the circuit under consideration is critically damped.

· · ·

By now you should be convinced of the usefulness of the Fourier transform. If not, then try to find the step responses of the circuits given in the most recent examples without the aid of transforms. Unfortunately, though, the Fourier transform is not the ultimate analytical tool. As an example, the Fourier transform of a ramp function does not exist. Analytically speaking, this is not a serious problem since from a time domain point of view we can employ the property that a ramp is the integral of a step.

The really important shortcoming of the Fourier transform is its inability to deal with initial conditions—assuming that such a situation is encountered. By making the appropriate modifications in the definition of the Fourier transform integral, we obtain another linear transformation—the Laplace transform. This topic is the subject discussed in the next chapter.

PROBLEMS   *13.19.   In the circuit in Fig. P13.19, the input is $v_1(t)$ and the output is $v_2(t)$.
  (a)   Find the system function for this network.
  (b)   Find the impulse response.
  (c)   Find the step response.
  (d)   Sketch the amplitude response.
  (e)   What type of filter is this circuit?

  Ans.   (a) $\dfrac{3}{j\omega + 3}$, (b) $3e^{-3t}u(t)$, (c) $(1 - e^{-3t})u(t)$

fig. P13.19

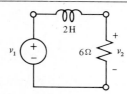

fig. P13.20

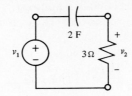

13.20. Repeat Problem 13.19 for the circuit shown in Fig. P13.20.

fig. P13.21

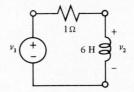

13.21. Repeat Problem 13.19 for the circuit shown in Fig. P13.21.

fig. P13.22

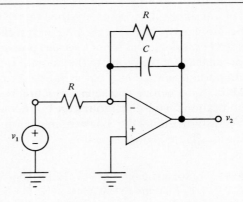

13.22. Repeat Problem 13.19 for the circuit shown in Fig. P13.22.

fig. P13.23

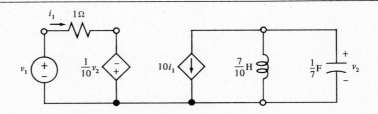

13.23. Repeat Problem 13.19 for the circuit in Fig. P13.23.

fig. P13.24

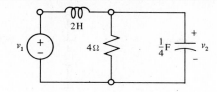

13.24.   Find the impulse response for the circuit in Fig. P13.24.

fig. P13.25

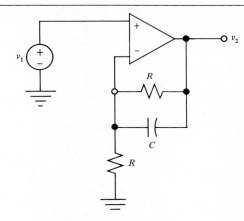

*13.25.   Consider the op amp circuit shown in Fig. P13.25.
   (a)   Find the impulse response.
   (b)   Find the step response.

Ans.   (a) $\delta(t) + \dfrac{1}{RC} e^{-t/RC} u(t)$, (b) $(2 - e^{-t/RC})\, u(t)$

fig. P13.26

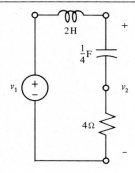

13.26.   Find the impulse response of the circuit in Fig. P13.26.

fig. P13.27

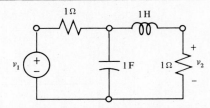

13.27.   Find the impulse response of the circuit shown in Fig. P13.27.

fig. P13.28

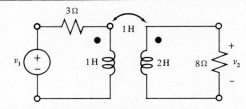

13.28.   Consider the transformer circuit in Fig. P13.28.
(a)   Find the impulse response.
(b)   Find the step response.

fig. P13.29

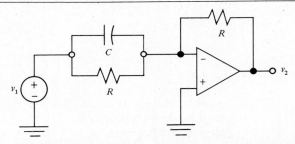

13.29.   Find the step response for the op amp circuit in Fig. P13.29.

fig. P13.30

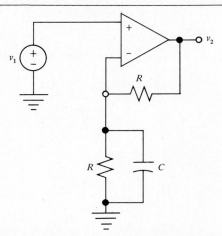

13.30.   Find the step response for the op amp circuit in Fig. P13.30.

fig. P13.31

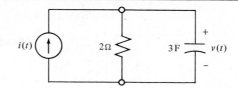

*13.31. In the circuit shown in Fig. P13.31, the input is the current $i(t) = \text{sgn}(t) + u(t)$. Find the output voltage $v(t)$.

 Ans. $2 \text{ sgn }(t) - 4e^{-t/6}u(t)$

*13.32. For the circuit given in Problem 13.19, find the output voltage $v_2(t)$ when the input voltage $v_1(t)$ is (a) $e^{-4t}u(t)$, (b) $e^{-3t}u(t)$, (c) $e^{-3t} \cos 4t \, u(t)$.

 Ans.   (b) $3te^{-3t}u(t)$

13.33. For the circuit given in Problem 13.20, find the output voltage $v_2(t)$ when the input voltage $v_1(t)$ is (a) $e^{-t/4}u(t)$, (b) $e^{-t/6}u(t)$, (c) $e^{-t/6} \cos 4t \, u(t)$.

13.34. For the circuit given in Problem 13.25, find the output voltage $v_2(t)$ when the input voltage $v_1(t)$ is (a) $e^{-t/RC}u(t)$, (b) $e^{-2t/RC}u(t)$, (c) $e^{-3t/RC}u(t)$, (d) $e^{-t/RC} \cos (1/RC)t \, u(t)$.

13.35. For the circuit given in Problem 13.28, find the output voltage $v_2(t)$ when the input voltage $v_1(t)$ is (a) $\delta(t) - 2e^{-4t}u(t)$, (b) $\delta(t) + 10e^{-2t}u(t)$.

13.36. For the circuit given in Problem 13.29, find the output voltage $v_2(t)$ when the input voltage $v_1(t)$ is (a) $e^{-t/RC} u(t)$, (b) $te^{-t/RC} u(t)$, (c) $e^{-2t/RC} u(t)$, (d) $e^{-t/RC} \sin t \, u(t)$.

*13.37. Given a linear system with impulse response $h(t)$ and output $y(t)$, find the corresponding input $x(t)$ for the following cases:

(a) $h(t) = e^{-at}u(t)$, $y(t) = e^{-at} \sin \omega_1 t \, u(t)$

(b) $h(t) = e^{-at} \cos \omega_1 t \, u(t)$, $y(t) = e^{-at}u(t)$

(c) $h(t) = (e^{-at} + e^{-bt})u(t)$, $y(t) = e^{-at}u(t)$

(d) $h(t) = e^{-at}u(t)$, $y(t) = \cos \omega_1 t \, u(t)$

 Ans.   (a) $\omega_1 e^{-at} \cos \omega_1 t \, u(t)$

*13.38. Given a linear system with input $x(t)$ and output $y(t)$, find the impulse response for the following cases:

(a) $x(t) = e^{-at} \sin \omega_1 t \, u(t)$, $y(t) = te^{-at}u(t)$

(b) $x(t) = e^{-at} \cos \omega_1 t \, u(t)$, $y(t) = e^{-at} \sin \omega_1 t \, u(t)$

(c) $x(t) = (e^{-at} + e^{-bt})u(t)$, $y(t) = te^{-at}u(t)$

(d) $x(t) = e^{-at}u(t)$, $y(t) = \sin \omega_1 t \, u(t)$

 Ans.   (b) $\omega_1 e^{-at} u(t)$

13.39. Given the system function of an ideal low-pass filter

$$\mathbf{H}(j\omega) = u(\omega + \omega_c) - u(\omega - \omega_c)$$

use the fact that

$$\mathscr{F} \left[u(t + a) - u(t - a) \right] = 2a \left(\frac{\sin a\omega}{a\omega} \right)$$

and the symmetry property (see Problem 13.18) of the Fourier transform to determine the impulse response $h(t)$. Sketch this function. Why is it not possible to construct a device with this impulse response?

Summary

1. The frequency spectrum of $f(t)$ consists of two continuous spectra. The amplitude spectrum is a plot of the magnitude of the Fourier transform of $f(t)$ versus frequency, and the phase spectrum is a plot of the angle of the Fourier transform versus frequency.
2. An energy relationship between a function and its Fourier transform is given by Parseval's theorem.
3. The Fourier transform is a linear transformation that can be used to solve differential equations or analyze circuits provided that the initial conditions are zero.
4. The use of Fourier transforms for circuit analysis results in the notion of impedance consistent with sinusoidal (phasor) analysis.
5. Multiplication of Fourier transforms corresponds to the convolution of the corresponding time functions.

THE LAPLACE TRANSFORM
14

The next stage in the evolution of techniques for circuit analysis is the generalization of the Fourier transform to the Laplace transform. One advantage of the latter over the former is the ability to deal with nonzero initial conditions.

Introduction

As was the case with the Fourier transform, the Laplace transform is a linear transformation. Also, multiplication of the Laplace transforms of two functions of time corresponds to the convolution of those two time functions.

With the use of the Laplace transform for circuit analysis, the concept of impedance as obtained for damped sinusoidal circuits results. Other important notions, such as Thévenin's theorem, are seen to conform easily to the general format. A proof of Thévenin's theorem is given in this chapter.

14.1 PROPERTIES OF THE LAPLACE TRANSFORM

In Chapter 13 we saw how to use the Fourier transform to analyze circuits, thereby eliminating the need to solve differential equations. However, we were able only to determine responses for circuits with zero initial conditions—that is, zero-state responses. By defining a new transform—the Laplace* transform—we shall be able to handle circuits with nonzero initial conditions.

The formula for the Fourier transform of $f(t)$ is

$$\mathscr{F}[f(t)] = \mathbf{F}(j\omega) = \int_{\infty}^{\infty} f(t)e^{-j\omega t}\, dt$$

When certain functions (e.g., a step function) are substituted into this formula, it cannot be evaluated since $e^{-j\omega t}$ has no limit as $t \to \infty$. However,

$$\lim_{t \to \infty} e^{-\sigma t}e^{-j\omega t} = \lim_{t \to \infty} e^{-(\sigma + j\omega)t} = 0$$

when $\sigma > 0$. Thus we define the complex variable $s = \sigma + j\omega$. In addition, from this

---

* Named for the French mathematician Marquis Pierre Simon de Laplace (1749–1827).

moment on, let us assume that unless otherwise specified, a function $f(t)$ is zero for time $t < 0$. Under this circumstance we define the *Laplace transform* of $f(t)$, denoted $\mathcal{L}[f(t)]$ or $\mathbf{F}(s)$, by

$$\mathcal{L}[f(t)] = \mathbf{F}(s) = \int_0^\infty f(t)e^{-st}\,dt$$

Although the lower limit on this integral is $t = 0$, because of possible discontinuities at this point there may be situations where the use of $t = 0$ and $t = -\varepsilon$ (where $\varepsilon > 0$ is arbitrarily small) yield different results. Under this circumstance, we shall use $t = -\varepsilon$, also denoted $t = 0^-$.

It can be shown that the *inverse Laplace transform* formula is given by

$$\mathcal{L}^{-1}[\mathbf{F}(s)] = f(t) = \frac{1}{2\pi j} \int_{c-j\infty}^{c+j\infty} \mathbf{F}(s)e^{st}\,ds$$

However, because the direct use of this formula requires some results from complex-variable theory, we shall avoid employing this integral. Instead, we shall form a table of Laplace transforms and the functions from which they come—and when we wish to find the inverse of a transform, we shall merely look up the inverse in the table.

· · ·

Example

It is a simple matter to determine the Laplace transform of a unit step function $u(t)$.

$$\mathcal{L}[u(t)] = \int_0^\infty u(t)e^{-st}\,dt = \int_0^\infty e^{-st}\,dt$$

$$= -\frac{1}{s}e^{-st}\Big|_0^\infty = -\frac{1}{s}(0-1) = \frac{1}{s}$$

It is even simpler to determine the Laplace transform of an impulse function. By the sampling property of the delta function,

$$\mathcal{L}[\delta(t)] = \int_0^\infty \delta(t)e^{-st}\,dt = 1$$

Generalizing the step function to the exponential $e^{-at}u(t)$ for $a \geq 0$, we have

$$\mathcal{L}[e^{-at}u(t)] = \int_0^\infty e^{-at}u(t)e^{-st}\,dt$$

$$= \int_0^\infty e^{-at}e^{-st}\,dt$$

$$= \frac{-1}{s+a}e^{-(s+a)t}\Big|_0^\infty$$

$$= \frac{-1}{s+a}(0-1) = \frac{1}{s+a}$$

· · ·

Suppose that $f(t) = f_1(t) + f_2(t)$. Then

$$\mathscr{L}[f(t)] = \int_0^\infty f(t)e^{-st}\, dt$$

$$= \int_0^\infty [f_1(t) + f_2(t)]e^{-st}\, dt$$

$$= \int_0^\infty f_1(t)e^{-st}\, dt + \int_0^\infty f_2(t)e^{-st}\, dt$$

so

$$\mathscr{L}[f_1(t) + f_2(t)] = \mathscr{L}[f_1(t)] + \mathscr{L}[f_2(t)]$$

Also,

$$\mathscr{L}[Kf(t)] = \int_0^\infty Kf(t)e^{-st}\, dt$$

$$= K \int_0^\infty f(t)e^{-st}\, dt$$

$$= K\mathscr{L}[f(t)]$$

Thus, the Laplace transform (like the Fourier transform) is a linear transformation.

· · ·

Example
By Equation (12.4),

$$\mathscr{L}[\sin \beta t\, u(t)] = \mathscr{L}\left[\frac{e^{j\beta t} - e^{-j\beta t}}{2j}\, u(t)\right]$$

$$= \frac{1}{2j}\, \mathscr{L}[e^{j\beta t}u(t)] - \frac{1}{2j}\, \mathscr{L}[e^{-j\beta t}u(t)]$$

But, just as $\mathscr{L}[e^{-at}u(t)] = 1/(s + a)$, it can be shown that

$$\mathscr{L}[e^{s_1 t}u(t)] = \frac{1}{s - s_1}$$

where s_1 is an arbitrary complex number. Thus

$$\mathscr{L}[\sin \beta t\, u(t)] = \frac{1}{2j}\frac{1}{s - j\beta} - \frac{1}{2j}\frac{1}{s + j\beta}$$

$$= \frac{1}{2j}\left[\frac{2j\beta}{(s - j\beta)(s + j\beta)}\right]$$

$$= \frac{\beta}{s^2 + \beta^2}$$

· · ·

Now suppose that $\mathscr{L}[f(t)] = \mathbf{F}(s)$. Then

$$\mathscr{L}\left[\frac{df(t)}{dt}\right] = \int_0^\infty \frac{df(t)}{dt}\, e^{-st}\, dt = \int_0^\infty e^{-st}\, \frac{df(t)}{dt}\, dt$$

To employ integration by parts

$$\int_a^b u\, dv = uv\Big|_a^b - \int_a^b v\, du$$

let

$$u = e^{-st} \qquad \text{and} \qquad dv = \frac{df(t)}{dt}\, dt = df(t)$$

then

$$du = -se^{-st}\, dt \qquad \text{and} \qquad v = f(t)$$

Thus

$$\mathscr{L}\left[\frac{df(t)}{dt}\right] = e^{-st}f(t)\Big|_0^\infty - \int_0^\infty f(t)[-se^{-st}]\, dt$$

and, provided that $\lim_{t\to\infty} f(t)$ does not approach infinity at a rate equal to or higher than the exponential $e^{\sigma t}$, then

$$\mathscr{L}\left[\frac{df(t)}{dt}\right] = -e^0 f(0) + s\int_0^\infty f(t)e^{-st}\, dt$$

$$= -f(0) + s\mathbf{F}(s)$$

This result is known as the *differentiation property* of the Laplace transform.

· · ·

Example

Since

$$\frac{d}{dt}[\sin \beta t\ u(t)] = \beta \cos \beta t\ u(t) + \sin \beta t\ \delta(t)$$

and since $\sin \beta t\ \delta(t) = 0\delta(t)$, then

$$\mathscr{L}\left[\frac{d}{dt}\sin \beta t\ u(t)\right] = \mathscr{L}[\beta \cos \beta t\ u(t)]$$

$$-\sin 0 + s\mathscr{L}[\sin \beta t\ u(t)] = \beta\mathscr{L}[\cos \beta t\ u(t)]$$

$$s\frac{\beta}{s^2 + \beta^2} = \beta\mathscr{L}[\cos \beta t\ u(t)]$$

Thus,

$$\mathscr{L}\left[\cos \beta t\, u(t)\right] = \frac{s}{s^2 + \beta^2}$$

. . .

We can apply the differentiation property to the second derivative of a function $f(t)$ as follows:

$$\mathscr{L}\left[\frac{d^2 f(t)}{dt^2}\right] = \mathscr{L}\left[\frac{d}{dt}\left(\frac{df(t)}{dt}\right)\right]$$

$$= -\frac{df(0)}{dt} + s\mathscr{L}\left[\frac{df(t)}{dt}\right]$$

$$= -\frac{df(0)}{dt} + s[-f(0) + s\mathbf{F}(s)]$$

$$= -\frac{df(0)}{dt} - sf(0) + s^2\mathbf{F}(s)$$

Extending this to the nth derivative, we get

$$\mathscr{L}\left[\frac{d^n f(t)}{dt^n}\right] = -\frac{d^{n-1} f(0)}{dt^{n-1}} - s\frac{d^{n-2} f(0)}{dt^{n-2}} - \cdots - s^{n-1} f(0) + s^n \mathbf{F}(s)$$

Note that $df(0)/dt$ is not the derivative of the constant $f(0)$, but is the notation for the derivative of $f(t)$ with t set equal to zero; that is,

$$\frac{df(0)}{dt} = \frac{df(t)}{dt}\bigg|_{t=0}$$

and therefore

$$\frac{d^n f(0)}{dt^n} = \frac{d^n f(t)}{dt^n}\bigg|_{t=0}$$

Having investigated the derivative of $f(t)$, let us consider its integral—that is, $\int_0^t f(t)\, dt$. If $\mathscr{L}[f(t)] = \mathbf{F}(s)$, then

$$\mathscr{L}\left[\int_0^t f(t)\, dt\right] = \mathscr{L}\left[\int_0^t f(x)\, dx\right]$$

$$= \int_0^\infty \left[\int_0^t f(x)\, dx\right] e^{-st}\, dt$$

Now let

$$u = \int_0^t f(x)\, dx \qquad \text{and} \qquad dv = e^{-st}\, dt$$

then

$$du = f(t)\, dt \qquad \text{and} \qquad v = -\frac{1}{s} e^{-st}$$

and

$$\mathscr{L}\left[\int_0^t f(t)\, dt\right] = \left[\int_0^t f(x)\, dx\right]\left[-\frac{1}{s} e^{-st}\right]\Big|_0^{\infty} - \int_0^{\infty} -\frac{1}{s} e^{-st} f(t)\, dt$$

If $\int_0^{\infty} f(x)\, dx$ does not approach infinity at a rate equal to or higher than the exponential $e^{\sigma t}$, then

$$\mathscr{L}\left[\int_0^t f(t)\, dt\right] = \left[\int_{0^-}^{0^-} f(x)\, dx\right]\left[\frac{1}{s}\right] + \frac{1}{s}\int_0^{\infty} f(t) e^{-st}\, dt$$

and since it is assumed that $f(t) = 0$ for $t < 0$,

$$\mathscr{L}\left[\int_0^t f(t)\, dt\right] = \frac{1}{s} \mathbf{F}(s)$$

This result is known as the *integration property* of the Laplace transform.

· · ·

Example
The unit ramp function $tu(t)$ is related to the unit step function $u(t)$ by

$$tu(t) = \int_{-\infty}^t u(t)\, dt = \int_0^t u(t)\, dt$$

Thus

$$\mathscr{L}\,[tu(t)] = \mathscr{L}\left[\int_0^t u(t)\, dt\right]$$

$$= \frac{1}{s}\left(\frac{1}{s}\right) = \frac{1}{s^2}$$

· · ·

Given $\mathscr{L}[f(t)] = \mathbf{F}(s)$, taking the derivative with respect to s of $\mathbf{F}(s)$ yields

$$\frac{d\mathbf{F}(s)}{ds} = \frac{d}{ds} \int_0^\infty f(t)e^{-st}\, dt$$

$$= \int_0^\infty f(t)\frac{de^{-st}}{ds}\, dt$$

$$= \int_0^\infty f(t)[-te^{-st}]\, dt$$

$$= -\int_0^\infty tf(t)e^{-st}\, dt$$

$$= -\mathscr{L}[tf(t)]$$

or

$$\mathscr{L}[tf(t)] = -\frac{d\mathbf{F}(s)}{ds}$$

Thus, multiplication by t in the time domain corresponds to (minus) differentiation in the transform (frequency) domain. This result is called the *complex differentiation property*.

· · ·

Example

$$\mathscr{L}[te^{-at}u(t)] = -\frac{d}{ds}\,\mathscr{L}[e^{-at}u(t)]$$

$$= -\frac{d}{ds}\left(\frac{1}{s+a}\right)$$

$$= \frac{1}{(s+a)^2}$$

· · ·

If $\mathscr{L}[f(t)] = \mathbf{F}(s)$, then by the definition of the Laplace transform

$$\mathscr{L}[e^{-at}f(t)] = \int_0^\infty e^{-at}f(t)e^{-st}\, dt$$

$$= \int_0^\infty f(t)e^{-(s+a)t}\, dt = \mathbf{F}(s+a)$$

which is the Laplace transform of $f(t)$ with s replaced by $s + a$. This property is known as *complex translation*.

$$\cdot \quad \cdot \quad \cdot$$

Example

Since $\mathscr{L}[tu(t)] = \dfrac{1}{s^2}$, then

$$\mathscr{L}[e^{-at}tu(t)] = \frac{1}{(s + a)^2}$$

which agrees with the result obtained in the previous example.
Furthermore, since

$$\mathscr{L}[\cos \beta t\, u(t)] = \frac{s}{s^2 + \beta^2}$$

then

$$\mathscr{L}[e^{-at} \cos \beta t\, u(t)] = \frac{s + \alpha}{(s + \alpha)^2 + \beta^2}$$

$$\cdot \quad \cdot \quad \cdot$$

If $\mathscr{L}[f(t)] = \mathbf{F}(s)$, then for $a > 0$,

$$\mathscr{L}[f(t - a)u(t - a)] = \int_0^\infty f(t - a)u(t - a)e^{-st}\, dt$$

$$= \int_a^\infty f(t - a)e^{-st}\, dt$$

Define the variable x by

$$x = t - a$$

Then

$$dx = dt$$

and

$$t = x + a$$

Thus,

$$t = a \Rightarrow x = 0$$

and

$$t = \infty \Rightarrow x = \infty$$

so

$$\mathscr{L}\left[f(t-a)u(t-a)\right] = \int_0^\infty f(x)e^{-s(x+a)}\,dt$$

$$= e^{-sa}\int_0^\infty f(x)e^{-sx}\,dx$$

$$= e^{-sa}\,\mathbf{F}(s)$$

and this result is known as the *real translation property* of the Laplace transform.

· · ·

Example
The Laplace transform of the function $f(t) = u(t) - u(t-a)$ can be determined as follows:

$$\mathscr{L}\left[f(t)\right] = \mathscr{L}\left[u(t) - u(t-a)\right]$$

$$= \mathscr{L}\left[u(t)\right] - \mathscr{L}\left[u(t-a)\right]$$

$$= \frac{1}{s} - e^{-sa}\frac{1}{s}$$

$$= \frac{1}{s}\left(1 - e^{-sa}\right)$$

· · ·

Now suppose that $f(t)$ is a periodic function for $t \geq 0$ and $f(t) = 0$ for $t < 0$. A sample of such a function is shown in Fig. 14.1. Since $f(t)$ is periodic for $t \geq 0$, it can be expressed as the sum of functions

$$f(t) = f_1(t) + f_2(t) + f_3(t) + \cdots$$

$$= \sum_{i=0}^\infty f_{i+1}(t)$$

where $f_{i+1}(t) = f_1(t - iT)$, as demonstrated in Fig. 14.2.
Thus

$$f(t) = \sum_{i=0}^\infty f_1(t - iT)$$

 fig. 14.1

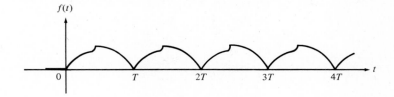

fig. 14.2

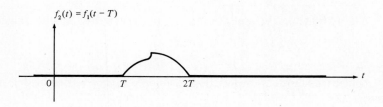

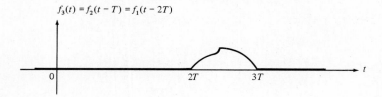

and

$$\mathscr{L}\left[f(t)\right] = \mathscr{L}\left[\sum_{i=0}^{\infty} f_1(t - iT)\right]$$

$$= \sum_{i=0}^{\infty} \mathscr{L}\left[f_1(t - iT)\right]$$

$$= \sum_{i=0}^{\infty} e^{-siT}\, \mathbf{F}_1(s)$$

where $\mathbf{F}_1(s) = \mathscr{L}\left[f_1(t)\right]$. Then $\mathbf{F}_1(s)$ is not dependent on i, so

$$\mathscr{L}\left[f(t)\right] = \mathbf{F}_1(s) \sum_{i=0}^{\infty} (e^{-sT})^i$$

But since the Taylor series expansion of the function $1/(1 - x)$ is given by

$$\frac{1}{1 - x} = 1 + x + x^2 + x^3 + \cdots = \sum_{i=0}^{\infty} x^i$$

then

$$\mathscr{L}\left[f(t)\right] = \frac{\mathbf{F}_1(s)}{1 - e^{-sT}}$$

Thus, knowing the Laplace transform of $f_1(t)$—the function obtained from the first period of $f(t)$—it is a simple matter to determine the Laplace transform of the "periodic" function $f(t)$.

$\cdot \quad \cdot \quad \cdot$

Example
Let us determine the transform of $f(t)$ shown in Fig. 14.3.

fig. 14.3

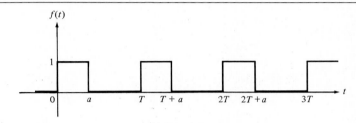

Since $f_1(t) = u(t) - u(t - a)$, from the last example,

$$F_1(s) = \frac{1}{s}(1 - e^{-sa})$$

and

$$F(s) = \frac{F_1(s)}{1 - e^{-sT}}$$

$$= \frac{1 - e^{-sa}}{s(1 - e^{-sT})}$$

$\cdot \quad \cdot \quad \cdot$

Table 14.1 summarizes the most important properties of the Laplace transform and the transforms of frequently used functions.

TABLE 14.1   Table of Laplace Transforms

$f(t)$	Property	$F(s)$
$f(t)$	Definition	$\displaystyle\int_0^\infty f(t)e^{-st}\, dt$
$f_1(t) + f_2(t)$	Linearity	$F_1(s) + F_2(s)$
$Kf(t)$	Linearity	$KF(s)$
$\dfrac{df(t)}{dt}$	Differentiation	$sF(s) - f(0)$
$\dfrac{d^n f(t)}{dt^n}$	Differentiation	$s^nF(s) - s^{n-1}f(0) - \cdots - \dfrac{d^{n-1}f(0)}{dt^{n-1}}$

TABLE 14.1 (Continued)

$f(t)$	Property	$F(s)$
$\displaystyle\int_0^t f(t)\,dt$	Integration	$\dfrac{1}{s}\,F(s)$
$tf(t)$	Complex differentiation	$-\dfrac{dF(s)}{ds}$
$e^{-at}f(t)$	Complex translation	$F(s+a)$
$f(t-a)u(t-a)$	Real translation	$e^{-sa}F(s)$
$f(t)$	Periodic function	$\dfrac{F_1(s)}{1-e^{-sT}}$
$\displaystyle\int_0^t x(\tau)h(t-\tau)\,d\tau$	Convolution	$H(s)X(s)$
$\delta(t)$		1
$u(t)$		$\dfrac{1}{s}$
$e^{-at}u(t)$		$\dfrac{1}{s+a}$
$\sin\beta t\,u(t)$		$\dfrac{\beta}{s^2+\beta^2}$
$\cos\beta t\,u(t)$		$\dfrac{s}{s^2+\beta^2}$
$e^{-\alpha t}\sin\beta t\,u(t)$		$\dfrac{\beta}{(s+\alpha)^2+\beta^2}$
$e^{-\alpha t}\cos\beta t\,u(t)$		$\dfrac{s+\alpha}{(s+\alpha)^2+\beta^2}$
$tu(t)$		$\dfrac{1}{s^2}$
$t^n u(t)$		$\dfrac{n!}{s^{n+1}}$
$te^{-at}u(t)$		$\dfrac{1}{(s+a)^2}$
$t^n e^{-at}u(t)$		$\dfrac{n!}{(s+a)^{n+1}}$

PROBLEMS *14.1.   Find the Laplace transform of each of the following functions:

(a) $te^{-t}u(t - a)$

(b) $(t - a)e^{-\alpha(t-a)}u(t - a)$

Ans.   (b) $\dfrac{e^{-as}}{(s + \alpha)^2}$

*14.2.   Repeat Problem 14.1 for the following:

(a) $\delta(t) + (a - b)e^{-bt}u(t)$

(b) $(t^3 + 1)e^{-2t}u(t)$

Ans.   (a) $\dfrac{s + a}{s + b}$

14.3.   Repeat Problem 14.1 for the following:

(a) $\sin(\beta t - \phi)u(t)$

(b) $\cos(\beta t - \phi)u(t)$

14.4.   Repeat Problem 14.1 for the following:

(a) $e^{-\alpha t}\sin \beta t\, u(t)$

(b) $e^{-\alpha t}\sin(\beta t - \phi)u(t)$

(c) $e^{-\alpha t}\cos(\beta t - \phi)u(t)$

14.5.   Repeat Problem 14.1 for the following:

(a) $\cos(t - \pi/4)u(t - \pi/4)$

(b) $\cos(t - \pi/4)u(t)$

*14.6.   Repeat Problem 14.1 for the following:

(a) $\sin t\,[u(t) - u(t - 2\pi)]$

(b) $\sin t\,[u(t) - u(t - \pi)]$

Ans.   (a) $\dfrac{1 - e^{-2\pi s}}{s^2 + 1}$

14.7.   Find the Laplace transform of $|\sin \pi t|\,u(t)$.

---

fig. P14.8

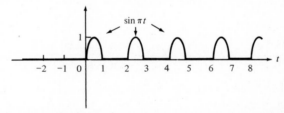

14.8.   Find the Laplace transform of the function shown in Fig. P14.8

---

fig. P14.9

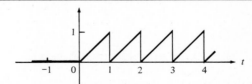

*14.9.   Find the Laplace transform of the function shown in Fig. P14.9.

Ans. $\dfrac{1 - (s + 1)e^{-s}}{(1 - e^{-s})s^2}$

14.10.   Find the Laplace transform of

(a) $\cos \alpha t \cos \beta t\, u(t)$

(b) $\sin \alpha t \cos \beta t\, u(t)$

14.2   PARTIAL FRACTION EXPANSIONS

A major application of the Laplace transform is in the solving of differential equations. Let us consider an example.

· · ·

Example

To determine the solution to the second-order differential equation

$$\frac{d^2 x(t)}{dt^2} + 3 \frac{dx(t)}{dt} + 2x(t) = 4e^{-3t}u(t)$$

subject to the initial conditions $x(0) = 2$ and $dx(0)/dt = -1$, take the Laplace transform of both sides of the equation. The result is the algebraic equation with complex coefficients

$$s^2 \mathbf{X}(s) - sx(0) - \frac{dx(0)}{dt} + 3[s\mathbf{X}(s) - x(0)] + 2\mathbf{X}(s) = \frac{4}{s+3}$$

or

$$s^2 \mathbf{X}(s) - 2s + 1 + 3s\mathbf{X}(s) - 6 + 2\mathbf{X}(s) = \frac{4}{s+3}$$

Thus

$$(s^2 + 3s + 2)\mathbf{X}(s) = \frac{4}{s+3} + 2s + 5$$

or

$$\mathbf{X}(s) = \frac{2s^2 + 11s + 19}{(s+1)(s+2)(s+3)}$$

which can be rewritten (we shall discuss the technique for accomplishing this step very soon) as follows:

$$\mathbf{X}(s) = \frac{5}{s+1} - \frac{5}{s+2} + \frac{2}{s+3}$$

From Table 14.1,

$$x(t) = (5e^{-t} - 5e^{-2t} + 2e^{-3t})u(t)$$

is the solution of the given differential equation subject to the boundary conditions stated.

· · ·

The Laplace transform can be used not only to solve a single differential equation, but also to solve a set of simultaneous differential equations.

• • •

Example

To solve the two simultaneous differential equations

$$2\frac{dx(t)}{dt} + 4x(t) + \frac{dy(t)}{dt} + 7y(t) = 5u(t)$$

$$\frac{dx(t)}{dt} + x(t) + \frac{dy(t)}{dt} + 3y(t) = 5\delta(t)$$

subject to the initial conditions $x(0) = 0$ and $y(0) = 0$, take the Laplace transform of both equations. The result is a pair of algebraic equations

$$2[sX(s) - x(0)] + 4X(s) + sY(s) - y(0) + 7Y(s) = \frac{5}{s}$$

$$sX(s) - x(0) + X(s) + sY(s) - y(0) + 3Y(s) = 5$$

which reduces to the pair of equations

$$(2s + 4)X(s) + (s + 7)Y(s) = \frac{5}{s}$$

$$(s + 1)X(s) + (s + 3)Y(s) = 5$$

which in matrix form is

$$\begin{bmatrix} 2s + 4 & s + 7 \\ s + 1 & s + 3 \end{bmatrix} \begin{bmatrix} X(s) \\ Y(s) \end{bmatrix} = \begin{bmatrix} \dfrac{5}{s} \\ 5 \end{bmatrix}$$

Forming the determinants

$$\Delta = \begin{vmatrix} 2s + 4 & s + 7 \\ s + 1 & s + 3 \end{vmatrix} = (2s + 4)(s + 3) - (s + 1)(s + 7) = s^2 + 2s + 5$$

$$\Delta_1 = \begin{vmatrix} \dfrac{5}{s} & s + 7 \\ 5 & s + 3 \end{vmatrix} = \frac{5}{s}(s + 3) - 5(s + 7) = \frac{-5s^2 - 30s + 15}{s}$$

$$\Delta_2 = \begin{vmatrix} 2s + 4 & \dfrac{5}{s} \\ s + 1 & 5 \end{vmatrix} = (2s + 4)5 - (s + 1)\frac{5}{s} = \frac{10s^2 + 15s - 5}{s}$$

By Cramer's rule, we have that

$$X(s) = \frac{\Delta_1}{\Delta} = \frac{-5s^2 - 30s + 15}{s(s^2 + 2s + 5)}$$

and

$$Y(s) = \frac{\Delta_2}{\Delta} = \frac{10s^2 + 15s - 5}{s(s^2 + 2s + 5)}$$

By the discussion following immediately, $\mathbf{X}(s)$ and $\mathbf{Y}(s)$ can be rewritten as

$$\mathbf{X}(s) = \frac{3}{s} - \frac{8s + 36}{s^2 + 2s + 5} = \frac{3}{s} - \frac{8(s + 1)}{(s + 1)^2 + 2^2} - \frac{14(2)}{(s + 1)^2 + 2^2}$$

$$\mathbf{Y}(s) = \frac{-1}{s} + \frac{11s + 17}{s^2 + 2s + 5} = -\frac{1}{s} + \frac{11(s + 1)}{(s + 1)^2 + 2^2} + \frac{3(2)}{(s + 1)^2 + 2^2}$$

From Table 14.1,

$$x(t) = (3 - 8e^{-t} \cos 2t - 14e^{-t} \sin 2t)u(t)$$

and

$$y(t) = (-1 + 11e^{-t} \cos 2t + 3e^{-t} \sin 2t)u(t)$$

$\bullet$ $\bullet$ $\bullet$

In the foregoing examples we encountered functions—in the form of ratios of polynomials in s—whose inverse transforms were desired. We obtained the inverses by expressing the function as a sum of simpler functions—fractions—and then looking up the inverse transform of each term. This decomposition of a function into a sum of simpler functions is known as a *partial fraction expansion*.

The process of taking a partial fraction expansion will be broken up into three cases. In each case we shall assume that

$$\mathbf{F}(s) = \frac{\mathbf{N}(s)}{\mathbf{D}(s)}$$

where $\mathbf{N}(s)$ and $\mathbf{D}(s)$ are polynomials in s, and deg $\mathbf{N}(s) <$ deg $\mathbf{D}(s)$. The situation that deg $\mathbf{N}(s) \geq$ deg $\mathbf{D}(s)$ will be discussed immediately following.*

Case I: Simple Real Poles

Suppose that $\mathbf{F}(s)$ can be written as

$$\mathbf{F}(s) = \frac{\mathbf{N}(s)}{(s + \alpha)\mathbf{D}_1(s)}$$

where the real number $-\alpha$ is not a root of $\mathbf{D}_1(s)$. Then we can write $\mathbf{F}(s)$ as

$$\mathbf{F}(s) = \frac{K}{s + \alpha} + \mathbf{F}_1(s)$$

To determine K, multiply both sides of this equation by $s + \alpha$, which yields

$$(s + \alpha)\mathbf{F}(s) = K + (s + \alpha)\mathbf{F}_1(s)$$

$$\frac{\mathbf{N}(s)}{\mathbf{D}_1(s)} = K + (s + \alpha)\mathbf{F}_1(s)$$

---

* The degree of polynomial $\mathbf{P}(s)$ is denoted by deg $\mathbf{P}(s)$.

Setting $s = -\alpha$, we obtain a formula for K; that is,

$$K = (s + \alpha)\mathbf{F}(s)\Big|_{s=-\alpha} = \frac{\mathbf{N}(s)}{\mathbf{D}_1(s)}\Big|_{s=-\alpha} = \frac{\mathbf{N}(-\alpha)}{\mathbf{D}_1(-\alpha)}$$

The problem of finding the partial fraction expansion of $\mathbf{F}(s)$ reduces to finding the partial fraction expansion of $\mathbf{F}_1(s)$. Clearly, $\mathbf{F}_1(s) = \mathbf{F}(s) - K/(s + \alpha)$. If $\mathbf{F}_1(s)$ has a simple real pole, we can repeat the above process.

· · ·

Example
The function

$$\mathbf{X}(s) = \frac{2s^2 + 11s + 19}{(s + 1)(s + 2)(s + 3)}$$

was encountered in a previous example. Since this function has simple poles and the degree of the denominator is greater than the degree of the numerator, the partial fraction expansion is

$$\mathbf{X}(s) = \frac{2s^2 + 11s + 19}{(s + 1)(s + 2)(s + 3)} = \frac{K_1}{s + 1} + \frac{K_2}{s + 2} + \frac{K_3}{s + 3}$$

where

$$K_1 = (s + 1)\mathbf{X}(s)\Big|_{s=-1} = \frac{2s^2 + 11s + 19}{(s + 2)(s + 3)}\Big|_{s=-1} = 5$$

$$K_2 = (s + 2)\mathbf{X}(s)\Big|_{s=-2} = \frac{2s^2 + 11s + 19}{(s + 1)(s + 3)}\Big|_{s=-2} = -5$$

$$K_3 = (s + 3)\mathbf{X}(s)\Big|_{s=-3} = \frac{2s^2 + 11s + 19}{(s + 1)(s + 2)}\Big|_{s=-3} = 2$$

Thus

$$\frac{2s^2 + 11s + 19}{(s + 1)(s + 2)(s + 3)} = \frac{5}{s + 1} - \frac{5}{s + 2} + \frac{2}{s + 3}$$

and

$$x(t) = (5e^{-t} - 5e^{-2t} + 2e^{-3t})u(t)$$

· · ·

Case II: Simple Complex Poles

Since the complex roots of a polynomial with real coefficients always appear in conjugate pairs, suppose that $\mathbf{F}(s)$ has the form

$$\mathbf{F}(s) = \frac{\mathbf{N}(s)}{(s + \alpha + j\beta)(s + \alpha - j\beta)\mathbf{D}_1(s)}$$

where $-\alpha - j\beta$ and $-\alpha + j\beta$ are not roots of $\mathbf{D}_1(s)$. Then we can write $\mathbf{F}(s)$ as

$$\mathbf{F}(s) = \frac{K}{s + \alpha + j\beta} + \frac{K^*}{s + \alpha - j\beta} + \mathbf{F}_1(s)$$

where we determine K, and hence its complex conjugate K^*, as was done for the case of simple real roots. In other words,

$$K = (s + \alpha + j\beta)\mathbf{F}(s)\bigg|_{s=-\alpha-j\beta} = \frac{\mathbf{N}(s)}{(s + \alpha - j\beta)\mathbf{D}_1(s)}\bigg|_{s=-\alpha-j\beta}$$

$$= \frac{\mathbf{N}(-\alpha - j\beta)}{(-2j\beta)\mathbf{D}_1(-\alpha - j\beta)}$$

Once K and K^* are determined, the corresponding two complex terms can be combined as follows: Suppose $K = a + jb$. Then $K^* = a - jb$, and

$$\frac{K}{s + \alpha + j\beta} + \frac{K^*}{s + \alpha - j\beta} = \frac{(a + jb)(s + \alpha - j\beta) + (a - jb)(s + \alpha + j\beta)}{(s + \alpha + j\beta)(s + \alpha - j\beta)}$$

$$= \frac{2a(s + \alpha) + 2b\beta}{(s + \alpha)^2 + \beta^2}$$

$$= \frac{2a(s + \alpha)}{(s + \alpha)^2 + \beta^2} + \frac{2b\beta}{(s + \alpha)^2 + \beta^2}$$

$$\cdots$$

Example
In a previous example we encountered the function

$$\mathbf{Y}(s) = \frac{10s^2 + 15s - 5}{s(s^2 + 2s + 5)} = \frac{10s^2 + 15s - 5}{s(s + 1 + j2)(s + 1 - j2)}$$

which has the partial fraction expansion

$$\frac{10s^2 + 15s - 5}{s(s + 1 + j2)(s + 1 - j2)} = \frac{K_0}{s} + \frac{K_1}{s + 1 + j2} + \frac{K_1^*}{s + 1 - j2}$$

where

$$K_0 = s\mathbf{Y}(s)\bigg|_{s=0} = \frac{10s^2 + 15s - 5}{s^2 + 2s + 5}\bigg|_{s=0} = -1$$

$$K_1 = (s + 1 + j2)\mathbf{Y}(s)\bigg|_{s=-1-j2} = \frac{10s^2 + 15s - 5}{s(s + 1 - j2)}\bigg|_{s=-1-j2}$$

$$= \frac{11}{2} + j\frac{3}{2} = a + jb$$

Thus

$$Y(s) = -\frac{1}{s} + \frac{\dfrac{11}{2} + j\dfrac{3}{2}}{s + 1 + j2} + \frac{\dfrac{11}{2} - j\dfrac{3}{2}}{s + 1 - j2}$$

$$= -\frac{1}{s} + \frac{11(s + 1)}{(s + 1)^2 + 2^2} + \frac{3(2)}{(s + 1)^2 + 2^2}$$

and therefore

$$y(t) = (-1 + 11e^{-t} \cos 2t + 3e^{-t} \sin 2t)u(t)$$

$$\bullet \quad \bullet \quad \bullet$$

Case III: Multiple Poles

Suppose that $\mathbf{F}(s)$ has the form

$$\mathbf{F}(s) = \frac{\mathbf{N}(s)}{(s + s_0)^n \mathbf{D}_1(s)}$$

where $-s_0$ is not a root of $\mathbf{D}_1(s)$ and in general is complex. Then we can write $\mathbf{F}(s)$ as

$$\mathbf{F}(s) = \frac{K_1}{s + s_0} + \frac{K_2}{(s + s_0)^2} + \cdots + \frac{K_{n-1}}{(s + s_0)^{n-1}} + \frac{K_n}{(s + s_0)^n} + \mathbf{F}_1(s)$$

Multiplying both sides of this equation by $(s + s_0)^n$ results in

$$(s + s_0)^n \mathbf{F}(s) = K_1(s + s_0)^{n-1} + K_2(s + s_0)^{n-2} + \cdots$$
$$+ K_{n-1}(s + s_0) + K_n + (s + s_0)^n \mathbf{F}_1(s)$$

Setting $s = -s_0$, we get

$$K_n = (s + s_0)^n \mathbf{F}(s)|_{s = -s_0}$$

To find K_{n-1}, after multiplication by $(s + s_0)^n$, take the derivative of both sides with respect to s. Then

$$\frac{d}{ds}[(s + s_0)^n \mathbf{F}(s)] = (n - 1)K_1(s + s_0)^{n-2} + (n - 2)K_2(s + s_0)^{n-3} + \cdots$$
$$+ 2K_{n-2}(s + s_0) + K_{n-1} + 0 + (s + s_0)^n \frac{d\mathbf{F}_1(s)}{ds}$$
$$+ n(s + s_0)^{n-1}\mathbf{F}_1(s)$$

Setting $s = -s_0$ in this equation, we get

$$K_{n-1} = \frac{d}{ds}[(s + s_0)^n \mathbf{F}(s)]\bigg|_{s = -s_0}$$

Repeating this process,

$$2K_{n-2} = \frac{d^2}{ds^2}[(s + s_0)^n \mathbf{F}(s)]\bigg|_{s = -s_0}$$

so

$$K_{n-2} = \frac{1}{2} \frac{d^2}{ds^2} \left[(s + s_0)^n \mathbf{F}(s) \right] \Bigg|_{s = -s_0}$$

In general,

$$K_{n-r} = \frac{1}{r!} \frac{d^r}{ds^r} \left[(s + s_0)^n \mathbf{F}(s) \right] \Bigg|_{s = -s_0}$$

for $r = 0, 1, 2, \ldots, n - 1$.

$\bullet \quad \bullet \quad \bullet$

Example
The function

$$\mathbf{F}(s) = \frac{s - 2}{s(s + 1)^3}$$

has the partial fraction expansion

$$\frac{s - 2}{s(s + 1)^3} = \frac{K_0}{s} + \frac{K_1}{s + 1} + \frac{K_2}{(s + 1)^2} + \frac{K_3}{(s + 1)^3}$$

where

$$K_0 = s\mathbf{F}(s) \Big|_{s=0} = \frac{s - 2}{(s + 1)^3} \Big|_{s=0} = -2$$

$$K_3 = (s + 1)^3 \mathbf{F}(s) \Big|_{s=-1} = \frac{s - 2}{s} \Big|_{s=-1} = 3$$

$$K_2 = \frac{d}{ds} \left[(s + 1)^3 \mathbf{F}(s) \right] \Big|_{s=-1} = \frac{d}{ds} \left[\frac{s - 2}{s} \right] \Big|_{s=-1} = \frac{2}{s^2} \Big|_{s=-1} = 2$$

$$K_1 = \frac{1}{2} \frac{d^2}{ds^2} \left[(s + 1)^3 \mathbf{F}(s) \right] \Big|_{s=-1} = \frac{1}{2} \frac{d}{ds} \left[\frac{2}{s^2} \right] \Big|_{s=-1} = -\frac{2}{s^3} \Big|_{s=-1} = 2$$

Thus

$$\mathbf{F}(s) = -\frac{2}{s} + \frac{2}{s + 1} + \frac{2}{(s + 1)^2} + \frac{3}{(s + 1)^3}$$

From Table 14.1,

$$\mathscr{L} \left[\frac{t^n}{n!} e^{-at} u(t) \right] = \frac{1}{(s + a)^{n+1}}$$

Hence

$$f(t) = \left(-2 + 2e^{-t} + 2te^{-t} + \frac{3}{2} t^2 e^{-t} \right) u(t)$$

$\bullet \quad \bullet \quad \bullet$

In the three cases of partial fraction expansions discussed above, it was assumed that the degree of the denominator polynomial is greater than the degree of the numerator polynomial. For a situation in which this condition does not hold, divide the numerator by the denominator, and then express $F(s) = N(s)/D(s)$ as

$$F(s) = \frac{N(s)}{D(s)} = Q(s) + \frac{R(s)}{D(s)}$$

where $Q(s)$ is the quotient and $R(s)$ is the remainder. Long division (called the *Euclidean division algorithm*) guarantees that deg $R(s) <$ deg $D(s)$. Thus we can take a partial fraction expansion of $R(s)/D(s)$ as described previously.

$\cdots$

Example
For the function

$$F(s) = \frac{2s^2 + s + 3}{(s + 1)(s + 2)} = \frac{2s^2 + s + 3}{s^2 + 3s + 2}$$

dividing the denominator into the numerator, we get the partial fraction expansion

$$F(s) = 2 - \frac{5s + 1}{(s + 1)(s + 2)} = 2 + \frac{4}{s + 1} - \frac{9}{s + 2}$$

from which

$$f(t) = 2\delta(t) + (4e^{-t} - 9e^{-2t})u(t)$$

$\cdots$

In the previous discussion we learned how to take partial fraction expansions algebraically. In addition, if $F(s)$ has simple poles, we can also take partial fraction expansions graphically. For suppose that

$$F(s) = \frac{N(s)}{D(s)} = \frac{K(s - z_1)(s - z_2)\cdots(s - z_n)}{(s - p_1)(s - p_2)\cdots(s - p_m)}$$

where $m > n$ and the poles and zeros are, in general, complex. Taking a partial fraction expansion of $F(s)$, we get

$$F(s) = \frac{K_1}{s - p_1} + \frac{K_2}{s - p_2} + \cdots + \frac{K_m}{s - p_m}$$

where

$$K_i = (s - p_i)F(s)|_{s = p_i}$$

is called the *residue* of the ith pole. (The term "residue" is used only in conjunction with simple poles.) Thus,

$$K_i = \frac{K(p_i - z_1)(p_i - z_2)\cdots(p_i - z_n)}{(p_i - p_1)\cdots(p_i - p_{i-1})(p_i - p_{i+1})\cdots(p_i - p_m)}$$

But, in the s-plane, $p_i - z_j$ is the vector from the jth zero to the ith pole; and $p_i - p_k$ is the vector from the kth pole to the ith pole. Thus

$$K_i = \frac{K(N_{i1}\underline{/\phi_{i1}})(N_{i2}\underline{/\phi_{i2}})\cdots(N_{in}\underline{/\phi_{in}})}{(M_{i1}\underline{/\theta_{i1}})\cdots(M_{i,i-1}\underline{/\theta_{i,i-1}})(M_{i,i+1}\underline{/\theta_{i,i+1}})\cdots(M_{im}\underline{/\theta_{im}})}$$

where

N_{ij} = the magnitude of $p_i - z_j$

ϕ_{ij} = the angle of $p_i - z_j$

M_{ik} = the magnitude of $p_i - p_k$

θ_{ik} = the angle of $p_i - p_k$

Therefore, from a pole-zero plot of $\mathbf{F}(s)$ we can determine the residues of the poles, and hence the partial fraction expansion of $\mathbf{F}(s)$, by graphical means.

$\cdots$

Example
Consider the function

$$\mathbf{F}(s) = \frac{s^2 + 6s + 8}{s^3 + 2s^2 + 4s} = \frac{(s+2)(s+4)}{s(s+1+j\sqrt{3})(s+1-j\sqrt{3})}$$

which has the partial fraction expansion

$$\mathbf{F}(s) = \frac{K_0}{s} + \frac{K_1}{s+1+j\sqrt{3}} + \frac{K_1{}^*}{s+1-j\sqrt{3}}$$

To determine K_0, draw vectors in the s-plane from each zero to the pole at the origin, as well as from each of the other poles to this pole. The result is shown in Fig. 14.4. Thus

$$K_0 = \frac{(2\underline{/0°})(4\underline{/0°})}{(2\underline{/60°})(2\underline{/-60°})} = 2$$

fig. 14.4

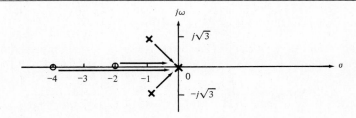

To determine K_1, draw vectors from each zero to the pole at $s = -1 - j\sqrt{3}$, as well as from each of the other poles. The resulting picture is shown in Fig. 14.5 and

$$K_1 = \frac{(2\underline{/-60°})(\sqrt{12}\underline{/-30°})}{(2\underline{/-120°})(2\sqrt{3}\underline{/-90°})} = 1\underline{/120°}$$

$$= -\frac{1}{2} + j\frac{\sqrt{3}}{2} = a + jb$$

so

$$K_1^* = -\frac{1}{2} - j\frac{\sqrt{3}}{2} = a - jb$$

fig. 14.5

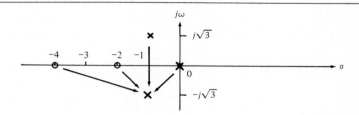

The resulting partial fraction expansion

$$F(s) = \frac{2}{s} + \frac{-\frac{1}{2} + j\frac{\sqrt{3}}{2}}{s + 1 + j\sqrt{3}} + \frac{-\frac{1}{2} - j\frac{\sqrt{3}}{2}}{s + 1 - j\sqrt{3}}$$

yields

$$F(s) = \frac{2}{s} + \frac{-(s + 1)}{(s + 1)^2 + 3} + \frac{\sqrt{3}(\sqrt{3})}{(s + 1)^2 + 3}$$

and hence

$$f(t) = (2 - e^{-t}\cos\sqrt{3}t + \sqrt{3}e^{-t}\sin\sqrt{3}t)u(t)$$

$\cdot \quad \cdot \quad \cdot$

PROBLEMS *14.11. Find the inverse Laplace transforms of the following functions:

(a) $\dfrac{600}{s(s + 10)(s + 30)}$ (b) $\dfrac{60(s + 4)}{s(s + 2)(s + 12)}$

Ans. (a) $(2 - 3e^{-10t} + e^{-30t})u(t)$

14.12. Repeat Problem 14.11 for the following:

(a) $\dfrac{s + a}{s + b}$ (b) $\dfrac{se^{-as}}{s + b}$

14.13.   Repeat Problem 14.11 for the following:

(a) $\dfrac{12s}{(s+3)(s^2+9)}$ (b) $\dfrac{4(s^2+1)}{s(s^2+4)}$

*14.14.   Repeat Problem 14.11 for the following:

(a) $\dfrac{6(s+1)^3}{s^4}$ (b) $\dfrac{(s+2)(s+3)}{s(s+1)^2}$ (c) $\dfrac{4(1-e^{-2s})}{s^2(s+2)}$

Ans. (b) $(6 - 5e^{-t} - 2te^{-t})u(t)$

*14.15.   Use Laplace transforms to solve the differential equation

$$\frac{d^2x}{dt^2} + 7\frac{dx}{dt} + 6x = 36u(t)$$

subject to the initial conditions $x(0) = -4$ and $dx(0)/dt = 0$.
Ans. $(6 - 12e^{-t} + 2e^{-6t})u(t)$

14.16.   Repeat Problem 14.15 for

$$\frac{d^2x}{dt^2} + 3\frac{dx}{dt} + 2x = 20 \cos 2t\, u(t)$$

subject to $x(0) = dx(0)/dt = 0$.

14.17.   Use Laplace transforms to solve the simultaneous differential equations

$$\frac{dx}{dt} + 3x + y = 0$$

$$x - \frac{dy}{dt} - y = 0$$

subject to the initial conditions $x(0) = 1$ and $y(0) = 2$.

*14.18.   Repeat Problem 14.17 for

$$\frac{dx}{dt} + 3x + \frac{dy}{dt} = 30u(t)$$

$$\frac{dx}{dt} + 2\frac{dy}{dt} + 8y = 0$$

subject to $x(0) = y(0) = 0$.
Ans. $20(1 - e^{-3t})u(t)$, $30(e^{-4t} - e^{-3t})u(t)$

14.19.   Use the graphical evaluation of residues to find the inverse transforms of the functions given in Problem 14.11.

14.20.   Repeat Problem 14.19 for the functions given in Problem 14.13.

14.3 APPLICATION TO LINEAR SYSTEMS

Consider the linear time-invariant system with input $x(t)$ and output $y(t)$ as depicted in Fig. 14.6. Since such a system can be described by a linear differential equation with constant coefficients, we can take the Laplace transform of this equation. If the system

fig. 14.6

$x(t)$ $y(t)$

is in the zero state (the initial conditions are zero), as was done in the similar discussion for the Fourier transform, there is a function $\mathbf{H}(s)$ such that

$$\mathbf{Y}(s) = \mathbf{H}(s)\mathbf{X}(s)$$

where $\mathbf{X}(s) = \mathscr{L}\,[x(t)]$ and $\mathbf{Y}(s) = \mathscr{L}\,[y(t)]$. From this equation,

$$\mathbf{H}(s) = \frac{\mathbf{Y}(s)}{\mathbf{X}(s)}$$

where $\mathbf{H}(s)$ is the *transfer function* of the system.

One simple but important system is shown in Fig. 14.7. Since

$$\mathbf{Y}(s) = \frac{1}{s}\,\mathbf{X}(s)$$

fig. 14.7

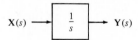

$\mathbf{X}(s)$ $\dfrac{1}{s}$ $\mathbf{Y}(s)$

then

$$y(t) = \mathscr{L}^{-1}\left[\frac{1}{s}\,\mathbf{X}(s)\right]$$

and, from Table 14.1,

$$y(t) = \int_0^t x(t)\,dt$$

That is, the output is the integral of the input. Thus this system is an *integrator*.

By using integrators and two other simple components, we can "build" or "simulate" transfer functions and thereby simulate more complex linear systems. This is the essence of the analog computer.

One of the components is a *scaler*, or *amplifier*, whose output is the input multiplied by a constant A. This device is designated as shown in Fig. 14.8. The other component is an *adder*, whose output is the sum of its inputs. For the case of three inputs, an adder is depicted as in Fig. 14.9.

fig. 14.8

x A Ax

fig. 14.9

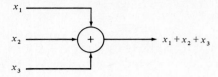

To simulate the transfer function

$$H(s) = \frac{Y(s)}{X(s)} = \frac{a_n s^n + a_{n-1} s^{n-1} + \cdots + a_1 s + a_0}{s^m + b_{m-1} s^{m-1} + \cdots + b_1 s + b_0}$$

where $m \geq n$, divide numerator and denominator by s^m. The result is

$$\frac{Y(s)}{X(s)} = \frac{\dfrac{a_n}{s^{m-n}} + \dfrac{a_{n-1}}{s^{m-n+1}} + \cdots + \dfrac{a_1}{s^{m-1}} + \dfrac{a_0}{s^m}}{1 + \dfrac{b_{m-1}}{s} + \cdots + \dfrac{b_1}{s^{m-1}} + \dfrac{b_0}{s^m}}$$

From this equation we get

$$Y(s) = \frac{a_n}{s^{m-n}} X(s) + \frac{a_{n-1}}{s^{m-n+1}} X(s) + \cdots + \frac{a_1}{s^{m-1}} X(s) + \frac{a_0}{s^m} X(s) - \frac{b_{m-1}}{s} Y(s) - \cdots$$

$$- \frac{b_1}{s^{m-1}} Y(s) - \frac{b_0}{s^m} Y(s)$$

which has the form $Y(s) = X_a(s) + Y_b(s)$. To realize the first $n + 1$ terms on the right-hand side of this equation, we need only take successive integrations of the input and scale them appropriately as shown in Fig. 14.10.

fig. 14.10

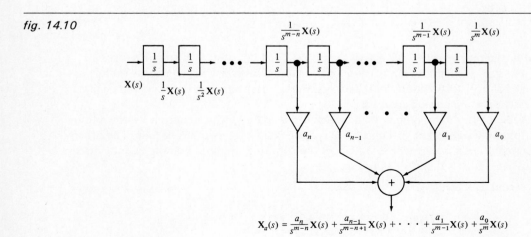

$$X_a(s) = \frac{a_n}{s^{m-n}} X(s) + \frac{a_{n-1}}{s^{m-n+1}} X(s) + \cdots + \frac{a_1}{s^{m-1}} X(s) + \frac{a_0}{s^m} X(s)$$

fig. 14.11

$$\mathbf{Y}_b(s) = -\frac{b_{m-1}}{s}\mathbf{Y}(s) - \cdots - \frac{b_1}{s^{m-1}}\mathbf{Y}(s) - \frac{b_0}{s^m}\mathbf{Y}(s)$$

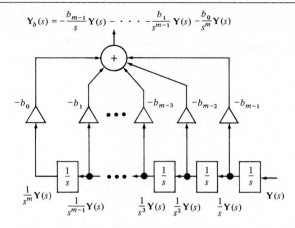

The remaining terms are obtained similarly from the output, as shown in Fig. 14.11. Since

$$\mathbf{Y}(s) = \mathbf{X}_a(s) + \mathbf{Y}_b(s)$$

we simply combine the separate realizations as shown in Fig. 14.12, to obtain the final result.

fig. 14.12

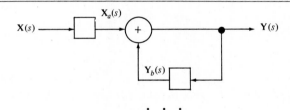

• • •

Example

To simulate the transfer function

$$\mathbf{H}(s) = \frac{4s^2 + 5s + 6}{s^2 + 2s + 3}$$

first divide numerator and denominator by s^2. Thus,

$$\frac{\mathbf{Y}(s)}{\mathbf{X}(s)} = \frac{4 + \dfrac{5}{s} + \dfrac{6}{s^2}}{1 + \dfrac{2}{s} + \dfrac{3}{s^2}}$$

From this,

$$\mathbf{Y}(s) = 4\mathbf{X}(s) + \frac{5}{s}\mathbf{X}(s) + \frac{6}{s^2}\mathbf{X}(s) - \frac{2}{s}\mathbf{Y}(s) - \frac{3}{s^2}\mathbf{Y}(s)$$

which can be realized by Fig. 14.13. Of course, the two three-input adders can be replaced by one five-input adder.

fig. 14.13

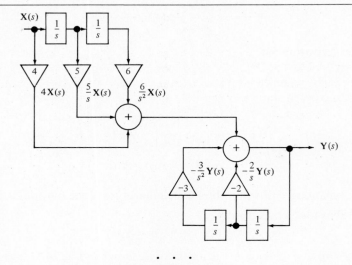

Consider the linear system shown in Fig. 14.14. When the system is in the zero state,

$$Y(s) = H(s)X(s)$$

fig. 14.14

For the case that the input is $x(t) = \delta(t)$, then $X(s) = 1$ and

$$Y(s) = H(s)$$

Thus the response to the unit delta function $\delta(t)$ is

$$y(t) = \mathscr{L}^{-1}[Y(s)] = \mathscr{L}^{-1}[H(s)]$$

Define the function $h(t)$ by

$$h(t) = \mathscr{L}^{-1}[H(s)]$$

For this reason, $h(t)$ is called the *impulse response* of the system.

The relationship between input and output transforms is a simple one:

$$Y(s) = H(s)X(s)$$

But what is the relationship between the input $x(t)$ and the output $y(t)$? It is not true that the output is the product of the input and impulse response; that is,

$$y(t) \neq h(t)x(t)$$

as you may be tempted to believe. The relationship is much more complicated.

By an approach similar to that done in the case of Fourier transforms, it is not difficult to show that

$$y(t) = \int_0^\infty x(\tau)h(t - \tau)\, d\tau$$

In the derivation of this formula, we use the fact that, for a real system, the impulse response $h(t) = 0$ for $t < 0$; that is, the response to an input cannot occur until the input is applied. Such systems are said to be *causal*. No one yet (aside from H. G. Wells) has been able to construct a noncausal system.

Furthermore, since

$$h(t) = 0 \qquad \text{for } t < 0$$

then

$$h(t - \tau) = 0 \qquad \text{for } t - \tau < 0 \,(\text{or } t < \tau)$$

Thus the above integral can be written as

$$\boxed{y(t) = \int_0^t x(\tau)h(t - \tau)\, d\tau}$$

When using the Laplace transform, it is assumed that a function $f(t) = 0$ for $t < 0$. Thus, interchanging the functions $x(t)$ and $h(t)$ results in the additional formula,

$$\boxed{y(t) = \int_0^t h(\tau)x(t - \tau)\, d\tau}$$

We refer to either of the previous two integrals as the *convolution integral*, and the process they describe is called *convolution*. Thus we see that multiplication in the frequency domain corresponds to convolution in the time domain.

Evaluating the convolution integral is not complicated, but it is quite subtle—it basically is a matter of bookkeeping. The form of bookkeeping, however, is best done in a graphical manner. The procedure will be illustrated with some examples.

· · ·

Example

The input $x(t) = tu(t)$ is applied to a system whose impulse response is $h(t) = e^{-2t}u(t)$. To determine the output by convolution, first sketch $x(\tau)$ versus τ and $h(\tau)$ versus τ. From $h(\tau)$ versus τ, we form $h(-\tau)$ versus $-\tau$, then $h(t - \tau)$ versus $-\tau$, and finally $h(t - \tau)$ versus τ. We then take the product $x(\tau)h(t - \tau)$ and integrate it with respect to τ to calculate $y(t)$. These steps are summarized in Fig. 14.15.

fig. 14.15

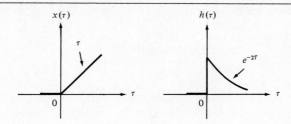

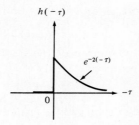

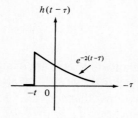

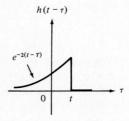

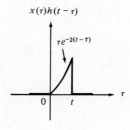

Therefore, for $0 \le t < \infty$,

$$y(t) = \int_0^t x(\tau)h(t-\tau)\,d\tau$$

$$= \int_0^t \tau e^{-2(t-\tau)}\,d\tau$$

$$= e^{-2t}\int_0^t \tau e^{2\tau}\,d\tau$$

$$= e^{-2t}\left(\frac{1}{2}\tau e^{2\tau}\Big|_0^t - \int_0^t \frac{1}{2}e^{2\tau}\,d\tau\right)$$

$$= e^{-2t}\left(\frac{1}{2}t e^{2t} - \frac{1}{4}e^{2t} + \frac{1}{4}\right)$$

$$= \frac{t}{2} - \frac{1}{4} + \frac{1}{4}e^{-2t}$$

Since $y(t) = 0$ for $t < 0$ (the output is zero before the input is applied), we have

$$y(t) = \left(\frac{t}{2} - \frac{1}{4} + \frac{1}{4}e^{-2t}\right)u(t)$$

which is the answer that is obtained when Laplace transform techniques are used to determine $y(t)$.

Suppose now that the input is changed to $x(t) = 2u(t) - 2u(t-1)$. Then we have the function shown in Fig. 14.16. In this case, though, in evaluating the convolution integral the limits will not be the same for all values of t. In particular, for $0 \le t < 1$, we get the plot shown in Fig. 14.17. Thus

$$y(t) = \int_0^t x(\tau)h(t-\tau)\,d\tau$$

$$= \int_0^t 2e^{-2(t-\tau)}\,d\tau$$

$$= 1 - e^{-2t} \qquad \text{for } 0 \le t < 1$$

fig. 14.16

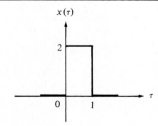

fig. 14.17

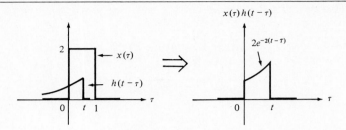

However, for $1 \le t < \infty$, we obtain the plot shown in Fig. 14.18. Therefore

$$y(t) = \int_0^t x(\tau)h(t - \tau) \, d\tau$$

$$= \int_0^1 2e^{-2(t-\tau)} \, d\tau$$

$$= e^{-2(t-1)} - e^{-2t} \qquad \text{for } 1 \le t < \infty$$

fig. 14.18

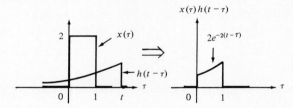

In summary, together with the fact that the output is zero before the input is applied, we have

$$y(t) = \begin{cases} 0 & \text{for } -\infty < t < 0 \\ 1 - e^{-2t} & \text{for } 0 \le t < 1 \\ e^{-2(t-1)} - e^{-2t} & \text{for } 1 \le t < \infty \end{cases}$$

An alternative, but equivalent, expression for $y(t)$ is

$$y(t) = (1 - e^{-2t})u(t) - (1 - e^{-2(t-1)})u(t - 1)$$

This form for $y(t)$ is obtained directly when we solve the same problem with the use of Laplace transforms.

Finally, suppose that the input is changed to $x(t) = 2u(t - 1) - 2u(t - 2)$. For demonstration purposes, we shall find $y(t)$ by using the alternative form of the convolution integral. We get the functions shown in Fig. 14.19.

Thus, for $-1 \le t - 1 < 0$ (i.e., for $0 \le t < 1$),

$$x(t - \tau)h(\tau) = 0$$

and

$$y(t) = \int_0^t x(t - \tau)h(\tau) \, d\tau = 0$$

fig. 14.19

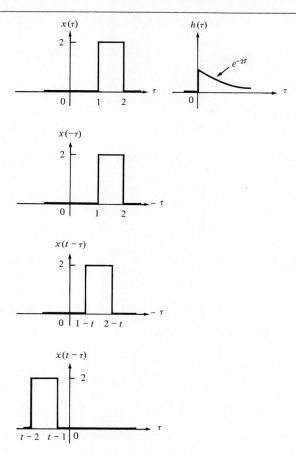

For $0 \le t - 1 < 1$ (or $1 \le t < 2$), we obtain the plots shown in Fig. 4.20. We therefore have

$$y(t) = \int_0^t x(t - \tau)h(\tau) \, d\tau$$

$$= \int_0^{t-1} 2e^{-2\tau} \, d\tau$$

$$= 1 - e^{-2(t-1)} \qquad \text{for } 1 \le t < 2$$

fig. 14.20

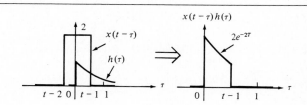

For $1 \leq t - 1 < \infty$ (or $2 \leq t < \infty$), the plots are as shown in Fig. 4.21. Hence

$$y(t) = \int_0^t x(t - \tau)h(\tau) \, d\tau$$

$$= \int_{t-2}^{t-1} 2e^{-2\tau} \, d\tau$$

$$= e^{-2(t-2)} - e^{-2(t-1)} \qquad \text{for} \qquad 2 \leq t < \infty$$

fig. 14.21

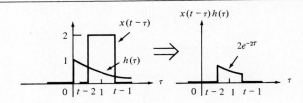

and therefore

$$y(t) = \begin{cases} 0 & \text{for} \quad -\infty < t < 1 \\ 1 - e^{-2(t-1)} & \text{for} \quad 1 \leq t < 2 \\ e^{-2(t-2)} - e^{-2(t-1)} & \text{for} \quad 2 \leq t < \infty \end{cases}$$

which can also be written in a single expression as

$$y(t) = (1 - e^{-2(t-1)})u(t - 1) - (1 - e^{-2(t-2)})u(t - 2)$$

Note that this response is equal to the previous response delayed by 1 second. This is due to the fact that the system is time-invariant and the present input is the previous input delayed by 1 second. Thus we could have determined this response simply by inspection.

• • •

PROBLEMS   14.21.   Using integrators, amplifiers, and adder(s), simulate the transfer function

$$H(s) = \frac{6s}{(s + 3)(s^2 + 9)}$$

14.22.   Repeat Problem 14.21 for

$$H(s) = \frac{12(s + 4)}{s(s + 2)(s + 12)}$$

*14.23.   Find the transfer function of the system shown in Fig. P14.23.

Ans. $\dfrac{s^2 + 1}{s^2 + 7s + 9}$

fig. P14.23

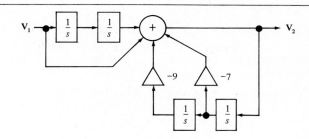

14.24. A system has the transfer function

$$\mathbf{H}(s) = \frac{s(s+1)}{(s+2)(s+3)}$$

Find the input $x(t)$ that will result in the output

$$y(t) = e^{-t}u(t) - e^{-(t-2)}u(t-2)$$

14.25. A system whose impulse response is $h(t) = \delta(t) - 11e^{-10t}u(t)$ has an output of $y(t) = (1 - 11t)e^{-10t}u(t)$. Find the input $x(t)$.

*14.26. Find the impulse response of the system whose output is $y(t) = e^{-t}u(t)$ when the input is $x(t) = e^{-2t}u(t)$.
Ans. $\delta(t) + e^{-t}u(t)$

14.27. Use convolution to determine the output of a system whose impulse response is $h(t) = e^{-2t}u(t)$ and whose input is (a) $x(t) = e^{-t}u(t)$ and (b) $x(t) = 2u(t) - 2u(t-1)$.

*14.28. Use convolution to find $y(t)$ when $h(t) = e^{-10t}u(t)$ and $x(t) = \delta(t) - 11e^{-10t}u(t)$.
Ans. $(1 - 11t)e^{-10t}u(t)$

fig. P14.29

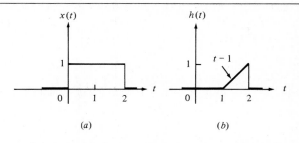

(a)　　　　　(b)

14.29. Repeat Problem 14.28 for the functions shown in Fig. P14.29.

14.30. Repeat Problem 14.28 for
(a) $h(t) = e^{-t}u(t)$, $x(t) = (1 - e^{-t})u(t)$
(b) $h(t) = e^{-at}u(t)$, $x(t) = 2e^{-at}u(t)$.

14.31. Repeat Problem 14.28 for $h(t) = 2u(t) - 2u(t-2)$ and $x(t) = \delta(t-1) - \delta(t-2)$.

14.32. Repeat Problem 14.28 for $h(t) = u(t) - 2u(t-1) + u(t-2)$ and $x(t) = 2u(t-1) - 2u(t-2)$.

14.4   NETWORK ANALYSIS

Since the Laplace transform can be used to solve linear differential equations with constant coefficients, it can be employed to analyze linear time-invariant circuits. One approach is to write the appropriate differential equations, and then take transforms. Even simpler is to take the transforms of individual components initially, and then apply Kirchhoff's laws and Ohm's law and proceed as before.

To begin with, consider the resistor (Fig. 14.22). Since

$$v(t) = Ri(t)$$

taking the Laplace transform yields

$$\mathscr{L}[v(t)] = \mathscr{L}[Ri(t)] = R\mathscr{L}[i(t)]$$

so

$$\mathbf{V}(s) = R\mathbf{I}(s)$$

fig. 14.22 (left)
fig. 14.23 (right)

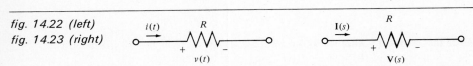

The transfer function $\mathbf{V}(s)/\mathbf{I}(s)$ is called the *impedance* of the element, while $\mathbf{I}(s)/\mathbf{V}(s)$ is the *admittance*. Thus, the impedance of the resistor, designated by $\mathbf{Z}_R(s)$, is

$$\mathbf{Z}_R(s) = \frac{\mathbf{V}(s)}{\mathbf{I}(s)} = R$$

Thus, in the frequency domain, we can model the resistor as shown in Fig. 14.23.

fig. 14.24

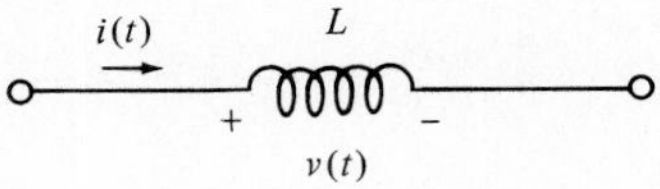

For the inductor, depicted in Fig. 14.24, since

$$v(t) = L\frac{di(t)}{dt}$$

then

$$\mathbf{V}(s) = L[s\mathbf{I}(s) - i(0)]$$

$$= Ls\mathbf{I}(s) - Li(0)$$

Remembering that the transfer function is defined for a system (component) in the zero state, for an inductor in the zero state (zero initial current), $i(0) = 0$ and

$$\mathbf{V}(s) = Ls\mathbf{I}(s)$$

Thus, the impedance of an inductor, denoted $\mathbf{Z}_L(s)$, is

$$\mathbf{Z}_L(s) = \frac{\mathbf{V}(s)}{\mathbf{I}(s)} = Ls$$

and an inductor with zero initial current, therefore, can be modeled by Fig. 14.25. For the case of nonzero initial current, since

$$\mathbf{V}(s) = Ls\mathbf{I}(s) - Li(0)$$

fig. 14.25 (left)
fig. 14.26 (right)

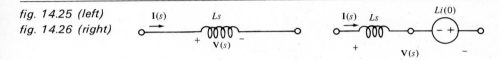

a general frequency domain model is given in Fig. 14.26. Rewriting the preceding equation as

$$\mathbf{I}(s) = \frac{\mathbf{V}(s)}{Ls} + \frac{i(0)}{s}$$

or by using a source transformation, we obtain the alternative general inductor model shown in Fig. 14.27.

fig. 14.27

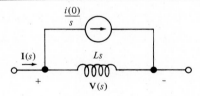

For the capacitor, depicted in Fig. 14.28, since

$$i(t) = C\frac{dv(t)}{dt}$$

then

$$\mathbf{I}(s) = C[s\mathbf{V}(s) - v(0)]$$
$$= Cs\mathbf{V}(s) - Cv(0)$$

For an initially uncharged capacitor, $v(0) = 0$, we have

$$\mathbf{I}(s) = Cs\mathbf{V}(s)$$

fig. 14.28 (left)
fig. 14.29 (right)

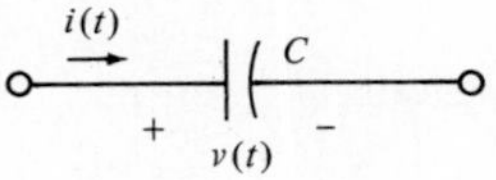

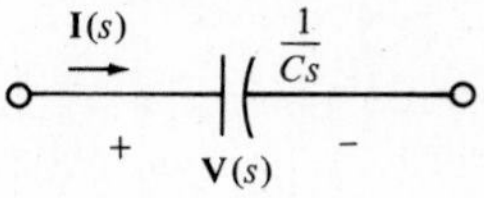

so that the impedance of a capacitor $\mathbf{Z}_C(s)$ is

$$\mathbf{Z}_C(s) = \frac{\mathbf{V}(s)}{\mathbf{I}(s)} = \frac{1}{Cs}$$

which is modeled as shown in Fig. 14.29.

From the general equation,

$$\mathbf{I}(s) = Cs\mathbf{V}(s) - Cv(0)$$

we get the frequency domain model shown in Fig. 14.30. Rearranging the previous equation, we get

$$\mathbf{V}(s) = \frac{\mathbf{I}(s)}{Cs} + \frac{v(0)}{s}$$

which yields the alternative model for the capacitor (Fig. 14.31), as does a source transformation.

fig. 14.30 (left)
fig. 14.31 (right)

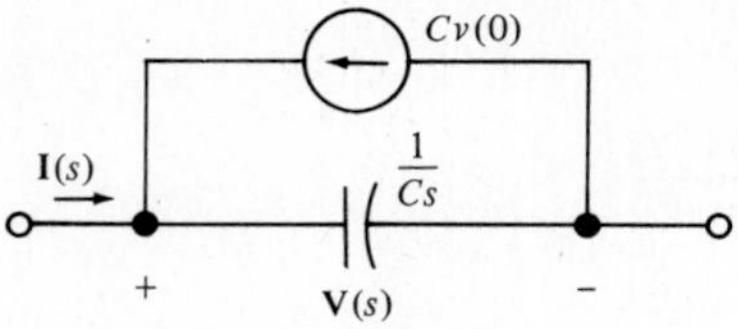

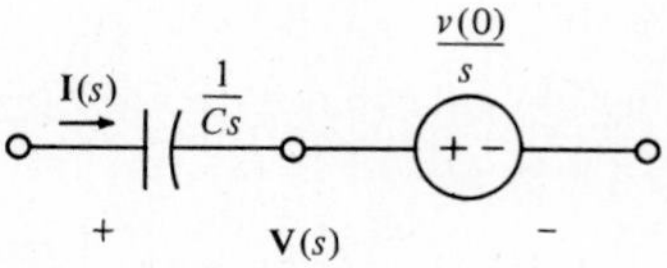

In summary, an element (R, L, or C) can be modeled in the frequency domain by an impedance and an appropriate independent source. The relationship between voltage $\mathbf{V}(s)$, current $\mathbf{I}(s)$, and impedance $\mathbf{Z}(s)$ [or admittance $\mathbf{Y}(s)$] in each case is given by the general forms of Ohm's law:

$$\mathbf{V}(s) = \mathbf{Z}(s)\mathbf{I}(s) = \frac{\mathbf{I}(s)}{\mathbf{Y}(s)}$$

$$\mathbf{I}(s) = \frac{\mathbf{V}(s)}{\mathbf{Z}(s)} = \mathbf{Y}(s)\mathbf{V}(s)$$

$$\mathbf{Z}(s) = \frac{\mathbf{V}(s)}{\mathbf{I}(s)} = \frac{1}{\mathbf{Y}(s)}$$

• • •

Example

Consider the circuit shown in Fig. 14.32, in which $i(0) = 2\,\text{A}$ and $v(0) = 2\,\text{V}$. Using the series models for the inductor and capacitor, in the frequency domain this circuit is as shown in Fig. 14.33.

fig. 14.32

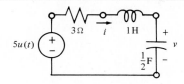

fig. 14.33

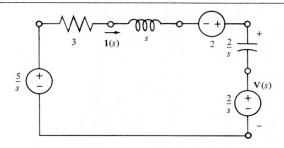

By KVL,

$$\frac{5}{s} = 3\mathbf{I}(s) + s\mathbf{I}(s) - 2 + \frac{2}{s}\mathbf{I}(s) + \frac{2}{s}$$

Solving for $\mathbf{I}(s)$, we obtain

$$\mathbf{I}(s) = \frac{2s + 3}{s^2 + 3s + 2} = \frac{2s + 3}{(s + 1)(s + 2)} = \frac{1}{s + 1} + \frac{1}{s + 2}$$

Thus

$$i(t) = e^{-t} + e^{-2t} \qquad \text{for} \qquad t \geq 0$$

In addition,

$$\mathbf{V}(s) = \frac{2}{s}\mathbf{I}(s) + \frac{2}{s} = \frac{2}{s}\left[\frac{2s + 3}{(s + 1)(s + 2)}\right] + \frac{2}{s}$$

$$= \frac{2(s^2 + 5s + 5)}{s(s + 1)(s + 2)} = \frac{5}{s} - \frac{2}{s + 1} - \frac{1}{s + 2}$$

so

$$v(t) = 5 - 2e^{-t} - e^{-2t} \qquad \text{for} \qquad t \geq 0$$

• • •

Example

In the zero-input circuit shown in Fig. 14.34, $i(0) = 1$ A and $v(0) = 1$ V. Although the use of the series inductor and capacitor models will require less manipulation, for the purpose of demonstration let us use the parallel models. The resulting frequency domain circuit representation is shown in Fig. 14.35.

fig. 14.34

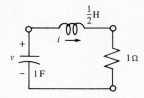

fig. 14.35

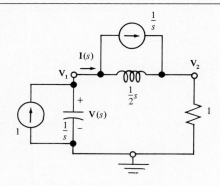

By nodal analysis, at node $\mathbf{V}_1$,

$$1 = s\mathbf{V}_1 + \frac{\mathbf{V}_1 - \mathbf{V}_2}{\frac{1}{2}s} + \frac{1}{s}$$

from which

$$(s^2 + 2)\mathbf{V}_1 - 2\mathbf{V}_2 = s - 1$$

At node $\mathbf{V}_2$,

$$\frac{1}{s} = \frac{\mathbf{V}_2 - \mathbf{V}_1}{\frac{1}{2}s} + \frac{\mathbf{V}_2}{1}$$

from which

$$-2\mathbf{V}_1 + (s + 2)\mathbf{V}_2 = 1$$

Solving for $\mathbf{V}_1$ yields

$$\mathbf{V}_1 = \frac{s + 1}{s^2 + 2s + 2}$$

Thus

$$V(s) = V_1 = \frac{s+1}{(s+1)^2 + 1}$$

so

$$v(t) = e^{-t} \cos t \qquad \text{for} \qquad t \geq 0$$

Furthermore, since

$$I(s) = 1 - sV_1 = 1 - \frac{s(s+1)}{s^2 + 2s + 2}$$

then

$$I(s) = \frac{s+2}{s^2 + 2s + 2} = \frac{s+1}{(s+1)^2 + 1} + \frac{1}{(s+1)^2 + 1}$$

and

$$i(t) = e^{-t} \cos t + e^{-t} \sin t$$
$$= e^{-t}(\cos t + \sin t) \qquad \text{for} \qquad t \geq 0$$

$$\bullet \quad \bullet \quad \bullet$$

For the transformer shown in Fig. 14.36, the voltages and currents are related by the pair of equations

$$v_1(t) = L_1 \frac{di_1(t)}{dt} + M \frac{di_2(t)}{dt}$$

$$v_2(t) = M \frac{di_1(t)}{dt} + L_2 \frac{di_2(t)}{dt}$$

fig. 14.36

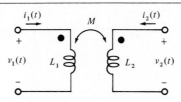

Taking Laplace transforms we get

$$V_1(s) = L_1[sI_1(s) - i_1(0)] + M[sI_2(s) - i_2(0)]$$
$$V_2(s) = M[sI_1(s) - i_1(0)] + L_2[sI_2(s) - i_2(0)]$$

from which

$$V_1(s) = L_1 sI_1(s) + MsI_2(s) - [L_1 i_1(0) + M i_2(0)]$$
$$V_2(s) = MsI_1(s) + L_2 sI_2(s) - [M i_1(0) + L_2 i_2(0)]$$

fig. 14.37

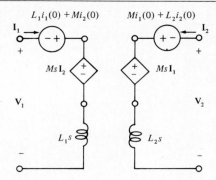

Thus the transformer can be modeled in the frequency domain as shown in Fig. 14.37.

As with our previous encounters, we can use Thévenin's theorem to replace a circuit by a voltage source in series with an impedance.

$$\cdot \quad \cdot \quad \cdot$$

Example

Consider the circuit shown in Fig. 14.38, in which the initial conditions are zero. To find the Thévenin equivalent circuit, first determine $\mathbf{V}_{oc}$ from Fig. 14.39. By voltage division,

$$\mathbf{V}_{oc} = \frac{2/s}{(2/s) + s + 3}\left(\frac{1}{2s}\right)$$

$$= \frac{1}{s(s^2 + 3s + 2)} = \frac{1/2}{s} - \frac{1}{s+1} + \frac{1/2}{s+2}$$

fig. 14.38

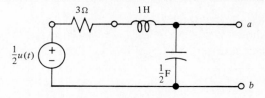

fig. 14.39

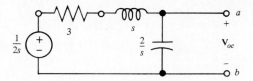

Thus

$$v_{oc}(t) = \left(\frac{1}{2} - e^{-t} + \frac{1}{2}e^{-2t}\right)u(t)$$

To determine $\mathbf{Z}_0$, set the independent voltage source to zero; a zero voltage source is equivalent to a short circuit. From the circuit shown in Fig. 14.40, it can be seen that $\mathbf{Z}_0$ is given by

$$\mathbf{Z}_0 = \frac{(2/s)(s+3)}{(2/s) + s + 3}$$

$$= \frac{2(s+3)}{s^2 + 3s + 2}$$

fig. 14.40

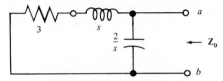

Since impedance is a frequency domain concept, we do not take the inverse transform of $\mathbf{Z}_0$, but we can "realize" or "build" $\mathbf{Z}_0$ with an R, L, and C as shown in Fig. 14.40. Therefore, the Thévenin equivalent of the circuit is given in Fig. 14.41 in both frequency and time-domain representations.

fig. 14.41

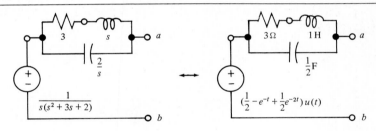

· · ·

Example
The circuit shown in Fig. 14.42 contains a dependent source. We determine $\mathbf{V}_{oc}$ from the circuit shown in Fig. 14.43.

fig. 14.42

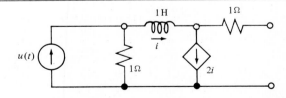

fig. 14.43

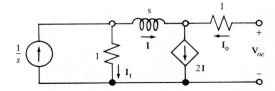

Since $\mathbf{I}_0 = 0$, then from

$$\mathbf{I} + \mathbf{I}_0 = 2\mathbf{I}$$

we have

$$\mathbf{I} = 0$$

Thus

$$\mathbf{V}_{oc} = 1\mathbf{I}_0 - s\mathbf{I} + 1\mathbf{I}_1$$

$$= 0 - 0 + 1\left(\frac{1}{s} - \mathbf{I}\right) = \frac{1}{s}$$

To determine $\mathbf{Z}_0$ we set the independent current source to zero; a current source of value zero is equivalent to an open circuit. However, because of the dependent current source, we cannot calculate $\mathbf{Z}_0$ as in the preceding example (a current source *has* an infinite impedance, but *is not* an infinite impedance). Instead, we apply a source (voltage or current) at the appropriate terminals and calculate the response (current or voltage, respectively). The ratio of voltage to current is $\mathbf{Z}_0$. For the given circuit applying a voltage source, we have the circuit shown in Fig. 14.44. Since

$$\mathbf{V}_0 = 1\mathbf{I}_0 - s\mathbf{I} - 1\mathbf{I}$$

---

fig. 14.44

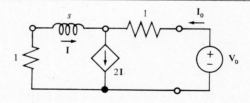

and

$$\mathbf{I} + \mathbf{I}_0 = 2\mathbf{I} \Rightarrow \mathbf{I} = \mathbf{I}_0$$

then

$$\mathbf{V}_0 = \mathbf{I}_0 - s\mathbf{I}_0 - \mathbf{I}_0 = -s\mathbf{I}_0$$

It follows that

$$\mathbf{Z}_0 = \frac{\mathbf{V}_0}{\mathbf{I}_0} = -s$$

That's right, we have a negative impedance (an inductor whose value is $L = -1$ H). The Thévenin equivalent circuit is shown in Fig. 14.45.

fig. 14.45

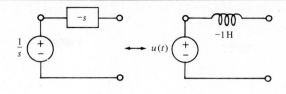

To prove Thévenin's theorem, consider the network shown in Fig. 14.46, where $\mathbf{V}_0$ is the voltage across the arbitrary load $\mathbf{Z}_L$.

fig. 14.46

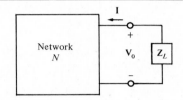

If $\mathbf{V}_0$ is the voltage across the load and $\mathbf{I}$ is the current through it, then with these values, KVL and KCL are satisfied throughout the entire network. If we replace the load by a voltage source whose value is $\mathbf{V}_0$, we have the circuit in Fig. 14.47. Clearly, all of the original voltages and currents in network N result in the satisfaction of KVL and

fig. 14.47

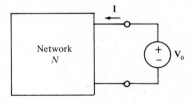

KCL throughout the entire new network. Thus the effect on network N in both configurations is the same. However, from the latter network, by the principle of superposition we have

$$\mathbf{I} = -\mathbf{I}_{sc} + \mathbf{I}_0 \qquad (14.1)$$

where $\mathbf{I}_{sc}$—the short-circuit current—is the current obtained when the source labeled $\mathbf{V}_0$ is set to zero as shown in Fig. 14.48, and $\mathbf{I}_0$ is the current obtained when all independent sources in network N are set to zero, as indicated in Fig. 14.49.

fig. 14.48 (left)
fig. 14.49 (right)

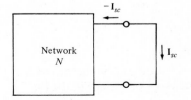
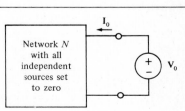

The output impedance (Thévenin equivalent impedance) of network N is defined by

$$\mathbf{Z}_0 = \frac{\mathbf{V}_0}{\mathbf{I}_0}$$

Then

$$\mathbf{I}_0 = \frac{\mathbf{V}_0}{\mathbf{Z}_0}$$

and, from Equation (14.1)

$$\mathbf{I} = -\mathbf{I}_{sc} + \frac{\mathbf{V}_0}{\mathbf{Z}_0}$$

Since $\mathbf{Z}_L$ is arbitrary, set $\mathbf{Z}_L = \infty$. For this special case, denote $\mathbf{V}_0$ by $\mathbf{V}_{oc}$—the open-circuit voltage. Under this condition $\mathbf{I} = 0$, so from the general equation (14.1)

$$\mathbf{I} = -\mathbf{I}_{sc} + \mathbf{I}_0$$

we have

$$0 = -\mathbf{I}_{sc} + \frac{\mathbf{V}_{oc}}{\mathbf{Z}_0}$$

Thus,

$$\mathbf{I}_{sc} = \frac{\mathbf{V}_{oc}}{\mathbf{Z}_0}$$

Substituting this into Equation (14.1) we get

$$\mathbf{I} = -\frac{\mathbf{V}_{oc}}{\mathbf{Z}_0} + \frac{\mathbf{V}_0}{\mathbf{Z}_0}$$

from which

$$\mathbf{I}\mathbf{Z}_0 = -\mathbf{V}_{oc} + \mathbf{V}_0$$

or

$$\mathbf{V}_0 = \mathbf{I}\mathbf{Z}_0 + \mathbf{V}_{oc}$$

and, since this equation can be modeled by the circuit shown in Fig. 14.50, we have proven Thévenin's theorem.

fig. 14.50

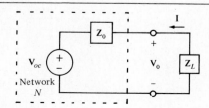

The proof above suggests an alternative for obtaining the output impedance $\mathbf{Z}_0$. Specifically, from the equation

$$\mathbf{I}_{sc} = \frac{\mathbf{V}_{oc}}{\mathbf{Z}_0}$$

we have that

$$\mathbf{Z}_0 = \frac{\mathbf{V}_{oc}}{\mathbf{I}_{sc}}$$

Thus, knowing $\mathbf{V}_{oc}$ and $\mathbf{I}_{sc}$, it is a simple matter to calculate $\mathbf{Z}_0$.

Suppose now that we take a source transformation of a Thévenin equivalent circuit. The result is shown in Fig. 14.51. However, since $\mathbf{V}_{oc}/\mathbf{Z}_0 = \mathbf{I}_{sc}$, then as in Thévenin's theorem, network N can be replaced by an independent current source $\mathbf{I}_{sc}$ in parallel with the impedance $\mathbf{Z}_0$ as in Fig. 14.52, where $\mathbf{I}_{sc}$ is the current flowing out of network N when the load is a short circuit. This result (Norton's theorem) can be proved in a dual manner to that of Thévenin's theorem.

fig. 14.51

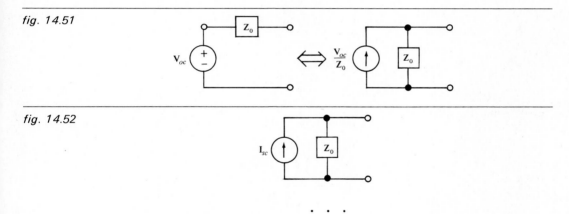

fig. 14.52

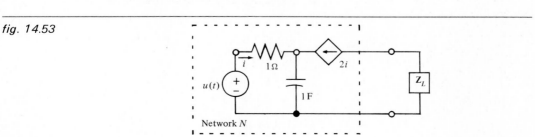

· · ·

Example

To find the Norton equivalent circuit of network N shown in Fig. 14.53, first determine $\mathbf{I}_{sc}$ from the circuit in Fig. 14.54.

fig. 14.53

fig. 14.54

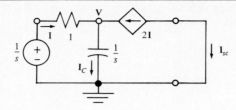

By nodal analysis,

$$I + 2I = I_C \qquad \text{or} \qquad 3I = I_C$$

so

$$3\left[\frac{(1/s) - V}{1}\right] = sV$$

from which

$$V = \frac{3}{s(s + 3)}$$

Then

$$I = \frac{(1/s) - V}{1} = \frac{1}{s + 3}$$

and

$$I_{sc} = -2I = \frac{-2}{s + 3}$$

To calculate Z_0, set the independent voltage source to zero and apply the source V_0 as shown in Fig. 14.55. By nodal analysis,

$$3I = I_C$$

fig. 14.55

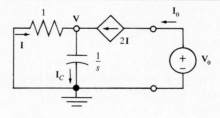

so

$$3\left(\frac{-V}{1}\right) = sV$$

from which

$$(s + 3)V = 0$$

Dividing both sides by $(s + 3)$ yields

$$\mathbf{V} = 0$$

Hence

$$\mathbf{I} = -\frac{\mathbf{V}}{1} = 0$$

Since

$$\mathbf{I}_0 = 2\mathbf{I} = 0$$

then

$$\mathbf{Z}_0 = \frac{\mathbf{V}_0}{\mathbf{I}_0} = \infty$$

That is, the output impedance is infinite. Thus the Norton equivalent circuit is as shown in Fig. 14.56.

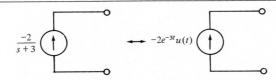

fig. 14.56

Hence the Norton equivalent is an ideal current source. Note that in this case there is no Thévenin equivalent circuit since an ideal current source cannot be equivalent to a voltage source in series with an impedance. For this example, consequently, it is not possible to obtain a unique value for $\mathbf{V}_{oc}$.

. . .

Needless to say, a circuit that has as its Thévenin equivalent circuit an ideal voltage source (zero output impedance) has no Norton equivalent.

There are networks whose Thévenin and Norton equivalents are identical. For such networks, $\mathbf{V}_{oc} = \mathbf{I}_{sc} = 0$; that is, the Thévenin and Norton equivalents consist solely of the output impedance $\mathbf{Z}_0$.

PROBLEMS   *14.33.   For the zero-input circuit in Fig. P14.33, find $v(t)$ for $t \geq 0$ given that $i(0) = 1$ A and $v(0) = 2$ V.
Ans. $2e^{-2t} + 3te^{-2t}$

fig. P14.33

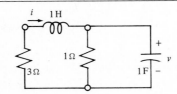

fig. P14.34

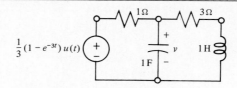

14.34.   Find the zero-state response $v(t)$ for the *RLC* circuit shown in Fig. P14.34.

fig. P14.35

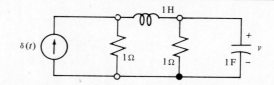

*14.35.   Repeat Problem 14.34 for the circuit in Fig. P14.35.
Ans. $e^{-t} \sin t \, u(t)$

fig. P14.36

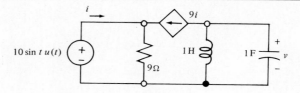

14.36.   Repeat Problem 14.34 for the circuit in Fig. P14.36.

fig. P14.37

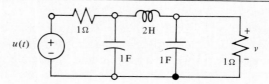

14.37.   Repeat Problem 14.34 for the circuit shown in Fig. P14.37.

fig. P14.38

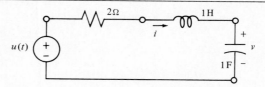

*14.38.   For the series *RLC* circuit in Fig. P14.38. Find $v(t)$ and $i(t)$ for $t \geq 0$ given that $i(0) = \frac{1}{2}$A and $v(0) = 0$.
Ans. $1 - e^{-t} - \frac{1}{2}te^{-t}$, $\frac{1}{2}e^{-t} + \frac{1}{2}te^{-t}$

fig. P14.39

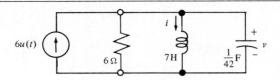

14.39.   Find $v(t)$ and $i(t)$ for $t \geq 0$ for the parallel *RLC* circuit in Fig. P14.39, given that $i(0) = -4$ A and $v(0) = 0$.

fig. P14.40

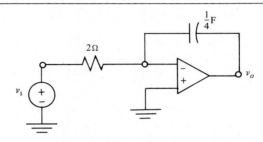

14.40.   For the op amp circuit shown in Fig. P14.40, find $v_o(t)$ for $t \geq 0$ given that $v_o(0) = -4$ V when (a) $v_1(t) = e^{-3t}u(t)$ and (b) $v_1(t) = \cos 2t\, u(t)$.

fig. P14.41

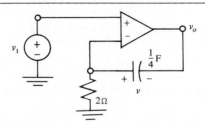

14.41.   For the op amp circuit in Fig. P14.41, given that $v(0) = 4$ V, find $v_o(t)$ for $t \geq 0$ when (a) $v_1(t) = e^{-2t}u(t)$, (b) $v_1(t) = e^{-3t}u(t)$, and (c) $v_1(t) = \cos 2t\, u(t)$.

fig. P14.42

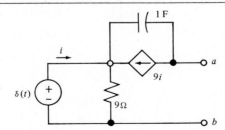

*14.42.   Given that the capacitor is initially uncharged, determine the Thévenin equivalent of the circuit in Fig. P14.42.

Ans. $\dfrac{s-1}{s}, \dfrac{10}{s}$

fig. P14.43

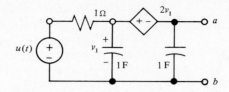

14.43.   The capacitors in the circuit shown in Fig. P14.43 are initially uncharged. Find the Thévenin equivalent of this circuit.

fig. P14.44

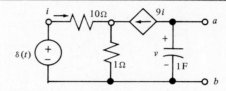

14.44.   In the circuit shown in Fig. P14.44, find the Thévenin equivalent for $t \geq 0$ given that $v(0) = 29/20\,\text{V}$.

Summary

1.   The Laplace transform is a linear transformation that can be used to solve differential equations or analyze circuits that have nonzero initial conditions.

2.   The inverse Laplace transform is found by using a table of transforms and various properties (and partial fraction expansions) rather than with the formula for the inverse transform.

3.   Certain transfer functions can be simulated by using three types of devices: adders, scalers, and integrators.

4.   Multiplication in the frequency domain corresponds to convolution in the time domain.

5.   Inductors and capacitors with nonzero initial conditions can be modeled with an appropriate independent source and zero-initial-condition element.

6.   Thévenin's theorem can be applied to circuits having zero or nonzero initial conditions.

MATRICES AND DETERMINANTS

A rectangular array of numbers having n rows and m columns, written as

$$\begin{bmatrix} a_{11} & a_{12} & a_{13} & \cdots & a_{1m} \\ a_{21} & a_{22} & a_{23} & \cdots & a_{2m} \\ a_{31} & a_{32} & a_{33} & \cdots & a_{3m} \\ \vdots & \vdots & \vdots & & \vdots \\ a_{n1} & a_{n2} & a_{n3} & \cdots & a_{nm} \end{bmatrix}$$

is called an $n \times m$ *matrix*. Note that the number in the ith row and jth column of the matrix is denoted by a_{ij}. We say that two $n \times m$ matrices are *equal* if all their corresponding entries are equal.

Given that two matrices are the same dimension (i.e., they are both $n \times m$ matrices), we can define their addition as follows:

Suppose that **A** and **B** are both $n \times m$ matrices, where a_{ij} and b_{ij} are the respective entries in row i and column j. Then the *sum* of **A** and **B**, denoted **A** + **B**, is the $n \times m$ matrix whose entry in row i and column j is $a_{ij} + b_{ij}$, for all $i = 1, 2, \ldots, n$ and $j = 1, 2, \ldots, m$. In other words, the sum of two matrices is the matrix obtained by adding the corresponding entries.

· · ·

Example
Let **A** and **B** be the following 2×3 matrices:

$$\mathbf{A} = \begin{bmatrix} 3 & 0 & -1 \\ 4 & -2 & -7 \end{bmatrix} \quad \text{and} \quad \mathbf{B} = \begin{bmatrix} 1 & 2 & 3 \\ -4 & 5 & -6 \end{bmatrix}$$

Then we have

$$\mathbf{A} + \mathbf{B} = \begin{bmatrix} 3+1 & 0+2 & -1+3 \\ 4-4 & -2+5 & -7-6 \end{bmatrix} = \begin{bmatrix} 4 & 2 & 2 \\ 0 & 3 & -13 \end{bmatrix}$$

· · ·

Note that as a consequence of ordinary arithmetic, it is true in general for two $n \times m$ matrices **A** and **B** that

$$\mathbf{A} + \mathbf{B} = \mathbf{B} + \mathbf{A}$$

In other words, matrix addition is *commutative*.

It is not only possible, but desirable, to multiply matrices. However, multiplication is a more complicated process than addition. To begin with, to take the matrix product **AB**, where **A** is an $n \times m$ matrix, it is required that **B** be an $m \times p$ matrix. (We can have that $p = n$ or $p = m$ or neither.)

Suppose that **A** is an $n \times m$ matrix and **B** is an $m \times p$ matrix. We can define multiplication as follows:

Let **C** be equal to the *product* of **A** and **B**; that is, $\mathbf{C} = \mathbf{AB}$. If the ith row of **A** is

$$a_{i1} \quad a_{i2} \quad a_{i3} \quad \cdots \quad a_{im}$$

and the jth column of **B** is

$$
\begin{matrix}
b_{1j} \\
b_{2j} \\
b_{3j} \\
\vdots \\
b_{mj}
\end{matrix}
$$

then the entry in row i and column j of **C** is

$$c_{ij} = a_{i1}b_{1j} + a_{i2}b_{2j} + a_{i3}b_{3j} + \cdots + a_{im}b_{mj}$$

Note that **C** is an $n \times p$ matrix.

· · ·

Example

Suppose that **A** is a 2×3 matrix and **B** is a 3×1 matrix as follows:

$$\mathbf{A} = \begin{bmatrix} 3 & 0 & -1 \\ 4 & -2 & -7 \end{bmatrix} \quad \text{and} \quad \mathbf{B} = \begin{bmatrix} 2 \\ -2 \\ 0 \end{bmatrix}$$

Then $\mathbf{C} = \mathbf{AB}$ is the following 2×1 matrix:

$$\mathbf{C} = \mathbf{AB} = \begin{bmatrix} 3 & 0 & -1 \\ 4 & -2 & -7 \end{bmatrix} \begin{bmatrix} 2 \\ -2 \\ 0 \end{bmatrix} = \begin{bmatrix} (3)(2) + (0)(-2) + (-1)(0) \\ (4)(2) + (-2)(-2) + (-7)(0) \end{bmatrix} = \begin{bmatrix} 6 \\ 12 \end{bmatrix}$$

Note that in this case we cannot even form the product **BA**.

Next consider the two 2×2 matrices

$$\mathbf{A} = \begin{bmatrix} 1 & 2 \\ 0 & -3 \end{bmatrix} \quad \text{and} \quad \mathbf{B} = \begin{bmatrix} 2 & 4 \\ -4 & 8 \end{bmatrix}$$

We can now form the products

$$\mathbf{AB} = \begin{bmatrix} 1 & 2 \\ 0 & -3 \end{bmatrix} \begin{bmatrix} 2 & 4 \\ -4 & 8 \end{bmatrix} = \begin{bmatrix} (1)(2) + (2)(-4) & (1)(4) + (2)(8) \\ (0)(2) + (-3)(-4) & (0)(4) + (-3)(8) \end{bmatrix}$$

$$= \begin{bmatrix} -6 & 20 \\ 12 & -24 \end{bmatrix}$$

and

$$\mathbf{BA} = \begin{bmatrix} 2 & 4 \\ -4 & 8 \end{bmatrix} \begin{bmatrix} 1 & 2 \\ 0 & -3 \end{bmatrix} = \begin{bmatrix} (2)(1) + (4)(0) & (2)(2) + (4)(-3) \\ (-4)(1) + (8)(0) & (-4)(2) + (8)(-3) \end{bmatrix}$$

$$= \begin{bmatrix} 2 & -8 \\ -4 & -32 \end{bmatrix}$$

Thus we can see, in general, for matrix multiplication that

$$\mathbf{AB} \neq \mathbf{BA}$$

In other words, matrix multiplication is not commutative in general.

$$\bullet \quad \bullet \quad \bullet$$

Having defined matrix equivalence, addition and multiplication, we are able to now talk about the concept of a matrix equation. First, consider the following set of simultaneous linear equations:

$$a_{11}x_1 + a_{12}x_2 + \cdots + a_{1m}x_m = b_1$$
$$a_{21}x_1 + a_{22}x_2 + \cdots + a_{2m}x_m = b_2$$
$$\vdots \qquad\qquad\qquad \vdots \qquad \vdots$$
$$a_{n1}x_1 + a_{n2}x_2 + \cdots + a_{nm}x_m = b_n$$

As a consequence of the definitions of matrix equivalence, addition, and multiplication, we can alternatively represent this set of equations in matrix form as

$$\begin{bmatrix} a_{11} & a_{12} & \cdots & a_{1m} \\ a_{21} & a_{22} & \cdots & a_{2m} \\ \vdots & \vdots & & \vdots \\ a_{n1} & a_{n2} & \cdots & a_{nm} \end{bmatrix} \begin{bmatrix} x_1 \\ x_2 \\ \vdots \\ x_m \end{bmatrix} = \begin{bmatrix} b_1 \\ b_2 \\ \vdots \\ b_n \end{bmatrix}$$

which is the matrix equation

$$\mathbf{AX} = \mathbf{B}$$

where

$$\mathbf{A} = \begin{bmatrix} a_{11} & a_{12} & \cdots & a_{1m} \\ a_{21} & a_{22} & \cdots & a_{2m} \\ \vdots & \vdots & & \vdots \\ a_{n1} & a_{n2} & \cdots & a_{nm} \end{bmatrix} \qquad \mathbf{X} = \begin{bmatrix} x_1 \\ x_2 \\ \vdots \\ x_m \end{bmatrix} \qquad \mathbf{B} = \begin{bmatrix} b_1 \\ b_2 \\ \vdots \\ b_n \end{bmatrix}$$

We have seen the difficulty of solving a set of n simultaneous equations. However, to solve the simple equation

$$\mathbf{AX} = \mathbf{B}$$

for $\mathbf{X}$, we need only "divide" both sides of the equation by $\mathbf{A}$. Since we are dealing with a matrix equation, though, ordinary division will not suffice. Thus we shall require the use of a few more new concepts.

We sometimes refer to an $n \times n$ matrix as a *square matrix*. Given a square matrix, the entries $a_{11}, a_{22}, a_{33}, \ldots, a_{nm}$ comprise the *diagonal* of the matrix. The $n \times n$ matrix in which all the diagonal entries are 1 and all the nondiagonal entries are 0 is called the *identity* or *unit matrix* and is denoted by $\mathbf{I}$.

· · ·

Example
The matrices

$$\mathbf{A} = \begin{bmatrix} 1 & -2 \\ 4 & 6 \end{bmatrix} \qquad \text{and} \qquad \mathbf{B} = \begin{bmatrix} 2 & -3 & -7 \\ 5 & 1 & 0 \\ -8 & 2 & 7 \end{bmatrix}$$

are both square matrices. $\mathbf{A}$ is 2×2 and $\mathbf{B}$ is 3×3. The 2×2 and 3×3 identity matrices are

$$\begin{bmatrix} 1 & 0 \\ 0 & 1 \end{bmatrix} \qquad \text{and} \qquad \begin{bmatrix} 1 & 0 & 0 \\ 0 & 1 & 0 \\ 0 & 0 & 1 \end{bmatrix}$$

respectively.

· · ·

Note that for any $n \times m$ matrix $\mathbf{A}$, it is true that

$$\mathbf{AI} = \mathbf{A}$$

where $\mathbf{I}$ is the $m \times m$ identity matrix, and

$$\mathbf{IA} = \mathbf{A}$$

where $\mathbf{I}$ is the $n \times n$ identity matrix. Specifically, if $\mathbf{A}$ is a square $n \times n$ matrix, then

$$\mathbf{AI} = \mathbf{IA} = \mathbf{A}$$

where $\mathbf{I}$ is the $n \times n$ identity matrix. Actually, for any $m \times n$ matrix $\mathbf{A}$ it is true that

$$\mathbf{AI} = \mathbf{IA} = \mathbf{A}$$

but it should be understood that two different dimensions of identity matrices appear in this equation.

Given an $n \times n$ matrix $\mathbf{A}$, then an $n \times n$ matrix, denoted $\mathbf{A}^{-1}$, is said to be the *inverse* of $\mathbf{A}$ if

$$\mathbf{A}\mathbf{A}^{-1} = \mathbf{A}^{-1}\mathbf{A} = \mathbf{I}$$

Furthermore, only a square matrix can have an inverse, and not every square matrix may have an inverse.

. . .

Example

For the matrices $\mathbf{A}$ and $\mathbf{B}$ given in the previous example, the corresponding inverses are

$$\mathbf{A}^{-1} = \begin{bmatrix} \dfrac{6}{14} & \dfrac{2}{14} \\ -\dfrac{4}{14} & \dfrac{1}{14} \end{bmatrix} \quad \text{and} \quad \mathbf{B}^{-1} = \begin{bmatrix} -1 & -1 & -1 \\ 5 & 6 & 5 \\ -\dfrac{18}{7} & -\dfrac{20}{7} & -\dfrac{17}{7} \end{bmatrix}$$

The matrix

$$\begin{bmatrix} 6 & 4 \\ 3 & 2 \end{bmatrix}$$

does not have an inverse. (Try to find one, if you don't believe it!)

. . .

The *transpose* of a matrix $\mathbf{A}$ is denoted $\mathbf{A}^t$ and is the matrix obtained from $\mathbf{A}$ by interchanging its rows with its columns. In other words, if a_{ij} is in row i and column j of $\mathbf{A}$, then a_{ij} is in row j and column i of $\mathbf{A}^t$.

. . .

Example

The following are examples of matrices and their transposes:

$$\mathbf{A}_1 = \begin{bmatrix} 3 & 0 & -1 \\ 4 & 1 & 2 \end{bmatrix} \quad \mathbf{A}_1{}^t = \begin{bmatrix} 3 & 4 \\ 0 & 1 \\ -1 & 2 \end{bmatrix}$$

$$\mathbf{A}_2 = \begin{bmatrix} 2 \\ -1 \\ 3 \end{bmatrix} \quad \mathbf{A}_2{}^t = \begin{bmatrix} 2 & -1 & 3 \end{bmatrix}$$

$$\mathbf{A}_3 = \begin{bmatrix} 1 & -1 & 0 \\ 2 & 5 & 7 \\ 8 & -3 & 3 \end{bmatrix} \quad \mathbf{A}_3{}^t = \begin{bmatrix} 1 & 2 & 8 \\ -1 & 5 & -3 \\ 0 & 7 & 3 \end{bmatrix}$$

. . .

Let us now discuss how to determine whether or not a square matrix has an inverse, and how to determine what that inverse is. For once we have done this, given a set of n simultaneous equations in n unknowns, written in matrix form as

$$\mathbf{AX} = \mathbf{B}$$

we can multiply both sides of this equation by $\mathbf{A}^{-1}$ to obtain

$$\mathbf{A}^{-1}\mathbf{AX} = \mathbf{A}^{-1}\mathbf{B}$$

$$\mathbf{IX} = \mathbf{A}^{-1}\mathbf{B}$$

$$\mathbf{X} = \mathbf{A}^{-1}\mathbf{B}$$

which is the solution.

Our first step is to introduce the concept of a "determinant." The *determinant* of a square matrix $\mathbf{A}$ is a number that is assigned to the matrix by the rule that is discussed below:

Given the square matrix

$$\mathbf{A} = \begin{bmatrix} a_{11} & a_{12} & a_{13} & \cdots & a_{1n} \\ a_{21} & a_{22} & a_{23} & \cdots & a_{2n} \\ \vdots & \vdots & \vdots & & \vdots \\ a_{n1} & a_{n2} & a_{n3} & \cdots & a_{nn} \end{bmatrix}$$

we denote the determinant of $\mathbf{A}$ by either of the three equivalent forms:

$$\det \mathbf{A} = \Delta_{\mathbf{A}} = \begin{vmatrix} a_{11} & a_{12} & a_{13} & \cdots & a_{1n} \\ a_{21} & a_{22} & a_{23} & \cdots & a_{2n} \\ \vdots & \vdots & \vdots & & \vdots \\ a_{n1} & a_{n2} & a_{n3} & \cdots & a_{nn} \end{vmatrix}$$

Note: The determinant of a 1×1 matrix $\mathbf{A} = [a_{11}]$ is simply the single entry; that is,

$$|a_{11}| = a_{11}$$

The *minor* of entry a_{ij} is the determinant of the matrix which is obtained from $\mathbf{A}$ by eliminating the ith row and jth column. The minor of a_{ij} is denoted by $\det \mathbf{A}_{ij}$.

The *cofactor* of a_{ij} is denoted by c_{ij} and is defined as

$$c_{ij} = (-1)^{i+j} \det \mathbf{A}_{ij}$$

· · ·

Example
Suppose that

$$\mathbf{A} = \begin{bmatrix} 2 & -3 & -7 \\ 5 & 1 & 0 \\ -8 & 2 & 7 \end{bmatrix}$$

Then the minor of a_{12} is

$$\det \mathbf{A}_{12} = \begin{vmatrix} 5 & 0 \\ -8 & 7 \end{vmatrix}$$

and the minor of a_{31} is

$$\det \mathbf{A}_{31} = \begin{vmatrix} -3 & -7 \\ 1 & 0 \end{vmatrix}$$

Furthermore, the cofactor of a_{12} is

$$c_{12} = (-1)^{1+2} \det \mathbf{A}_{12}$$
$$= -\det \mathbf{A}_{12}$$

and the cofactor of a_{31} is

$$c_{31} = (-1)^{3+1} \det \mathbf{A}_{31}$$
$$= \det \mathbf{A}_{31}$$

· · ·

We calculate a determinant as follows: Pick either any row or any column. If row i is selected, then

$$\det \mathbf{A} = a_{i1}c_{i1} + a_{i2}c_{i2} + \cdots + a_{in}c_{in}$$

If column j is selected, then

$$\det \mathbf{A} = a_{1j}c_{1j} + a_{2j}c_{2j} + \cdots + a_{nj}c_{nj}$$

In either case, the same number is obtained.

· · ·

Example
Given that

$$\mathbf{A} = \begin{bmatrix} 1 & -2 \\ 4 & 6 \end{bmatrix}$$

suppose that we select the first row. Then

$$\det \mathbf{A} = a_{11}c_{11} + a_{12}c_{12}$$
$$= 1c_{11} + (-2)c_{12}$$
$$= 1(-1)^{1+1}|6| + (-2)(-1)^{1+2}|4|$$
$$= 6 + 8$$
$$= 14$$

Next suppose that

$$\mathbf{A} = \begin{bmatrix} 2 & -3 & -7 \\ 5 & 1 & 0 \\ -8 & 2 & 7 \end{bmatrix}$$

Let us now choose the first column. We then have

$$\det \mathbf{A} = a_{11}c_{11} + a_{21}c_{21} + a_{31}c_{31}$$

$$= 2c_{11} + 5c_{21} - 8c_{31}$$

$$= 2(-1)^{1+1}\begin{vmatrix} 1 & 0 \\ 2 & 7 \end{vmatrix} + 5(-1)^{2+1}\begin{vmatrix} -3 & -7 \\ 2 & 7 \end{vmatrix} - 8(-1)^{3+1}\begin{vmatrix} -3 & -7 \\ 1 & 0 \end{vmatrix}$$

$$= 2\begin{vmatrix} 1 & 0 \\ 2 & 7 \end{vmatrix} - 5\begin{vmatrix} -3 & -7 \\ 2 & 7 \end{vmatrix} - 8\begin{vmatrix} -3 & -7 \\ 1 & 0 \end{vmatrix}$$

We now see that in evaluating the determinant of a 3×3 matrix we must evaluate the determinants of three 2×2 matrices. Arbitrarily selecting the first row for each case, we have

$$\det \mathbf{A} = 2[1(-1)^{1+1}|7| + 0(-1)^{1+2}|2|]$$

$$- 5[-3(-1)^{1+1}|7| - 7(-1)^{1+2}|2|]$$

$$- 8[-3(-1)^{1+1}|0| - 7(-1)^{1+2}|1|]$$

$$= 2(7) - 5(-21 + 14) - 8(7)$$

$$= 14 + 35 - 56$$

$$= -7$$

· · ·

Because of the frequent appearance of 2×2 and 3×3 matrices in this book, let us develop formulas for calculating the corresponding determinants.

Suppose that

$$\mathbf{A} = \begin{bmatrix} a_{11} & a_{12} \\ a_{21} & a_{22} \end{bmatrix}$$

and suppose that we calculate the determinant of $\mathbf{A}$ by selecting the first column. We then have

$$\det \mathbf{A} = a_{11}(-1)^{1+1}|a_{22}| + a_{21}(-1)^{2+1}|a_{12}|$$

$$= a_{11}a_{22} - a_{21}a_{12}$$

A simple way to remember this formula is to make an "X" inside the determinant as shown in Fig. A.1.

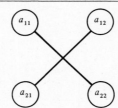

to indicate that the products $a_{11}a_{22}$ and $a_{21}a_{12}$ are formed. In addition, the signs associated with these products can be remembered as shown in Fig. A.2.

Thus for the matrix **A** given in the last example we have

$$\begin{vmatrix} 1 & -2 \\ 4 & 6 \end{vmatrix} = (1)(6) - (4)(-2) = 6 + 8 = 14$$

For a 3×3 matrix

$$\mathbf{A} = \begin{bmatrix} a_{11} & a_{12} & a_{13} \\ a_{21} & a_{22} & a_{23} \\ a_{31} & a_{32} & a_{33} \end{bmatrix}$$

let us select the first row. Then

$$\det \mathbf{A} = a_{11}(-1)^{1+1}\begin{vmatrix} a_{22} & a_{23} \\ a_{32} & a_{33} \end{vmatrix} + a_{12}(-1)^{1+2}\begin{vmatrix} a_{21} & a_{23} \\ a_{31} & a_{33} \end{vmatrix} + a_{13}(-1)^{1+3}\begin{vmatrix} a_{21} & a_{22} \\ a_{31} & a_{32} \end{vmatrix}$$

$$= a_{11}(a_{22}a_{33} - a_{32}a_{23}) - a_{12}(a_{21}a_{33} - a_{31}a_{23}) + a_{13}(a_{21}a_{32} - a_{31}a_{22})$$

$$= a_{11}a_{22}a_{33} - a_{11}a_{32}a_{23} - a_{12}a_{21}a_{33} + a_{12}a_{31}a_{23} + a_{13}a_{21}a_{32} - a_{13}a_{31}a_{22}$$

$$= a_{11}a_{22}a_{33} + a_{12}a_{31}a_{23} + a_{13}a_{21}a_{32} - (a_{11}a_{32}a_{23} + a_{12}a_{21}a_{33} + a_{13}a_{31}a_{22})$$

At first glance it may appear that this formula is just too long to commit to memory. However, if you can remember the pattern associated with the three positively signed terms, which is shown in Fig. A.3,

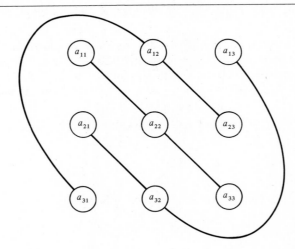

and the pattern associated with the three negatively signed terms, which is shown in Fig. A.4, then in effect you have remembered the formula.

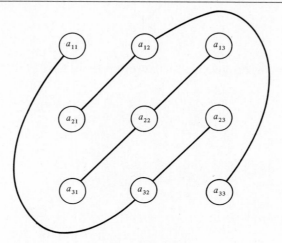

Consider the 3×3 matrix given in the last example. Applying the above formula to this matrix we have

$$\begin{vmatrix} 2 & -3 & -7 \\ 5 & 1 & 0 \\ -8 & 2 & 7 \end{vmatrix} = (2)(1)(7) + (-3)(-8)(0) + (-7)(5)(2)$$
$$\qquad\qquad - [(2)(2)(0) + (-3)(5)(7) + (-7)(-8)(1)]$$
$$= 14 + 0 - 70 - [0 - 105 + 56]$$
$$= -56 + 49 = -7$$

Unfortunately, there is no simple formula for the determinant of an $n \times n$ matrix, where n is larger than 3.

Having discussed the concept of the determinant of a square matrix $\mathbf{A}$, we can now determine whether or not $\mathbf{A}$ has an inverse, and if so, what it is. We begin by stating a classical result from linear algebra: A square matrix $\mathbf{A}$ has an inverse if and only if det $\mathbf{A} \neq 0$.

Given that det $\mathbf{A} \neq 0$, form the matrix $\mathbf{B}$, where b_{ij} the entry in row i and column j is defined (recalling that c_{ij} is the cofactor of a_{ij}) as

$$b_{ij} = \frac{c_{ij}}{\det \mathbf{A}}$$

Then the inverse of $\mathbf{A}$ is equal to the transpose of $\mathbf{B}$; that is,

$$\mathbf{A}^{-1} = \mathbf{B}^t$$

· · ·

Example

Reconsider the 2×2 matrix

$$\mathbf{A} = \begin{bmatrix} 1 & -2 \\ 4 & 6 \end{bmatrix}$$

Since det $\mathbf{A} = 14$, this matrix has an inverse $\mathbf{A}^{-1}$. We form the matrix

$$\mathbf{B} = \begin{bmatrix} b_{11} & b_{12} \\ b_{21} & b_{22} \end{bmatrix}$$

where

$$b_{11} = \frac{c_{11}}{\det \mathbf{A}} = \frac{(-1)^{1+1}|6|}{14} = \frac{6}{14}$$

$$b_{12} = \frac{c_{12}}{\det \mathbf{A}} = \frac{(-1)^{1+2}|4|}{14} = \frac{-4}{14}$$

$$b_{21} = \frac{c_{21}}{\det \mathbf{A}} = \frac{(-1)^{2+1}|-2|}{14} = \frac{2}{14}$$

$$b_{22} = \frac{c_{22}}{\det \mathbf{A}} = \frac{(-1)^{2+2}|1|}{14} = \frac{1}{14}$$

Thus we have

$$\mathbf{A}^{-1} = \mathbf{B}^t = \begin{bmatrix} b_{11} & b_{21} \\ b_{12} & b_{22} \end{bmatrix} = \begin{bmatrix} \dfrac{6}{14} & \dfrac{2}{14} \\ -\dfrac{4}{14} & \dfrac{1}{14} \end{bmatrix}$$

For the 3×3 matrix

$$\mathbf{A} = \begin{bmatrix} 2 & -3 & -7 \\ 5 & 1 & 0 \\ -8 & 2 & 7 \end{bmatrix}$$

we have seen that det $\mathbf{A} = -7$. Thus, this matrix also has an inverse. To find $\mathbf{A}^{-1}$ we form the matrix

$$\mathbf{B} = \begin{bmatrix} b_{11} & b_{12} & b_{13} \\ b_{21} & b_{22} & b_{23} \\ b_{31} & b_{32} & b_{33} \end{bmatrix}$$

where

$$b_{11} = \frac{c_{11}}{\det \mathbf{A}} = \frac{(-1)^{1+1}\begin{vmatrix} 1 & 0 \\ 2 & 7 \end{vmatrix}}{-7} = -1$$

$$b_{12} = \frac{c_{12}}{\det \mathbf{A}} = \frac{(-1)^{1+2}\begin{vmatrix} 5 & 0 \\ -8 & 7 \end{vmatrix}}{-7} = 5$$

$$b_{13} = \frac{c_{13}}{\det \mathbf{A}} = \frac{(-1)^{1+3}\begin{vmatrix} 5 & 1 \\ -8 & 2 \end{vmatrix}}{-7} = -\frac{18}{7}$$

$$b_{21} = \frac{c_{21}}{\det \mathbf{A}} = \frac{(-1)^{2+1}\begin{vmatrix} -3 & -7 \\ 2 & 7 \end{vmatrix}}{-7} = -1$$

$$b_{22} = \frac{c_{22}}{\det \mathbf{A}} = \frac{(-1)^{2+2}\begin{vmatrix} 2 & -7 \\ -8 & 7 \end{vmatrix}}{-7} = 6$$

$$b_{23} = \frac{c_{23}}{\det \mathbf{A}} = \frac{(-1)^{2+3}\begin{vmatrix} 2 & -3 \\ -8 & 2 \end{vmatrix}}{-7} = -\frac{20}{7}$$

$$b_{31} = \frac{c_{31}}{\det \mathbf{A}} = \frac{(-1)^{3+1}\begin{vmatrix} -3 & -7 \\ 1 & 0 \end{vmatrix}}{-7} = -1$$

$$b_{32} = \frac{c_{32}}{\det \mathbf{A}} = \frac{(-1)^{3+2}\begin{vmatrix} 2 & -7 \\ 5 & 0 \end{vmatrix}}{-7} = 5$$

$$b_{33} = \frac{c_{33}}{\det \mathbf{A}} = \frac{(-1)^{3+3}\begin{vmatrix} 2 & -3 \\ 5 & 1 \end{vmatrix}}{-7} = -\frac{17}{7}$$

Therefore, we have

$$\mathbf{A}^{-1} = \mathbf{B}^t = \begin{bmatrix} b_{11} & b_{21} & b_{31} \\ b_{12} & b_{22} & b_{32} \\ b_{13} & b_{23} & b_{33} \end{bmatrix} = \begin{bmatrix} -1 & -1 & -1 \\ 5 & 6 & 5 \\ -\dfrac{18}{7} & -\dfrac{20}{7} & -\dfrac{17}{7} \end{bmatrix}$$

$$\cdot \quad \cdot \quad \cdot$$

Now that we know how to take matrix inverses, it is a routine matter to solve a set of n simultaneous linear equations having n unknowns. We simply write the equations in the matrix form

$$\mathbf{AX} = \mathbf{B}$$

Then, if det $\mathbf{A} \neq 0$, we multiply both sides of the equation by $\mathbf{A}^{-1}$ to obtain the unique solution

$$\mathbf{X} = \mathbf{A}^{-1}\mathbf{B}$$

If det $\mathbf{A} = 0$, then there is no unique solution of the equations.

$\bullet \quad \bullet \quad \bullet$

Example
Given the two equations

$$x_1 - 2x_2 = 28$$
$$4x_1 + 6x_2 = 21$$

we can rewrite them in matrix form $\mathbf{AX} = \mathbf{B}$ as

$$\begin{bmatrix} 1 & -2 \\ 4 & 6 \end{bmatrix} \begin{bmatrix} x_1 \\ x_2 \end{bmatrix} = \begin{bmatrix} 28 \\ 21 \end{bmatrix}$$

The solution is simply $\mathbf{X} = \mathbf{A}^{-1}\mathbf{B}$, or

$$\begin{bmatrix} x_1 \\ x_2 \end{bmatrix} = \begin{bmatrix} 1 & -2 \\ 4 & 6 \end{bmatrix}^{-1} \begin{bmatrix} 28 \\ 21 \end{bmatrix}$$

and having calculated $\mathbf{A}^{-1}$ previously, we have

$$\begin{bmatrix} x_1 \\ x_2 \end{bmatrix} = \begin{bmatrix} \dfrac{6}{14} & \dfrac{2}{14} \\ -\dfrac{4}{14} & \dfrac{1}{14} \end{bmatrix} \begin{bmatrix} 28 \\ 21 \end{bmatrix}$$

and, by matrix multiplication,

$$\begin{bmatrix} x_1 \\ x_2 \end{bmatrix} = \begin{bmatrix} 15 \\ -\dfrac{13}{2} \end{bmatrix}$$

In other words, $x_1 = 15$ and $x_2 = -13/2$.

$\bullet \quad \bullet \quad \bullet$

Example

Next consider the three simultaneous linear equations

$$2x_1 - 3x_2 - 7x_3 = 5$$
$$5x_1 + x_2 \quad\quad = -2$$
$$-8x_1 + 2x_2 + 7x_3 = 1$$

In matrix form this becomes

$$\begin{bmatrix} 2 & -3 & -7 \\ 5 & 1 & 0 \\ -8 & 2 & 7 \end{bmatrix} \begin{bmatrix} x_1 \\ x_2 \\ x_3 \end{bmatrix} = \begin{bmatrix} 5 \\ -2 \\ 1 \end{bmatrix}$$

which is of the form $AX = B$ and, since we have already seen that det $A \neq 0$, we have

$$\begin{bmatrix} x_1 \\ x_2 \\ x_3 \end{bmatrix} = \begin{bmatrix} 2 & -3 & -7 \\ 5 & 1 & 0 \\ -8 & 2 & 7 \end{bmatrix}^{-1} \begin{bmatrix} 5 \\ -2 \\ 1 \end{bmatrix}$$

$$= \begin{bmatrix} -1 & -1 & -1 \\ 5 & 6 & 5 \\ \dfrac{18}{7} & -\dfrac{20}{7} & -\dfrac{17}{7} \end{bmatrix} \begin{bmatrix} 5 \\ -2 \\ 1 \end{bmatrix} = \begin{bmatrix} -4 \\ 18 \\ -\dfrac{67}{7} \end{bmatrix}$$

Thus $x_1 = -4$, $x_2 = 18$, $x_3 = -67/7$ is the unique solution to the given set of simultaneous equations.

$$\cdot \quad \cdot \quad \cdot$$

We have just seen that we can solve a set of simultaneous equations for all the unknowns by utilizing the inverse of a matrix. However, if the value of only one of the unknowns is desired, this can be accomplished without the necessity of first calculating an inverse. The technique for doing this is known as *Cramer's rule*, which can be stated as follows:

Cramer's Rule

Given a set of simultaneous, linear equations written in matrix form as $AX = B$, let $\Delta = \det A$. Define Δ_i to be the determinant of the matrix that is obtained by replacing the ith column of A with the column comprising B. Then

$$x_i = \frac{\Delta_i}{\Delta}$$

· · ·

Example

For the matrix equation of the form $\mathbf{AX} = \mathbf{B}$ given by

$$\begin{bmatrix} 1 & -2 \\ 4 & 6 \end{bmatrix} \begin{bmatrix} x_1 \\ x_2 \end{bmatrix} = \begin{bmatrix} 28 \\ 21 \end{bmatrix}$$

we have

$$\Delta = \det \mathbf{A} = \begin{vmatrix} 1 & -2 \\ 4 & 6 \end{vmatrix} = 14$$

Replacing the first column of $\mathbf{A}$ by the column comprising $\mathbf{B}$, we get

$$\Delta_1 = \begin{vmatrix} 28 & -2 \\ 21 & 6 \end{vmatrix} = 168 + 42 = 210$$

Thus

$$x_1 = \frac{\Delta_1}{\Delta} = \frac{210}{14} = 15$$

In addition, we have

$$\Delta_2 = \begin{vmatrix} 1 & 28 \\ 4 & 21 \end{vmatrix} = 21 - 112 = -91$$

and therefore

$$x_2 = \frac{\Delta_2}{\Delta} = \frac{-91}{14} = -\frac{13}{2}$$

· · ·

Example

For the matrix equation

$$\begin{bmatrix} 2 & -3 & -7 \\ 5 & 1 & 0 \\ -8 & 2 & 7 \end{bmatrix} \begin{bmatrix} x_1 \\ x_2 \\ x_3 \end{bmatrix} = \begin{bmatrix} 5 \\ -2 \\ 1 \end{bmatrix}$$

we have seen that

$$\Delta = \begin{vmatrix} 2 & -3 & -7 \\ 5 & 1 & 0 \\ -8 & 2 & 7 \end{vmatrix} = -7$$

Since

$$\Delta_1 = \begin{vmatrix} 5 & -3 & -7 \\ -2 & 1 & 0 \\ 1 & 2 & 7 \end{vmatrix} = 35 + 0 + 28 - (-7 + 0 + 42) = 28$$

we have

$$x_1 = \frac{\Delta_1}{\Delta} = \frac{28}{-7} = -4$$

Furthermore,

$$\Delta_2 = \begin{vmatrix} 2 & 5 & -7 \\ 5 & -2 & 0 \\ -8 & 1 & 7 \end{vmatrix} = -28 + 0 - 35 - (-112 + 0 + 175) = -126$$

and thus

$$x_2 = \frac{\Delta_2}{\Delta} = \frac{-126}{-7} = 18$$

Finally,

$$\Delta_3 = \begin{vmatrix} 2 & -3 & 5 \\ 5 & 1 & -2 \\ -8 & 2 & 1 \end{vmatrix} = 2 - 48 + 50 - (-40 - 8 - 15) = 67$$

and therefore

$$x_3 = \frac{\Delta_3}{\Delta} = \frac{67}{-7} = -\frac{67}{7}$$

INDEX

fig. 3.32

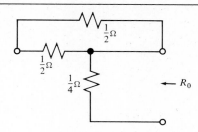

Note that the two $\frac{1}{2}$-Ω resistors are connected in series—and thus is equivalent to a 1-Ω resistor. However, this 1-Ω resistor is in parallel with a short circuit; the parallel combination has a value of zero ohms. This 0-Ω resistor is in series with the $\frac{1}{4}$-Ω resistor, which results in the conclusion that

$$R_0 = \tfrac{1}{4} \ \Omega$$

The Thévenin equivalent of the given circuit along with the $\frac{1}{3}$-Ω load resistance is shown in Fig. 3.33. By voltage division,

$$v = \frac{\dfrac{1}{3}\left(\dfrac{5}{2}\right)}{\dfrac{1}{3} + \dfrac{1}{4}} = \tfrac{10}{7} \ \text{V}$$

fig. 3.33

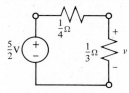

Clearly, in order to absorb the maximum amount of power, the load resistance should be changed to $\frac{1}{4}$ Ω.

• • •

So far, we have worked examples of finding the Thévenin equivalent of circuits consisting of resistors and independent sources. Let's now consider circuits having dependent sources.

• • •

Example
Find the Thévenin equivalent of the circuit in Fig. 3.34 with no load, by nodal analysis.

fig. 3.34

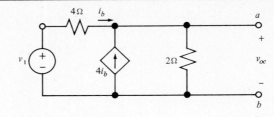

At node *a*,

$$i_b + 4i_b = \frac{v_{oc}}{2}$$

or

$$5i_b = \frac{v_{oc}}{2}$$

But

$$i_b = \frac{v_1 - v_{oc}}{4}$$

Substituting this into the previous equation yields

$$5\left(\frac{v_1 - v_{oc}}{4}\right) = \frac{v_{oc}}{2}$$

from which

$$v_{oc} = \frac{5}{7}v_1$$

We determine R_0 from Fig. 3.35. Note that this circuit is different from the previous one. For this reason, the current through the 4-Ω resistor (and hence, the current of the dependent source) has been labeled with a different variable since

fig. 3.35

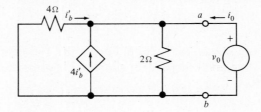

this current and the corresponding current in the former circuit do not, in general, have the same values. Remember this in the future when, because of laziness, we use the same labels in different circuits.

By nodal analysis, at node a,

$$i_b' + 4i_b' + i_0 = \frac{v_0}{2}$$

or

$$5i_b' + i_0 = \frac{v_0}{2}$$

Since

$$i_b' = -\frac{v_0}{4}$$

then

$$5\left(-\frac{v_0}{4}\right) + i_0 = \frac{v_0}{2}$$

so

$$i_0 = \frac{v_0}{2} + \frac{5v_0}{4}$$

$$i_0 = \frac{7}{4}v_0$$

Thus,

$$R_0 = \frac{v_0}{i_0} = \tfrac{4}{7}\ \Omega$$

and the Thévenin equivalent of the circuit given is shown in Fig. 3.36.

---

fig. 3.36

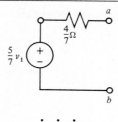

$\cdot\ \cdot\ \cdot$

Suppose that for the Thévenin equivalent in Fig. 3.37 we perform a source transformation. The result is shown in Fig. 3.38. Since, in the application of Thévenin's theorem, circuit B can be any load, consider the case that the load is a short circuit. If

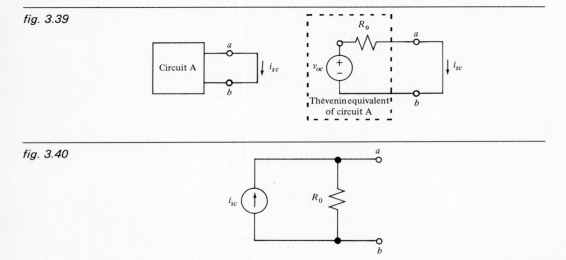

fig. 3.37 (left)
fig. 3.38 (right)

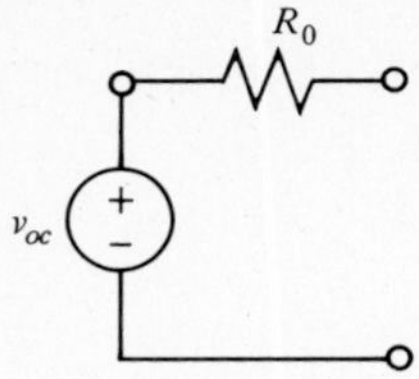

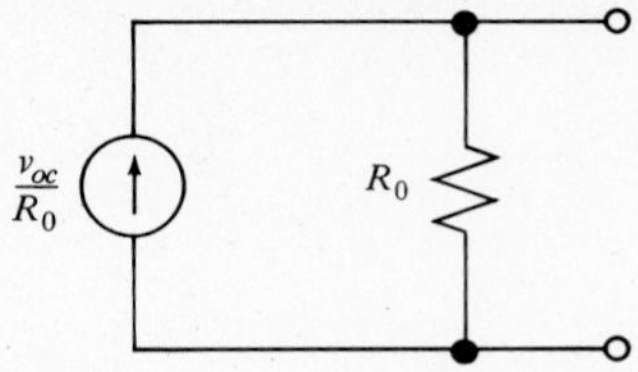

the resulting current is denoted i_{sc}, then the two equivalent situations shown in Fig. 3.39 yield i_{sc}. From the circuit on the right, we see that

$$i_{sc} = \frac{v_{oc}}{R_0}$$

Thus a circuit equivalent to the Thévenin equivalent is shown in Fig. 3.40, where i_{sc} is the current that results when a short circuit is placed between nodes a and b. This result

fig. 3.39

fig. 3.40

is known as *Norton's theorem* since it appeared in a memo sent by Edward L. Norton in 1926.

From the equation

$$i_{sc} = \frac{v_{oc}}{R_0}$$

we see that

$$R_0 = \frac{v_{oc}}{i_{sc}}$$

Hence, this is an alternative for obtaining R_0. Instead of calculating R_0 directly as described above, both v_{oc} and i_{sc} are determined individually, and then the ratio of v_{oc} to i_{sc} is found.

• • •

Example

For the circuit given in the preceding example, the short-circuit current is found from Fig. 3.41. Since the voltage across the 2-Ω resistor is zero, the current through it is zero. Thus, by KCL,

$$i_b + 4i_b = 0 + i_{sc}$$

or

$$i_{sc} = 5i_b = 5\left(\frac{v_1}{4}\right) = \frac{5}{4}v_1$$

fig. 3.41

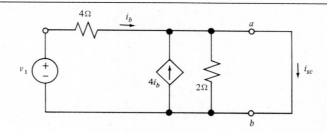

In the preceding example it was found that

$$v_{oc} = \frac{5}{7}v_1$$

Thus

$$R_0 = \frac{v_{oc}}{i_{sc}} = \frac{(5/7)v_1}{(5/4)v_1} = \frac{4}{7}\ \Omega$$

as obtained previously, and the Norton equivalent circuit is as shown in Fig. 3.42.

fig. 3.42

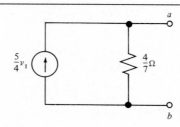

• • •

Example

Let us find the Norton equivalent of the circuit in Fig. 3.43, which contains no independent source.

To find i_{sc}, place a short circuit across terminals a and b. The result is shown in Fig. 3.44. By KCL,

$$\frac{v}{5} + 2i + \frac{v}{2} = 0$$

fig. 3.43

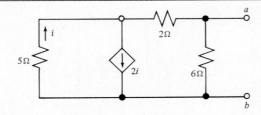

fig. 3.44

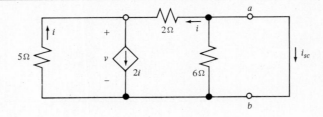

or

$$\frac{v}{5} + 2\left(-\frac{v}{5}\right) + \frac{v}{2} = 0$$

from which

$$3v = 0$$

and

$$v = 0$$

Thus

$$i = -\frac{v}{5} = 0$$

Since the voltage across the 6-Ω resistor is zero (due to the short circuit), the current through it is zero; therefore

$$i_{sc} = -i = 0$$

Actually, this result could have been arrived at intuitively since there is no independent source in the circuit that would initially institute a current.

To find R_0, we need to find the ratio of v_0/i_0 in the circuit shown in Fig. 3.45. By KCL at node c,

$$\frac{v}{5} + 2i + \frac{v - v_0}{2} = 0$$

fig. 3.45

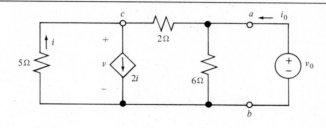

or

$$\frac{v}{5} + 2\left(-\frac{v}{5}\right) + \frac{v - v_0}{2} = 0$$

from which

$$-2v + 5v - 5v_0 = 0$$

so

$$v = \frac{5}{3}v_0$$

At node a,

$$i_0 = \frac{v_0}{6} + \frac{v_0 - v}{2}$$

$$= \frac{v_0}{6} + \frac{v_0 - \frac{5}{3}v_0}{2}$$

$$= \frac{v_0}{6} + \frac{v_0}{2} - \frac{5}{6}v_0 = -\frac{1}{6}v_0$$

Hence,

$$R_0 = \frac{v_0}{i_0} = -6 \; \Omega$$

That's right, a negative resistance! We've constructed a negative resistance from positive resistors and a dependent source. What is a negative resistor? It's just like a positive resistor in that voltage and current are proportional ($v = Ri$). However, in a positive resistor (positive-valued) current flows from a given potential to a lower potential, whereas for a negative resistor (positive-valued) current flows from a given potential to a higher potential.

We thus have that the Norton equivalent of the given circuit is as shown in Fig. 3.46.

fig. 3.46

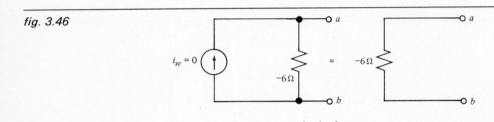

• • •

In the foregoing example, the Norton equivalent is simply a resistance and, because of source transformations, so is the Thévenin equivalent. This result is due to the fact that the original circuit contained no independent source. However, it is possible to obtain a similar result for some circuits containing independent sources—but this is a very special case indeed!

• • •

Example
To find the Norton equivalent of the circuit in Fig. 3.47, first place a short circuit between nodes *a* and *b*. The resulting circuit is shown in Fig. 3.48.

fig. 3.47

fig. 3.48

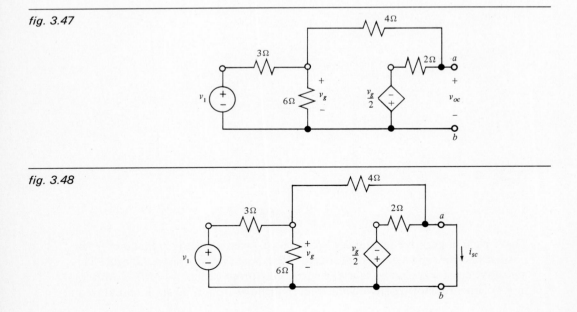

At node a

$$\frac{v_g}{4} - \frac{v_g/2}{2} = i_{sc}$$

Thus,

$$0 = i_{sc}$$

Since

$$R_0 = \frac{v_{oc}}{i_{sc}}$$

it is tempting to conclude that $R_0 = \infty$. However, for the original circuit,

$$\frac{v_g - v_{oc}}{4} = \frac{v_{oc} + v_g/2}{2}$$

so

$$\frac{v_g}{4} - \frac{v_{oc}}{4} = \frac{v_{oc}}{2} + \frac{v_g}{4}$$

or

$$0 = \frac{v_{oc}}{2} + \frac{v_{oc}}{4}$$

and

$$v_{oc} = 0$$

Thus, the formula $R_0 = v_{oc}/i_{sc}$ results in an indeterminate value for R_0.
However, we can determine R_0 from the circuit shown in Fig. 3.49. At node a,

$$i_0 = \frac{v_0 - v_g}{4} + \frac{v_0 + v_g/2}{2}$$

$$= \frac{v_0}{4} - \frac{v_g}{4} + \frac{v_0}{2} + \frac{v_g}{4}$$

$$= \frac{3}{4}v_0$$

fig. 3.49

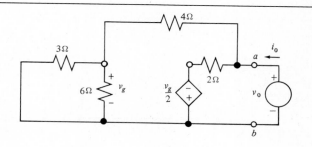

Therefore,

$$R_0 = \frac{v_0}{i_0} = \tfrac{4}{3} \; \Omega$$

and the Norton equivalent circuit is as shown in Fig. 3.50.

fig. 3.50

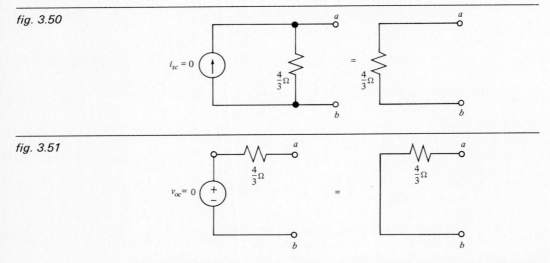

fig. 3.51

Also, the Thévenin equivalent is as shown in Fig. 3.51, and thus we see that the Thévenin and Norton equivalent circuits are identical for this example, as was true for the preceding example as well.

· · ·

Most circuits have both a Thévenin and Norton equivalent circuit (in each of the last two examples, the Thévenin and Norton equivalents were identical) and in such a case knowing one is tantamount to knowing the other, since one can be obtained from the other simply by a source transformation. However, there are special cases where a circuit may have a Thévenin equivalent but not a Norton equivalent, or vice versa.

· · ·

Example
Let us find the Thévenin equivalent of the circuit shown in Fig. 3.52.

fig. 3.52

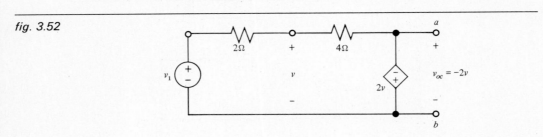

Applying KCL at the node common to the two resistors, we get

$$\frac{v_1 - v}{2} = \frac{v + 2v}{4}$$

from which

$$2v_1 - 2v = 3v$$

so that

$$v = \frac{2}{5}v_1$$

Hence

$$v_{oc} = -2v = -\frac{4}{5}v_1$$

Having obtained v_{oc}, let us now determine R_0. If we want to use the formula $R_0 = v_{oc}/i_{sc}$, we need to calculate i_{sc}. But to do this we must place a short circuit across the dependent voltage source—a definite no-no. This suggests that i_{sc}, and hence the Norton equivalent, does not exist. Let's see whether this is the case.

We obtain R_0 by setting $v_1 = 0$ and taking the ratio of v_0/i_0 for the circuit in Fig. 3.53. By KCL, at the node common to the resistors,

$$\frac{v}{2} + \frac{v + 2v}{4} = 0$$

fig. 3.53

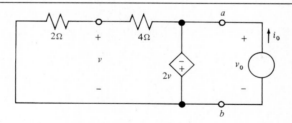

from which

$$2v + 3v = 0$$
$$5v = 0$$
$$v = 0$$

Since

$$v_0 = -2v = 0$$

in order not to have a contradictory situation, this means that the independent source must be a current source whose value is i_0. Thus

$$R_0 = \frac{v_0}{i_0} = 0 \; \Omega$$

Hence the Thévenin equivalent of the original circuit is the ideal independent voltage source as shown in Fig. 3.54.

fig. 3.54

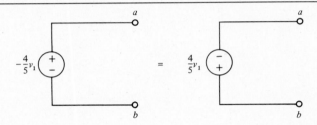

Needless to say, therefore, the original circuit does not have a Norton equivalent (as we suspected), for there is no way we can model an ideal voltage source as a practical current source—that is, an ideal current source in parallel with a resistance.

In this example we stated that i_{sc} did not exist. However, the skeptic may have refused to accept this statement at face value and try to calculate i_{sc} from the circuit in Fig. 3.55. Since $2v = 0$, then $v = 0$. Thus, $i = v_1/2$, and therefore $v_2 = 4i = 2v_1$.

fig. 3.55

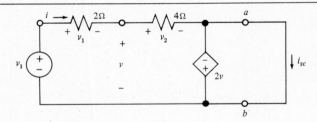

But by KVL, $v = v_2$. Hence, as long as v_1 is a nonzero voltage source, $v = v_2$ is a contradiction. We deduce that such a circuit cannot be analyzed (because of the paradoxical nature of placing a short circuit across a voltage source) and i_{sc} does not exist.

· · ·

PROBLEMS *3.5.   (a)   Use source transformations to find the Thévenin equivalent of the circuit to the left of terminals a and b in the circuit in Fig. P3.5.
(b)   Use the result of part (a) to find i.
Ans.   (a) $\frac{112}{9}$ V, $\frac{29}{9}$ Ω; (b) 2 A

fig. P3.5

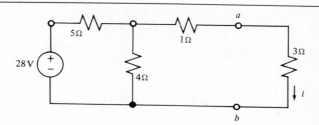

fig. P3.6

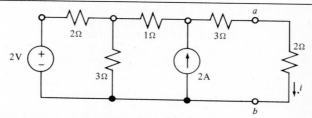

3.6.   Repeat Problem 3.5 for the circuit in Fig. P3.6.

fig. P3.7

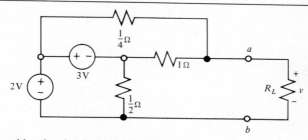

3.7.   Consider the circuit shown in Fig. P3.7.
   (a) Find the Thévenin equivalent of the circuit to the left of terminals a
        and b.
   (b) Find v when $R_L = \frac{1}{3}\ \Omega$.
   (c) What value of R_L absorbs the maximum power?
   (d) For the value of R_L determined in part (c), find the power absorbed by
        R_L.

fig. P3.8

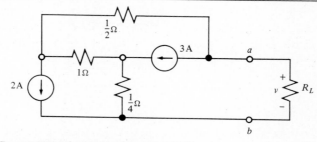

*3.8.   Repeat Problem 3.7 for the circuit in Fig. P3.8.
   Ans. (a) -7 V, $\frac{7}{4}\ \Omega$; (b) $-\frac{28}{25}$ V; (c) $\frac{7}{4}\ \Omega$; (d) 7 W

fig. P3.9

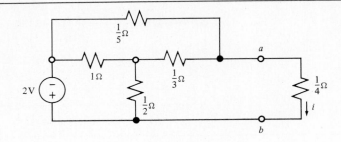

3.9.   (a)   Find the Norton equivalent to the left of terminals *a* and *b* of the circuit shown in Fig. P3.9.
       (b)   Use the result of part (a) to find *i*.

fig. P3.10

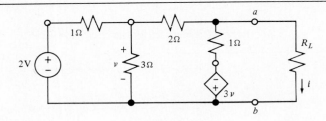

3.10.   Consider the circuit shown in Fig. P3.10.
        (a)   Find the Thévenin equivalent of the circuit to the left of terminals *a* and *b*.
        (b)   Find *i* when $R_L = 2 \, \Omega$.
        (c)   What value of R_L absorbs the maximum power?
        (d)   For the value of R_L determined in part (c), find the power absorbed by R_L.

fig. P3.11

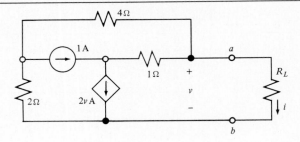

3.11.   Using the Norton equivalent instead of the Thévenin equivalent, repeat Problem 3.10 for the circuit shown in Fig. P3.11.
3.12.   Repeat Problem 3.10 for the circuit in Fig. P3.12.

fig. P3.12

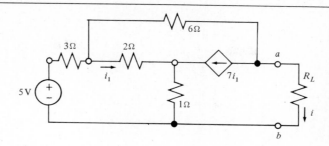

fig. P3.13

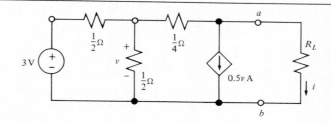

3.13. Repeat Problem 3.11 for the circuit in Fig. P3.13.

fig. P3.14

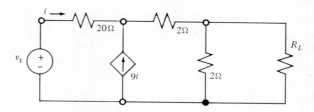

*3.14. In the circuit shown in Fig. P3.14, what value of R_L will absorb the maximum power?

Ans. $\frac{4}{3} \Omega$

fig. P3.15

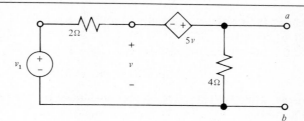

3.15. Find the Thévenin equivalent of the circuit in Fig. P3.15.

3.16. Show that the Thévenin equivalent of the circuit in Fig. P3.16 is an ideal voltage source.

fig. P3.16

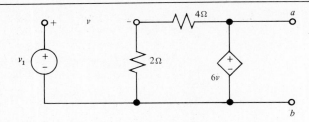

fig. P3.17

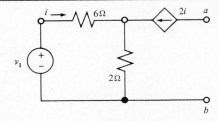

3.17. Show that the Norton equivalent of the circuit shown in Fig. P3.17 is an ideal current source. (*Hint*: In determining R_0, apply voltage source v_0 and calculate the resulting current i_0.)

3.4 *THE PRINCIPLE OF SUPERPOSITION*

As we have seen, given a circuit that contains two or more independent sources (voltage or current or both), one way to determine a specific variable (either voltage or current) is by the direct use of nodal or mesh or loop analysis. An alternative, however, is to find those portions of the variable due to each independent source and then sum these up. This concept is known as the *principle of superposition.*

The justification for the principle of superposition is based on the concept of linearity. Specifically, let us first consider the relationship between voltage and current for a resistor (i.e., Ohm's law). Suppose that a current i_1 (the input) is applied to a resistor R. Then the resulting voltage v_1 (the response or output) is $v_1 = Ri_1$. Similarly, if i_2 is applied to R, then $v_2 = Ri_2$ results. But if $i = i_1 + i_2$ is applied then the response is

$$v = Ri = R(i_1 + i_2) = Ri_1 + Ri_2 = v_1 + v_2$$

In other words, the response to a sum of inputs is equal to the sum of the individual responses (condition I).

In addition, if v is the response to i (i.e., $v = Ri$), then the response to Ki is

$$R(Ki) = K(Ri) = Kv$$

In other words, if the input is scaled by the constant K, then the response is also scaled by K (condition II).

Because conditions I and II are satisfied, we say that the relationship between current (input) and voltage (response) is *linear* for a resistor. Similarly, by using the alternative form of Ohm's law, $i = v/R$, we can show that the relationship between voltage (input) and current (response) is also linear for a resistor.

Although the relationships between voltage and current for a resistor are linear, the power relationships $p = Ri^2$ and $p = v^2/R$ are not. For instance, if the current flowing through a resistor is i_1, then the power absorbed by the resistor R is

$$p_1 = Ri_1^2$$

while if the current is i_2, then the power absorbed is

$$p_2 = Ri_2^2$$

However, the power absorbed due to the current $i_1 + i_2$ is

$$p_3 = R(i_1 + i_2)^2$$
$$= Ri_1^2 + Ri_2^2 + 2Ri_1i_2$$
$$\neq p_1 + p_2$$

Hence, the relationship $p = Ri^2$ is *nonlinear*.

Since the relationships between voltage and current are linear for resistors, we say that a resistor is a *linear element*. A dependent source (either current or voltage) whose value is directly proportional to some voltage or current is also a linear element. Because of this we say that a circuit consisting of independent sources, resistors, and linear dependent sources (as well as other linear elements to be introduced later) is a *linear circuit*. All of the circuits dealt with so far and those to be encountered in the remainder of this text are linear circuits.

Return now to the simple case of a resistor R connected to a voltage source v. If the voltage is v_1, then the current is $i_1 = v_1/R$; if v_2 is applied, then $i_2 = v_2/R$ results. By linearity, we have that for the input $v_1 + v_2$, the response is $i_1 + i_2$. We can represent this situation as in Fig. 3.56. The individual components (i_1 and i_2) of the response can be determined as shown in Fig. 3.57.

fig. 3.56

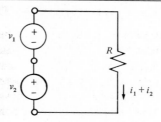

fig. 3.57

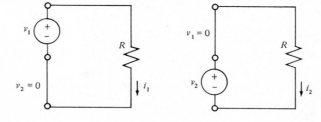

This is a very special case showing how it is possible to obtain the response due to two independent voltage sources by calculating the response to each one separately and then summing the responses. However, for any linear circuit we can take the same approach: Find the response to each independent source (both voltage and current) separately, and then sum the responses. Needless to say, a formal general proof is much more involved, but it utilizes analogous ideas.

For demonstration purposes, consider the linear circuit shown in Fig. 3.58, which contains three independent sources. By KCL at node v,

$$\frac{v}{6} + \frac{v - v_g}{3} + i_y - i_x = 0$$

from which

$$3v = 2v_g + 6i_x - 6i_y \qquad (3.1)$$

---

fig. 3.58

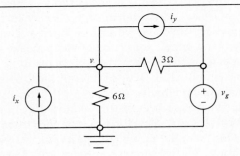

Suppose that when $i_x = i_y = 0$, the resulting node voltage is v_1. Thus, we have the equality

$$3v_1 = 2v_g + 6(0) - 6(0) \qquad (3.2)$$

For the case that $v_g = 0$ and $i_y = 0$, let the resulting node voltage be v_2. This yields the equality

$$3v_2 = 2(0) + 6i_x - 6(0) \qquad (3.3)$$

Finally, let v_3 be the node voltage that results when $v_g = 0$ and $i_x = 0$. This gives us the equality

$$3v_3 = 2(0) + 6(0) - 6i_y \qquad (3.4)$$

Adding Equations (3.2), (3.3), and (3.4) produces the equality

$$3(v_1 + v_2 + v_3) = 2v_g + 6i_x - 6i_y$$

which means that $v_1 + v_2 + v_3$ is the solution to Equation (3.1). We now formally state the principle of superposition:

Given a linear circuit with independent sources $s_1, s_2, \ldots, s_m$, let r be the response (either voltage or current) of this circuit. If r_i is the response of the circuit to source s_i with all independent sources set to zero (dependent sources are left as is), then $r = r_1 + r_2 + \cdots + r_n$.

$\cdot$ $\cdot$ $\cdot$

Example

The circuit shown in Fig. 3.59 contains three independent sources. Find the voltage v by using the principle of superposition.

fig. 3.59

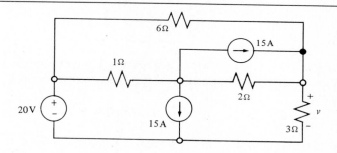

We first find that portion of v, call it v_a, due to the independent voltage source. Thus, we set the two independent current sources to zero as shown in Fig. 3.60. Note that the 1-Ω and 2-Ω resistors are in series, and the resulting effective 3-Ω resistance is in parallel with the 6-Ω resistor. These three resistors, therefore, are equivalent to a 2-Ω resistance. By voltage division,

$$v_a = \frac{3}{3+2}(20) = 12 \text{ V}$$

fig. 3.60

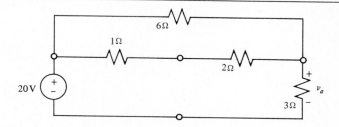

To find the voltage v_b due solely to the lower 15-A current source, set the other current source and the voltage source to zero as shown in Fig. 3.61. In this circuit the 6-Ω resistor is in parallel with the 3-Ω resistor. We can redraw the circuit as in Fig. 3.62. The 3-Ω and 6-Ω resistors in parallel are equivalent to a 2-Ω

fig. 3.61

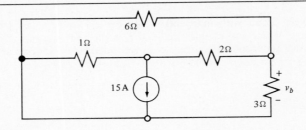

fig. 3.62

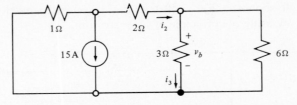

resistor—and this is in series with a 2-Ω resistor. The combination is effectively a 4-Ω resistor in parallel with the 1-Ω resistor. By current division,

$$i_2 = \frac{1}{1 + 4}(-15) = -3 \text{ A}$$

Having determined i_2, we can use current division again to get

$$i_3 = \left(\frac{6}{6 + 3}\right)i_2 = -2 \text{ A}$$

Thus

$$v_b = 3i_3 = -6 \text{ V}$$

To find the voltage v_c due to the upper 15-A current source, set the other two independent sources to zero. The resulting circuit is shown in Fig. 3.63.

fig. 3.63

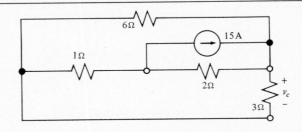

Since the 6-Ω and 3-Ω resistors are in parallel, we can redraw the circuit as shown in Fig. 3.64. Suppose for this circuit that we perform a source transformation on the 15-A source in parallel with the 2-Ω resistor. The result is a 15(2) = 30-V source (with the + toward node n) in series with a 2-Ω resistor.

fig. 3.64

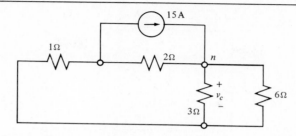

Since the 3-Ω and 6-Ω resistors in parallel is an effective 2-Ω resistance, by voltage division we find that

$$v_c = \frac{2}{2 + 1 + 2} \, (30) = 12 \text{ V}$$

By the principle of superposition,

$$v = v_a + v_b + v_c$$

$$= 12 - 6 + 12$$

$$= 18 \text{ V}$$

$$\bullet \quad \bullet \quad \bullet$$

In the example just completed, the circuit contained only independent sources. Let us now consider a situation in which dependent sources are also included.

$$\bullet \quad \bullet \quad \bullet$$

Example
The symmetrical network shown in Fig. 3.65 is in the form of what is known as a "differential amplifier."

fig. 3.65

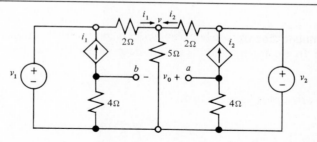

To find v_0 for this circuit by the principle of superposition, first find $v_0{}'$ when v_2 is set to zero as in Fig. 3.66. At the node labeled v,

$$i_1 + i_2 = \frac{v}{5}$$

$$\frac{v_1 - v}{2} - \frac{v}{2} = \frac{v}{5}$$

fig. 3.66

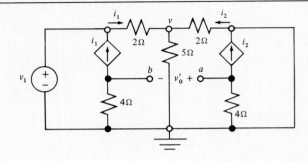

from which

$$v = \frac{5}{12}v_1$$

Since

$$i_1 = \frac{v_1 - v}{2} = \frac{7}{24}v_1$$

and

$$i_2 = -\frac{v}{2} = -\frac{5}{24}v_1$$

then

$$v_0' = -4i_2 + 4i_1$$

$$= \frac{5}{6}v_1 + \frac{7}{6}v_1$$

$$= 2v_1$$

Next, find v_0'', the voltage between nodes a and b when v_2 is operating and v_1 is set to zero. In a manner identical to that described above, we get

$$v_0'' = -2v_2$$

By the principle of superposition,

$$v_0 = v_0' + v_0''$$

$$= 2v_1 - 2v_2$$

$$= 2(v_1 - v_2)$$

$$\cdot \quad \cdot \quad \cdot$$

PROBLEMS *3.18. Use the principle of superposition to find v in the circuit shown in Fig. P3.18.

Ans. $\frac{4}{11}$ V

fig. P3.18

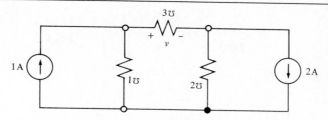

3.19.   For the circuit in Problem 3.18, find the power absorbed by the 3-℧ conductance. Show that the principle of superposition does not hold for power.

fig. P3.20

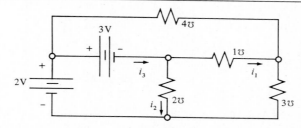

*3.20.   For the circuit shown in Fig. P3.20, use the principle of superposition to find i_1, i_2, and i_3.
Ans. $-\frac{15}{8}$ A, -2 A, $-\frac{31}{8}$ A

fig. P3.21

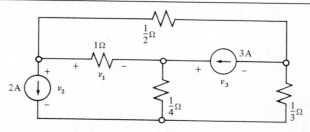

3.21.   Use the principle of superposition to find v_1, v_2, and v_3 in the circuit shown in Fig. P3.21.

fig. P3.22

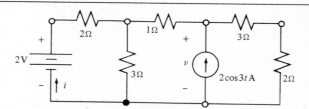

3.22.   For the circuit in Fig. P3.22 use the principle of superposition to find i and v.

fig. P3.23

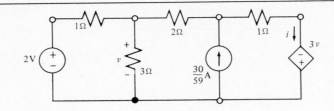

3.23. Repeat Problem 3.22 for the circuit in Fig. P3.23.

fig. P3.24

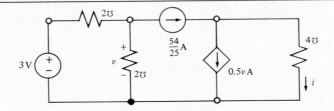

*3.24. Repeat Problem 3.22 for the circuit in Fig. P3.24.
   Ans. $\frac{42}{25}$ A, $\frac{24}{25}$ V

fig. P3.25

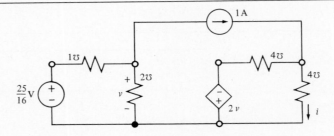

3.25. Repeat Problem 3.22 for the circuit in Fig. P3.25.

fig. P3.26

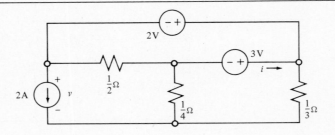

3.26. Repeat Problem 3.22 for the circuit in Fig. P3.26.

fig. P3.27

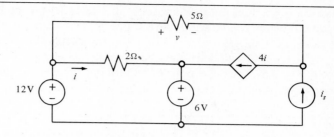

3.27. Repeat Problem 3.22 for the circuit in Fig. P3.27 when (a) $i_s = 4$ A, (b) $i_s = 12$ A.

fig. P3.28

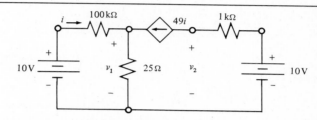

3.28. Use the principle of superposition to find v_1 and v_2 in the circuit shown in Fig. P3.28.

3.5 THE OPERATIONAL AMPLIFIER

Of fundamental importance in the study of electric circuits is the *ideal voltage amplifier*. Such a device, in general, has two inputs, v_1 and v_2, and one output, v_o. The relationship between the output and the inputs is given by $v_o = K(v_2 - v_1)$, where K is called the *gain* of the amplifier. The ideal amplifier is modeled by the circuit shown in Fig. 3.67, which contains a dependent source.

fig. 3.67

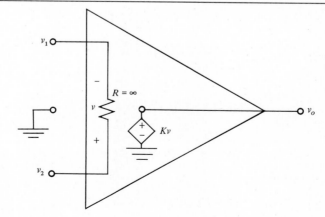

Note that since the input resistance $R = \infty$, when such an amplifier is connected to any circuit no current will flow into the input terminals. Also, since the output v_o is the voltage across an ideal source, $v_o = K(v_2 - v_1)$ regardless of what is connected to the output. For the sake of simplicity, the ideal amplifier having gain K is often represented as shown in Fig. 3.68. We refer to the input terminal labeled " $-$ " as the *inverting input* and the input terminal labeled " $+$ " as the *noninverting input*.

---

fig. 3.68

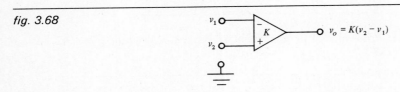

$$v_o = K(v_2 - v_1)$$

---

fig. 3.69

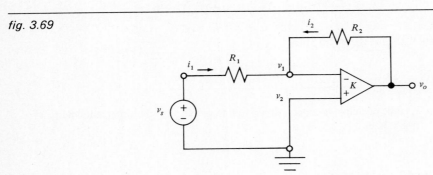

In the circuit shown in Fig. 3.69 the noninverting input is at the reference potential; that is, $v_2 = 0$. Furthermore, since the input node v_s and the output node v_o are constrained by voltage sources (independent and dependent, respectively), using nodal analysis we need only sum currents at node v_1. Because the amplifier draws no current, by KCL,

$$i_1 + i_2 = 0$$

or

$$\frac{v_s - v_1}{R_1} + \frac{v_o - v_1}{R_2} = 0$$

from which

$$R_2 v_s - R_2 v_1 + R_1 v_o - R_1 v_1 = 0$$

or

$$R_2 v_s = (R_1 + R_2) v_1 - R_1 v_o \qquad (3.5)$$

But, because of the amplifier,

$$v_o = K(v_2 - v_1) = -K v_1$$

so

$$v_1 = -\frac{v_o}{K} \tag{3.6}$$

and substituting this into Equation (3.5) we get

$$R_2 v_s = (R_1 + R_2)\left(-\frac{v_o}{K}\right) - R_1 v_o$$

$$= -\frac{(R_1 + R_2) + KR_1}{K} v_o$$

Thus

$$v_o = \frac{-KR_2}{(K+1)R_1 + R_2} v_s$$

For such a circuit let us consider the case in which the gain K becomes arbitrarily large. We can rewrite the last expression as

$$v_o = \frac{-R_2}{\left(1 + \dfrac{1}{K}\right)R_1 + \dfrac{R_2}{K}} v_s$$

Therefore, as $K \to \infty$,

$$v_o = \frac{-R_2}{R_1} v_s$$

We see that although the gain of the amplifier is infinite, for a finite input voltage v_s the output voltage v_o is finite (provided, of course, that $R_1 > 0$). Inspection of Equation (3.6) indicates why the output voltage remains finite—as $K \to \infty$, then $v_1 = -v_o/K \to 0$. This result occurs because there is a resistor connected between the output and one of the inputs. Such a connection is called *feedback*.

An ideal amplifier having gain $K = \infty$ is known as an *operational amplifier* or *op amp*, for short. In an op amp circuit, because of the infinite-gain property, we must have a feedback resistor and must not connect a voltage source directly between the amplifier's input terminals. For the circuit of Fig. 3.69, the corresponding op amp circuit is usually drawn as in Fig. 3.70. Using the fact that $v_1 = 0$, summing the currents

fig. 3.70

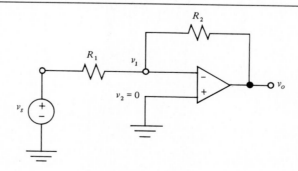

at that node, we have

$$\frac{v_s}{R_1} = -\frac{v_o}{R_2}$$

from which

$$v_o = -\frac{R_2}{R_1} v_s$$

and the gain of the overall circuit is

$$\frac{v_o}{v_s} = -\frac{R_2}{R_1}$$

Notice how simple the analysis of the op amp circuit is when we use the fact that $v_1 = 0$. Although this result was originally deduced from Equation (3.6), the combination of infinite gain and feedback constrains the voltage applied to the op amp (between terminals v_1 and v_2) to be zero. In other words, we must have that $v_1 = v_2$. This conclusion is confirmed in the following example.

· · ·

Example
Consider the op amp circuit with feedback shown in Fig. 3.71.

fig. 3.71

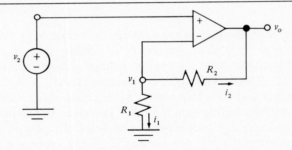

Again, the amplifier draws no current so, applying KCL at node v_1, we find that

$$i_1 + i_2 = 0$$

so

$$\frac{v_1}{R_1} + \frac{v_1 - v_o}{R_2} = 0$$

or

$$R_2 v_1 + R_1 v_1 - R_1 v_o = 0$$

from which

$$(R_1 + R_2)v_1 = R_1 v_o$$

Thus

$$v_o = \frac{R_1 + R_2}{R_1} v_1 \qquad (3.7)$$

But for an ideal amplifier having gain K,

$$v_o = K(v_2 - v_1) = Kv_2 - Kv_1$$

from which

$$Kv_1 = Kv_2 - v_o$$

so

$$v_1 = v_2 - \frac{v_o}{K} \qquad (3.8)$$

Substituting this into Equation (3.7) yields

$$v_o = \frac{R_1 + R_2}{R_1} \left(v_2 - \frac{v_o}{K} \right)$$

$$= \frac{R_1 + R_2}{R_1} v_2 - \frac{R_1 + R_2}{R_1 K} v_o$$

and

$$\left(1 + \frac{R_1 + R_2}{R_1 K} \right) v_o = \frac{R_1 + R_2}{R_1} v_2$$

so that

$$v_o = \frac{\dfrac{R_1 + R_2}{R_1}}{1 + \dfrac{R_1 + R_2}{R_1 K}} v_2$$

$$= \frac{R_1 + R_2}{R_1 + \dfrac{R_1 + R_2}{K}} v_2$$

But, for an operational amplifier, $K = \infty$. Hence

$$v_o = \frac{R_1 + R_2}{R_1} v_2 = \left(1 + \frac{R_2}{R_1} \right) v_2$$

and Equation (3.8) becomes

$$v_1 = v_2$$

and again the input voltage to the op amp is zero.

In summary, given a circuit containing an op amp with feedback, the input voltage to the amplifier is zero, or equivalently, both input terminals must be at the same potential.

Using this fact for the circuit in Fig. 3.71, we have the following simple nodal analysis:

$$v_1 = v_2$$

$$i_1 + i_2 = 0$$

$$\frac{v_1}{R_1} + \frac{v_1 - v_o}{R_2} = 0$$

$$\frac{v_2}{R_1} + \frac{v_2 - v_o}{R_2} = 0$$

$$R_2 v_2 + R_1 v_2 - R_1 v_o = 0$$

$$(R_1 + R_2)v_2 = R_1 v_o$$

$$v_o = \frac{R_1 + R_2}{R_1}\, v_2 = \left(1 + \frac{R_2}{R_1}\right)v_2$$

The overall gain of this circuit is therefore

$$\frac{v_o}{v_2} = \frac{R_1 + R_2}{R_1} = 1 + \frac{R_2}{R_1}$$

Note that if $R_2 = 0$, then $v_o = v_2$. Under this circumstance, R_1 is superfluous and may be removed. The resulting op amp circuit, shown in Fig. 3.72, is known as a *voltage follower*. This configuration is used to isolate or buffer one circuit from another.

---

fig. 3.72

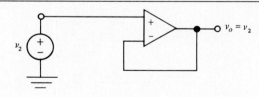

· · ·

Example

The op amp circuit shown in Fig. 3.73 can be used as an electronic ohmmeter, where R is the resistance whose value is to be determined.

The device connected between the output and the reference potential is a dc voltmeter. Since the positive terminal of the voltmeter is connected to the reference, the voltmeter displays $-v_2$. By KCL,

$$\frac{12}{12,000} + \frac{v_2}{R} = 0$$

fig. 3.73

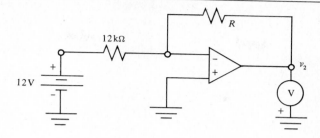

from which

$$\frac{v_2}{R} = -\frac{1}{1000}$$

so

$$R = -1000v_2$$

Suppose that the voltmeter "reads" 10 V. Then $R = 10 \text{ k}\Omega$. If the voltmeter "reads" 27 V, then $R = 27 \text{ k}\Omega$.

• • •

Example

Let us now consider the circuit shown in Fig. 3.74, which contains two operational amplifiers.

fig. 3.74

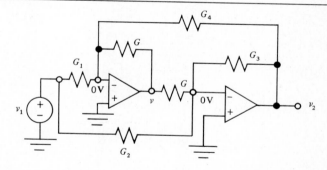

Summing currents at the inverting input of the op amp on the left we get

$$G_1 v_1 + G_4 v_2 = -Gv$$

and, at the inverting input of the other op amp,

$$G_2 v_1 + G_3 v_2 = -Gv$$

Combining these two equations, we get

$$G_1 v_1 + G_4 v_2 = G_2 v_1 + G_3 v_2$$

or

$$(G_1 - G_2)v_1 = (G_3 - G_4)v_2$$

from which

$$v_2 = \frac{G_1 - G_2}{G_3 - G_4}\, v_1$$

$$\cdot \quad \cdot \quad \cdot$$

Operational amplifier circuits can be quite useful for some situations in which there is more than a single input.

$$\cdot \quad \cdot \quad \cdot$$

Example
Consider the op amp circuit shown in Fig. 3.75.

fig. 3.75

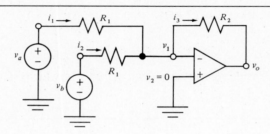

Since $v_2 = 0$, then $v_1 = 0$. Thus from

$$i_1 + i_2 = i_3$$

we get

$$\frac{v_a}{R_1} + \frac{v_b}{R_1} = -\frac{v_o}{R_2}$$

or

$$v_o = -\frac{R_2}{R_1}(v_a + v_b)$$

Since the output v_o is the sum of the voltages v_a and v_b (multiplied by the constant $-R_2/R_1$), this circuit is called an *adder* or *summer*. A common application of such a circuit is as an "audio mixer," where, for example, the inputs are voltages due to two separate microphones and the output voltage is their sum.

$$\cdot \quad \cdot \quad \cdot$$

Example

Another op amp circuit having two inputs is depicted in Fig. 3.76.

fig. 3.76

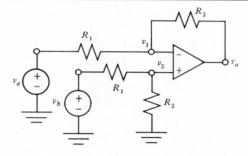

Using the fact that $v_1 = v_2 = v$, by KCL at the inverting input,

$$\frac{v_a - v}{R_1} = \frac{v - v_o}{R_2}$$

from which

$$R_1 v_o + R_2 v_a = (R_1 + R_2)v \tag{3.9}$$

Applying KCL at the noninverting input, we get

$$\frac{v_b - v}{R_1} = \frac{v}{R_2}$$

from which

$$R_2 v_b = (R_1 + R_2)v \tag{3.10}$$

Combining Equations (3.9) and (3.10) results in

$$R_1 v_o + R_2 v_a = R_2 v_b$$

and hence

$$v_o = \frac{R_2}{R_1}(v_b - v_a)$$

Since the output is the difference of the inputs (multiplied by the constant R_2/R_1), such a circuit is called a *difference amplifier* or *differential amplifier*.

$\cdot\quad\cdot\quad\cdot$

For a physical operational amplifier the gain and the input resistance are large, but not infinite. Nor is the output (Thévenin equivalent) resistance zero; that is, a more precise model of an actual op amp is to have a resistance in series with the dependent voltage source. Yet, assuming that an op amp is ideal often yields a simple analysis with very accurate results.

Although their usefulness is great, operational amplifiers are typically small in size. This is a result of the fact that op amps are commonly available as integrated circuits that normally contain between one and four op amps. In our discussions above, we have depicted the operational amplifier as a three-terminal device. In actuality, a package containing one or more op amps typically has between 8 and 14 terminals. As we have seen, three of the terminals are the inverting input, the noninverting input and the output. However, there are also terminals for applying dc voltages ("power supplies") to "bias" the numerous transistors comprising an op amp, for "frequency compensation," and for other details that should be discussed in your electronics courses. One of the important practical electronic considerations for proper op amp operation is that when feedback is between the output and only one input terminal, that input terminal should be the inverting input.

Figure 3.77 shows a simple practical amplifier that uses the popular 741 op amp. In this circuit, terminal ("pin") 4 is the inverting input, terminal 5 is the noninverting input, and terminal 10 is the output. Furthermore, pin 6 has a dc voltage of -15 V and pin 11 has a dc voltage of $+15$ V applied to them. The 741 op amp typically has a gain of $K = 200,000$, input resistance $R = 2$ MΩ and an output resistance of $R_0 = 75$ Ω.

fig. 3.77

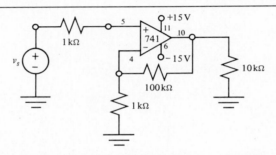

Suppose that v_j is the voltage at terminal j. Assuming that the op amp is ideal, then $v_5 = v_s$ as the (ideal) op amp draws no current. Also, due to feedback, $v_4 = v_5 = v_s$, and by KCL at terminal 4,

$$\frac{v_{10} - v_s}{100,000} = \frac{v_s}{1000}$$

from which

$$v_{10} - v_s = 100v_s$$

Thus,

$$v_{10} = 101v_s$$

We can get a more accurate result by not assuming that the op amp is ideal. Doing so, which requires a much greater analysis effort (try it and see), yields

$$v_{10} \approx 100v_s$$

which is not significantly different from our simple approach given above.

PROBLEMS   3.29.   Find the Thévenin equivalent of the ideal amplifier circuit in Fig. P3.29. (*Hint:* To find R_0, apply a current source i_0 and calculate v_0.)

fig. P3.29

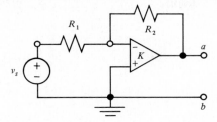

3.30.   For the circuit given in Problem 3.29, find the Thévenin equivalent of the corresponding op amp circuit ($K = \infty$).

fig. P3.31

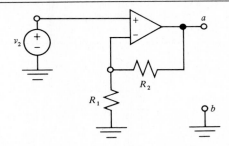

*3.31.   Find the Thévenin equivalent of the op amp circuit in Fig. P3.31. (*Note:* Terminal b is the reference node. See the hint given in Problem 3.29.)

Ans. $\left(1 + \dfrac{R_2}{R_1}\right) v_2, 0\ \Omega$

3.32.   Find the resistance seen by the voltage source v_s for the circuit given in (a) Problem 3.29; (b) Problem 3.30; and (c) Problem 3.31.

fig. P3.33

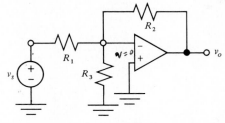

*3.33.   Given the op amp circuit in Fig. P3.33.
(a)   Find the voltage gain v_o/v_s.
(b)   Find the resistance seen by the voltage source v_s.
(c)   How do the results of (a) and (b) differ from the case that $R_3 = \infty$? Why?

Ans.   (a) $-\dfrac{R_2}{R_1}$, (b) R_1, (c) same results

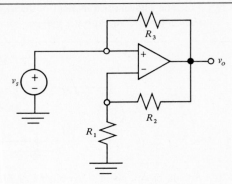

3.34.   Repeat Problem 3.33 for the op amp circuit shown in Fig. P3.34.

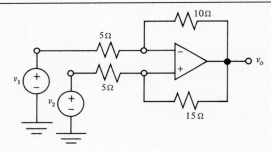

3.35.   For the op amp circuit shown in Fig. P3.35 find v_o.

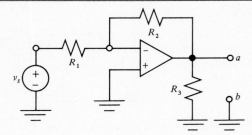

3.36.   Repeat Problem 3.31 for the op amp circuit in Fig. P3.36. What happens
        for the case that $R_3 = \infty$? (See Problem 3.30.) Why?
3.37.   Given the op amp circuit in Fig. P3.37, choose values for resistors R_1 and
        R_2 such that the 100-Ω resistor absorbs 10 mW when $v_s = 4$ V.

fig. P3.37

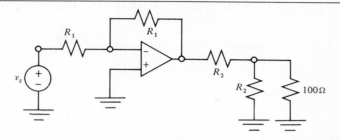

fig. P3.38

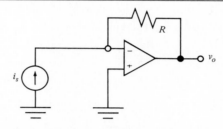

*3.38. Consider the op amp circuit shown in Fig. P3.38.
 (a) Find the ratio v_o/i_s.
 (b) Find the resistance seen by the current source i_s.
 Ans. (a) $-R$, (b) 0 Ω

fig. P3.39

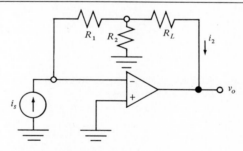

3.39. Repeat Problem 3.38 for the circuit in Fig. P3.39.
3.40. For the circuit given in Problem 3.39, find the current gain i_2/i_s.

fig. P3.41

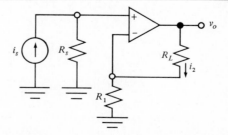

3.41. Repeat Problem 3.38 for the op amp circuit shown in Fig. P3.41.

3.42.   For the op amp circuit given in Problem 3.41, (a) find the current gain i_2/i_s and (b) find the resistance seen by the current source.

fig. P3.43

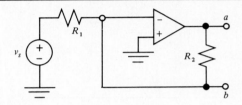

*3.43.   Find the Thévenin equivalent of the op amp circuit in Fig. P3.43.

Ans. $-\dfrac{R_2}{R_1}\, v_s,\ R_2$

fig. P3.44

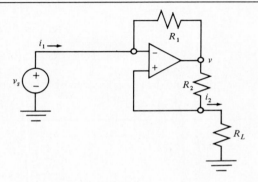

3.44.   Consider the op amp circuit in Fig. P3.44.
   (a)   Find the voltage v.
   (b)   Find the current gain i_2/i_1.

fig. P3.45

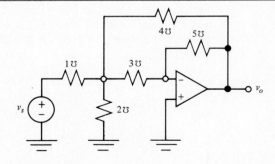

3.45.   (a)   Find the voltage gain v_o/v_s for the op amp circuit in Fig. P3.45.
   (b)   Find the conductance seen by the voltage source v_s.
3.46.   Repeat Problem 3.45 for the op amp circuit shown in Fig. P3.46.

fig. P3.46

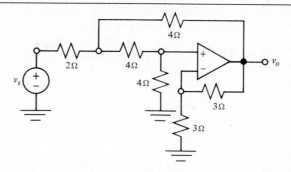

fig. P3.47

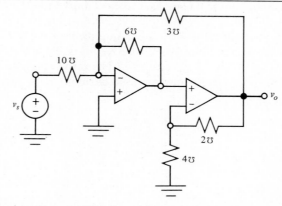

*3.47. Repeat Problem 3.45 for the op amp circuit in Fig. P3.47.
$Ans.$ (a) $- 2$, (b) 10 ℧

fig. P3.48

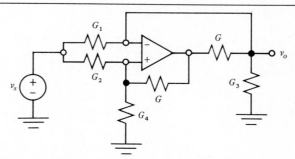

3.48. Find v_o/v_s for the op amp circuit in Fig. P3.48.

Summary

1. Some nonideal voltage sources can be modeled as an
    ideal voltage source in series with a resistance, and certain
    nonideal current sources can be modeled as an ideal
    current source in parallel with a resistance.

2. A voltage source in series with a resistance behaves as a current source in parallel with that resistance, and vice versa.

3. A nonideal source with internal resistance R_s delivers maximum power to a resistive load R_L when $R_L = R_s$.

4. The effect of an arbitrary circuit on a load is equivalent to an appropriate voltage source in series with an appropriate resistance (Thévenin's theorem) or an appropriate current source in parallel with that resistance (Norton's theorem).

5. The response of a circuit having n independent sources equals the sum of the n responses to each individual independent source (the principle of superposition).

6. An operational amplifier is an ideal amplifier with infinite gain. When used in conjunction with feedback, the input voltage to an op amp is constrained to be zero volts. The input terminals of an op amp draw no current.

ENERGY STORAGE ELEMENTS

4

The applications of electric circuits consisting of sources and resistors is quite limited indeed. Furthermore, sources that produce voltages or currents that vary with time can be extremely useful. Because of this, in the present chapter we introduce two circuit elements, the inductor and the capacitor, that have voltage-current relationships that are not simple direct proportionalities. Instead, the behavior of each element can be described by either a differential or an integral relationship. In addition, we define some basic, but important, nonconstant functions of time and examine their derivatives and integrals. Furthermore, we see how simple functions can be combined into more complicated functions and vice versa.

Introduction

Although resistors dissipate power (or energy), inductors and capacitors store energy. The relationships describing this property are derived in this chapter. We shall also see how to deal with series and parallel connections of inductors and capacitors and how to write mesh and node equations for circuits containing inductors and capacitors. By using the concept of "duality," analysis of a planar circuit results in the automatic analysis of a second circuit. We shall also see how to handle inductors that have initial currents flowing through them and capacitors that have initial charge (voltage across them).

4.1 THE INDUCTOR

In our high school science courses (or earlier) we learned that a current flowing through a conductor, such as a wire, produces a magnetic field around that conductor. Furthermore, winding such a conductor into a coil strengthens the magnetic field. For the resulting element, known as an *inductor*, the voltage across it is directly proportional to the time rate of change of the current flowing through it. This fact is credited to the independent work of the American inventor Joseph Henry (1797–1878) and the English physicist Michael Faraday (1791–1867).

To signify a coil of wire, we represent an *ideal inductor* as shown in Fig. 4.1. The relationship between voltage and current for this element is given by

$$v = L \frac{di}{dt}$$

fig. 4.1

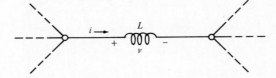

where *L* is the *inductance* of the element, and its unit is the *henry* (abbreviated H) in honor of Joseph Henry. Remember that in the differential relationship just given, *v* and *i* represent the functions of time $v(t)$ and $i(t)$, respectively.

· · ·

Example

Consider the circuit of Fig. 4.2 in which the ideal source produces a constant current, called a *direct current* (abbreviated dc). From the equation

$$v = L \frac{di}{dt}$$

fig. 4.2

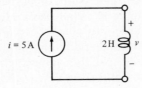

we have that

$$v = 2 \frac{d}{dt} [5]$$

$$= 0 \text{ V}$$

This example is an illustration of the fact that an (ideal) inductor behaves as a short circuit to dc.

· · ·

In the example just presented, the dc circuit analyzed was quite simple. Let us therefore consider a more complicated situation.

. . .

Example

Let us determine the current i in the circuit shown in Fig. 4.3. This circuit has one independent voltage source whose value is constant. For a resistive circuit we would naturally anticipate that all voltages and currents are constant. However,

fig. 4.3

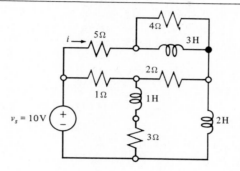

this is not a resistive circuit. Yet, our intuition suggests that the constant-valued voltage source (even if there were additional constant-valued independent sources, and dependent sources, too) produces constant-valued responses. This fact will be confirmed more rigorously in future chapters. In the meantime, we shall use the result that a circuit containing only constant-valued sources is a dc circuit.

In the dc circuit given in Fig. 4.3 the inductors behave as short circuits. Thus, as far as the resistors are concerned, we have the circuit shown in Fig. 4.4. Note that in this circuit the 4-Ω resistor does not appear. The reason is that a 4-Ω resistor

fig. 4.4

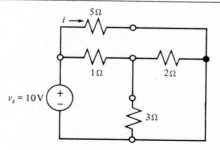

in parallel with a short circuit (zero resistance) is equivalent to a short circuit. By KVL, the voltage across the 5-Ω resistor is 10 V (the plus is on the left). Therefore,

$$i = \frac{10}{5} = 2 \text{ A}$$

. . .

Because of the dc source in the preceding example, we were able to reduce the problem to that of analyzing a resistive circuit. But don't get the impression that the analysis of a circuit containing an inductor or inductors is a snap. Generally circuits containing inductors have sources that are not simply constant-valued.

. . .

Example
For the simple circuit given in Fig. 4.2, let the current $i(t)$ that is produced by the source be described by the function of time shown in Fig. 4.5, which is described in analytical terms as follows:

$$i(t) = \begin{cases} 0 & \text{for } -\infty < t < 0 \\ t & \text{for } 0 \le t < 1 \\ -(t-2) & \text{for } 1 \le t < 2 \\ 0 & \text{for } 2 \le t < \infty \end{cases}$$

fig. 4.5

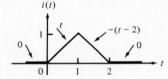

Since

$$v = L\frac{di}{dt} = 2\frac{di}{dt}$$

then

$$v(t) = \begin{cases} 0 & \text{for } -\infty < t < 0 \\ 2 & \text{for } 0 \le t < 1 \\ -2 & \text{for } 1 \le t < 2 \\ 0 & \text{for } 2 \le t < \infty \end{cases}$$

A sketch of $v(t)$ is shown in Fig. 4.6.

fig. 4.6

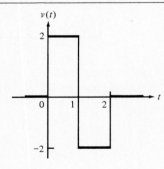

We know that a resistor always absorbs power and the energy absorbed is dissipated as heat. But how about an inductor? For the inductor in the circuit shown in Fig. 4.2, the instantaneous power $p(t) = v(t)i(t)$ absorbed by the inductor is given by

$$p(t) = \begin{cases} 0 & \text{for } -\infty < t < 0 \\ 2t & \text{for } 0 \le t < 1 \\ 2(t-2) & \text{for } 1 \le t < 2 \\ 0 & \text{for } 2 \le t < \infty \end{cases}$$

A sketch of $p(t)$ is shown in Fig. 4.7, in which we see that the power absorbed by the inductor is zero for $-\infty < t \le 0$ and $2 \le t < \infty$. For $0 < t < 1$, since $p(t)$ is a positive quantity, the inductor is absorbing power (which is produced by the source). However, for $1 < t < 2$, since $p(t)$ is a negative quantity, the inductor is actually supplying power (to the source)!

fig. 4.7

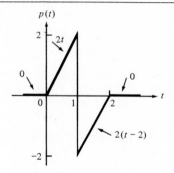

To get the energy absorbed by the inductor, we simply integrate over time the power absorbed. For this example, the energy absorbed increases from 0 to $\frac{1}{2}(1)(2) = 1$ J as time goes from $t = 0$ to $t = 1$ s. However, from $t = 1$ to $t = 2$ s, the inductor supplies energy such that at time $t = 2$ s and thereafter, the net energy absorbed by the inductor is zero. Since all of the energy absorbed by the inductor is not dissipated but is eventually returned, we say that the inductor *stores* energy. The energy is stored in the magnetic field that surrounds the inductor.

· · ·

In the example just given, we obtained an expression for $p(t)$, the instantaneous power absorbed by the inductor. From this we can obtain an expression for $w_L(t)$, the energy stored in the inductor; we just sum up the power over time. In other words, the energy stored in the inductor at time t is

$$w_L(t) = \int_{-\infty}^{t} p(t)\, dt$$

However, rather than find the particular expression $w_L(t)$ for the above example, let us derive a simple formula for the energy stored in an inductor. We proceed as follows:

For the inductor illustrated in Fig. 4.8, since

$$p(t) = i(t)v(t)$$

and

$$v(t) = L\frac{di(t)}{dt}$$

then

$$p(t) = i(t)\left[L\frac{di(t)}{dt}\right]$$

$$= Li(t)\frac{di(t)}{dt}$$

fig. 4.8

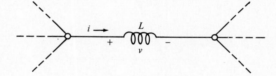

Thus, the energy stored in the inductor at time t is $w_L(t)$, where

$$w_L(t) = \int_{-\infty}^{t} p(t)\, dt$$

$$= \int_{-\infty}^{t} Li(t)\frac{di(t)}{dt}\, dt$$

Using the chain rule of calculus, the variable of integration can be changed from t to $i(t)$. Changing the limits of integration appropriately results in

$$w_L(t) = \int_{i(-\infty)}^{i(t)} Li(t)\, di(t)$$

$$= L\left.\frac{i^2(t)}{2}\right|_{i(-\infty)}^{i(t)}$$

$$= \frac{1}{2}L[i^2(t) - i^2(-\infty)]$$

Assuming that you can go back far enough in time, you will eventually return to that time when there was no current flowing in the inductor. Thus, we adopt the convention that $i(-\infty) = 0$. The result is the following expression for the energy stored at time t by an inductor whose value is L henries:

$$w_L(t) = \frac{1}{2} Li^2(t)$$

$\cdot$ $\cdot$ $\cdot$

Example

The energy stored in the inductor given in the previous example is

$$w_L(t) = \frac{1}{2} Li^2(t) = i^2(t) = \begin{cases} 0 & \text{for } -\infty < t < 0 \\ t^2 & \text{for } 0 \leq t < 1 \\ (t-2)^2 & \text{for } 1 \leq t < 2 \\ 0 & \text{for } 2 \leq t < \infty \end{cases}$$

A sketch of the energy stored in the inductor versus time is shown in Fig. 4.9.

fig. 4.9

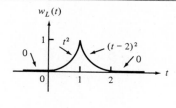

$\cdot$ $\cdot$ $\cdot$

Unlike an ideal inductor, an actual inductor (an automobile ignition coil is an example) has some resistance associated with it. This is due to the fact that a physical inductor is constructed from real wire, which may have a very small (but still nonzero) resistance. For this reason, we model a physical inductor as an ideal inductor in series with a resistance. Practical inductors (also called "coils" and "chokes") range in value from about 10^{-6} H $= 1$ μH all the way up to around 100 H. To construct inductors having large values requires many turns of wire and iron cores, and consequently results in larger values of series resistance. The series resistance is typically in the range from a fraction of an ohm to several hundred ohms.

$\cdot$ $\cdot$ $\cdot$

Example

In the circuit shown in Fig. 4.10 suppose that $i(t)$ is the function given in the previous two examples. Then the voltage across the inductor

$$v = L \frac{di}{dt} = 2 \frac{di}{dt}$$

fig. 4.10

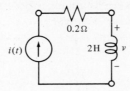

is the same as was obtained previously, as is the energy stored in the inductor:

$$w_L(t) = \frac{1}{2} Li^2(t) = i^2(t)$$

In addition, however, the resistor dissipates energy in the form of heat. The power dissipated is

$$p_R(t) = Ri^2(t) = 0.2i^2(t)$$

Hence, the total energy dissipated is

$$w_R = \int_{-\infty}^{\infty} p_R(t)\, dt = 0.2 \int_{-\infty}^{\infty} i^2(t)\, dt$$

$$= 0.2 \left[\int_0^1 t^2\, dt + \int_1^2 (t-2)^2\, dt \right]$$

$$= 0.2 \left[\left. \frac{t^3}{3} \right|_0^1 + \left. \frac{(t-2)^3}{3} \right|_1^2 \right]$$

$$= 0.2 \left[\frac{1}{3} + \frac{1}{3} \right] = \frac{1}{5} \left[\frac{2}{3} \right] = \tfrac{2}{15} \text{ J}$$

$$\cdot \ \cdot \ \cdot$$

4.2   THE CAPACITOR

Another extremely important circuit element is obtained when two conducting surfaces (called *plates*) are placed in proximity to one another, with a nonconducting material called a *dielectric* between them. For the resulting element, known as a *capacitor* (formerly, *condenser*), a voltage across the plates results in an electric field between them and the current through the capacitor is directly proportional to the time rate of change of the voltage across it. The *ideal capacitor* is depicted in Fig. 4.11 and the relationship between current and voltage is given by*

$$i = C \frac{dv}{dt}$$

*For those who remember from freshman physics that $q = Cv$, where q is charge, C is capacitance, and v is voltage, taking the derivative we get $C \dfrac{dv}{dt} = \dfrac{dq}{dt} = i$.

fig. 4.11

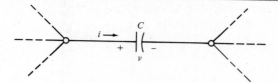

where *C* is the *capacitance* of the element. Its unit is the *farad* (abbreviated F) in honor of Michael Faraday.

· · ·

Example

Consider the simple dc circuit shown in Fig. 4.12.

fig. 4.12

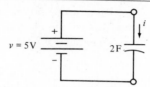

From the differential relationship for the capacitor we have

$$i = C \frac{dv}{dt} = 2 \frac{d}{dt} [5] = 0 \text{ A}$$

· · ·

We deduce, therefore, that a capacitor (ideal) is an open circuit to dc. A more elaborate dc circuit is presented in the following example.

· · ·

Example

Suppose that we wish to find *i* in the dc circuit shown in Fig. 4.13.

fig. 4.13

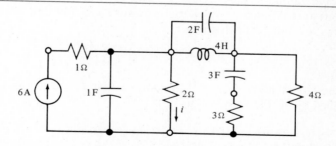

Using the fact that a capacitor behaves as an open circuit to dc—as well as the fact that an inductor behaves as a short circuit—we can determine the current *i*

from the equivalent resistive circuit shown in Fig. 4.14. By current division, we find that

$$i = \frac{4}{4 + 2}(6) = 4 \text{ A}$$

fig. 4.14

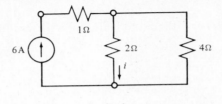

· · ·

Perhaps the fact that a capacitor acts as an open circuit to dc does not seem surprising since such an element has a nonconducting dielectric (material which will not allow the flow of charge) between its (conducting) plates. But how, then, can there ever be any current flow* through a capacitor? To answer this question, consider the simple circuit shown in Fig. 4.15.

fig. 4.15

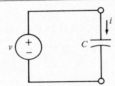

For the case that v is a constant, then indeed $i = 0$ (that is, there is no current through the capacitor). However, there is a voltage v across the capacitor and a charge q on the plates. From elementary physics, the relationship is $q = Cv$. If the voltage is not constant with time—let's explicitly write $v(t)$—the charge also varies with time; that is, $q(t) = Cv(t)$. Since the net charge on the plates fluctuates—and no charge crosses the dielectric—there must be a transfer of charge throughout the remaining portion of the circuit. Hence, when $v(t)$ is nonconstant, there is a nonzero current $i(t) = C\, dv(t)/dt$. In summary, even though there is no flow of charge across the dielectric of a capacitor, the effect of alternate charging and discharging due to a nonconstant voltage results in an actual current outside of the dielectric.

· · ·

Example
For the circuit given in Fig. 4.15 suppose that $C = 2$ F and the voltage v is described by the function in Fig. 4.16.

---

*Grammatically speaking, the phrase "current flow," though frequently used, may seem incorrect. Since current is a flow of charge, then such a phrase is tantamount to saying "a flow of charge flow"—which is redundant, to say the least. The same comment can be made about the oft-encountered phrase "dc current"—which really means "direct current current." However, because of their wide usage, try to be understanding when you run across such phrases in this and other books.

fig. 3.26

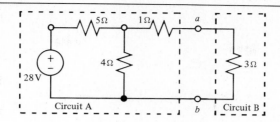

Circuit A   Circuit B

and calculate v_{oc} from Fig. 3.27. Since the current through the 1-Ω resistor in this circuit is zero, the voltage across it is also zero. Thus, by voltage division,

$$v_{oc} = \frac{4}{4+5}(28) = \tfrac{112}{9} \text{ V}$$

fig. 3.27

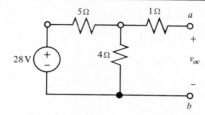

The Thévenin equivalent resistance is found from the circuit in Fig. 3.28, which is obtained by setting the 28-V independent source to zero. Thus

$$R_0 = 1 + \frac{(4)(5)}{4+5} = 1 + \frac{20}{9}$$

$$= \tfrac{29}{9} \ \Omega$$

fig. 3.28

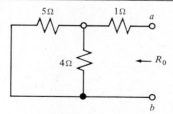

Replacing circuit A by its Thévenin equivalent, we get the circuit of Fig. 3.29. By voltage division, the voltage across the 3-Ω resistor is

$$v = \frac{3}{3 + \dfrac{29}{9}} \left(\frac{112}{9}\right) = \frac{3}{\dfrac{56}{9}} \left(\frac{112}{9}\right)$$

$$= 6 \text{ V}$$

fig. 3.29

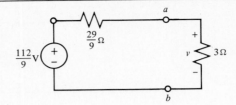

Direct analysis of the original circuit, of course, yields the same result (see Fig. 1.70). Thus, in order for it to absorb the maximum amount of power, the 3-Ω load resistor should be changed to $\frac{29}{9}$ Ω.

. . .

Example
Let us find the voltage across the $\frac{1}{3}$-Ω load resistor in Fig. 3.30 by replacing the remainder of the circuit by its Thévenin equivalent.

fig. 3.30

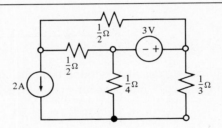

First we remove the $\frac{1}{3}$-Ω resistor and determine v_{oc} (see Fig. 3.31). Note that for this circuit, by KCL, $i = -2$ A. Thus,

$$v_{oc} = 3 + \frac{1}{4}i = 3 - \frac{1}{2} = \tfrac{5}{2} \text{ V}$$

fig. 3.31

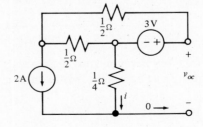

To find R_0, remove the $\frac{1}{3}$-Ω resistor and set the independent sources to zero. The result is the circuit shown in Fig. 3.32.